I. Burc Oral
Jan 1992, MIT

International Association of Geodesy Symposia

Ivan I. Mueller, Series Editor

International Association of Geodesy Symposia

Ivan I. Mueller, Series Editor

Symposium 101: Global and Regional Geodynamics
Symposium 102: Global Positioning System: An Overview
Symposium 103: Gravity, Gradiometry, and Gravimetry
Symposium 104: Sea Surface Topography and the Geoid
Symposium 105: Earth Rotation and Coordinate Reference Frames

Global Positioning System: An Overview

Symposium No. 102

Edinburgh, Scotland, August 7–8, 1989

Convened and Edited by
Yehuda Bock
Norman Leppard

Springer-Verlag
New York Berlin Heidelberg
London Paris Tokyo Hong Kong

Yehuda Bock
IGPP A-025
Scripps Institute of Oceanography
La Jolla, CA 92093
USA

Norman Leppard
Elmwood Avenue
Feltham, Middlesex TW13 7AE
United Kingdom

Series Editor
Ivan I. Mueller
Department of Geodetic Science & Surveying
The Ohio State University
Columbus, OH 43210-1247
USA

For information regarding previous symposia volumes contact:
Secretáire Général
Bureau Central de l'Association Internationale de Géodésie
138, rue de Grenelle
75700 Paris
France

Library of Congress Cataloging-in-Publication Data
Global positioning system : an overview / Yehuda Bock, Norman Leppard, editors.
p. cm. — (International Association of Geodesy symposia ; symposium 102)
Proceedings from a symposium held Aug. 7–8, 1989, in Edinburgh, Scotland, as part of the International Association of Geodesy's 125th anniversary General Meeting.
Includes bibliographical references.
1. Global Positioning System — Congresses. I. Bock, Yehuda. II. Leppard, Norman. III. International Association of Geodesy. General Meeting (1989 : Edinburgh, Scotland) IV. Series.
VK562.G58 1990
526′.6 — dc20 90-9531

Printed on acid-free paper.

ISBN 0-387-97266-8/1990 $0.00 + 0.20

Camera-ready copy provided by the editors.
Printed and bound by Edwards Brothers, Ann Arbor, Michigan.
Printed in the United States of America.

9 8 7 6 5 4 3 2 1

ISBN 0-387-97266-8 Springer-Verlag New York Berlin Heidelberg
ISBN 3-540-97266-8 Springer-Verlag Berlin Heidelberg New York

Foreword

A General Meeting of the IAG was held in Edinburgh, Scotland, to commemorate its 125th Anniversary. The Edinburgh meeting, which attracted 360 scientific delegates and 80 accompanying persons from 44 countries, was hosted jointly by the Royal Society, the Royal Society of Edinburgh and the University of Edinburgh.

The scientific part of the program, which was held in the Appleton Tower of the University, included the following five symposia:

Symposium 101	Global and Regional Geodynamics
Symposium 102	GPS and Other Radio Tracking Systems
Symposium 103	Gravity, Gradiometry and Gravimetry
Symposium 104	Sea Surface Topography, the Geoid and Vertical Datums
Symposium 105	Earth Rotation and Coordinate Reference Frames

All together there were 90 oral and 160 poster presentations. The program was arranged to prevent any overlapping of oral presentations, and thus enabled delegates to participate in all the sessions.

The 125th Anniversary Ceremony took place on August 7, 1989, in the noble surroundings of the McEwan Hall where, 53 years earlier, Vening-Meinesz gave one of the two Union Lectures at the 6th General Assembly of the IUGG. The Ceremony commenced with welcome speeches by the British hosts. An interlude of traditional Scottish singing and dancing was followed by the Presidential Address given by Professor Ivan Mueller, on 125 years of international cooperation in geodesy. The Ceremony continued with greetings from representatives of sister societies, and was concluded by the presentation of the Levallois Medal to Professor Arne Bjerhammar. The 125th Anniversary was also commemorated by an exhibition entitled *The Shape of the Earth*, which was mounted in the Royal Museum of Scotland. An abbreviated version of the President's speech and the list of all participants are included as appendices in this volume.

A social program enabled delegates to experience some of the hospitality and culture of both Edinburgh and Scotland, as well as provided an opportunity to explore the beautiful City of Edinburgh and the surrounding countryside.

A Scottish Ceilidh on the last night concluded a pleasant week, which was not only scientifically stimulating, but also gave delegates and accompanying persons an opportunity to renew *auld acquaintances* and make new ones.

The International Association of Geodesy and the UK Organizing Committee express their appreciation to the local organizers of the General Meeting, especially to Dr. Roger G. Hipkin and Mr. Wm. H. Rutherford, for their tireless efforts in running the meeting to its successful conclusion.

Commencing with these symposia the proceedings of IAG-organized scientific meetings will be published by Springer Verlag Inc., New York from author-produced camera-ready manuscripts. Although these manuscripts are reviewed and edited by IAG, their contents are the sole responsibility of the authors, and they do not reflect official IAG opinion, policy or approval.

V. Ashkenazi
A. H. Dodson
UK Organizing Committee

Ivan I. Mueller
President, International
Association of Geodesy

Preface

Space geodesy has evolved in the last twenty years into one of the most exciting disciplines in the earth sciences. In particular, the last five years have seen unprecedented growth in this field with no end in sight. This development is due to a large extent to the versatility of applications provided by the radio interferometric technique called GPS. It is fair to say that the Global Positioning System has already revolutionized or will soon revolutionize almost all conceivable positioning applications (precise or otherwise, large or small scale, scientific or commercial) on land, in the air, in space, and at sea. And the GPS constellation is only beginning to approach its full operational phase with the launching of the first five Block II satellites by the end of 1989.

When we were asked to convene Symposium 102 on *The Global Positioning System and Other Radio Tracking Systems* at the IAG General Meeting in Edinburgh, we thought it appropriate to present a broad review of GPS geodesy, achievements to date and prospects for the future. A natural way to accomplish this was through the active participation of the IAG special study groups and subcommissions whose charters were related to the scope of the symposium, in particular SSG's 1.104—Static and Geodynamic Positioning with GPS—Y. Bock, president; 1.105—GPS Kinematic Positioning Methods and Applications —G. Lachapelle, president; 2.54—Determination of Orbits to Cm-Accuracy— O. Colombo, president; 4.117—Optimization of Modern Positioning Techniques— D. Delikaraoglou, president; and the CSTG Global Positioning System Subcommission— G. Mader, president. The presidents of these groups were asked to serve on the organizing committee and were invited to deliver keynote presentations and to invite other speakers. We decided on this approach because of the limited time (a day and a half) allocated to a symposium whose scope covers such an active research area, and the large number of contributed papers that were anticipated. We felt that review papers of a half-hour length would provide the participants with a good overview of the field at a reasonable, nonrushed pace, with current research activity covered in detail at the poster boards. The interactions at the poster boards between the authors and participants have been found to be more conducive to fruitful exchange of ideas than a flurry of ten-minute oral presentations that leave all participants exhausted.

Symposium 102 was held on August 7–8, 1989, in Edinburgh, Scotland, as part of the International Association of Geodesy's 125th Anniversary General Meeting. The IAG call for papers produced 63 abstracts, which were all accepted for the poster sessions. Of these, 43 were eventually presented. The structure of the Symposium was a series of nine keynote papers presented in plenary session supported by the poster presentations. The posters were grouped to correspond with the keynote papers so that the plenary and poster sessions could be linked. The Symposium had the advantage of having all the posters displayed before, during and after the plenary sessions with the formal poster presentations after their plenary lead papers. Reducing the number of oral presentations in the Symposium enabled discussion periods to be meaningful. The Symposium attracted considerable attention with much interest in the poster presentations.

In this volume the papers are grouped into sections according to the relevant IAG special study groups and subcommission. In each section the first papers are the keynote presentations. They are intended as a comprehensive review of the current status of GPS geodesy with a look towards the future, from the perspective of the presidents of the GPS-related IAG study groups and subcommissions, and their colleagues.

The rest of the sections contain a collection of 29 papers based on the poster presentations. They provide an overview of current research activities in GPS geodesy and thereby complement the keynote presentations.

We would like to thank all the authors for their contributions to the realization of this volume and to the IAG Executive Committee for giving us the opportunity to organize Symposium 102 and prepare this document.

Yehuda Bock　　　　Norman Leppard

Contents and Program

Orbit Determination — IAG Special Study Group 2.106 & CSTG Subcommission

Optimization and Design — IAG Special Study Group 4.117

Dynamic, Kinematic GPS/INS — IAG Special Study Group 1.105

Radio Tracking Systems — IAG Special Study Group 2.54

Campaigns

Appendix 1 — Presidential Address

RELATIVE STATIC POSITIONING WITH THE GLOBAL POSITIONING SYSTEM: BASIC TECHNICAL CONSIDERATIONS

G. Beutler, W. Gurtner, M. Rothacher, U. Wild
Astronomical Institute
CH - 3012 Berne, Switzerland

E. Frei
Wild Leitz AG
CH - 9435 Heerbrugg, Switzerland

1. INTRODUCTION

We give an overview of some of the important everyday-problems of using the Global Positioning System (GPS) for relative positioning in networks of small to medium size (let us say from 5×5×2 km to 100×100×2 km). We assume that the surveys are performed with single- and dual-frequency receivers. We allow for different receiver types and we assume that at least some of the dual frequency instruments are of C/A-code type (recovering the L_2-phase using some squaring technique). The questions we address are the following: (1) What linear combination should be used for a special problem type (e.g., for pre-processing, ambiguity resolution, final solutions)? (2) How are mathematical correlations handled correctly in the general case; are we ready to process upcoming events with 50-100 simultaneously operating receivers in a correct way? (3) What is the optimum way of using "basic" atmosphere information (surface meteorological data, use of dual frequency data for single frequency receivers working in the same area)? (4) What is the state of the art of combining receivers of different type? (5) If we are interested in moderate accuracies only: how long do we have to stay on a survey point to get a satisfactory answer. Or, more generally, how to perform a kinematic survey using the methods of static positioning?

When we browse through the proceedings of conferences on satellite positioning (e.g., Rockville 1985, Austin 1986) and the three major journals (*Journal of Geophysical Research*, *Bulletin Géodésique*, *Manuscripta Geodaetica*) for papers on the geodetic use of the Global Positioning System (GPS) we find that these may be grouped as follows:

(1) Accuracy demonstration of relative static positioning
(2) Orbit determination/satellite clock modeling
(3) Papers dealing with technical aspects:
  - Mathematical correlations between simultaneous GPS observations
  - Special linear combinations of the L_1 and L_2 carrier phases
  - Ambiguity resolution
(4) Influence of unmodeled biases on geodetic results (e.g., orbits, atmosphere)
(5) Kinematic surveys

Let us have a closer look at some of these contributions:

(i) The papers of group (1) tell us that there are remarkable improvements since 1982: From a few parts per million and a few millimeters on very short baselines (Bock *et al.*, 1984; *Goad et al.*, 1984) the accuracy in the horizontal position came down to a few parts per 10 millions (e.g., Beutler *et al.*, 1987) to a few parts in 100 millions (e.g., Lichten and Border, 1987). The step from 1 ppm to 0.1 ppm was possible due to the availability of dual frequency receivers and orbit improvement techniques, the step from 0.1 to 0.01 ppm due to the consequent application of the fiducial point concept (Kroger *et al.*, 1985) and refined modeling techniques (Lichten and Border, 1987) to determine the orbits of the GPS satellites. It also became obvious that high relative accuracies are more easily obtained on long baselines (the ideal situation would be an accuracy not depending on the baseline length). We found (Beutler *et al.*, 1989]) that at present the baseline accuracy obtainable with GPS is quite well reflected by the "law"

$$\frac{db}{b} = \sqrt{\frac{1}{2b}} \quad \text{mm/km} \tag{1.1}$$

where: b is the baseline length in km
db is the error in (one of) the baseline components (in mm).

(2) Orbit determination using the observations of a regional fiducial tracking network is an absolute prerequisite for obtaining GPS results of the order of 0.01 ppm in a regional or continental network. A very careful analysis of this problem is given by Lichten and Border (1987), a review of the same topic is included in these proceedings by Lichten. It seems worthwhile to mention that this problem may become significantly easier to deal with when high precision code measurements become available at the fiducial sites (Jet Propulsion Laboratory's ROGUE receivers promise code accuracies of 5 cm (see, e.g., Blewitt *et al.*, 1988): The time consuming preprocessing and ambiguity resolution steps will be superfluous. This is an important aspect for routine procedures.

Kinematic surveys which are outside the scope of the present article will be dealt with in detail in these proceedings by Lachapelle and Schwarz. It is interesting however that this topic received so much attention. Obviously the goal of kinematic surveys, to obtain a moderate point accuracy (of the order of 1–3 cm) in a short time, is most attractive to many people. For us this was the reason to include section 6, where we try to solve the same problem, using the methods of static positioning only. This problem has promising solutions if, and only if, we manage to resolve the initial phase ambiguities using a short observation time span.

Let us have a quick look at the remaining sections:

In section 2 we summarize the essential knowledge concerning linear combinations of dual frequency (code and) phase observations.

In section 3 we review the state of the art of modeling mathematical correlations between simultaneously observed phase (or code) observations. For the sake of completeness we include the background necessary to handle this problem in the most general case when single- and dual-band receivers are combined in the same adjustment process (there are users doing this on a routine basis).

In section 4 we look into the problem of atmospheric refraction. We start with a brief literature survey, then we put the emphasis on elementary, but as we think, important practical aspects.

Experienced GPS users may have noticed that in small networks baseline- or network-repeatability is usually better when actual meteorological measurements are discarded and not used. We show that there are better ways of using these measurements than to introduce them into the well known Saastamoinen or Hopfield models.

We will also present practical ways to estimate models of ionospheric refraction using the dual band phase (or code) observations of one or more receiver(s). These models, when used for single or dual frequency surveys in the neighborhood, will in most cases reduce ionospheric biases well below the residual level of the observations..

In section 5 we present the current knowledge of combining receivers and antennae from different manufacturers. This problem will become more and more important because there undoubtedly will be many more products on the market. From the economical point of view it is absolutely mandatory to combine observations from all these receivers.

2. FORMING LINEAR COMBINATIONS

Linear combinations of the basic GPS observables

- phases ϕ_1 and ϕ_2
- (P- or C/A-) codes P_1 and P_2

on the two carriers L_1 and L_2 are used for partial or complete elimination of biases and nuisance parameters in the functional model. Such linear combinations can be formed in several ways

- using observations on the same carrier and of the same type but stemming from different receivers, satellites, and epochs (single-, double-, triple-differences)

- using observations of the same type on the two carriers

- using observations of different type (code and phase).

The main purpose of forming single-, double-, and triple-differences is the elimination of satellite clock errors, receiver clock errors (largest part) and initial phase ambiguities, respectively. If the mathematical correlations introduced by such differencing techniques are handled rigorously this procedure leads to the same results as processing undifferenced observations introducing the respective biases as additional unknown parameters (see section 3).

We could even speak of linear combinations of the primary unknowns (the station coordinates), namely the "differential application of GPS." This differencing is usually done implicitly by keeping the coordinates of one station fixed yielding a great reduction of the influences of orbit errors and (to a certain extent) atmospheric refraction.

The main topic of this section is the linear combination of observations on different carriers. The main reasons to form these combinations are the following:

- Elimination of ionospheric refraction to avoid biased solutions.

- Ionospheric refraction and unmodeled orbit errors may make ambiguity resolution on longer baselines difficult or even impossible when using L_1 and/or L_2 observations.

In the following, we describe linear combinations that either lead to solutions unbiased by the ionosphere or help ambiguity resolution on medium range baselines.

Let us start with the basic single-difference phase observation equation involving two stations i and k (the phase observations ϕ_f, f=1,2 have been replaced by their metric representation $\ell_f = \phi_f \cdot \lambda_f$, λ_f being the wavelength of the carrier L_f, f=1,2):

$$\Delta\ell_f + v_f = \Delta\rho + (c-\dot{\rho}_i)\,\Delta t + \Delta\delta_f^{ION} + \Delta\delta^{TROP} + \lambda_f \cdot N_f \qquad (2.1)$$

where

Δ	:	single difference operator (stations i and k)
v_f	:	residual of observation ℓ_f
N_f	:	integer ambiguity
ρ	:	range to the satellite
$\dot{\rho}_i$	:	range rate with respect to station i
Δt	:	synchronization error between receivers
δ_f^{ION}	:	ionospheric refraction
δ^{TROP}	:	tropospheric refraction
c	:	speed of light

Note that by forming the double-difference observation equation (difference of eqns. of type (2.1) for two different satellites j, m) the receiver clock synchronization error still remains in the equation with a coefficient of :

$$\dot{\rho}_i^j - \dot{\rho}_i^m$$

That means that the synchronization error has to be known to about 1 microsecond, which is easily obtainable (even in real-time) using the code measurements. The frequency-dependent ionospheric refraction (meters) is given by

$$\Delta\delta_f^{ION} = -\frac{41 \cdot (E_i - E_k)}{\nu_f^2} = \frac{K}{\nu_f^2}; \qquad f = 1, 2 \qquad (2.2)$$

where: E_i and E_k are the total electron contents of the ionosphere in a cylinder of 1 m^2 cross section along the lines of sight from the two stations i, k to the satellite,

ν_f is the frequency of the carrier f.

The general linear combination of two single-differences (same station, same satellite) observed in L_1 and L_2 may be written as

$$\Delta\ell_j = \chi_{j,1} \cdot \Delta\ell_1 + \chi_{j,2} \cdot \Delta\ell_2 \qquad (2.3)$$

Using the error propagation law (assuming no correlation between $\Delta\ell_1$ and $\Delta\ell_2$) we may compute the corresponding noise

$$m_{\Delta\ell_j} = \sqrt{\chi_{j,1}^2 \cdot m_{\Delta\ell_1}^2 + \chi_{j,2}^2 \cdot m_{\Delta\ell_2}^2} \qquad (2.4)$$

Or, if we postulate identical standard deviations in the phases (measured in cycles)

$$m_{\Delta\phi_1} = m_{\Delta\phi_2} = m_{\Delta\phi}$$

we obtain

$$m_{\Delta\ell_1} = \lambda_1 \cdot m_{\Delta\phi}$$

$$m_{\Delta\ell_2} = \lambda_2 \cdot m_{\Delta\phi} = \frac{\lambda_2}{\lambda_1} \cdot m_{\Delta\ell_1} = \frac{\nu_1}{\nu_2} \cdot m_{\Delta\ell_1} \tag{2.5}$$

and finally:

$$m_{\Delta\ell_j} = \sqrt{\chi_{j,1}^2 + \frac{\nu_1^2}{\nu_2^2}\chi_{j,2}^2} \cdot m_{\Delta\ell_1} \tag{2.6}$$

Starting from the observations $\Delta\ell_1$ and $\Delta\ell_2$ we present in Table 2.1 a summary of the coefficients of the five most important linear combinations, together with the influence of frequency-independent biases (e.g., troposphere, orbits), frequency-dependent biases (ionosphere), and error propagation of the observation noise, all expressed in "length units" as well as cycles, relative to an assumed size of 1 in $\Delta\ell_1$.

LC	λ	LC-Factors		Tropospheric/Orbit Biases		Ionospheric Biases		Observation Noise	
	[m]	$x_{\ell,1}$	$x_{\ell,2}$	[length]	[cycles]	[length]	[cycles]	[length]	[cycles]
L_1	c/ν_1 0.19	1	0	1	1	1	1	1	1
L_2	c/ν_2 0.24	0	1	1	$\frac{\nu_2}{\nu_1}$ 0.78	$\frac{\nu_1^2}{\nu_2^2}$ 1.6	$\frac{\nu_1}{\nu_2}$ 1.3	$\frac{\nu_1}{\nu_2}$ 1.3	1
L_3	0	$\frac{\nu_1^2}{\nu_1^2-\nu_2^2}$ 2.5	$\frac{-\nu_2^2}{\nu_1^2-\nu_2^2}$ -1.5	1	-	0	-	$\frac{\nu_1\sqrt{\nu_1^2+\nu_2^2}}{\nu_1^2-\nu_2^2}$ 3.1	-
L_4	∞	1	-1	0	-	$-\frac{\nu_1^2-\nu_2^2}{\nu_2^2}$ -0.6	-	$\sqrt{1+\frac{\nu_1^2}{\nu_2^2}}$ 1.6	-
L_5	$\frac{c}{\nu_1-\nu_2}$ 0.86	$\frac{\nu_1}{\nu_1-\nu_2}$ 4.5	$\frac{-\nu_2}{\nu_1-\nu_2}$ -3.5	1	$\frac{\nu_1-\nu_2}{\nu_1}$ 0.22	$-\frac{\nu_1}{\nu_2}$ -1.3	$\frac{\nu_2-\nu_1}{\nu_2}$ -0.28	$\frac{\nu_1}{\nu_1-\nu_2}\sqrt{2}$ 6.4	$\sqrt{2}$ 1.4
L_5'	$\frac{c}{\nu_1-2\nu_2}$ -0.34	$\frac{\nu_1}{\nu_1-2\nu_2}$ -1.8	$\frac{-2\nu_2}{\nu_1-2\nu_2}$ 2.8	1	$\frac{\nu_1-2\nu_2}{\nu_1}$ -0.56	$\frac{(\nu_2-2\nu_1)\nu_1}{(\nu_1-2\nu_2)\nu_2}$ 2.8	$\frac{\nu_2-2\nu_1}{\nu_2}$ -1.6	$\frac{\nu_1}{\nu_1-2\nu_2}\sqrt{5}$ -4.0	$\sqrt{5}$ 2.2

Table 2.1
Summary of the linear combinations (factors to apply to metric observations ℓ_1 and ℓ_2)

$L'_3 \rightarrow LC$

Comments on Table 2.1:

(1) The linear combination L_3 does not contain any ionospheric refraction, yielding therefore solutions unbiased by the ionosphere. However, the corresponding linear combination of the integer ambiguities will be a real-valued expression. The noise level of L_3 is about 3 times larger than the one of L_1 The latter fact may be a source of concern, especially on short baselines. Bock *et al.* (1986) and Schaffrin and Bock (1988) discuss this topic in some detail. They state that the "obvious" combination of L_1 and L_2 (eliminating ionospheric refraction) is not always the best choice because "... any source of noise which is dispersive as, e.g., multipath, will be amplified in the ionosphere free linear combination ... ." This is certainly true for small networks where the differential influence of the ionosphere may be very well accounted for by a simple model. Another possibility, also pointed out by Bock *et al.* (1986), would be to introduce into the processing, instead of the L_3 linear combination, the original L_1 and L_2 equations plus a weighted equation for the ionospheric contribution. Depending on the adopted weight the equation system may be varied between complete neglect and a rigorous elimination of the ionosphere.

$L_3 \rightarrow LC$

$L_5 \rightarrow WL$

(2) The linear combination L_4 contains the contribution of the ionosphere plus a real-valued linear combination of the ambiguities in L_1 and L_2 only. It is unaffected by any other influence like orbits, receiver clocks, troposphere. It may be used, e.g., for studies of ionospheric behavior.

(3) The linear combination L_5 is formed by differencing of the L_1 and L_2 phases (expressed in cycles). Its wavelength is about 86 cm ("wide-lane"), and its corresponding ambiguity $N_5 = N_1 - N_2$ keeps its integer nature. It can be used to determine the difference of the basic ambiguities $N_1 - N_2$ on medium range baselines. In a further step we can introduce the known $N_1 - N_2$ into the observation equations for L_3 to resolve the remaining N_1 ambiguity.

(4) If we want to process dual frequency observations from instruments which measured L_2 in units of half-cycles, we had two possibilities to form a "wide-lane type" of linear combination:

- We just form the L_5 linear combination of the observations ℓ_1 and ℓ_2 ($\ell_1 = \phi_1 \cdot \lambda_1$, $\ell_2 = \phi_2 \cdot \lambda_2 / 2$). The resulting linear combination of the ambiguities may be written as

$$(N_1 - \frac{1}{2} N_2) \cdot \lambda_5 = (2N_1 - N_2) \cdot (\lambda_5/2) = N_5 \cdot (\lambda_5/2)$$

which shows that we have to look for an integer multiple of $(\lambda_5/2) = 43$ cm.

- We use another linear combination (L'_5 of Table 2.1), i.e., the difference of the original phase measurements (units of cycles, 1 cycle in L_1 corresponding to $\approx$ 19 cm) with a resulting wavelength of about 34 cm (Allison, 1988).

The latter linear combination however has distinct disadvantages for the resolution of the ambiguities: The influences of wavelength-independent error sources is (expressed in units of the ambiguities) about 1.3, the influence of ionospheric refraction about 2.8 times larger than in the former combination.

After having successfully resolved the ambiguities either of the linear combination L_5 (in multiples of $\lambda_5/2 = 43$ cm) or of the linear combination L'_5 (in multiples of $\lambda'_5 = 34$ cm), we may now use either the ionosphere-free linear combination or the original carriers L_1

and L_2 to solve for the N_1 ambiguities of the L_1 carrier. It is important that for this purpose we are looking for multiples of λ_1 and not, as one might think, for multiples of $\lambda_1/2$.

Linear Combinations of Phase and Code in L_1 and L_2: There were two remarkable papers about this subject at the 1985 Rockville Symposium (Melbourne, 1985; Wübbena, 1985). Both authors independently came up with the same idea: to use a special linear combination w of the two code measurements P_1 and P_2 (on L_1 and L_2) and the corresponding phases ℓ_1 and ℓ_2 to resolve the difference of the L_1 and the L_2 initial carrier phase ambiguities ($N_1 - N_2$). For a specific epoch they put

$$w = \chi_1 \cdot \ell_1 + \chi_2 \cdot \ell_2 + \chi_3 \cdot P_1 + \chi_4 \cdot P_2 \qquad (2.7)$$

where

P_1, P_2, ℓ_1, ℓ_2 are expressed in meters

$\chi_1 = \nu_1/(\nu_1-\nu_2), \quad \chi_2 = -\nu_2/(\nu_1-\nu_2)$

$\chi_3 = -\nu_1/(\nu_1+\nu_2), \quad \chi_4 = -\nu_2/(\nu_1+\nu_2)$

ν_1, ν_2 are the frequencies of the L_1, L_2 carriers

Ionosphere, geometry, satellite- and receiver-clocks cancel out completely when forming the right hand side of eqn. (2.7). What remains is a term proportional to the difference of the initial carrier phase ambiguities N_1-N_2. By collecting all eqns. (2.7) for all epochs (for a special satellite) it is therefore possible to give an estimate for N_1-N_2 unbiased by orbits, station coordinates, atmospheric refraction! The approach has proved to be very successful when processing TI-4100 observations; it will be even more powerful when in the future antennae less affected by multipath will be used.

3. MATHEMATICAL CORRELATION

Forming differences between (quasi-) simultaneous GPS observations is an essential, if not the most important, aspect of all GPS processing techniques: When forming the difference between observations of the same satellite made by two receivers (usually called single difference), the satellite clock error (and the orbit error) are greatly reduced. When forming the difference between the observations of two satellites recorded by the same receiver (let us call this an inter-satellite difference) the receiver clock error is greatly reduced. Therefore the functional model of the so-called double difference (inter-satellite difference of two single differences made by the same receivers) is much simpler than that of the original phase observations: satellite clocks usually do not have to be modeled (compare however section 5), receiver clocks have to be modeled only on the 1 microsecond level, which may be done easily using C/A- or P-code. As usual such advantages have their price. In this case the price is mathematical correlation between $(m-1)\cdot(n-1)$ linearly independent double differences formed from $m\cdot n$ uncorrelated phase observations (m is the number of stations, n the number of satellites at the observation time considered).

Correlations Within Single Baselines

Correlations are easily handled when processing data from a single baseline It is then possible to decorrelate the double difference observations of one epoch using, e.g., a

Gram-Schmidt scheme (Remondi, 1984) or to compute the weight matrices of all simultaneous double differences as the inverse of the corresponding variance-covariance matrix (Beutler *et al.*, 1984). These weight matrices may be computed once and for all if the between-satellite differences are always formed using the same pattern (e.g., satellite 1–2, 2–3, ..., or 1–2, 1–3, ...), because then the weight matrix is uniquely a function of the number of the satellites observed.

Correlations Within a Network

The problem is much more complex if data from more than two receivers are processed. Several algorithms were discussed for that purpose. It is important to state that the result will be identical in all cases:

(1) It is attractive to look at the problem from the physical point of view (Goad, 1985). The reason for forming differences are satellite- and receiver-clocks. An equivalent approach to differencing therefore is to use undifferenced phases (which are uncorrelated) and to introduce per epoch one clock parameter for each satellite and one for each station (but one). These epoch-specific parameters have to be pre-eliminated after each epoch, otherwise the number of parameters in the final adjustment would be far too large.

(2) It is also possible to keep the double difference as basic observable and, as in the single difference case, to compute the weight matrix pertaining to one epoch as the inverse of the corresponding variance-covariance matrix (Beutler *et al.*, 1986; Bock *et al.*, 1986). Very efficient algorithms could be developed when (at a specific epoch) the observation scheme is identical for each receiver. The analysis (and the algorithms) are becoming more and more tedious when more and more data outages occur. If the number of outages equals about 30% of the number of double differences of the complete observation scheme, it becomes more economical to compute the weight matrix starting from the corresponding variance-covariance matrix.

(3) Goad (1988) proposes to decorrelate the double difference measurements of one epoch (Cholesky decomposition of the contribution to the normal equation system at the epoch considered), which may lead to a more efficient updating of the normal equation system with the measurements of one epoch.

General Case: Different Linear Combinations (LC's) of L_1 and L_2 for Different Baselines

Imagine the following two scenarios:

Scenario 1: To measure a network you use both single and dual frequency GPS receivers (an example is given in Figure 3.1). When processing the data you would like to combine the L_1 data of the single (and dual) frequency instruments and the ionosphere-free linear combination (L_3, see Table 2.1) of L_1 and L_2 for the dual frequency receivers in one adjustment process.

Scenario 2: You have measured a network with a mixture of very short (a few kilometers) and long baselines. For the short baselines you introduce the L_1 and L_2 observations (because of the smaller noise) and for the long baselines L_3 (because of the ionosphere bias) into the parameter estimation program.

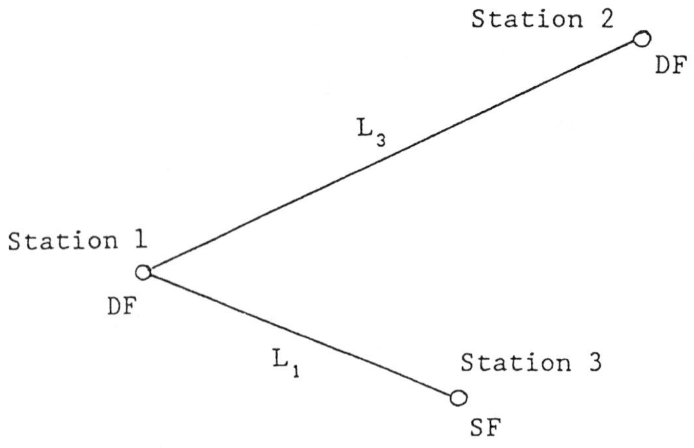

Figure 3.1 Baselines of Scenario 1

SF: Single Frequency Receiver
DF: Dual Frequency Receiver
L_3: Ionosphere free LC of L_1 and L_2

In both cases, if you want to model correlations correctly (not just within the same LC), you have to take into account that the L_1 (and L_2) observations of one baseline are correlated with the LC of another baseline (see Figure 3.1).

Let us therefore derive the covariance matrix for the double difference observations of one epoch in the general case of using different LC's. We may write

$$\nabla\Delta\ell\,| = C \cdot \ell\,| \tag{3.1}$$

with

$$\ell\,| = \begin{pmatrix} \ell_1\,| \\ \ell_2\,| \end{pmatrix} \tag{3.2}$$

where

$\ell_f\,|$: $\ell_f\,| = (\ldots, \ell^k_{i,f}, \ldots)^T$, f=1,2, is the vector containing the undifferenced phase measurements $\ell^k_{i,f}$, of frequency f, station i to satellite k, for one epoch. We assume that the phases are already expressed in meters.

C : matrix to form the double difference observations starting from the original phases $\ell^k_{i,f}$, (C may be called a double difference operator).

$\nabla\Delta\ell\,| : \nabla\Delta\ell\,| = (..., \nabla\Delta\ell^{k_1,k_2}_{i_1,i_2,\,j}\,, ...)^T$

is the vector with a complete set of linearly independent double difference observations of one epoch. Each double difference in $\nabla\Delta\ell\,|$ may be an arbitrary LC j of L_1 and L_2 (e.g., the ionosphere-free LC or wide-lane LC):

$$\nabla\Delta\ell^{k_1,k_2}_{i_1,i_2,\,j} = \chi_{j,1}\left[\left(\ell^{k_1}_{i_1,1} - \ell^{k_2}_{i_1,1}\right) - \left(\ell^{k_1}_{i_2,1} - \ell^{k_2}_{i_2,1}\right)\right]$$

$$+\ \chi_{j,2}\left[\left(\ell^{k_1}_{i_1,2} - \ell^{k_2}_{i_1,2}\right) - \left(\ell^{k_1}_{i_2,2} - \ell^{k_2}_{i_2,2}\right)\right]$$

stand for the double difference observation of stations i_1 and i_2 and satellites k_1, k_2 forming the LC j. $\chi_{j,1}$ and $\chi_{j,2}$ are the coefficients defining the LC of L_1 and L_2 (e.g., for the ionosphere free LC we have (compare Table 2.1)

$$\chi_{j,1} = \frac{v_1^2}{v_1^2 - v_2^2}\,, \quad \chi_{j,2} = \frac{-v_2^2}{v_1^2 - v_2^2}$$

Example: Let us assume that the three stations given in scenario 1 are observing the same three satellites 1, 2, 3. We then have:

$$\ell_1\,| = (\ell^1_{1,1}, \ell^2_{1,1}, \ell^3_{1,1}, \ell^1_{2,1}, \ell^2_{2,1}, \ell^3_{2,1}, \ell^1_{3,1}, \ell^2_{3,1}, \ell^3_{3,1})^T$$

$$\ell_2\,| = (\ell^1_{1,2}, \ell^2_{1,2}, \ell^3_{1,2}, \ell^1_{2,2}, \ell^2_{2,2}, \ell^3_{2,2})^T$$

and matrix C might look as follows:

The covariance matrix of the double difference observations is then given by:

$$\text{cov}\,(\nabla\Delta\ell\,|\,) = C \cdot \text{cov}\,(\ell\,|\,) \cdot C^T \tag{3.3}$$

Assuming a measurement noise of σ_1 for L_1, and σ_2 for L_2 observations the covariance matrix cov (ℓ |) has the following form:

$$\text{cov}(\ell\,|\,) = \begin{pmatrix} \sigma_1^2 E_1 & 0 \\ 0 & \sigma_2^2 E_2 \end{pmatrix} \tag{3.4}$$

where

E_f, f=1,2 are the unit matrices with dimensions equal to the number of L_f phase observations.

σ_f, f=1,2 are the rms errors of the original L_1, L_2 observations.

Finally, the weight matrix P to be introduced into the normal equation system is computed as:

$$P = (\text{cov}(\nabla\Delta\ell\,|\,))^{-1} \tag{3.5}$$

Eqn. (3.3) is the basic matrix equation handling correlations in the general case. Starting from there it is, in principle, possible to proceed in ways analogous to the methods developed for processing single frequency data (or, more precisely, for processing only one frequency (L_1 or L_2 or one linear combination of L_1 and L_2 in the same program run)). It is, e.g., possible to use methods analogous to those presented by (Beutler *et al.*, 1987) to compute matrix P (see eqn. (3.5)) in an efficient way, or to use a Cholesky Decomposition algorithm (Goad *et al.*, 1988) or, what might be an optimum, to develop an algorithm combining the advantages of both methods.

Problem Areas. Two problems may show up when modeling correlations correctly:

(a) *Coordinates:* Let us have a closer look at the coordinates we might expect from a scenario of type 1 (combination of single and dual frequency receivers). It is clear that, if we take into account the correlations within the observations of one LC only (no correlations between L_1 and L_3) the L_1-baselines will be too short due to the ionosphere (see Section 4) whereas the L_3-baselines will remain unbiased by the ionosphere. The covariance matrix of the estimated coordinates will be separated into totally decorrelated parts (L_1 and L_3 parts), which is unrealistic. On the other hand, when modeling correlations correctly between L_1 and L_3, the L_3-baselines, too, will be biased by the ionosphere due to the correlations with L_1-baselines. But the covariance matrix of the resulting coordinates will be fully populated. Introducing an ionosphere model or estimating ionosphere model parameter may help to reduce the ionospheric influence (see Section 4).

(b) *Computer time*: As the number of stations and the number of satellites simultaneously observed increase, correct modeling of mathematical correlations becomes more and more laborious. Whereas the memory needed to compute the weight matrix P may be kept small by just storing the positions of the nonzero elements of the C-matrix (one always has the same pattern: $\chi_1, -\chi_1, -\chi_1, \chi_1, \chi_2, -\chi_2, -\chi_2, \chi_2$), there seems to be no way to avoid the time-consuming matrix multiplications necessary to update the normal equation system at each epoch. Let us give an example: During the EUREF Campaign this year in Europe approximately 60 dual frequency receivers were operating simultaneously tracking up to 6 satellites. The dimension of the weight matrix of such a scenario will be of the order of 300 (= 60·5). Who is going to model correlations correctly in this case?

4. ATMOSPHERE MODELING IN SMALL NETWORKS

We have to distinguish between ionospheric and tropospheric refraction. Ionospheric refraction is caused by free electrons in the upper atmosphere (essential contributions from 200-500 km), tropospheric refraction originates from the lowest layers of the earth's atmosphere (essential contributions from 0-10 km). Tropospheric refraction does not depend on the wavelength whereas ionospheric refraction is wavelength-dependent. This dependence allows us to form a linear combination of the L_1 and the L_2 GPS observables (phase or code) which does not depend on ionospheric refraction (see section 2 for measurement-noise considerations). It is another essential difference that the survey area (antenna points) is lying in the troposphere, whereas ionosphere and survey area are well separated (more than 100 km) from each other. This difference leads to a completely different picture of ionosphere- and troposphere-induced errors into small networks: whereas troposphere biases usually almost uniquely show up in the heights of the estimated (relative) sites, ionosphere biases show up as a scale factor of the network (Beutler *et al.*, 1988a) There is one more important aspect of tropospheric refraction: Standard meteorological equipment in the survey area is a low-cost tool to model the troposphere in the area of the survey, which is moreover independent of the GPS observable. (Other tools like, e.g., water vapor radiometers are even more promising, they are not exactly low cost however; for GPS surveys in small areas they probably will not find widespread use.)

Much of what was said above is, or should be, common knowledge in the GPS community today. Many of the research results reported are not GPS specific, nor have they been obtained using GPS data (alone). It is, e.g., clear that the GPS observable, after ambiguity resolution, is very closely related to the VLBI observable. It is therefore not surprising that the strong correlation between atmospheric zenith delays and vertical position biases was also discovered in the analysis of VLBI data (Herring, 1986). In this reference we also find an empirical description of the vertical position error as a function of the minimum elevation angle observed. The use of water vapor radiometers, which we will not consider in this article (where we confine ourselves to small networks), is also common to the VLBI and the GPS community (see, e.g., Tralli *et al.*, 1988 for a GPS-oriented article).

The GPS is the successor of the Navy Navigation Satellite System (NNSS), for a detailed analysis see e.g. [Wells, 1974]. Many of the basic aspects of the two systems are similar. In particular the NNSS as the GPS space-vehicles transmit two frequencies (150 MHz, 400 MHz for the NNSS). Exactly as in the GPS the two frequencies of the NNSS were used to eliminate the first-order terms of ionospheric refraction. It is therefore not surprising that research concerning the ionosphere was stimulated by some of the institutions working with NNSS data. In this context we would like to mention the order of magnitude prediction for the ionosphere-induced scale error when using GPS L_1 phase observations only (Campbell, 1984) and the research concerning ionospheric models using NNSS data (see, e.g., Leitinger and Putz, 1988). Due to the advent of relatively cheap dual frequency GPS receivers and due to the limited lifetime of the NNSS this approach probably will be of little practical importance in the future.

Ionosphere Modeling for Small Networks

Due to the clear separation between the survey area and the ionosphere the so-called single layer model, where we assume that all free electrons are concentrated in a spherical shell in height H above the surface of the earth, makes sense. Although it is well known that the number of free electrons shows a strong day-time variation (see, e.g., Spilker, 1980) the assumption of a constant electron density is not too bad for order-of-magnitude considerations for a small network. Several authors have studied the biases introduced into a GPS network by a single layer model using constant electron density (Georgiadou and Kleusberg, 1988; Beutler *et al.*, 1988a). Although the actual formulae differ considerably, they all give the same order of magnitude. We reproduce the formula derived by Beutler *et al.* (1988a), where a very simple geometrical approach was used. It is a good approximation for $z_{max} \leq 70°$,

$$\frac{d\ell}{\ell} = -41 \frac{E}{\nu^2} \frac{1}{R} \frac{1}{\cos(z_{max})} \tag{4.1}$$

where:

ℓ	is the baseline length
$d\ell$	is the error in baseline length
E	is the (constant) electron density
ν	is the frequency of the observed carrier
R	is the radius of the earth
z_{max}	is the maximum zenith distance observed.

Writing formula (4.1) for the L_1 frequency of the GPS, using z_{max}=70°, and expressing the scale factor $d\ell/\ell$ in mm/km (or ppm) we obtain

$$\frac{d\ell}{\ell} = -\,0.7 \cdot 10^{-17}\, E \quad \text{(mm/km)} \tag{4.2}$$

The same order of magnitude is found by Santerre (1989). In this excellent analysis one may also find the latitude dependence of biases and rms errors due to the GPS satellite distribution.

Nobody will be surprised that the single layer model with constant electron density is somewhat too simple for real-life situations. It was demonstrated, however, (Georgiadou and Kleusberg, 1988; Wild *et al.*, 1989) that if the electron density in the single layer is modeled as a function of latitude and the hour angle of the sun (using Taylor series developments in these variables), and if the observations of one or more dual frequency GPS receiver(s) near the survey area (let us say within 100 km) are used to estimate the coefficients of these Taylor series, the resulting ionosphere models are good enough to eliminate systematic effects due to the ionosphere from single frequency GPS analyses in small networks.

It seems that this approach gives acceptable results even if the ionosphere shows a very irregular behavior which may be the case in the auroral zones. This aspect was discussed recently by Beutler *et al.* (1988b), Delikaraoglou (1989), and Heroux and Kleusberg (1989). It is clear today that GPS surveys (may) become very difficult in these zones. It is true that single frequency data may become completely useless, since no objective pre-processing will be possible. Heroux and Kleusberg (1989) and Beutler *et al.* (1988b) independently conclude that, with carefully designed pre-processing procedures, dual frequency surveys will be more or less unaffected by these problems. Another important aspect showing up in these investigations is a very close relationship between GPS dual frequency data and the short period variations of the earth's magnetic field (Beutler *et al.*, 1988b). It may very well be that GPS offers an excellent tool to explore the latter phenomenon. We therefore strongly suggest that activities along these lines should be coordinated. We think that the moment has come to bring the specialists in the field (coming from GPS, NNSS, theory) together with the goal to work out processing schemes which as soon as possible should be used for routine processing on permanent tracking sites (let us mention that, e.g., Onsala (Sweden) and Yellowknife (Canada) are tracking sites in the auroral zone).

Troposphere Modeling for Small Networks

When processing GPS observations it was (and is) very tempting to copy the procedure that was so successful in the NNSS processing algorithms, namely to take the Hopfield or Saastamoinen models (Hopfield, 1969; Saastamoinen, 1972) to account for tropospheric refraction, using surface meteorological data (SM) recorded at the receiver sites.

The results of this approach never were really exciting (see, e.g., Tralli *et al.*, 1988 for an example in a regional network; Rothacher *et al.*, 1986, for a small network), they often turned out to be disastrous when processing data from small networks: Instead of having height repeatabilities of a few millimeters (as for the horizontal baseline components), only centimeters were obtained. This is why usually SM were either not observed or not used in processing. What was done instead was to feed the Saastamoinen or Hopfield models not with real data, but with SM values derived from some standard atmosphere (characterized by one set of SM data, e.g., at sea level and by assumption of height gradients for the SM values). This procedure usually gave excellent repeatabilities, especially when the height differences in the network considered were small (100 m or less) and when the weather conditions were comparable throughout the campaign. Does this mean that, at least in small networks, the SM do not contain valuable information? The answer will be a clear "NO." SM do contain very important information when properly used. It has to be accepted however that the classical approach of using SM outlined above is inappropriate for small networks. How can we do better? Let us briefly look at the influence of a troposphere bias on relative positioning before answering this question.

In order to get an impression of the order of magnitude we may very well use the Saastamoinen model. This model tells us that tropospheric refraction correction for an observation made at zenith distance z may be written (approximately) as

$$dr = dr_0/\cos(z) \tag{4.3}$$

where dr_0 is the tropospheric zenith path delay. It consists of a dry delay (main contribution of about 2.2 m, depending on atmospheric pressure only) and a wet delay (depending on temperature and humidity, up to 20 cm for tropical areas). Very critical from the point of view of the model is temperature (and, to some extent, humidity) because the model has to assume gradients, which were the result of some statistical studies, which may be completely wrong for the special case under consideration. The same quantities are also critical from the point of view of the measurement: Some elementary computations (Beutler *et al.*, Table 1, 1988a) show that an error of only one degree Celsius will produce an error in dr_0 of 3 to 27 mm (depending on temperature and humidity). Everybody having worked in practical meteorology must be concerned about these orders of magnitude. Errors of one degree Celsius are certainly not exceptional (instrument calibrations, topography effects). It has to be pointed out that no cancellation on short baselines will occur. The plain difference Δdr_0 between the errors in $dr_{0,i}$, i=1,2 at both ends of the baseline will enter as a bias into computation. Beutler *et al.* (1988a) gives the bias dh in the estimated station height h as a function of the bias Δdr_0 in the tropospheric zenith path delay and the maximum zenith distance z_{max} observed:

$$dh = \frac{\Delta dr_0}{\cos(z_{max})} \tag{4.4}$$

For $z_{max} = 70°$, we obtain:

$$dh = 3 \cdot \Delta dr_0 \tag{4.5}$$

Santerre (1989), using a refined analysis, finds the same order of magnitude. This means, e.g., that an error of one degree in the temperature difference between the two ends of a baseline will, depending on actual temperature and pressure, generate a height bias between 10 and 80 mm!

Let us now repeat the question asked above: How can we make better use of measured SM data: The answer comes when analyzing networks with large height differences (of the order of 1000 m) under very different atmospheric conditions. The puzzle looks as follows: (1) The repeatabilities when using SM in the classical sense are bad (a few centimeters) when comparing results obtained under similar or very different weather conditions. (2) The repeatabilities are excellent when comparing results taken under similar weather conditions using one and the same standard atmosphere in processing. (3) When comparing results taken under very different weather conditions but processed using one and the same standard atmosphere, the two sets of heights may differ considerably (up to 10 cm let us say). The differences however are not random, but simple functions of height (usually, if no special situations like temperature inversions occurred, we see linear functions of height).

Fact (1) is easily explained when looking at eqn. (4.5). Fact (2) is only explainable (at least if we are not analyzing a degenerate network), if in no case horizontal gradients are present. This is the key to the answer of our problem! Fact (3) just shows that the underlying standard atmosphere does not reflect the troposphere in the area between the lowest and the highest point of the survey for one or both of the weather conditions. It is the motivation to use the measured SM data!

How do we make use of surface meteorological data?

- The entire set of surface meteorological data is used to derive a height dependent profile of the atmosphere (temperature, pressure, and humidity) in the layer between the highest and the lowest point of the survey. Usually linear modeling should be good enough, two different linear models may be needed in inversion situations.

- The signal path between a station A (see Figure 4.1) and the satellite S is divided into two sections: section 1 from point A to point A_0, the intersection of the signal path with the horizontal plane in the height h_{A0} of the highest SM station, section 2 from point A_0 to the satellite S.

- Atmospheric refraction is computed as the sum of refraction along AA_0 using the standard formulae for terrestrial refraction (Kahmen, 1978), and refraction along A_0S, where we use the standard Saastamoinen or Hopfield models, assuming a hypothetical station at A_0 and (for all stations) the same SM derived data for the height h_{A0}.

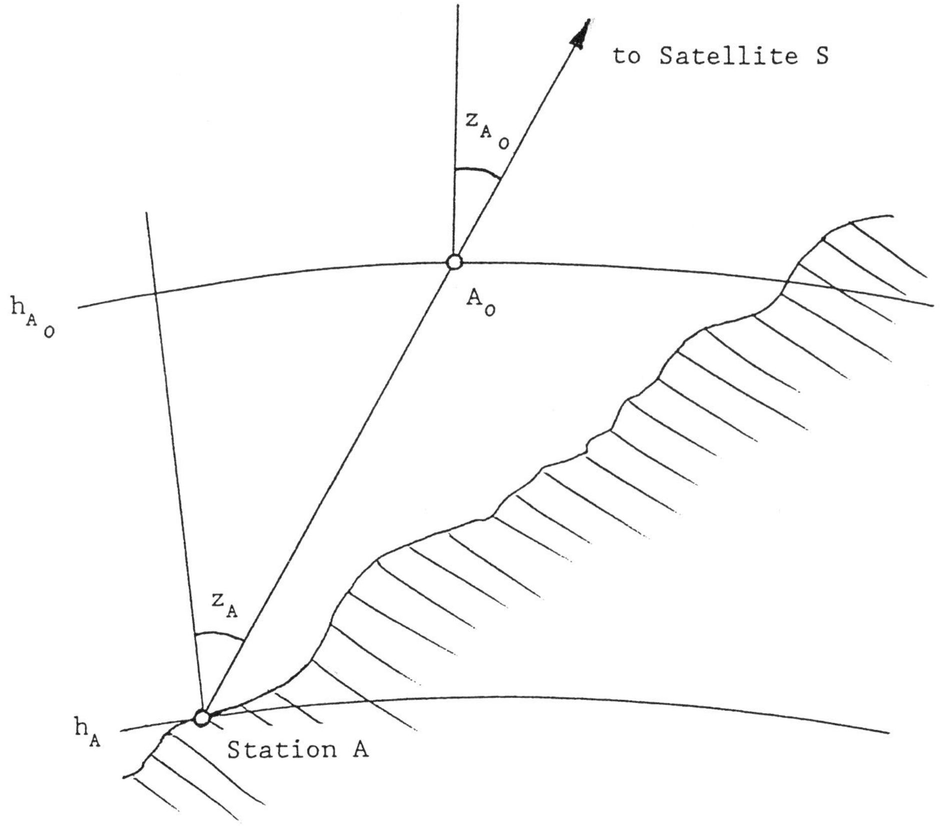

Figure 4.1 Differential Tropospheric Correction

With this procedure we get height repeatabilities comparable to those using standard atmospheres under similar weather conditions. As opposed to the latter procedure we do not have to be afraid of systematic height errors when using the, as we think, only correct procedure. More information concerning this topic and examples may be found in Gurtner *et al.* (1989b).

5. RECEIVER / ANTENNA MIXING

With the availability of several different receiver types and the improvement in the antenna design more and more geodetic users of GPS would like to have the possibility of

- using different receiver types in one and the same survey
- selecting among the available antenna types the one with, e.g., the smallest multipath effects of the smallest phase center variations.

One of the most serious obstacles in the routine mixing of different receiver types is the shortage of commercially available software able to process simultaneously more than one receiver type.

There are several problems to be solved when we want to combine different receiver types on the observation level:

(1) Each receiver type has its own raw data format. Moreover, the basic observable is often defined in a way that is not directly usable by the processing software (e.g., time tag in transmission time rather than in receiver time, phase measurements in fractional parts of cycles only, different time tags for code and phase, no simultaneous observations to the different satellites).
(2) Different sampling times and intervals between the receivers.
(3) Usually each receiver uses its own antenna type with its specific phase center definition, phase center variation and sensitivity to multipath.

Ad (1):
One possibility to overcome the first problem would be to define a unique data interface between the different receivers and the processing software. A first step in this direction is the "Receiver Independent Exchange Format" (RINEX) as it has been proposed by Gurtner *et al.* (1989c) at the Fifth International Geodetic Symposium on Satellite Positioning in Las Cruces.

Ad (2):
When the receivers observe at different times, we have to take into account the behavior of the satellite clocks between the individual sampling times. One possibility is the use of the clock polynomial coefficients as they are transmitted by the satellites in the broadcast message. However, these coefficients only give the overall (and predicted) drift and drift rate: Short term variations cannot be accounted for. It seems that time differences of up to about one second can be bridged by these coefficients without loss of accuracy in the results (Wübbena, 1988; Gurtner *et al.*, 1989a). Some receivers give the operator the possibility to set the observation times and intervals to predefined values. Obviously a synchronization of the observation times is the preferable way to handle this problem.

Ad (3):
Each antenna type (or even each antenna of one and the same type) has its own phase center definition. Strictly spoken each phase observation is biased by a value which may depend on the current direction to the satellite. As long as both antennae at the two ends of a (short) baseline show identical patterns, the resulting baseline will be unaffected. However, as soon as we use different antenna types (with identical or different receiver types), we have to take into account the differences in the patterns. Assuming the existence of a constant (or "mean") phase center, we could try to determine experimentally the

difference of the phase centers of the antenna types involved. If we could assume at least an identical pattern between antennae of the same type it would make sense to determine in the laboratory a correction function (e.g., spherical harmonics in azimuth and zenith angle) for each type, which could then be used to correct the individual phase observations.

Until now many papers have been published dealing with theoretical and practical problems of the definition of the antenna phase centers (e.g., Tranquilla , 1986; Geiger, 1988). Reports about combinations of different receiver and/or antenna types however are not so numerous. Experiences reported by Gurtner *et al.* (1989b) and Rocken *et al.* (1989) on short baseline tests show that such combinations are possible, but that the results do not yet show the same accuracy and reliability as classically observed baselines (on the level of a few mm up to about 1 cm, only). The former paper also contains a first summary of observed phase center differences between several receiver/antenna pairs. The latter paper additionally shows possibilities to improve the performance of a specific receiver by connecting it to better (modern) antenna types.

6. FAST AMBIGUITY RESOLUTION

Since GPS-based surveying techniques have been used in detail surveys, the aspect of productivity has become very important. All techniques proposed so far to increase productivity have one basic objective in common, namely to reduce the observation time required to determine site coordinates. The first published methods and results were mainly based on B. Remondi's kinematic GPS approach, which utilizes the fact that the initial double difference phase ambiguity is a constant in time even if the receiver is in motion (Remondi, 1985). This technique is based on the assumption that receivers do not lose satellites when travelling from one site to another. Considering built-up areas, the potential drawbacks of this approach become evident.

Recently, a fast ambiguity resolution technique has been proposed (Frei *et al.* (1989), which is based on the methods of static positioning techniques. Due to the fact that this technique does not require the receiver to maintain lock to the satellites while moving from one site to another, there are no limitations put on the operational flexibility using this approach. Its performance is almost completely a function of satellite geometry. This allows for an accurate and reliable prediction of what will take place in an actual survey in terms of required site occupation time and achievable accuracy.

We know from experience that measurements to five satellites over an observation period of one hour guarantee ambiguity resolution for short baselines (up to ~10 km). We know furthermore that ambiguity resolution becomes critical (for a five satellite constellation) with only 10 minutes worth of data. In order to study the principles and mechanisms of the ambiguity resolution process some simulations have been carried out. Figure 6.1 shows such a simulation. It presents the development of formal accuracies of the unknown parameters (site coordinates and ambiguities) of a baseline determination with eight satellites. The satellite constellation used to process the simulation shown in Figure 6.1 is a predicted constellation for Los Angeles in 1991 as part of a global 18 satellite constellation.

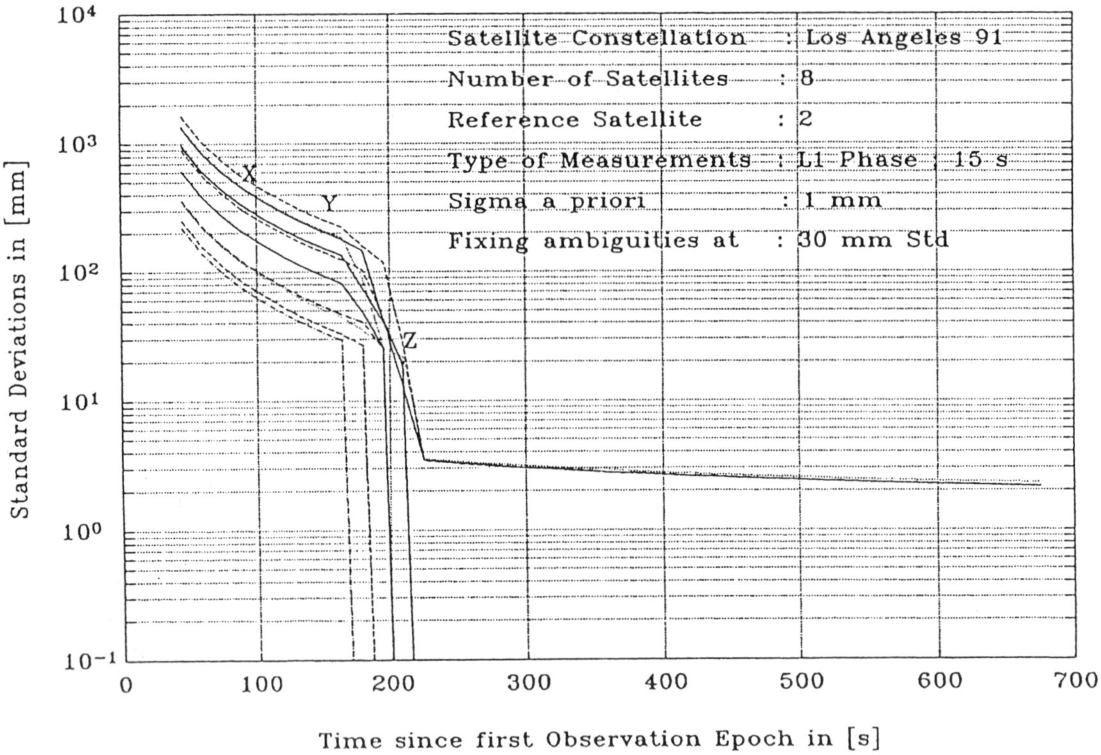

Figure 6.1. Development of formal accuracy as a function of time

To produce Figure 6.1 we have assumed that the ambiguity parameters can be safely resolved as soon as the formal accuracies reach the 30 mm level. As shown in Fig. 6.1 a short baseline can then be determined within 2 to 3 minutes with all ambiguities resolved. Clearly, this can only be achieved in the absence of systematic effects.

The ambiguity resolution approach employed in the simulation above is only suboptimal, because it does not utilize the entire geometrical information contained in the inverted normal equation system. The exploitation of this additional information leads to what we call fast ambiguity resolution technique. It consists of an ambiguity search on an epoch by epoch basis up to the point where all ambiguities can be reliably fixed to integer values. The information in the inverted normal equation system is used to narrow the search-width so that only a small number of integer ambiguity tuples has actually to be analyzed for the determination of the final tuple.

Unfortunately, today there are no eight satellite constellations available. So, the question arises whether there are methods to strengthen the current constellations. Let us consider an observation time span of a few minutes with, let us say, five satellites which did not provide sufficient information to resolve the ambiguities using the above technique. Let us split this observation time span into two halves. What is the contribution of the second half to the determination of a short baseline? Obviously, satellite geometry for both observation periods is almost identical. Therefore, the second half mainly helps to improve the formal accuracies but does hardly provide new geometrical information. Let us move one step further and let us now consider the second half of the observations (still with its original length) to be separated from the first one by, let us say, twenty minutes. Owing to the fact that the satellite geometry has considerably changed during these 20 minutes, the second half of observations does now not only improve the formal accuracies, it also strengthens the determination of the baseline by providing additional geometrical information. This is in a sense as if one would add additional satellites to the first half of the observations. Exactly this additional information is needed for a reliable and fast determination of the "correct" tuple of integer ambiguities.

Simulations and processing of real data with this fast ambiguity resolution technique have shown that it is well possible to resolve ambiguities within short observation periods, especially if repeated short occupations of the same sites are considered. The ability to resolve ambiguities is strongly dependent on the quality of the available satellite geometry. The selection of the "correct" tuple of integer ambiguities becomes critical, if the available satellite constellation does not provide sufficient geometrical information. The techniques outlined above involve ambiguity search strategies which utilize the entire information contained in the inverted normal equation system. Further investigations will be necessary in order to improve the ambiguity search algorithms and to learn more about the operational side of this technique.

REFERENCES

Allison, T., Westfall, B., Eschenbach, R., and Hyatt , R. (1988). Status of dual frequency GPS development at Trimble Navigation, *Lecture Notes in Earth Sciences*, **19**, Springer Verlag.

Beutler, G., Davidson, D.A., Langley, R., Santerre, R., Vanicek, P., and Wells, D.E. (1984). Some theoretical and practical aspects of geodetic positioning using carrier phase difference observations of GPS satellites. *Mitteilungen der Satelliten-Beobachtungsstation Zimmerwald*, **14** and *Department of Surveying Engineering, Technical Report*, **109**, University of New Brunswick, Fredericton, Canada.

Beutler, G., Gurtner, W., Bauersima, I., and Rothacher, M. (1986). Efficient computation of the inverse of the covariance matrix of simultaneous GPS carrier phase difference observations, *Manuscripta Geodaetica*, **11**, 249–255.

Beutler, G., Bauersima, I., Gurtner, W., Rothacher, M., and Schildknecht, T. (1987). Evaluation of the Alaska global positioning system campaign with the Bernese software, *J. Geophys. Res.*, **92**, 1295–1303.

Beutler, G., Bauersima, I., Gurtner, W., Rothacher, M., Schildknecht, T., and Geiger, A. (1988a). Atmospheric refraction and other important biases in GPS carrier phase observations. Monograph 12, School of Surveying, University of New South Wales, Australia.

Beutler, G., Gurtner, W., Hugentobler, U., Rothacher, M., Schildknecht, T., and Wild, U. (1988b). Ionosphere and GPS processing techniques, 1988 *Chapman Conference on the Use of GPS for Geodynamics*, Ft. Lauderdale, USA.

Beutler, G., Bauersima, I., Botton, S., Gurtner, W., Rothacher, M., and Schildknecht, T. (1989). Accuracy and biases in the geodetic application of the global positioning system, *Manuscripta Geodaetica*, **14**, 28–35.

Blewitt, G., Melbourne, W.G., Bertiger, W.I., Dixon, T.H., Kroger, P.M., Lichten, S.M., Meehan, T.K., Neilan, R.E., Skrumeda, L.L., Thornton, C.L., Wu, S.C., and Young, L.E. (1988). GPS geodesy with centimeter accuracy, *Lecture Notes in Earth Sciences,* **19**,Springer Verlag, 30–40.

Bock, Y., Abbot, R.I., Counselman, C.C., Gourevitch, S.A., King, R.W., and Paradis, A.R. (1984). Geodetic accuracy of the Macrometer Model V-1000, *Bulletin Géodésique*, **58**, 211–221.

Bock, Y., Gourevitch, S.A., Counselman, C.C., King, R.W., and Abbot, R.I. (1986). Interferometric analysis of GPS phase observations, *Manuscripta Geodaetica*, **11**, 282–288.

Campbell, J., Cloppenburg, H., and Lohmar, F.-J. (1984). Estimating the ionospheric refraction effect on interferometric GPS measurements, *International Symposium on Space Techniques for Geodynamics*, Sopron, Hungary, 196–206.

Delikaraoglou, D. (1989). On the stochastic modeling of GPS ionospheric delays, *Manuscripta Geodaetica*, **14**, 100–109.

Frei, E., and Beutler, G. (1989). Some considerations concerning an adaptive, optimized technique to resolve the initial phase ambiguities for static and kinematic GPS surveying techniques, *Fifth International Symposium on Satellite Positioning*, Las Cruces, New Mexico, 671–686.

Geiger, A. (1988). Einfluss und bestimmung der variabilität des phasenzentrums von GPS-antennen, *Mitteilung*, **43**, Institut für Geodäsie und Photogrammetrie, ETH Zürich.

Georgiadou, Y., Kleusberg, A. (1988). On the effect of ionospheric delay on geodetic relative positioning, *Manuscripta Geodaetica*, **13**, 1–8.

Goad, C.C., and Remondi, B.W. (1984). Initial relative positioning results using the global positioning system, *Bulletin Géodésique*, **58**, 193–210.

Goad, C.C. (1985). Precise relative position determination using global positioning system carrier phase measurements in a nondifference mode, *First International Symposium on Precise Positioning with the Global Positioning System,* Rockville, 347–356.

Goad, C.C., and Müller, A. (1988). An automated procedure for generating an optimum set of independent double difference observables using global positioning system carrier phase measurements, *Manuscripta Geodaetica*, **13**, 365–369.

Gurtner, W. Beutler, G., and Rothacher, M. (1989a). Combination of GPS observations made with different receiver types, *Fifth International Symposium on Satellite Positioning,* Las Cruces, New Mexico, 362–374.

Gurtner, W., Beutler, G., Botton, S., Rothacher, M., Geiger, A., Kahle, H.-G., Schneider, D., and Wiget, A. (1989b). The use of GPS in mountainous areas, *Manuscripta Geodaetica*, **14**, 53–60.

Gurtner, W., Mader, G., and MacArther, D. (1989c). A common exchange format for GPS data, *GPS Bulletin*, **2**, No. 3, Commission VIII, International Coordination of Space Techniques for Geodesy and Geodynamics (CSTG), and *Fifth International Symposium on Satellite Positioning*, Las Cruces, New Mexico, 920–931,.

Heroux, P., and Kleusberg, A. (1989). GPS precise relative positioning and ionosphere in auroral regions, *Fifth International Symposium on Satellite Positioning*, Las Cruces, New Mexico, 475–486.

Hopfield, H.S. (1969). Tropospheric effect on electromagnetically measured range: Prediction from surface weather data, *Radio Science*, **6**, 357–367.

Kahmen, H. (1978). *Elektronische Messverfahren in der Geodäsie*, Wichmann, Karlsruhe.

Kroger, P.M., Thornton, C.L., Davidson, J.M., Stephens, S.A., and Beckman, B.C. (1985). Sensitivity of GPS Caribbean baseline performance to the location of a southerly fiducial station, *First International Symposium on Precise Positioning with the Global Positioning System*, Rockville, USA, 447–456.

Leitinger, R., and Putz, E. (1988). Ionospheric refraction errors and observables, *Monograph* **12**, School of Surveying, University of New South Wales, Australia.

Lichten, S.M., and Border, J.S. (1987). Strategies for high-precision global positioning system orbit determination, *J. Geophys. Res.*, **92**, 12751–12762.

Melbourne, W.G. (1985). The case for ranging in GPS based geodetic systems, *First International Symposium on Precise Positioning with the Global Positioning System,* Rockville, USA, 373–386.

Remondi, B.W. (1984). Using the global positioning system (GPS) phase observable for relative geodesy: Modeling, processing, and results. *Ph.D. Thesis*, University of Texas at Austin.

Remondi, B.W. (1985). Global positioning system carrier phase: Description and use, *Bulletin Géodésique*, **59**, 361–377.

Röcken, C., Meertens, C.M. (1989). GPS antenna and receiver tests: Multipath reduction and mixed receiver baselines, *Fifth International Symposium on Satellite Positioning*, Las Cruces, New Mexico, 375–385.

Rothacher, M., Beutler, G., Gurtner, W., Geiger, A., Kahle, H.-G., and Schneider, D. (1986). The Swiss 1985 GPS campaign, *Fourth International Geodetic Symposium on Satellite Positioning*, Austin, Texas, 979–991.

Saastamoinen, J. (1972). Atmospheric correction for the troposphere and stratosphere in radio ranging of satellites, *The Use of Artificial Satellites for Geodesy*, Geophysical Monograph 15, American Geophysical Union, Washington, D.C., 247–251.

Santerre, R. (1989). GPS satellite sky distribution: Impact on the propagation of some important errors in precise relative positioning, *Ph.D. Thesis*, Dept. of Surveying Engineering, University of New Brunswick.

Schaffrin, B., and Bock , Y. (1988). A unified scheme for processing GPS dual-band phase observations, *Bulletin Géodésique*, **67**, 142–160.

Spilker, J.J. (1980). GPS signal structure and performance characteristics, *Journal of the Institute of Navigation (U.S.),* **25**, 121–146.

Tralli, D.M., Dixon, T.H. and Stephens, S.A. (1988). Effect of wet tropospheric path delays on estimation of geodetic baselines in the Gulf of California using the global positioning system, *J. Geophys. Res.*, **93**, 6545–6557.

Tranquilla, J.M. (1986). Multipath and imaging problems in GPS receiver antennas, *Fourth International Geodetic Symposium on Satellite Positioning*, Austin, Texas, 557–571.

Wells, D.E. (1974). Doppler satellite control, *Department of Surveying Engineering Technical Report,* **29**, University of New Brunswick, Fredericton, Canada.

Wübbena, G. (1985). Software developments for geodetic positioning with GPS using TI-4100 code and carrier measurements, *First International Symposium on Precise Positioning with the Global Positioning System*, Rockville, USA, 403–412.

Wübbena, G. (1988). GPS carrier phases and clock modeling, *Lecture Notes in Earth Sciences*, **19**, Springer Verlag, 381–392.

Wild, U., Beutler, G., Gurtner, W., Rothacher, M. (1989). Estimating the ionosphere using one or more dual frequency GPS receivers, *Fifth International Symposium on Satellite Positioning*, Las Cruces, New Mexico, 724–736.

NNSS

PRESENT CAPABILITIES OF GPS FOR HIGH-PRECISION REGIONAL SURVEYS

Robert W. King
Department of Earth, Atmospheric, and Planetary Sciences
Massachusetts Institute of Technology
Cambridge, MA 02139 USA

Geoffrey Blewitt
Jet Propulsion Laboratory, California Institute of Technology
Pasadena, CA 91109 USA

INTRODUCTION

Within the next few years the efficiency of high-precision Global Positioning System (GPS) surveys will increase by nearly an order of magnitude with the introduction of lighter-weight and lower-power instruments, the application of kinematic techniques, and the availability of an improved satellite constellation for most areas of the world. There may also be a significant improvement in accuracy from higher data rates and the removal of setup errors with permanently mounted antennas. Much of our knowledge of the capabilities of the system has derived from analysis of static surveys in the southwestern United States, a region for which the Block I constellation has been optimized. This region also includes a large number of sites whose relative positions have been measured independently by the NASA Crustal Dynamics Project and the National Geodetic Survey (NGS) using fixed and mobile very long baseline interferometry (VLBI) systems. By reviewing what has been learned from the five major campaigns and a similar number of smaller ones in California, we attempt to provide a context for surveys which have been and will be carried out in other parts of the world. To complement the review by Beutler and Gurtner (this volume), we concentrate on regional networks, with baselines between 50 and 1000 km.

ACCURACY OF GPS SURVEYS

Figure 1 shows a comparison of GPS and VLBI determinations of baselines in the western U.S., presented by several analysis groups at a recent Chapman Conference sponsored by the American Geophysical Union (Herring and King, 1989). The VLBI values are taken from the "GLB223" analysis of fixed and mobile observations at the Goddard Space Flight Center (J.W. Ryan, C. Ma, and E. Himwich, unpublished results, 1988). The GPS values are a composite of estimates obtained by the Jet Propulsion Laboratory (JPL), NGS, and the United States Geological Survey (USGS) using three different computer codes. Most of these are averages of a few sessions, each of 6-8 hours duration. Some of the VLBI baselines have been measured only a few times and have uncertainties at least as large at those from GPS; for others which have been measured dozens of times, local ties may introduce uncertainties of several mm. Hence, there may be significant contributions to these difference from sources other than errors in the GPS measurements. We infer from Figure 1 that the accuracy of GPS measurements in the southwestern U.S. is no worse than 5 mm + 1 part in 10^8 of the baseline length in the horizontal components, and 20 mm in the vertical, from the equivalent of a few days of observations.

Another measure of the utility of GPS for high-precision geodesy is the long-term repeatability of the measurements. Figure 2 compares GPS and VLBI estimates horizontal components of the 400-km baseline between the Mojave and Vandenberg VLBI sites (K.. M. Larson, personal communication, 1989). Although the GPS measurements span less than three years, compared with six for the VLBI, the estimates of relative station velocity differ by less than 2 mm/yr in either component. The rms scatter of the GPS points about the best-fit slope is at the 1 cm level, comparable to the precision of VLBI.

The accuracy and repeatability results illustrated in Figures 1 and 2 provide clear evidence of the utility of GPS for monitoring crustal deformation in plate boundary regions. In the following sections examine the influence of hardware, experiment design, and analysis software in obtaining results of this quality.

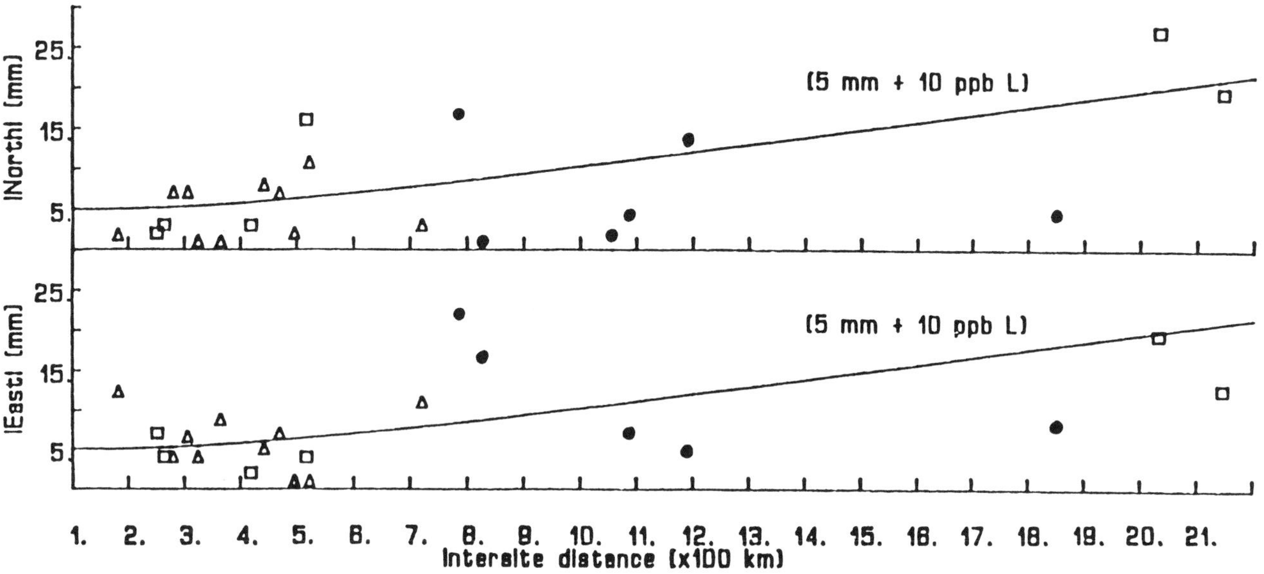

Fig. 1. Differences between GPS and VLBI determined baselines in the U. S. The figure is a composite of the results presented at the Chapman Conference on GPS Measurements for Geodynamics (Herring and King, 1989) by G. Blewitt, S. M. Lichten, U. J. Lindqwister, P. M. Kroger, L. L Skrumeda, M. S. Kornreich, and W. I. Bertiger of JPL; J. L. Davis and W. H. Prescott of USGS; and W. E. Strange of NGS. The different symbols used refer to the number of 3- to 8-hour sessions which have been averaged before being differenced from the VLBI estimates: dots, 1-3; triangles, 4-5; squares, >5.

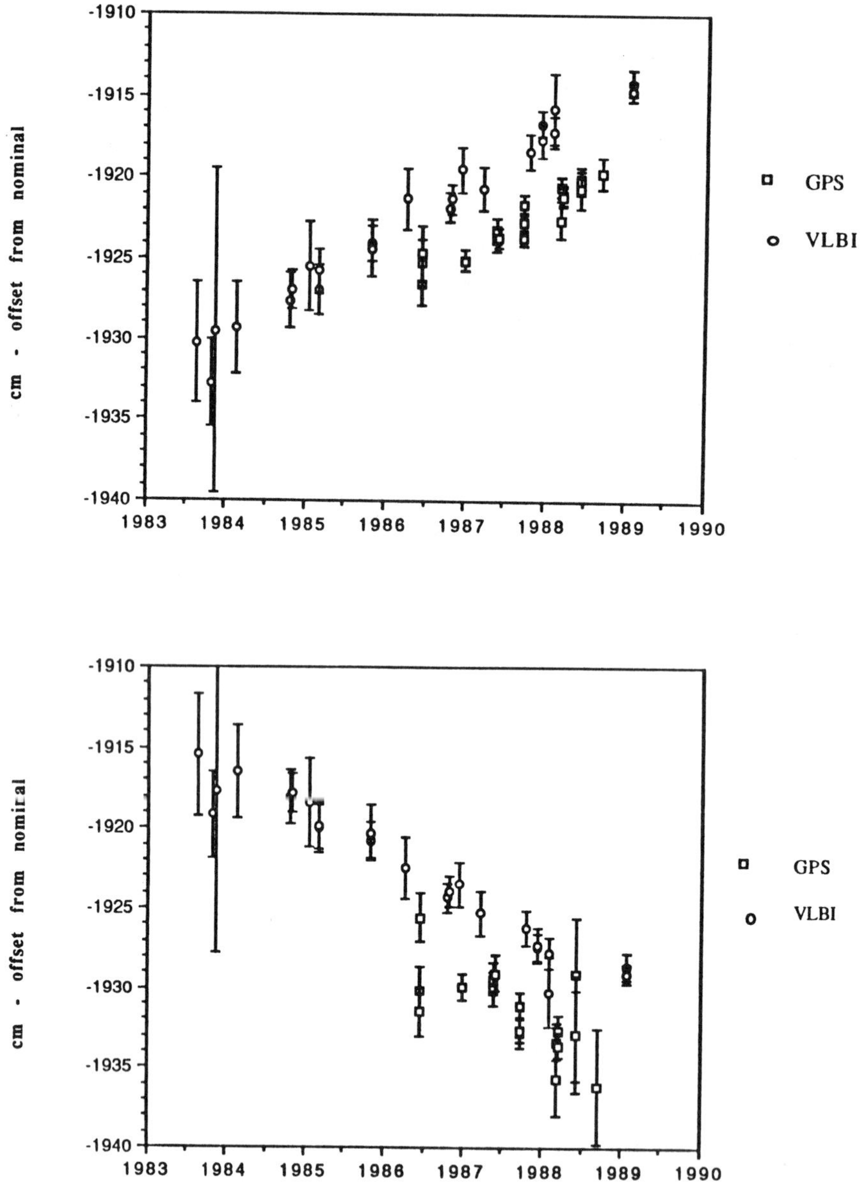

Fig. 2. GPS and VLBI estimates of the north (top) and east (components) of the 400-km baseline between Mojave and Vandenberg. The VLBI values are from the NASA Goddard GLB363 solution (J. W. Ryan, C. Ma, and E. Himwich, unpublished results, 1989). The Mojave GPS observations were obtained by the University of Texas TI-4100 tracker operated continuously by NGS; the Vandenberg observations were made by a USGS field crew, also using a TI-4100 (Davis et al., 1989) The GPS results were obtained by K. M. Larson using the GIPSY (GPS-Inferred Positioning System) software developed at JPL. Each GPS value represents an estimate from 7-8 hours of observation, each VLBI value from 24 hours. The offset between GPS and VBLI points reflects a difference in reference frames and does not affect the slope estimates.

FACTORS IN EXPERIMENT DESIGN

Hardware

The one essential ingredient for high-precision GPS geodesy is the recording of carrier-phase measurements at both the L1 (1575.42 MHz) and L2 (1227.6 MHz) frequencies. Although some investigators have achieved accuracies of a few parts in 10^7 of the baseline length using one or two dual-frequency receivers to calibrate the ionosphere for a small network of single-band receivers (Georgiadou and Kleusberg, 1988) there is no evidence that this technique will work at the level of a few parts in 10^8 for networks several hundred kilometers in extent.

The absence of pseudo-range need not compromise accuracy in a regional network, but it makes both the acquisition and analysis of the observations more difficult. Virtually all state-of-the-art receivers provide at least C/A code pseudo-range for clock synchronization, thus avoiding the requirement of bringing receivers together before and after an observation session to compare their clocks. As we discuss later, P-code pseudo-range can aid both data editing and the resolution of phase ambiguities, particularly in large networks and in the presence of strong ionospheric activity. In the TI 4100, the P-code receiver most commonly used thus far for regional surveys, the usefulness of the pseudo-range measurements has been hampered by high receiver noise and meter-level multipath effects. Nevertheless, Blewitt (1989a) has used it effectively to aid in ambiguity resolution over long baselines. With the Rogue receiver, and perhaps others in the future, pseudo-range can be obtained with a precision of 10-cm or better, enhancing its usefulness for ambiguity resolution and providing additional strength to orbit estimation in cases where phase ambiguities cannot be resolved (Lichten, this volume).

The number of channels in the receiver will become increasingly important as more satellites become available. Although we cannot cite a systematic study, it is clear that the redundancy provided by additional satellites could at times be critical, particularly for short sessions. The receiver manufacturers have obviously appreciated this point since most new receivers can track eight or more satellites. Simultaneous observations from a large number of satellites may also improve the estimation of tropospheric parameters.

Another important factor is the antenna. The phase residuals from most observations, even with good antennas, are dominated by signal multipath effects. The antennas that have been used in controlled tests by several groups show differences of up to a factor of two in suppression of high-frequency multipath, and at certain sites all can show rms phase fluctuations of a centimeter or more (Rocken and Meertens, 1989). If matched antennas are used, the effect on baseline estimates is less than 2 mm in the horizontal components, but up to 5 mm for some antennas in the vertical (Rocken and Meertens, 1989; Davis et al., 1989). It may be that recent tests have not been sufficiently rigorous and the results are unduly pessimistic. There is strong evidence that some antennas, in single-frequency observations over short baselines, can achieve three-dimensional baseline accuracies of 1 mm (Y. Bock, personal communication, 1989). More tests are needed to establish the magnitude of antenna-related effects for dual-frequency observations.

The value of using meteorological instruments in most GPS experiments is slight at best. Measurements of surface pressure can (with some effort) be made accurately enough to calibrate the "dry" component of the tropospheric delay, but surface conditions provide a poor measure of the water-vapor delay. Brightness-temperature measurements with radiometers and solar hygrometers allow the water-vapor delay to be inferred, but not with sufficient accuracy to improve upon the total excess path-delay estimated stochastically in the data analysis. Our covariance studies indicate that 5-mm accuracy in the determination of the zenith tropospheric path delay is required before these measurements can improve GPS baseline estimates. This conclusion is corroborated by analysis of VLBI (Elgared et al., 1989) and GPS (Dixon and Kornreich Wolf, 1989) observations from a limited number of sites.

Reference Frame

Establishing and maintaining a consistent reference frame is the key to accuracy in using GPS for regional surveys. Operationally, the reference frame is defined by the set of regional, continental, and/or global sites whose observations are used to determine the orbits of the satellites. Since the primary observations of the satellites are phase differences between ground stations, they determine most strongly a satellite's motion (or position) along a path parallel to the baseline vector between the stations. At a minimum, one would need a non-degenerate triangle of "fiducial" sites, whose coordinates are known *a priori*, to determine the satellites' trajectories in the plane of the sky. The radial component is determined less strongly through parallax, but also through the dynamics of the satellites' motions. A useful way of estimating the necessary size of the fiducial network is to think of the satellites simply as intermediaries transferring the accuracy of the fiducial-site coordinates to a denser network of stations. The *relative* accuracy maintained will be the same, so the larger the fiducial network the more accurate will be the relative positions of the survey network. In North America, for example, we can use VLBI or SLR sites which are separated by 2000 km and have three-dimensional relative-position accuracies of about 2 cm, or 1 part in 10^8 (Figure 3). Thus, for a regional survey covering 200 km, the North American fiducial network is sufficient for an accuracy of 2 mm.

In practice a network of four or more fiducial sites is desirable to provide redundancy and good tracking coverage of all satellites used by receivers in the region network. The geometric strength provided by additional sites can be important even if their coordinates are not well known *a priori*. In one regional survey in central and southern California, for example, the only VLBI sites outside of California for which GPS observations were available were Westford, Algonquin, Platteville. Adding observations from Churchill to the analysis reduced the regional baseline uncertainties by a factor of two, even though the Churchill coordinates were unconstrained. Similar improvements were obtained for the CASA (Central And South America) UNO experiment (Kellogg et al., 1989) when data from GPS tracking sites at unconstrained locations in the South Pacific were used in addition to North American tracking sites (Kornreich et al., 1989). The use of a global tracking network, with long satellite arcs, is discussed in more detail by Lichten (this volume). Before performing any survey, it is wise to perform computer simulations to determine the tradeoffs between the number of sites simultaneously occupied within the regional network and the number of continental- or global-scale fiducial sites which might be occupied.

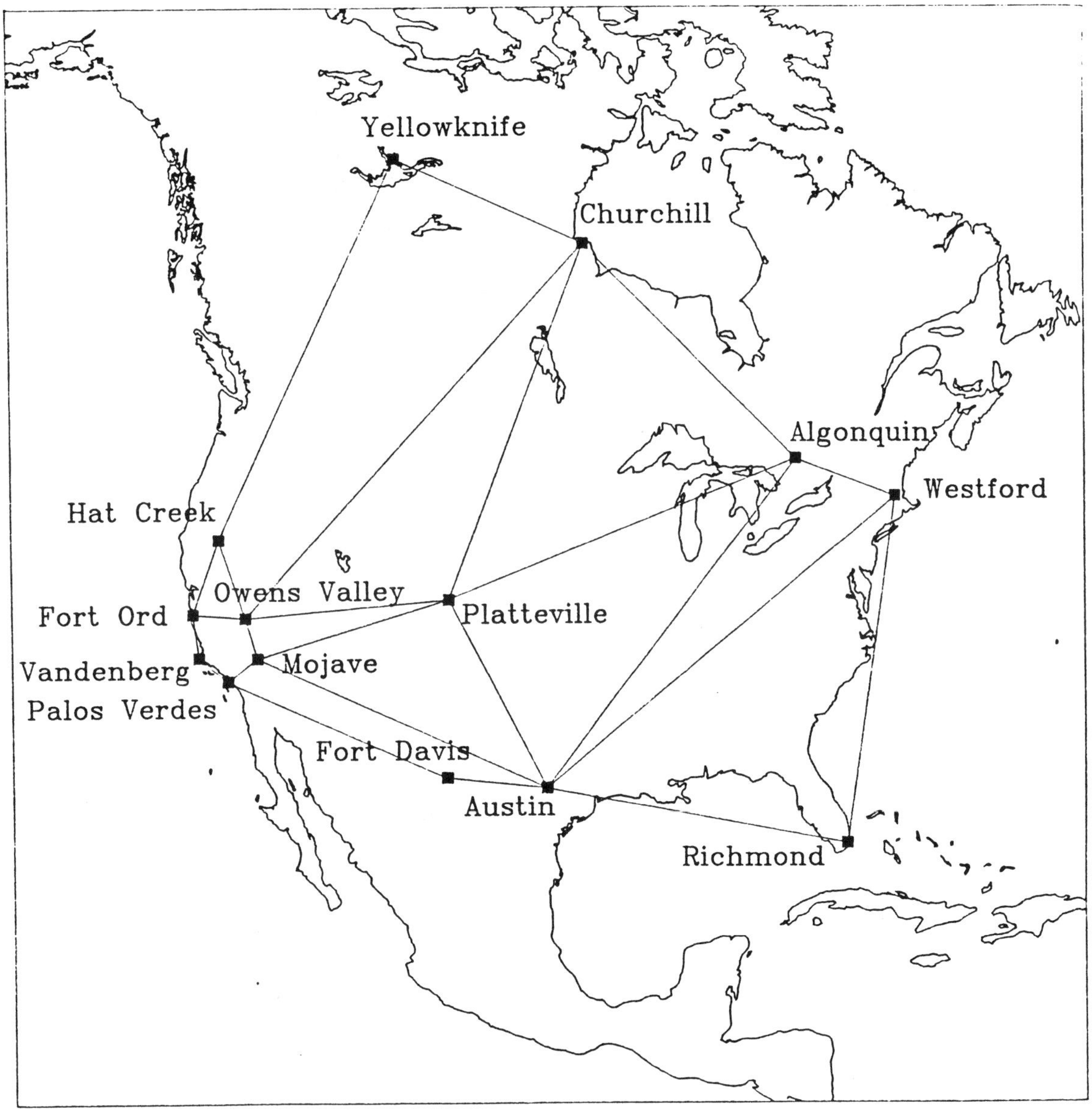

Fig. 3. Superset of the North American sites used for orbit determination in the campaigns described in this review. All except Austin and Churchill have also been used for VLBI observations. Mojave, Yellowknife, Richmond, and Westford now collect GPS data regularly as part of the Cooperative International GPS Network (CIGNET).

Network Design for Ambiguity Resolution

With a fixed observation span, the highest precision is always achieved when the cycle ambiguity in the (doubly differenced) carrier phases is resolved. Ambiguity resolution is significantly enhanced if the network is designed with a wide spread in baseline lengths (Blewitt, 1989a, Counselman and Abbot, 1989; Dong and Bock, 1989). Fiducial baselines of several thousand kilometers are required to establish good orbit control, and regional baselines as short as 50 km may be required in order to initiate the bootstrapping process of ambiguity resolution. As ambiguities associated with the shorter baselines are resolved, the uncertainties of all the estimated parameters in the fit are reduced, including the orbital state vectors, baseline components, and ambiguity parameters for the longer baselines. It is important to overdesign an experiment because one cannot anticipate data outages and/or high ionospheric activity which may make ambiguity resolution more difficult. Moreover, there is no clear answer to how one maintains consistency and reliability in the analysis for those cases where resolution is marginal. We shall address the topic of ambiguity resolution later in the context of data processing considerations.

Length of Observing Session

Like the choice of fiducial stations, developing a criterion for the length of observing sessions requires simulation of actual scenarios for a survey. There may also be practical constraints such as battery capacity, operator fatigue, and travel times between stations. A good rule of thumb is to observe as long as possible on a site without coming up against a limit that would require the expenditure of significantly greater resources. In our California surveys using Block I satellites, sessions encompassed the full time during which at least two satellites were visible, typically between 7.5 and 8 hours. We have tested our results for sensitivity to reducing the length of track. For example, if we reduce the span of data used in the analysis to 6.5 hours, eliminating periods when only two satellites are in view, the formal errors change by only a few percent; however, the baseline estimates often change by 5-15 mm, suggesting that the redundancy provided by the extra tracking time is important in the influence of unmodeled systematic errors.

Number of Sessions per Occupation

Similar arguments can be used in favor of more redundancy in the number of sessions per occupation. Scientific common sense says that observing for only one day risks a complete loss; two days provides a backup but no means to assess the significance of differences between the parameter estimates obtained from each session. Thus, a three-session occupation is the minimum to obtain a useful error analysis. In California, we have used four or five days, with four being a reasonable minimum number from which to derive single-occupation statistics and five providing a cushion to obtain four. For ambiguity resolution, the use of a two-day orbital arc is sometimes superior to a one-day arc since the number of parameters and the correlations among them are both reduced. However, extending the length of orbital arc may in some cases lead to undesirable systematic errors unless care is taken to model satellite forces optimally. This is especially true if the GPS satellites move into the Earth's shadow twice per day (Lichten, this volume).

Our analyses suggest that the errors in the horizontal components of baselines are dominated by fiducial network and multipath, which are not averaged by multi-day occupations. Thus the uncertainty in the estimate from one good session is reduced by less than a factor of the square root of n by adding additional days with the same site geometry. For the vertical component of the baselines, however, the more random tropospheric errors are more nearly equal to fiducial network and multipath effects, resulting in greater improvement from multi-day occupations. Figure 4 compares with VLBI values GPS estimates of the vertical components of 11 baselines (Blewitt et al., 1988). The differences are significantly smaller when the data from four or more days is used in the analysis.

If ionospheric constraints, rather than precise P-code pseudo-range, are used to resolve phase ambiguities, then multiple sessions can be critical to obtaining maximum accuracy. Ionospheric activity fluctuates greatly with time of day, and from day to day. Successful resolution of all the ambiguities may depend on having one period, out of several days of observation, when ionospheric activity is low.

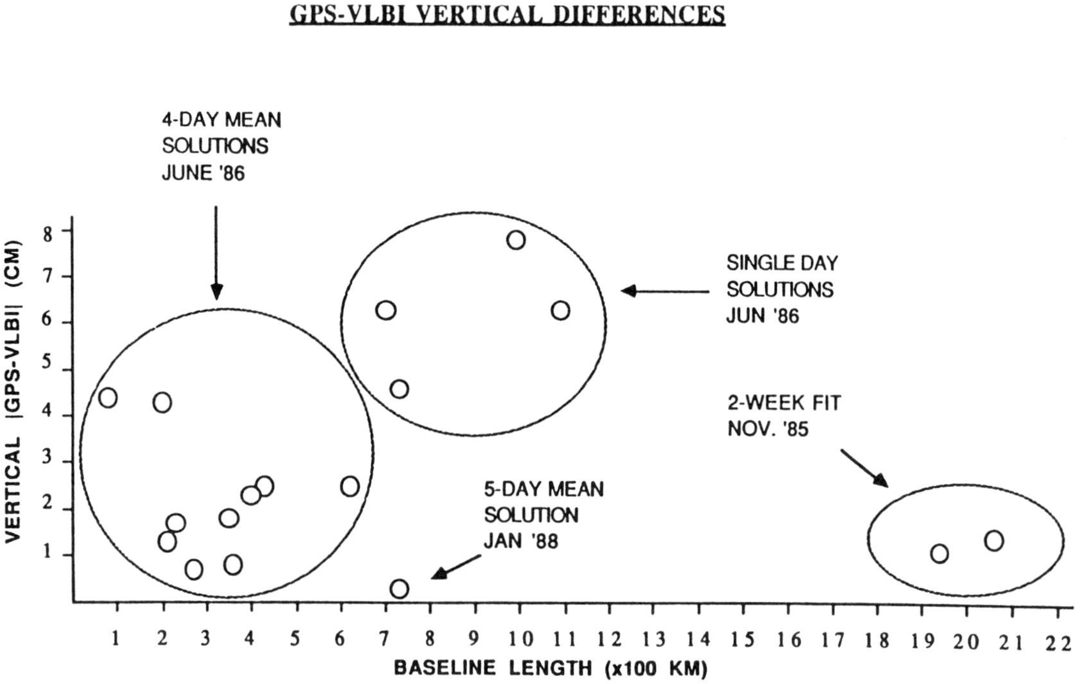

Fig. 4 Differences (absolute value) in GPS and VLBI estimates of the vertical components of 11 baselines. VLBI values are from the NASA Goddard Global 223 solution (J. W. Ryan, C. Ma, and E. Himwich, unpublished results, 1988). The GPS values were obtained from analyses at JPL of data from three experiments in 1985, 1986, and 1988 (Blewitt et al., 1988).

FACTORS IN DATA PROCESSING

Much of the complexity in network design becomes clearer if we look more carefully at factors in data processing. We have identified four that are central to the analysis of regional surveys.

Orbital estimation

The most direct method available for the analyst to place a survey in a well-defined reference frame is by including in the analysis orbital tracking data and a particular set of constrained coordinates of the fiducial stations. There are many different ways that orbital adjustment can be accomplished. If numerical integration codes with complex models for satellite motion are available, then they can be used with little concern for modeling errors except for radiation pressure and other non-gravitational accelerations (Schutz and Tapley, 1980; Abbot et al., 1985; Beutler et al., 1985; Krakiwsky et al., 1985; Rizos and Stolz, 1985; Lichten and Border, 1987; Schutz et al, 1989). But mathematically simpler schemes can work well over short arcs (Beutler et al., 1987; Columbo, 1989). At the other extreme, global tracking coupled with a stochastic model for non-gravitational accelerations can provide geometric strength where good regional fiducial sites are not available (Lichten and Bertiger, 1989). More care is required in this case to assure that re-occupations of a network can be tied to the same effective global reference frame.

Ambiguity resolution

Resolving the integer ambiguity in carrier phase biases effectively converts carrier phase data into very precise data, significantly strengthening the estimates of geodetic parameters. For observation sessions of six or more hours using the Block I satellites, the improvement in both the formal uncertainty and accuracy of satellite orbital parameters and at least one of the horizontal components of a baseline is a factor of 1.5 to 3 (Blewitt, 1989a; Counselman and Abbot, 1989; Dong and Bock, 1989). Figure 5 compares the differences between GPS and VLBI estimates of 15 baselines in the western U.S. obtained with and without ambiguity resolution. In the ambiguity-fixed GPS analysis, both horizontal components show better agreement with VLBI, by a factor of 1.25 in the north and 2.8 in the east. component. The improved geometry proved by the Block II constellation will lessen the impact of ambiguity resolution for long observing sessions, but it will allow significant reductions in the amount time one has to spend on a site. Therefore, the tools needed to resolve ambiguities, obtained with either code-correlating or codeless receivers, should be incorporated into GPS analysis software.

Two recent papers explore in some detail most of the approaches that have been suggested for ambiguity resolution. One by Blewitt (1989a) emphasizes the use of pseudo-ranges; the other by Dong and Bock (1989) emphasizes constraint of the ionosphere. With either approach the objective is to constrain the effect of all sources of error on the L1 and L2 phase observations to a small fraction of a wavelength, roughly speaking a few centimeters.

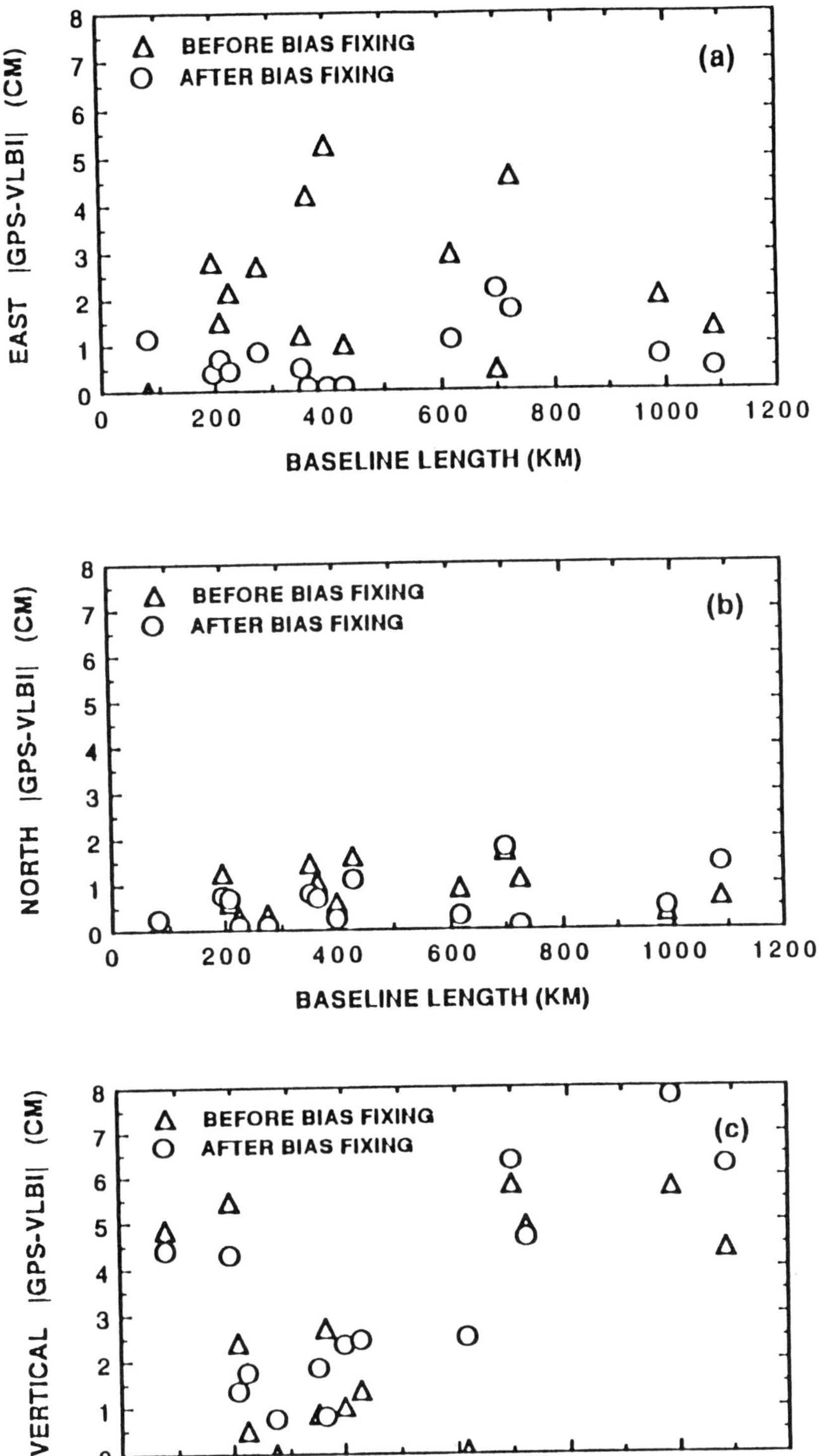

Fig. 5 Magnitude of differences between GPS and VLBI baseline estimates versus baseline length, with and without ambiguity resolution (Blewitt, 1989). The VLBI values are from the NASA Goddard Global 223 solution (J. W. Ryan, C. Ma, and E. Himwich, unpublished results, 1988) evaluated at June 1986, the epoch of the GPS observations.

If sub-decimeter pseudo-range is available for both the L1 and L2 signals, then ambiguity resolution is straightforward. Unfortunately, only the Rogue receiver, not yet available as a true field instrument, comes close to meeting this requirement. In his paper, Blewitt describes the use of much less precise P-code ranges from the TI 4100 to resolve the differences of L2 and L1 ambiguities, the so-called "wide-lane biases" (Counselman et al., 1979; Bender and Larden, 1985; Hatch and Larson, 1985) with an effective wavelength of 86 cm. Since ionospheric effects can be removed for any length baseline, this method is preferred if you have at least decimeter-level pseudo-range at both L1 and L2. It is likely that future receivers will meet this requirement without knowledge of the P-code by employing codeless tracking of the difference between the P2 and P1 pseudo-range.

Without precise pseudo-range, the contribution of the ionosphere must be constrained in the estimation of the wide-lane phase ambiguities. Except at times of peak ionosphere that contribution has been of the order of a few parts in 10^6 even for near-equatorial day-time observations. Thus, on a 50-km baseline, for example, one could reasonably constrain the ionospheric contribution to 10-20 centimeters.

However one resolves the wide-lane ambiguities, a challenge remains in resolving the "narrow-lane" ambiguities in the presence of errors other than the ionosphere. The effective wavelength of the narrow-lane, ionosphere-free combination of L1 and L2 phase observable is 10.7 cm. To resolve this ambiguity for a given baseline and satellite pair, it is necessary to reduce to a few centimeters the effect of errors or uncertainties in fiducial-site coordinates, multipath, tropospheric delays, and satellite motion. Success depends both on good experiment design and accurate modeling of these effects in the analysis software. In repeated surveys intended to measure site motion, it may be possible to "bootstrap" the analysis by initially constraining the coordinates of the entire network at the level of the expected tectonic motion, usually only a few centimeters. Once the ambiguities are resolved, then the *a priori* constraint is removed so that the final result is unbiased by tectonic assumptions. This tool would be particularly useful if not enough data were collected at a particular site to enable ambiguity resolution without constraints. In this way, certain experiments with field problems may be "rescued."

Modeling Atmospheric Effects

Errors in modeling the excess path delay introduced by the neutral atmosphere affect GPS baseline estimates in several different ways. The most direct is the mapping of zenith delay errors into estimates of the vertical component of the baseline vector. However, the atmosphere can introduce errors into the horizontal components through azimuthal asymmetries, due to spatial or temporal variations in the atmosphere, and through unmodeled atmospheric effects at the fiducial sites, resulting in errors in estimates of orbital parameters.

Two strategies have been employed for calibrating the effects of atmospheric delays. One relies primarily on auxillary measurements of meteorological parameters to infer the path delay as a function of time. Another uses the GPS phase measurements themselves to estimate the delay. The advantage of using independent meteorological measurements is that they provide a means to break the correlation between atmospheric delay and station height in the estimation process. The disadvantage is that the reliability and accuracy of the meteorological instruments may not be adequate. Centimeter-level calibration is possible but depends on accurate (<1 mb) measurements of surface pressure, a good temperature-height profile, and reliable water-vapor radiometer (WVR) measurements from which to estimate the usually dominant "wet" component of the delay (e.g., Tralli et al., 1988). In principal, meteorological and GPS measurements can be combined to obtain the optimal estimate of atmospheric and geodetic parameters. Thus far, the information added to GPS analyses by meteorological measurements has produced little or no improvement in accuracy compared with the best estimation strategies using GPS data alone.

The simplest common estimation strategy is to model the atmospheric delay at the zenith by a single parameter for each day and for each site in the network. A more sophisticated approach is to model the zenith delay at each site as a stochastic process, estimating values at each epoch, subject to the constraint implied by the stochastic model. Figure 6 shows the repeatabilities obtained for five daily estimates of the components of a 270-km baseline in Costa Rica using four classes of models for the atmospheric delay: a) WVR calibration with no estimation of atmospheric parameters; b) WVR calibration with estimation of a constant zenith-delay parameter; c) WVR calibration with stochastic estimation of the residual zenith delay; and d) stochastic estimation of the zenith delay with no WVR calibration (Dixon and Kornreich Wolf, 1989). For cases c) and d), the repeatabilities are shown for six different values of the parameter governing how rapidly the zenith delay was allowed to change in the stochastic model. In this example, stochastic estimation improved the baseline repeatabilities by a factor of two compared with WVR calibration with or without estimation of a constant zenith delay. Moreover, adding WVR calibration for the two sites produced no improvement. over stochastic estimation alone.

Data Editing and Compression

One part of the GPS data processing that is particularly susceptible to the whims of the analyst is the data editing, which encompasses both outlier deletion and cycle-slip detection and correction. To ensure reproducible results, and to speed up this historically labor-intensive task, efforts have been made to automate the data editing procedure with robust algorithms. These algorithms rely on frequent sampling of the carrier phase (10–120 seconds, depending on ionospheric conditions) and/or P-code pseudo-range to distinguish cycle slips from short-period fluctuations of the satellite and receiver oscillators, the atmosphere, and the ionosphere (e.g., Beutler et al., 1987; Blewitt, 1989b). These algorithms should tend to err on the conservative side, i.e., by deleting more data points than is perhaps necessary, and by introducing extra phase-ambiguity parameters when there is the slightest probability of incorrectly repairing a cycle-slip. Crude yet robust algorithms (e.g., utilizing triple differences) are very useful in providing "first-cut" solutions.

In order to check the integrity of the data editing process, various display packages have been developed. They usually include the display of phase residuals for various linear combinations of data, e.g., double-differences, the ionospheric combination, etc.

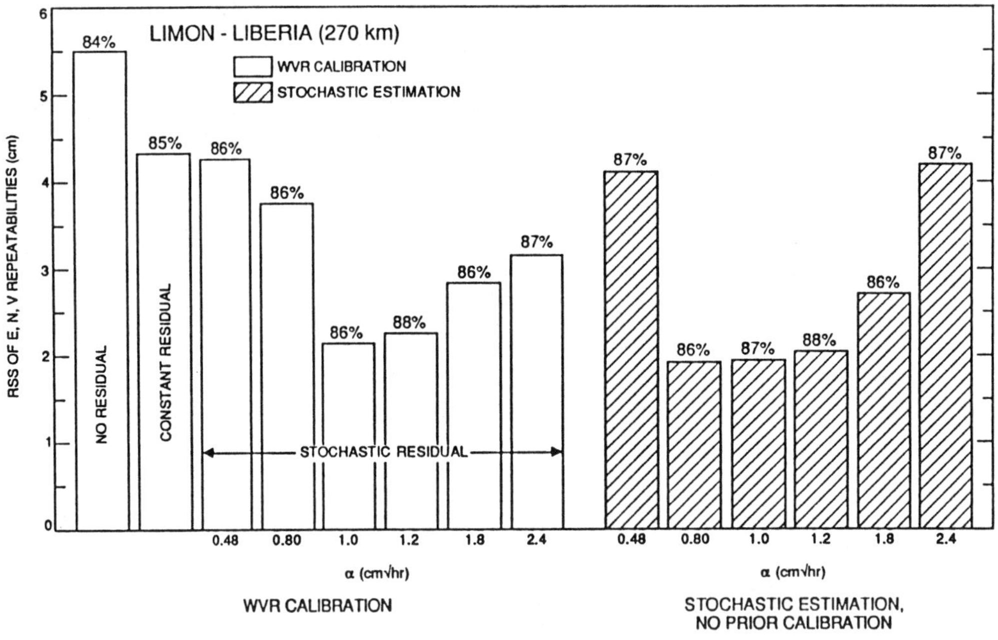

Fig. 6 Repeatability of five independent daily estimates for the baseline between sites on the Caribbean (Limon) and Pacific (Liberia) coasts of Costa Rica. (From Dixon and Kornreich Wolf, 1990) The unshaded bars indicate cases that used WVR calibration, estimating no residual delay (left-most bar), a constant zenith delay at each site (second from left), or a stochastic residual delay (cases 3 - 8). The shaded bars indicate cases using stochastic estimation with no WVR calibration. The best results were obtained with a value of the process-noise rate parameter (α) that allows the residual delay to vary by about 3 cm over the 8-hour period of the observations. Data from the two Costa Rican sites were analyzed with an 8-station subset of the CASA Uno network (Kellogg et al., 1989) using the GIPSY software developed at JPL. In all cases shown, a stochastic model for the atmospheric delay was used for the other six sites. The numbers above each bar indicate the percentage of phase ambiguities resolved in each analysis.

Since no algorithm is completely fail-safe, graphics packages have proved convenient to correct obvious errors, and to simply look at the data for routine checking purposes.

The process of "data compression" serves two main functions: (1) to relieve the computational burden of the data reduction, and (2) to align the time-tags of data from different types of receivers to common epochs. Data compression is usually achieved in one of two ways: (1) data decimation, or (2), fitting the data in short segments to obtain a "normal point" near the center of each segment. The latter option allows alignment of time tags so that data from different receivers can be successfully reduced together.

Generally speaking, a data compression algorithm should be selected which does not render the final solution signal-to-noise limited. Our studies indicate that negligible precision is lost by forming 6 min normal points from 30-second samples, yet it results in an order of magnitude decrease in the required computational time. Strictly speaking, the statistics used in the final least-squares adjustment should account for temporal correlation of the observations (e.g., by multipathing), yet to our knowledge, no software does this. It appears to be adequate to reduce compressed data under the white noise assumption, adjusting the assumed level of noise so that the resulting reduced chi-square of the fit be close to unity. Typical carrier phase noise levels determined by this process are at the level of 5 to 10 mm for compressed 6 minute points.

CONCLUSIONS

Our experience with GPS experiments in North American indicates that 5 to 10 mm horizontal and 20 mm vertical accuracy is currently possible for baselines up to 1000 km. The primary factors in achieving this accuracy are (1) the use of dual frequency receivers, (2) a continental network with at least 3 good fiducial sites, (3) estimation of orbital parameters, (4) long (> 7 hrs) tracks of a good satellite constellation, and (5) the resolution of carrier phase ambiguities.

Increased accuracy can be obtained from the use of pseudo-range receivers, low-multipath antennas, and multi-day site occupations. Data reduction software should include the capability for stochastic troposphere estimation, automatic and manual data editing, and data compression.

ACKNOWLEDGEMENTS

We wish to thank Tim Dixon, Susan Kornreich Wolf, Kristine Larson, Steve Lichten, and Mark Murray for their efforts in providing analysis results and figures used in this review. We also appreciate comments on the text provided by Gerhard Beutler, Yehuda Bock, Tom Herring, Mark Murray, and Catherine Thornton.

REFERENCES

Abbot, R. I., Bock, Y., Counselman III, C. C., and King, R. W. (1985). Interferometric determination of GPS satellite orbits, *Proc. 1st Int. Sym. on Precise Positioning with GPS*, NOAA, Rockville, Maryland, 63-72.

Agnew, D. C., Bock, Y., Jordan, T. H., King, R. W., Dixon, T. H., Hager, B. H., Jackson, D. D., Prescott, W. H., Stowell, J. L., Schutz, B. E., and Strange, W. E. (1987). GPS measurements in Central and Southern California (abstract), *Eos Trans. AGU*, **68**, 282.

Bender, P. L., and Larson, D. R. (1985). GPS carrier phase ambiguity resolution over long baselines, *Proc. 1st Int. Sym. on Precise Positioning with GPS*, NOAA, Rockville, Maryland, 357-362.

Beutler, G., Bauersima, I., Gurtner, W., Rothacher, M., and Schildknecht, T. (1987). Evaluation of the 1984 Alaska Global Positioning System Campaign with the Bernese GPS Software, *J. Geophys. Res.* **92**, 1295-1303.

Beutler, G., Gurtner, W., Bauersima, I., and Langley, R. (1985). Modelling and estimating the orbits of GPS satellites, *Proc. 1st Int. Sym. on Precise Positioning with GPS*, NOAA, Rockville, Maryland, 99-111.

Blewitt, G., Lichten, S. M., Kroger, P. M., Kornreich, M. S., Linqwister, U. J., Skrumeda, L. L, and Bertiger, W. I. (1988). Accuracy and long-term repeatability of GPS baseline estimates, *Eos Trans. AGU*, **69**, 1151.

Blewitt, G. (1989a). Carrier phase ambiguity resolution for the Global Positioning System applied to baselines up to 2000 km, *J. Geophys. Res.* **94**, 10,187-10,203.

Blewitt, G. (1989b). An automatic editing algorithm for GPS data, submitted to *Geophys. Res. Lett.*

Columbo, O. (1989). The dynamics of Global Positioning System orbits and the determination of precise ephemerides, *J. Geophys. Res.*, **94**, 9167-9182.

Counselman, C. C., III, Shapiro, I. I., Greenspan, R. L., and Cox, D. B., Jr. (1979). Backpack VLBI terminal with subcentimeter capability, *Proc. Radio Interferometry Techniques for Geodesy, NASA Conf. Publ. 2115*, 409-414.

Counselman, C. C., III, and Abbot, R. I. (1989). Method of resolving radio ambiguity in satellite orbit dertermination., *J. Geophys. Res.*, **94**, 7058-7064.

Davis, J. L., Prescott, W. H., Svarc, J. L., and Wendt, K. J. (1989). Assessment of Global Positioning System measurements for studies of crustal deformation, *J. Geophys. Res.*, 94, 13,635-13,650.

Dixon, T. H., and Kornreich Wolf, S. (1989). Some tests of wet tropospheric calibration for the CASA Uno Global Positioning System experiment, submitted to *Geophys. Res. Lett.*.

Dong, D., and Bock, Y. (1989). GPS network analysis with phase ambiguity resolution applied to crustal deformation studies in California, *J. Geophys. Res.*, **94**, 3949-3966.

Elgared, G., Davis, J. L., Herring, T. A., and Shapiro, I. I., (1989). Geodesy by radio interferometry: water vapor radiometry for estimation of the wet delay, *J. Geophys. Res.*, in press.

Georgiadou, Y., and Kleusberg, A. (1988). On the effect of ionospheric delay on geodetic relative GPS positioning, *Man. Geod.*, **13**, 1-8.

Hatch, R. and Larson, K. (1985). MAGNET-4100 GPS survey program processing techniques and test results, *Proc. 1st Int. Sym. on Precise Positioning with GPS*, NOAA, Rockville, Maryland, 285-298.

Kellogg, J., Dixon, T., and Neilan, R (1989). CASA: Central and South America GPS geodesy, *Eos Trans. AGU*, **70**, 649-656.

Kornreich, M. S., Dixon, T., and Freymueller, J. T., (1989). The effect of tracking network configuration on GPS baseline estimates for the CASA UNO experiment, *Geophys. Res. Lett.* (in press).

Krakiwsky, E. J., Wanless, B., Buffett, B., Schwartz, K. P., and Nakiboglu, M. (1985). GPS orbit improvement and precise positioning, *Proc. 1st Int. Sym. on Precise Positioning with GPS*, NOAA, Rockville, Maryland, 73-86.

Lichten, S. M., and Border, J. S. (1987). Strategies for high-precision Global Positioning System orbit determination, *J. Geophys. Res.*, **92**, 12,751-12762.

Lichten, S., Towards GPS orbit accuracy of tens of centimeters, submitted to *Geophys. Res. Lett.,* 1989.

Lichten, S. M., and Bertiger, W. I. (1989). Demonstration of sub-meter GPS orbit determination and 1.5 parts in 10^8 three-dimensional baseline accuracy, *Bull. Geod.*, **63**, 167-189.

Murray et al., 1988, 89?

Rizos, C., and Stolz, A. (1985). Force modelling for GPS satellite orbits, *Proc. 1st Int. Sym. on Precise Positioning with GPS*, NOAA, Rockville, Maryland, 87-96.

Rocken, C., and Meertens, C. M. (1989). GPS antenna and receiver tests: multipath reduction and mixed receiver baselines, *Proc. 5th Int. Geod. Sym. on Satellite Positioning*, Physical Science Laboratory, New Mexico State Univ., Las Cruces, 375-385.

Schutz, B. E., and Tapley, B. D. (1980). UTOPIA: University of Texas Orbit Processor, TR 80-1, Center for Space Research, Univ. of Texas, Austin.

Schutz, B. E., Tapley, B. D., Ho, C. S., Rim, H. J., and Abusali, P. A. M. (1989). GPS orbit determination: experiments and results, *Proc. 5th Int. Geod. Sym. on Satellite Positioning*, Physical Science Laboratory, New Mexico State Univ., Las Cruces, 201-209.

Tralli, D. M., Dixon, T. H., and Stephens, S. (1988). The effect of wet tropospheric path delays on estimation of geodetic baselines in the Gulf of California using the Global Positioning System, *J. Geophys. Res.*, **93**, 6545-6557.

Tralli, D. M., and Lichten, S. M. (1989). Stochastic estimation of tropospheric path delays in Global Positioning System geodetic measurements, *Bull Geod..* (submitted).

CONTINUOUSLY MONITORING GPS NETWORKS FOR DEFORMATION MEASUREMENTS

Yehuda Bock[1]
IGPP A-025
Scripps Institution of Oceanography
La Jolla, CA 92093 USA

Seiichi Shimada
National Research Center for Disaster Prevention
3-1 Tennodai, Tsukuba
Ibaraki, 305 JAPAN

INTRODUCTION

A powerful rationale for monitoring the deformations of the earth's crust and of man-made structures is to understand the underlying physics of phenomena that have historically caused major loss of life and destruction of property, most notably earthquakes, volcanic eruptions, and collapse of structures such as dams. We are entering the International Decade of Natural Hazard Reduction. Space geodetic techniques, in particular the Global Positioning System (GPS), will contribute valuable data to the alleviation of natural hazards.

Crustal deformation, earthquakes, and volcanic eruptions are driven by complex interactions between the crust and upper mantle. The transmission of stress in the lithosphere is fundamental to understanding how earthquakes occur and how volcanic eruptions evolve, but the underlying mechanisms are poorly understood. Continuous monitoring with GPS will enable us to characterize and understand the spatial distribution and time dependence of deformation within tectonic regions, from which constraints on the physics of the deformation process can be inferred (Mueller and Zerbini, 1989). Particularly valuable are continuous records of deformation associated with seismic and volcanic activity before and after these events occur. In addition, continuous monitoring networks will provide reference grids for detailed GPS static and kinematic surveys of the earth's crust at regional and local scales. At this stage in GPS development, continuous monitoring will also provide invaluable data for understanding and modeling the GPS error spectrum over a wide range of spatial and temporal scales.

Large engineering structures have failed in the past, sometimes catastrophically, and will continue to do so particularly in a climate of an aging infrastructure. Hydro-electric dams have collapsed causing loss of life in Rempen (1924), Glenau (1932), Stava (1985) in Europe and Saint Francis (1928) and Teton (1976) in the U.S.A. (Krakiwsky, 1986). Traditional geodetic monitoring techniques are too labor intensive and expensive to justify continuous monitoring of critical facilities. Continuous measurements using GPS should prove to be a cost effective method to monitor the integrity of large structures.

We describe the concept and theory of continuous monitoring and present three examples of continuous monitoring systems, two for seismotectonic studies of crustal deformation and one for monitoring the structural deformation of dams.

---

[1] Also at Jet Propulsion Laboratory, Pasadena, California.

CONCEPT OF CONTINUOUS MONITORING

The concept of continuous monitoring with GPS is straightforward. A fixed network of receivers continuously tracks the GPS constellation (whenever visible) for an indefinite period of time, searching for deformation at temporal scales ranging from as short as several minutes to as long as several years. A central facility monitors the performance of the network remotely, automatically and periodically and collects data from all the sites via a high-speed communications link. The data are continuously analyzed at the central facility to obtain accurate "snapshots" of the relative positions of the network stations. Significant variations in these positions indicate deformation within the network. The data are archived for subsequent more refined analysis and disseminated to interested users. The basic building blocks of a continuous monitoring system are depicted in Figure 1. We distinguish continuous monitoring from global tracking for GPS orbit determination which is also performed "continuously." Eventually, this distinction will become obsolete.

The implementation of continuous monitoring with GPS is less than straightforward, at least at this stage of GPS technology development. The challenge is to collect, process, comprehend, archive and disseminate large volumes of data, and to accomplish these tasks efficiently and rapidly. A backlog of data is generally unacceptable because the goal of continuous monitoring is near real-time monitoring of deformation. Of course, there are applications where the data need not be analyzed immediately, for example, in detailed studies of pre-seismic deformation. The most fundamental challenge is to rapidly analyze and correctly interpret the data collected so as to be able to distinguish between deformation signal and measurement noise.

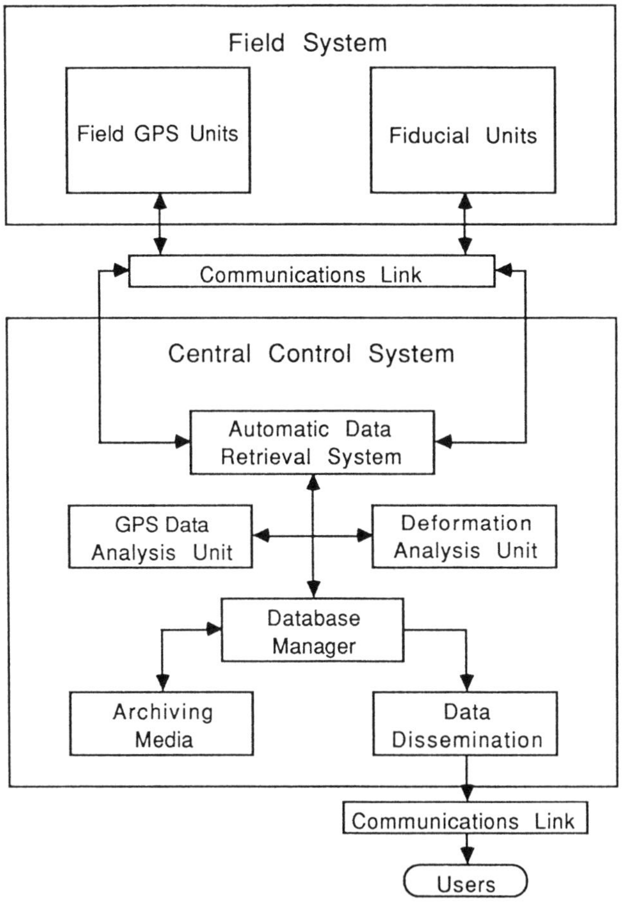

Fig. 1. Components of a continuous monitoring system

THEORY OF CONTINUOUS MONITORING

GPS Error Sources

The models for continuous monitoring with GPS are a direct extension of the static GPS positioning formulation. For satellite i and station j at epoch t_j, the (linearized) observation equations for one set of dual-frequency carrier-beat phase and pseudorange measurements, expressed in units of cycles, are

$$\phi_{1ij}(t_j) = -f_1\tau_{ij}(t_j) + \kappa_{ij}(t_j)/f_1 + \phi_{1ij}^{trop}(t_j) + \phi_{1ij}^{mult}(t_j) + \phi_{1ij}^{clock}(t_j) + b_{1ij} + v_{\phi 1ij}(t_j) \quad (1)$$

$$\phi_{2ij}(t_j) = -f_2\tau_{ij}(t_j) + \kappa_{ij}(t_j)/f_2 + \phi_{2ij}^{trop}(t_j) + \phi_{2ij}^{mult}(t_j) + \phi_{2ij}^{clock}(t_j) + b_{2ij} + v_{\phi 2ij}(t_j) \quad (2)$$

$$P_{1ij}(t_j) = -f_1\tau_{ij}(t_j) - \kappa_{ij}(t_j)/f_1 + P_{1ij}^{trop}(t_j) + P_{1ij}^{mult}(t_j) + P_{1ij}^{clock}(t_j) + v_{P1ij}(t_j) \quad (3)$$

$$P_{2ij}(t_j) = -f_2\tau_{ij}(t_j) - \kappa_{ij}(t_j)/f_2 + P_{2ij}^{trop}(t_j) + P_{2ij}^{mult}(t_j) + P_{2ij}^{clock}(t_j) + v_{P2ij}(t_j) \quad (4)$$

where

f_1 = 1575.42 MHz (L1 frequency), f_2 = 1227.6 MHz (L2 frequency)

τ_{ij} is the geometric travel time of the radio signal between satellite and receiver

κ_{ij} represents the ionospheric refraction effect,which is proportional to the total electron density along the signal propagation path

$\phi_{ij}^{clock}(t_j)$ indicates the time-dependent variations of the transmitter and receiver oscillator phases

ϕ_{ij}^{trop} is the tropospheric refraction delay along the signal propagation path

ϕ_{ij}^{mult} represents the effects of antenna multipath

b_{ij} is the mixture of the integer ambiguity ("bias") n_{ij} and the unknown initial phase offset of the satellite and receiver oscillators

v_{ij} represents the measurement error due to random effects and remaining unmodeled effects

and similarly for the P-code pseudorange terms. The geometric delay is given by

$$\tau_{ij}(t_j) = |\, \mathbf{s}_i(t_j - \tau_{ij}(t_j)) - \mathbf{s}_j(t_j)\,| / c \quad (5)$$

where $\mathbf{s}_i$ and $\mathbf{s}_j$ are the three-dimensional position vectors of the satellite and station, respectively, and c is the speed of light.

The primary utility of P-code pseudoranges for continuous monitoring is in the detection and repair of cycle-slips in the phase data. The procedure for repairing cycle slips is to compute the wide-lane slip at each observation epoch:

$$n_2 - n_1 = \phi_2 - \phi_1 + \frac{f_1 - f_2}{f_1 + f_2}(P_1 + P_2) + v_{n_2 - n_1} \quad (6)$$

P-code pseudorange, because of its longer wavelength (30 m), is much more sensitive to multipath interference than phase with wavelengths of 19.0 cm and 24.4 cm. Therefore, it is necessary to average over a period of time long enough to reduce the effects of multipath to well below the wide-lane wavelength of 86 cm. Once the wide-lane slip is resolved, polynomial fitting to a narrow-lane combination (either the ionosphere combination or the ionosphere-free combination) is used to separate n_1 and n_2 (e.g., Lindqwister et al., 1989). Once the cycle-slips are eliminated, the pseudorange measurements are of limited utility and we can drop them from the discussion. Clock synchronization can be readily accomplished using only S-code pseudorange and ambiguity resolution using pseudoranges is not a critical issue, as described below.

Differencing of the observables between receivers and between satellites (i.e., double differencing) removes the time-dependent variations of both the transmitter and receiver oscillators. Moreover, it cancels the non-integer part of the phase bias term, b_{ij}, leaving the integer part, n_{ij}. The advantage of continuous monitoring is that a static network is being repeatedly surveyed so that the station coordinates are very well known. It is then a simple matter to resolve the phase ambiguities, and assuming that phase lock can be maintained, the ambiguity terms drop out as parameters in the estimation process and the observables from that time on become doubly-differenced unambiguous phase (i.e., range):

$$\Delta^2 r_1 = -f_1 \Delta^2 \tau + \Delta^2 \kappa / f_1 + \Delta^2 r_1^{trop} + \Delta^2 r_1^{mult} + \Delta^2 v_{r1} \quad (7)$$

$$\Delta^2 r_2 = -f_2 \Delta^2 \tau + \Delta^2 \kappa / f_2 + \Delta^2 r_2^{trop} + \Delta^2 r_2^{mult} + \Delta^2 v_{r2} \quad (8)$$

where the scalar double difference operator, Δ^2, denotes a differencing of observables between two receivers and two satellites. These are the basic observation equations for GPS dual-frequency processing of continuous monitoring data. The first step in processing continuous data is to resolve the integer cycle ambiguities from initial network solutions along the lines of Blewitt (1989), Counselman and Abbot (1989) and Dong and Bock, (1989).

Any estimation scheme for continuous monitoring should naturally be sequential, e.g., sequential least squares estimation, Kalman filtering, or sequential robust estimation. Kalman filtering allows a convenient sequential algorithm for modeling the various GPS error sources with temporal as well as spatial correlations. Efficient algorithms are available to update the system of normal equations incrementally (in the least squares case) as well as the solution vector as new observations are collected. The estimation process will differ with the dimensions of the network because of the different error sources that must be modeled.

For small aperture networks (e.g., monitoring deformations on a dam or locally across a fault) the residual effects of tropospheric and ionospheric refraction are so small that they can be neglected and so can errors in the GPS broadcast ephemeris. In addition, once cycle slips are repaired the second frequency is redundant, which reduces the system to the very simple equation

$$\Delta^2 r_1 = -f_1 \Delta^2 \tau + \Delta^2 r_1^{mult} + \Delta^2 v_{r1} \quad (9)$$

The only parameters to be estimated are the three coordinates of each monitoring site for each observation period, keeping fixed the coordinates of several reference marks outside the region of deformation. The only remaining systematic error sources for small aperture networks are multipath effects and instabilities in the site monuments. Multipath errors can be minimized by proper selection of sites (this may be difficult in certain situations) and antennae. Multipath effects tend to average out over observation schedules of several hours. However, for near real-time monitoring of deformations this is not possible and so multipath effects must be considered the main source of error for small aperture networks. Research into the mapping and modeling of the multipath signatures at each site should receive high priority. On the other hand, since the network geometry is fixed and the satellite constellation repeats from day to day (this may change with the Block II orbits), it is not clear how much effect multipath will have on the analysis of deformation. Geodetic monument stability (at the sub-millimeter level) is still an area of active research (e.g., Wyatt et al., 1989). Ultimately, the accuracy obtainable with small aperture networks will depend on the geometric strength of the satellite constellation and the network configuration. With the full Block II constellation, sub-millimeter horizontal and vertical accuracies should be obtainable in near real-time.

For larger networks, we form the ionospheric-free linear combination of unambiguous phase (range):

$$\Delta^2 r_c = -f_c \Delta^2 \tau + \Delta^2 r_c^{trop} + \Delta^2 r_c^{mult} + \Delta^2 v_{rc} \tag{10}$$

where

$$\Delta^2 r_c = \Delta^2 r_1 - \frac{g}{1-g^2} (\Delta^2 r_2 - g\Delta^2 r_1) \; ; \; g = \frac{60}{77} \tag{11}$$

The main error sources for larger networks are due to the satellite ephemerides and to tropospheric refraction. The errors in the satellite ephemerides can be essentially eliminated by well-proven orbit improvement techniques (Lichten and Border, 1987; Blewitt, 1989; Counselman and Abbot, 1989; Dong and Bock, 1989). Depending on the application and the scale of the network, orbit improvement may involve the continuous array sites and possibly GPS fiducial tracking sites well away from the region of interest. In the latter case, the GPS tracking data need to be rapidly available to be effective in near real-time applications, although accurate extrapolations of orbital elements are possible for at least several hours. Another option is to constrain the coordinates of the continuous tracking stations to a priori values which will be well determined from the continuous analysis of the data. A general rule for any scale continuous network is, if possible, to only fix coordinates of stations outside of the expected region of deformation.

Tropospheric refraction, particularly the wet delay caused by water vapor in the atmosphere, is the most poorly controlled of the GPS error sources (Davis, 1986; Treuhaft and Lanyi, 1987). However, except for large-scale networks, the primary effect of this error is in the vertical component. Modeling the tropospheric delay in the zenith direction by a stochastic process has been somewhat successful in improving vertical baseline precision and in orbit determination and is an area of active research (Tralli et al., 1988; Lichten and Bertiger, 1989). For precise continuous array applications, lateral inhomogeneities in the tropospheric delay may be a significant source of error.

Deformation analysis

The analysis of deformation involves comparing repeated estimates of the station positions of the network at some chosen temporal resolution. Consider a network of regional extent or smaller: the phase ambiguities of the network are resolved, cycle-slips are fixed and phase lock is maintained, ionospheric refraction effects are eliminated (at least to first order) with the ionosphere-free linear combination, orbital errors are modeled using data from tracking stations outside of the area of deformation, multipath effects are minimized to a great extent by selecting proper sites and antennae, and tropospheric refraction errors are modeled stochastically (and minimally effect the horizontal). Then, a "snapshot" of the positions of the network stations can be taken every few minutes with respect to a reference frame defined by fixed stations outside of the region of deformation (i.e, the orbit tracking sites), and deformation estimated with respect to the initial configuration of the network with phase ambiguities resolved. The analogy to kinematic surveying with GPS is very clear. With the above scenario, this snapshot should be accurate to the several millimeter level in the horizontal and several centimeters in the vertical and only limited by high-frequency multipath effects and the geometric strength of the network and the satellite constellation. The stations are allowed to move from epoch to epoch, an epoch being the time over which new positions are estimated. The longer the epoch, the more accurate are the snapshots but the coarser the temporal resolution of deformation. The network can be thought of as sampling a deforming body at small, regular intervals of time.

Complications

The worst complications in the scenario mentioned above are due to unexpected changes in the network observing scheme, e.g., when a station drops a satellite or fails altogether. The additions of stations to the network and satellites to the constellation require the estimation of new phase ambiguity parameters. In practice, it may be complicated algorithmically to maintain continuity of the ambiguity parameters because of such changes.

Over the long term any monument instabilities would bias the deformation analysis. Therefore, it is imperative to establish a set of at least three reference monuments near each site and to monitor monument stability on a regular basis, using conventional geodetic instruments or with other GPS receivers. Ideally two sets of reference marks are needed, one set within 100 meters or less of the site to monitor local stability and to recover a damaged mark, and a second set within several kilometers to monitor geological stability.

Solid earth tides and ocean loading tidal effects must be modeled carefully for applications at the several parts in 10^{-8} level. Secular deformations of the network need to modeled in the deformation analysis. The addition of new fiducial stations or updating the coordinates of the old fiducial stations will change the external reference frame. Finally, an increasing number of stations in the network will require a combination of increasingly efficient processing algorithms and more powerful computing facilities.

Continuous Arrays as Reference Frames

Continuous monitoring networks can serve as accurate and convenient reference frames for regional and local GPS static and kinematic deformation surveys. A person with only one receiver could perform at any time a positioning survey with respect to these reference sites. Furthermore, sufficiently dense continuous networks with well known relative positions and well modeled satellite orbits could be used to generate regional maps of ionospheric and tropospheric refraction, that could be used in densification surveys. These applications are meant for offline analysis so the data from the permanent array need to be recorded and made easily available to interested users.

THREE ONGOING CONTINUOUS MONITORING PROJECTS

GPS Fixed-Point Network in Kanto-Tokai District in Japan

A GPS fixed-point network was established in April, 1988 in the Kanto-Tokai district of central Japan for the purpose of earthquake prediction research. Ten GPS receivers were installed at tracking sites shown in Figure 2. In this area, the Philippine and the Pacific plates converge with the main part of Japan. Two large earthquakes of magnitude 7 or greater are expected to occur along the plate boundary of the Philippine Sea plate in the near future (Ishibashi, 1984). The major objective of the network is to monitor the crustal movements within the network and to detect any precursory movements of large inter- and intra-plate earthquakes.

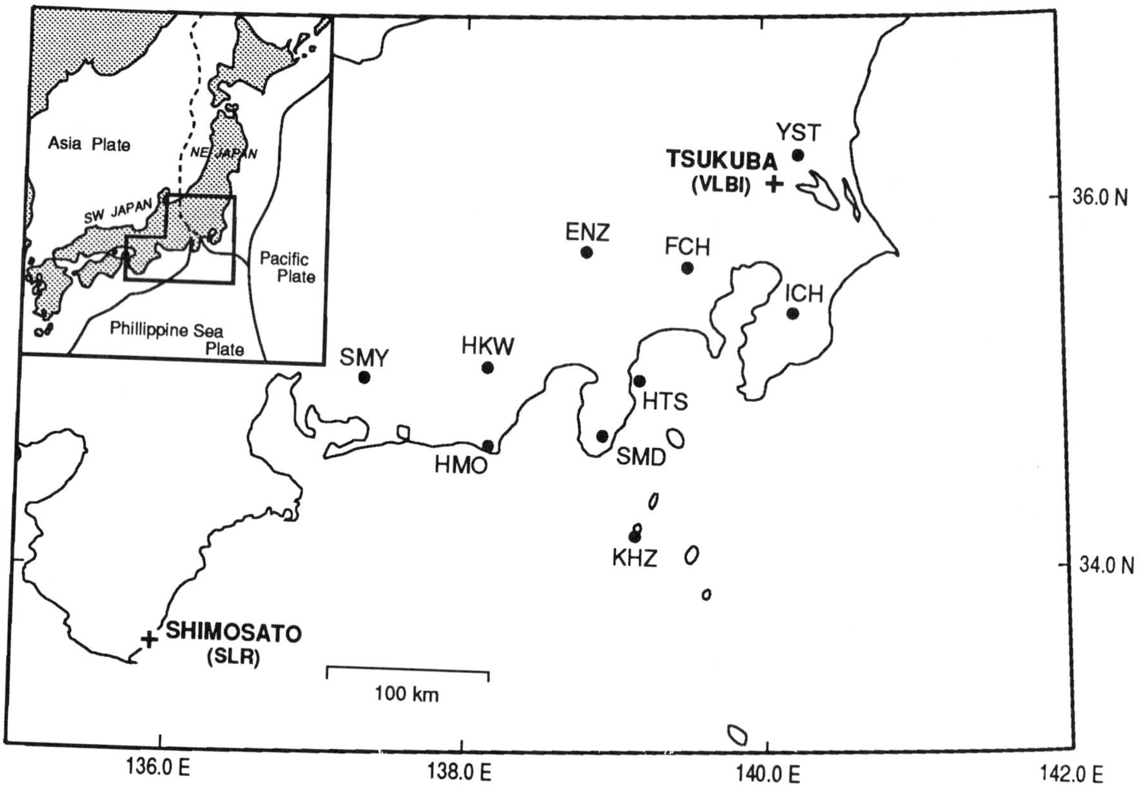

Fig. 2. Locations of the fixed-point GPS tracking sites in the Kanto-Tokai district. NRCDP sites are denoted by solid circles and fiducial points by a cross.

At each tracking site, a Mini-Mac™ 2816 AT GPS receiver automatically tracks GPS satellites every day[2]. The Mini-Mac 2816 AT is a dual-frequency C/A code L1 and codeless L2 receiver which tracks eight satellites simultaneously. It has the capability to maintain more than one month of data continuously tracked with a 30-second sampling rate, as well as surface meteorological data collected for the correction of the tropospheric delay. Usually, tracked data are collected automatically every day at the National Research Center for Disaster Prevention (NRCDP) in Tsukuba, using dial-up communication over public

[2] Mini-Mac is a trademark of Aero Service Division, Western Atlas Company

telephone lines. Details of the hardware composition are described by Shimada et al. (1988). The GPS data are analyzed daily with the SONAP™ (Simultaneous Orbit and Network Adjustment Program) package (Aero Service, 1988). The basic components of the NRCDP fixed-point network are depicted in Figure 3.

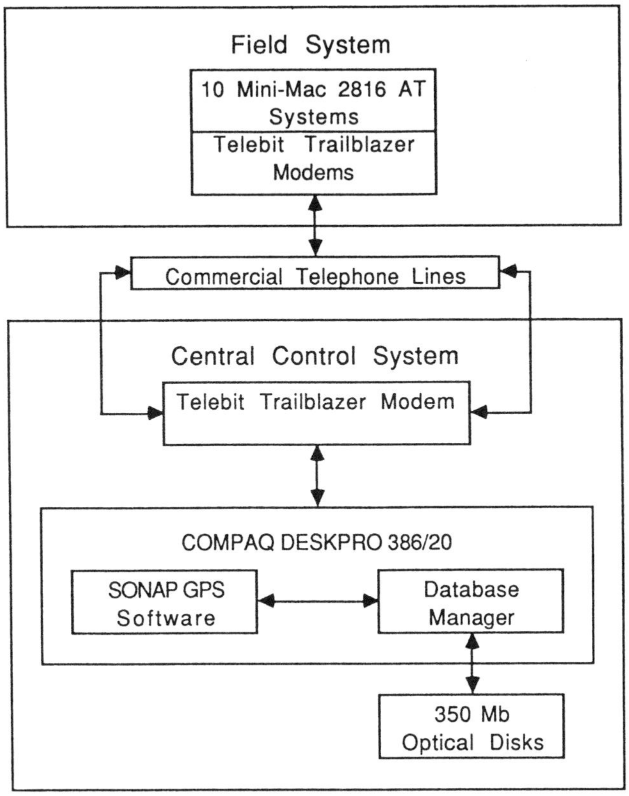

Fig. 3. Components of NRCDP fixed-point network.

In its first year of operation, the satellite constellation over the network was not favorable, and only four satellites in the NW direction were available for the period between the beginning of the observation in 1988 and the spring of 1989 (Shimada et al., 1989). In addition, there are insufficient numbers of fiducial points available in Japan. The initial coordinates of each site of the network were determined by simultaneous GPS measurements in December, 1988 with the Tsukuba VLBI site of the Geographical Survey Institute and Shimosato SLR site of the Hydrographic Department. The errors of the obtained coordinates of the network sites in WGS84 are thought to be about 1 m or less for each component (Sekiguchi et al., in preparation).

Using the initial tracking data for about one month, 28 baselines were obtained between 8 sites in the network using the SONAP package. The positions of 2 sites could not be obtained because of the obstacles in the NW direction. As in our daily analyses, soft constraints were applied to both orbital parameters and site coordinates. The distribution of the deviation of 372 normalized baseline vectors are shown in Figure 4. The vectors are normalized by their baseline lengths, and projected on the XY, YZ, and ZX planes of the WGS 84 geocentric coordinate system. The ellipses are the projection of the 2 sigma error ellipsoid (Fujinawa et al., 1990).

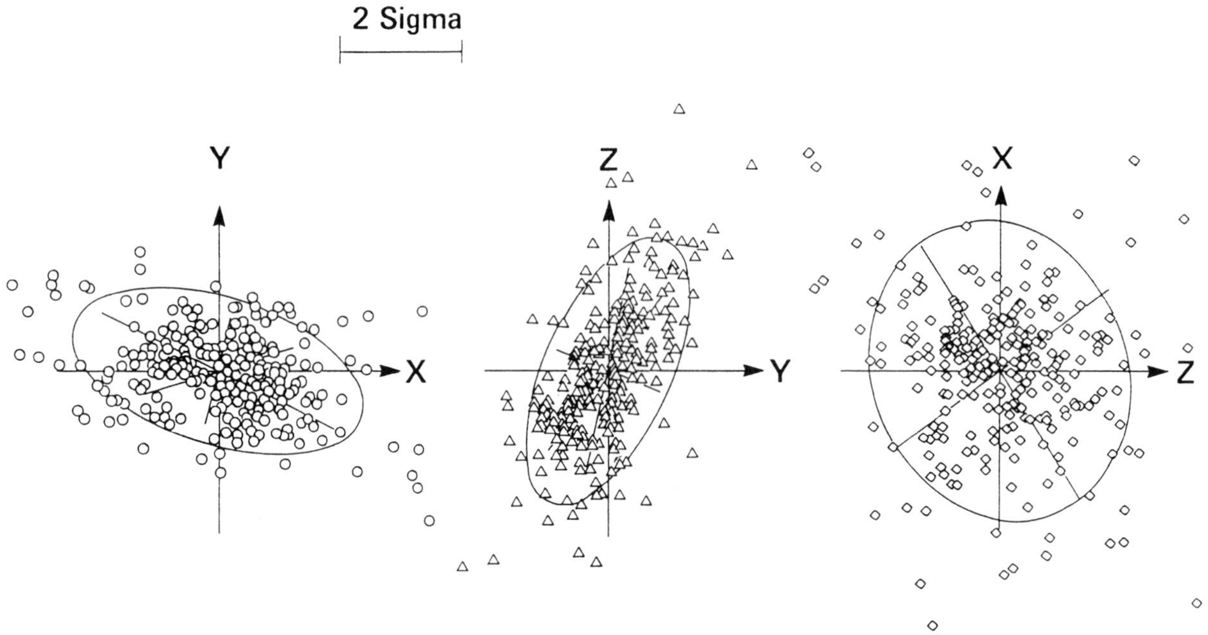

Fig. 4 Distribution of the deviations of the baseline vectors observed by the NRCDP network. Vectors are projected on the XY, YZ, and ZX plane of the WGS 84 geocentric coordinate system. The ellipses indicate projections of the 2-sigma error ellipsoid.

The Gaussian error distribution suggests a statistical approach to estimate the site coordinates. The site coordinates are revised using the average of the obtained coordinates for all sites, except HKW where the position is fixed in the analysis. Then the same data set is calculated again, using the revised site coordinates. This iteration is repeated three times. The error distribution of the site coordinates for each iteration is shown in Table 1.

Table 1. Error of the site positions of the NRCDP network for each iteration. Values indicate the 90% confidence limit obtained from the error distribution.

Site	Component	Iteration 1	Iteration 2	Iteration 3
YST	N-S	15 cm	5 cm	2cm
	E-W	9	3	1
	U-D	15	7	5
ICH	N-S	22	5	3
	E-W	18	4	2
	U-D	19	6	3
ENZ	N-S	3	2	1
	E-W	8	4	2
	U-D	9	10	11
SMD	N-S	14	9	7
	E-W	10	4	3
	U-D	10	7	5
HMO	N-S	5	2	1
	E-W	9	3	2
	U-D	11	4	2

We only show five sites among the seven sites whose coordinates are determined. The tendency of the other two sites is similar. The distribution values in the table are given as those for the 90% confidence limit calculated from the standard deviations using the *t*-distribution. After the last iteration, the root-mean-square repeatability of the component of the baseline vector which is least affected by the unfavorable satellite constellation is 1.9 cm for baseline lengths between 40 and 270 km (Fujinawa et al., 1989). The results suggest that convergence to good initial site positions yields good repeatability in the GPS baseline estimation. Convergence to precise site positions will be more efficient when the full GPS constellation is available and continuous tracking become possible.

A least three fiducial points are necessary to provide a regional or nation-wide terrestrial reference. Although the position of the fiducial points should be determined from VLBI or SLR coordinate solutions, in most regions of the world (except North America and Western Europe) it is very difficult to utilize more than three VLBI or SLR fiducial points with good geometry. It will therefore be convenient to use regional GPS fixed-points as fiducial sites for GPS surveying. The experience gained with the NRCDP network suggests that orbit relaxation techniques can be successfully applied for regional networks. If the deviations of the initial positions of GPS fixed-points are equal or less than 1 m, the NRCDP experience suggests that the relative coordinates of the fixed-points can be determined from repeated measurements with an accuracy of 3 cm and those GPS fixed-points present good fiducials for the GPS measurements within the network. Even if such a regional datum deviates by 1m from the world-wide terrestrial frame, the deviation does not considerably affect the GPS baseline measurements, and is correctable by connecting the regional network with the world-wide network.

In our experience at NRCDP, the GPS measurements often have cycle-slips and data outliers. Cleaning data is the most inefficient aspect of our system, at present, since manual editing is sometimes required after an initial automatic analysis. After data editing and after initial coordinates are estimated, "clean" data sets are saved on mass storage devices. These data sets are usually re-analyzed several times, to improve initial coordinates using the scheme described above, and when improved analyzing techniques or satellite ephemerides become available. In the case of the NRCDP network, 30-second sampled continuous data with eight satellite channels of L1 and L2 carrier-beat phases and S-code pseudo-ranges amount to 450 Kb per site per day, using binary formatted files. One-month of data storage requires more than 130 Mb for a 10 station network. At NRCDP, we utilize 350 Mb per side optical disks for mass storage of raw and "clean" data sets. These have the advantage of reliable long-term storage and random and rapid access to files. A database is necessary to maintain the data and the results of the various analysis schemes.

A comparison of the positions of the NRCDP network in April, 1988 and March, 1989, indicates crustal deformation in agreement with the anticipated motion of the tectonic plates colliding in this area. In addition, a precursory crustal deformation was observed prior to the 1989 volcanic eruption east of the Izu peninsula. A deformation of about 14 cm was observed over a 10 kilometer baseline one week before the eruption. (Fujinawa et al., 1989). This is the first recorded precursory deformation observed with GPS.

Continuously Monitoring of Crustal Deformation in Southern California

The Jet Propulsion Laboratory (JPL) and a consortium of four universities (University of California San Diego (UCSD), Massachusetts Institute of Technology, California Institute of Technology and University of California Los Angeles) are implementing in 1990-1991 a continuously monitoring network in central and southern California (Figure 5). This network will span the entire width of the California plate boundary zone separating the Pacific and North American plates and will provide a reference grid for seismotectonic studies using smaller scale GPS static and kinematic surveys. Any GPS receiver operating in the southern half of the state will be within 100 kilometers of a continuous array station.

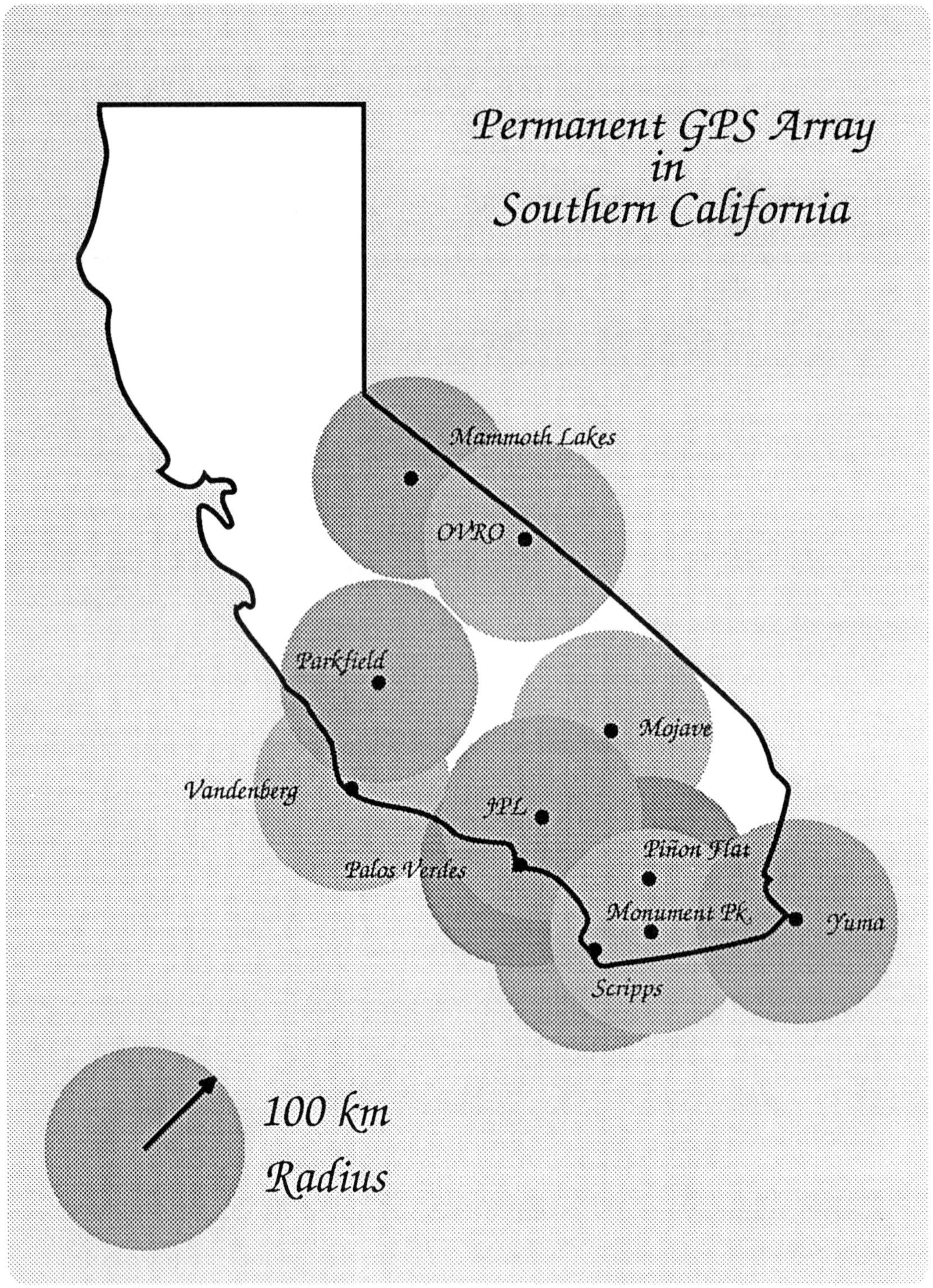

Fig. 5: Distribution of continuously monitoring GPS sites in southern California

The network will deploy SNR-8 Rogue receivers which are capable of both P-code and codeless dual-frequency operation to eight satellites simultaneously. The precise measurement of P-code pseudorange should allow for automatic cycle-slip detection and repair in undifferenced phase measurements which will greatly enhance the efficiency of data processing. JPL engineers are designing automatic in-receiver cycle-slip repair, and data compression schemes to reduce the quantities of data that must be transferred over communication links to the central data facility (Lindqwister et al., 1989). The measurement of precise pseudorange requires a benign local multipath environment. The Dorn-Margolin antenna and choke ring (designed at JPL) have been engineered to reduce multipath effects.

A central control system is being implemented at UCSD. The components of this system are illustrated in Figure 6.

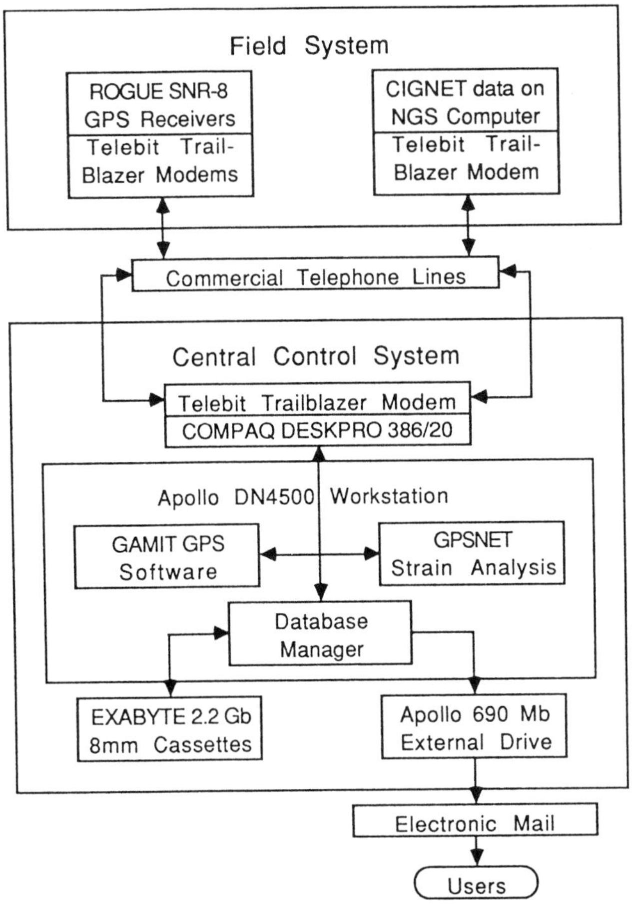

Fig. 6. Components of Permanent GPS Geodetic Array in Southern California

Data will be downloaded daily from each field site. In addition, fiducial data from Mini-Mac 2816 AT receivers tracking at the North American CIGNET stations in Massachusetts, Florida, and California will be collected from the National Geodetic Survey within 24 hours of observation. All data will initially be transferred over commercial telephone lines via high-speed (maximum 19,200 baud rate) Telebit Trailblazer® T2SAA modems[3]. The data transfer will be performed automatically by a COMPAQ DESKPRO 386/20 computer using Co/Session Application Version Communication Software. The COMPAQ is connected to a local Apollo workstation ring as well as an ETHERNET Local Area Network. The data will be transferred to an Apollo DN4500 workstation on the Apollo ring for daily automatic analysis using the GAMIT GPS software package. Initial plans call for the data to be managed using the ORACLE data base management system and stored on 2.2 Gb EXABYTE magnetic cassettes. The data will be maintained on-line in RINEX format (Gurtner et al., 1989) for dissemination to users via electronic mail (initially, anonymous FTP over TELENET).

A stable GPS monument has been designed for use in the network (Wyatt et al., 1989) Two have been already installed at the Scripps and Pinõn Flat sites. The basic form of the monument is: (1) a ground-level base that is anchored to depth, and decoupled from the surface, and (2) an antenna-mount that can be precisely positioned on this base, and can also support other types of measurements, by being able to be removed and precisely reset.

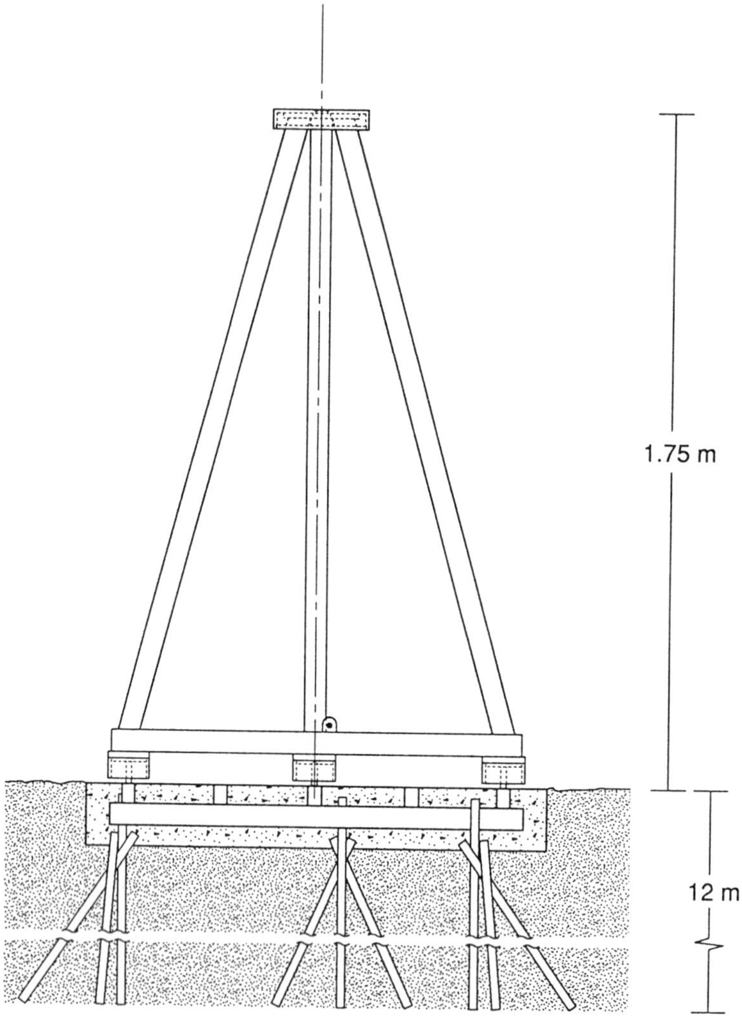

Fig. 7. GPS tripod with deep anchoring for continuous monitoring

[3] Data transfer via communications satellite is being investigated since commercial telephone lines are often unreliable after large earthquakes as was the case after the Loma Prieta earthquake (1989, M7.1).

The initially daily position solutions for the network stations will provide a reference grid (and satellite ephemerides) for GPS surveys of all types in central and southern California, particularly accurate seismotectonic surveys. The data will be used to develop daily ionospheric and tropospheric models for southern California which would enhance rapid densification surveys with highly portable receivers.

The long range goal is a dense network covering the entire state of California to support a reference grid of several millimeter accuracy for the detection and analysis of both steady and transient strains over a broad spectrum of spatial and temporal scales. With a combination of the GPS network with seismic and gravimetric measurements, models of inter- and intra-plate deformation can be developed to clarify the stress accumulation, strain propagation, and the tectonic features of the plate boundary zone, associated with the occurrences of large earthquakes. The reference network would span the entire state with a 100 km grid spacing and a densification to 10 km spacing along a number of active faults and in areas of particular seismic hazard. This reference network would then allow periodic surveys of the entire state at 10 km spacing to be accomplished by a pool of portable receivers and would allow rapid (within hours) deployment of receivers for critical surveys of, for example, post-seismic or inter-seismic strain events.

Continuous Monitoring of Dam Deformations

The behavior of large engineering structures such as dams has been traditionally monitored geodetically by triangulation and trilateration techniques which provide direct, precise measurements of position displacement as a function of time. However, these techniques are quite expensive and are therefore only repeated periodically and generally in cases where there is suspicion of impending structural distress. Continuous monitoring with GPS can provide automatic millimeter-level monitoring of large structures, that would yield a better understanding of their deformations over a wide range of temporal scales.

The U.S. Army Corps of Engineers is responsible for monitoring the integrity of many of the dams in the United States. It is not surprising that they have turned to continuous deformation monitoring with GPS as a practical and cost-effective solution to this problem (DeLoach, 1989). An automatic continuous monitoring GPS system for local scale applications has been developed by TRIMBLE Navigation, Inc. under contract to the U.S. Army Corps of Engineers (Potterfield et al., 1989). The system managed automatically by two near-site personal computers supports a network of up to ten GPS receivers connected to the computers with dual twisted-pair wire or fiber optic cables. Two receivers are intended for occupation of two stable reference monuments in the vicinity of the structure, and up to eight receivers are used on the structure itself. Each computer operates in a continuously looping cycle of epoch observations and processing. An epoch is the basic unit of the deformation analysis during which the structure is assumed to be static. In the current system, epoch length can be set to one hour, at a minimum. The deformation analysis is based on a comparison of the positions between two epochs, using standard static GPS positioning techniques to analyze the data collected during each epoch. A third personal computer at a remote location allows an individual to access the system through a telephone modem.

In tests of the technique, induced motions of 3 mm to 60 mm were imparted to GPS antennae on short baselines. These test results indicate measurement precision in the subcentimeter range (S. DeLoach, personal communication).

A demonstration of the continuous deformation monitoring system (CDMS) was held in August 1989 at the U.S. Army Corps of Engineers Dworshak Dam Project on the north fork of the Clearwater River near Orofino, Idaho. The Dworshak Dam is the highest straight-axis, concrete gravity dam in the Western world with a height of 218 m. Four antennas were placed on the crest of the dam and four receivers just below the roadway along the crest. Each receiver was supplied with AC power, and a fiber optics cable was the data link from the receivers to a multiport data switch. Another fiber optics cable ran from the data switch to the two personal computers operating in the powerhouse at the base of the dam. Two receivers were placed at stable reference sites, one 1.5 km upstream and one 3 km downstream from the dam. Each of these units was installed in a secure building with AC power and telephone lines. Auto-dial modems were used to transmit data to the personal computers in the powerhouse. The CDMS operated for a three-month period taking surveys of the structure daily. At the time the reservoir was undergoing drawdown. The CDMS reported that the structure was creeping upstream which was confirmed from optical plummet readings (S. DeLoach, personal communication). CDMS precision analysis is still ongoing.

FUTURE PROSPECTS FOR CONTINUOUS MONITORING

Continuous monitoring with GPS is still in its very early stages of development. In the next few years we can expect to see a proliferation of continuous monitoring networks not only to monitor deformation on land but also at sea, for example, to monitor sea level variations and to support sea floor positioning experiments. An exciting new development in continuous monitoring is the proposed FLINN network.

A Global Strain Network

The FLINN network (Fiducial Laboratories for an International Natural science Network) was proposed by a panel of scientists at the July, 1989 NASA planning workshop at Coolfont, West Virginia (Minster at al., 1989). It would comprise about 200 GPS receivers distributed world-wide with an intersite spacing of approximately 1,000 km, and would therefore provide a framework for the study of plate motion and deformation on a global scale. This density of sites would mean, for example, that 30 of them would be located in North America. A subset of these sites will be co-located with VLBI/SLR sites to insure accurate relative positions of widely separated monuments. The stated goal of FLINN is to maintain a knowledge of relative site positions with a precision of 1 cm over a 1 day interval, and 1 mm over three months, on a continuous basis. Selected sites would be equipped with an expanded set of geodetic, geophysical, and environmental sensors.

FLINN should result in better measurements of crustal deformation, improved understanding of the earthquake cycle, and much improved constraints on sea-level change. Its existence would lead to highly accurate orbits for GPS, and of course to the establishment of a very accurate fiducial network for local and regional GPS geodesy.

Other Geodetic Systems for Continuous Monitoring

Other radio tracking systems have been proposed for continuous monitoring applications but are still many years away. The "Super-GPS" concept includes a satellite constellation similar to GPS but designed specifically for geodetic and precise orbit determination applications. By broadcasting a well chosen set of tones, data analysis would be vastly simplified and instantaneous sub-centimeter accuracy could be determined from a strictly geometric solution (Mueller and Zerbini, 1989). A second system is called "Geo-Beacon." In this concept, many low power radio transmitters are placed on points to be monitored. Signals from all transmitters in view are relayed by small repeater satellites to a central site where detection, measurement and data processing take place. To enable precise positioning signals are emitted at multiple frequencies (Cangahuala and Counselman, 1989).

CONCLUSIONS

The deployment of the full Block II GPS satellite constellation will allow true continuous monitoring of deformations within two to three years. The widespread deployment of continuous GPS arrays will be driven primarily by technological developments: the availability of low-cost, robust GPS receivers, the development of inexpensive, reliable, and high-rate telemetry links, and the development of tools (hard and soft) for manipulating, storing and comprehending large volumes of data. The payoff from continuous monitoring in understanding the physics of phenomena driving deformations and reducing hazards will, of course, ultimately drive the widespread use of this technique. Projects such as described in this paper will contribute to the development of continuous monitoring and the assessment of its utility.

Acknowledgement. We would like to acknowledge individuals who have contributed in assorted ways to the early stages of GPS continuous monitoring development: Miranda Chin, Gerald Mader and Bill Strange (National Geodetic Survey), Jonathan Ladd and Bob Welsh (Aero Service Division), Steve Deloach (U.S. Army Corps of Engineers), Bruce Peetz (TRIMBLE Navigation), Duncan Agnew, Bernard Minster, Frank Wyatt (Scripps Institution of Oceanography), Brad Hager, Kurt Feigl, Tom Herring, Tom Jordan, Bob King (Massachusetts Institute of Technology), Dave Jackson (University of California Los Angeles), Ken Hudnut and Kerry Sieh (California Institute of Technology), Geoff Blewitt, Steve Dinardo, Tim Dixon, Steve Fisher, Ulf Lindqwister, Tom Meehan, Bill Melbourne, John Scheid, Dave Tralli and Larry Young (Jet Propulsion Laboratory), Yukio Fujinawa, Shoji Sekiguchi, Takao Eguchi and Yoshimitsu Okada (National Research Center for Disaster Prevention), Ted Flinn, Miriam Baltuck and Herb Frey (National Aeronautics and Space Administration), Bob Burford and Will Prescott (U.S. Geological Survey).

We would like to thank Bernard Minster for his useful suggestions and for preparing Figure 5, and Frank Wyatt and Duncan Agnew for preparing Figure 7. We appreciate the information supplied to us by Steve DeLoach on continuous monitoring of dam deformations.

This work was supported by United States Geological Survey Grant 14-08-0001-G1673 and the National Aeronautics and Space Administration.

REFERENCES

Aero Service Division (1988). SONAP™ Processing Manual, Houston, Texas.

Blewitt, G. (1989). Carrier phase ambiguity resolution for the Global Positioning System applied to geodetic baselines up to 2000 km, *J. Geophys. Res.*, ***94***, 10187-10203.

Cangahuala, L.A. and Counselman C.C. (1989). Feasibility of millimeter-accuracy geodetic positioning and vehicle tracking with repeater satellites, *Eos, Trans. Amer. Geophys. Union*, ***70***, 1062.

Counselman, C.C. and Abbot, R.I. (1989). Method of resolving radio phase ambiguity in satellite orbit determination, *J. Geophys. Res.*, ***94***, 7058-7064.

DeLoach, S.R. (1989). Continuous deformation monitoring with GPS, *J. Surv. Eng.*, **115**, 93-110.

Davis, J.L. (1986). Atmospheric propagation effects on radio interferometry, PhD thesis, Mass. Inst. of Technol., Cambridge, 1986.

Dong, D., and Bock, Y. (1989). Global Positioning System network analysis with phase ambiguity resolution applied to crustal deformation studies in California, *J. Geophys. Res.*, ***94***, 3949-3966.

Fujinawa, Y., Shimada, S., Kishii, T., Sekiguchi, S., Eguchi, T., Okada, Y. and Watada, S. (1989). Some results in the preliminary data analysis of the fixed-point GPS baseline determination network in central Japan, presented at IAG General Meeting, Edinburgh, Scotland, 3-12 August 1989, this volume.

Gurtner, W., Mader, G., and MacArthur, D. (1989). A common exchange format for GPS data, *GPS Bulletin, 2*, Global Positioning System Subcommission, Int. Assoc. of Geodesy.

Ishibashi, K. (1984). Possibility of large earthquake near Odawara, Central Japan, preceding the Tokai earthquake, *Earthquake Prediction Res., 3*, 319-344.

Krakiwsky, E.J. (1986). An overview of deformations, measurement technologies, and mathematical modeling and analysis, Deformation Measurements Workshop, Y. Bock (ed.), Massachusetts Institute of Technology, Cambridge, 7-33.

Ladd J. (1986). Continuous monitoring of deformation with GPS, proc. Deformation Measurements Workshop, Y. Bock (ed.), Massachusetts Institute of Technology, Cambridge, 416-436.

Lichten, S.M. and Border, J.S. (1987). Strategies for high-precision GPS orbit determination, *J. Geophys. Res., 92,* 12751-12762.

Lichten, S.M. and Bertiger, W.I. (1989). Demonstration of sub-meter GPS orbit determination and 1.5 parts in 10 three-dimensional baseline accuracy, *Bull. Geod., 63,* 167-190.

Lindqwister, U., Blewitt, G., Pogorelc, S., Roth, M., Caissy, M. and Tetreault (1989). Design and testing of a continuously monitoring GPS-based system, *Eos, Trans. Amer. Geophys. Union, 70*, 1054.

Minster, B., Prescott W., Royden, L., Bock, Y., Kastens, K., McNutt, M., Peltzer, G., Reilinger, R., Rundle, J., Sauber, J., Scheid, J., and Zuber, M. (1989). Report of the Plate Motion and Deformation Panel, NASA Coolfont Workshop, August, 1989.

Mueller, I. and S. Zerbini (eds.) (1989). The Interdisciplinary Role of Space Geodesy, *Lecture Notes in Earth Sciences, Vol. 22*, Springer-Verlag, Berlin.

Potterfield M., Peetz, B., Frohring, B. and Dew, D. (1989). Continuous Monitoring System, Phase II, Final Report, Trimble Navigation, Sunnyvale, California.

Sekiguchi,S., K.Hirahara, H.Tsuji, A.Sengoku, S.Watada, M.Satomura, and S.Shimada, Simultaneous measurements by GPS interferometers at VLBI and SLR sites in Japan, in preparation.

Shimada,S., Sekiguchi S., Eguchi T., Okada Y., and Fujinawa Y. (1988). Simultaneous baseline determination network in Kanto-Tokai area, central Japan, Proc. Japanese Symp. on GPS, Jan. 27-29, 1988, 260-270.

Shimada,S., S.Sekiguchi, T.Eguchi, Y.Okada, and Y.Fujinawa (1989). Preliminary results of the observation by fixed-point GPS simultaneous baseline determination network in Kanto-Tokai district, *J. Geod. Soc. Japan, 35*, 85-95.

Tralli, D.M., Dixon, T.H., and Stephens S.A. (1988). Effect of wet tropospheric path delays on estimation of geodetic baselines in the Gulf of California using the Global Positioning System, *J. Geophys. Res., 93*, 6545-6557.

Treuhaft, R.N. and Lanyi, G.E. (1987). The effect of the dynamic wet troposphere on radio interferometric measurements, *Radio Sci., 22*, 251-265.

Wyatt, F.K., Bolton H., Bralla S., and D.C. Agnew (1989). New Designs of Geodetic Monuments for Use with GPS, *Eos, Trans. Amer. Geophys. Union, 70*, 1054-1055.

GPS-DATA PREPROCESSING FOR CYCLE-SLIP DETECTION

by
H. Lichtenegger and B. Hofmann-Wellenhof
Technical University Graz, Austria

Abstract

Cycle-slip detection is one important aspect for GPS data. There are several methods known depending on the various data sets available, e.g. code-pseudoranges, phase-pseudoranges. Generally speaking, the methods known so far contain advantages and disadvantages as well.

Preprocessing methods are presented permitting cycle-slip detection in a two-step procedure. After setting-up a testing quantity its behaviour is checked with respect to smoothness. Discontinuities shown up reflect cycle slips.

1. Introduction

To assure a routinely achievement of the GPS-accuracy of 1 ppm or even better it is necessary to free the measurement data from cycle slips. Since the trend for data processing aims to be fully automatic, a check and a correction of the measurements becomes a necessary step. The most desirable case would be a computer programm where the data are fed in separately, i.e. phases, pseudoranges station by station, and all cycle slips are detected and repaired without any additional information as e.g. coordinates of satellites or stations. As it is shown later, this could become true in the near future. Today different methods are used based on different data sets available (one or two frequencies). The goal of this presentation are proposals for investigating the data for cycle slips in a preprocess. This means that all methods based on preliminary baseline computations are omitted. Therefore e.g. one of the "classical methods" originally proposed by Remondi (1985) based on the triple-difference technique for fixing cycle-slips is not mentioned.

Many of the subsequent sections show a list of advantages and disadvantages where, certainly, the judgement is subjective for each plus or minus. However, basic criteria were put upon the target of kinematic applications with the final goal of precise real-time applications. In addition the question of one or two frequencies was taken into account since still many hardware companies struggle with managing this problem.

The implementation of the following methods on a computer should be easily achievable.

2. Definition of cycle-slips

Assuming phase measurements, then

$$\lambda \cdot \Phi_i^j(t) = \rho_i^j(t) + \lambda \cdot N_i^j + c \cdot \Delta\tau_i^j(t) - \frac{A_i^j(t)}{f^2} + \ldots \qquad (2\text{-}1)$$

can be used as a model for the phase equation, where λ is the wavelength of the satellite signal, $\Phi_i^j(t)$ is the measured beat-phase of the satellite j with respect to station i at epoch t. On the right-hand side of eq. (2-1) $\rho_i^j(t)$ denotes the geometric distance between the satellite j and the station i at epoch t, N_i^j is the time-independent integer ambiguity, c is the speed of light, $\Delta\tau_i^j(t)$ contains errors of the satellite clock and the receiver clock as well. Finally, f is the frequency of the satellite signal and $A_i^j(t)$ models the influence of the ionosphere. It depends on the electron content in the ionosphere, therefore it varies with location and with time. In principle $A_i^j(t) = A(E(t), z(t))$ meaning that $A_i^j(t)$ is a function of the electron content E and the zenith angle z. Some influences being almost frequency and time invariant have been omitted in eq. (2-1) as e.g. the troposphere. Note that λ in eq. (2-1) is a function of the frequency f.
For later use the relation

$$f \cdot \lambda = c \qquad (2\text{-}2)$$

is needed with the dimensions $[\text{cycle} \cdot \text{sec}^{-1}]$ for the frequency, $[\text{m} \cdot \text{cycle}^{-1}]$ for the wavelength λ and $[\text{m} \cdot \text{sec}^{-1}]$ for c. Thinking of GPS and the carrier waves L1 and L2, from the equation above the formula

$$\frac{\lambda_{L1}}{\lambda_{L2}} = \frac{f_{L2}}{f_{L1}} \qquad (2\text{-}3)$$

can be deduced.

One additional remark should be considered for eq. (2-1). Exactly the same equation holds for pseudo-code distance measurements if the term containing the integer ambiguity is omitted and the sign of the ionospheric term is changed.

Starting from eq. (2-1), a cycle-slip can be defined as a change of N_i^j by an unknown integer number from one epoch to the next. In addition, this change must stay permanently thus the original integer number of N_i^j never appears again. As a consequence, a jump of the integer number N_i^j by an amount of $+\Delta N_i^j$ for a few epochs and jumping back by $-\Delta N_i^j$ is not considered to be a cycle-slip, cf. Hofmann-Wellenhof and Lichtenegger (1988).

The reasons for cycle-slips can be

- receiver dependent as e.g.
  - low signal strength (by others than just low elevation),
  - possibly from rotation and inclination of the antenna in the kinematic mode,
  - signal processing;

- observation dependent as e.g.
  - obstructions,
  - signal noise (e.g. ionosphere, multipath)
  - low satellite elevation.

For reasons of completeness the so-called half-cycle-slips are mentioned. These are receiver dependent and are caused by temporary confusion in the tracking loop between the 0^0 phase and the 180^0 phase zero crossings.

3. Methods for cycle-slip elimination

3.1. General remarks

The following principle holds for all methods: a testing quantity is obtained from observation data (sometimes supplied by satellite and station coordinates). The variation with time of the testing quantity should be smooth. Looking at eq. (2–1), which terms may disturb the smooth behaviour of the testing quantity? In the sequence of the strongest influence due to variation with time there are

- $\rho_i{}^j(t)$ because of the satellite motion and, to a lower degree, because of the receiver motion in the kinematic mode,
- $c \cdot \Delta\tau_i{}^j(t)$ because of varying clock errors,
- $A_i{}^j(t)$ because of changes in the ionosphere.

3.2. Testing quantity

First of all one statement given by Mader et al. (1986) is important: "There is no analytic method to determine the integer correction at the cycle-slips." This is the reason why always combinations of different quantities must be performed. What can be combined? There are available phase measurements, pseudo-code measurements, satellite coordinates, station coordinates. Depending on the kind of combination, at least four different methods can be found in the literature:

Method with double-differences where phases, satellite and station coordinates are used, cf. e.g. Beutler et al. (1989).

Method with ionospheric residuals where phases of L1 and of L2 are used, cf. e.g. Goad (1986).

Method with kinematic mode where phases, satellite coordinates and precise coordinates of one station with known integer ambiguities are used, cf. e.g. Remondi (1988).

Method with phase-range combinations where phases and pseudo-code measurements are used without any further information, cf. e.g. Landau (1989).

3.2.1. Method with double-differences

The testing quantity is obtained from the double-difference equation. Backing on the notation of eq. (2–1), then

$$DD_{12}^{jk}(t) = \frac{1}{\lambda}\cdot[\rho_2{}^k(t) - \rho_2{}^j(t) - \rho_1{}^k(t) + \rho_1{}^j(t)] + N_{12}^{jk} - \frac{1}{f^2}\cdot[A_2{}^k(t) - A_2{}^j(t) - A_1{}^k(t) + A_1{}^j(t)] \tag{3–1}$$

is the double-difference equation for the two satellites j and k and the two stations 1 and 2 where the influence of the troposphere has been neglected. Having available approximate values for the satellite and station coordinates, all the geometric distances ρ can be computed and the first bracketed term in eq. (3–1) may be subtracted from the measurement value $DD_{12}^{jk}(t)$. Compared to the geometric distances ρ the influence of the ionosphere is much smaller, say in the amount of meters. By forming the difference as in eq. (3–1), the influence decreases further. Therefore two possibilities arise:

- either to neglect the whole term resulting from the ionosphere, i.e. the second bracketed term on the right-hand side of eq. (3–1),
- or, e.g. for long time gaps, to calculate it by a simple layer model, cf. Gervaise et al. (1985).

Anyway, it may be assumed that instead of eq. (3–1) now

$$\ell DD_{12}^{jk}(t) = N_{12}^{jk} \tag{3–2}$$

can be used, where

$$\ell DD_{12}^{jk} = DD_{12}^{jk}(t) - \frac{1}{\lambda}\cdot[\rho_2{}^k(t) - \rho_2{}^j(t) - \rho_1{}^k(t) + \rho_1{}^j(t)] + \frac{1}{f^2}\cdot[A_2{}^k(t) - A_2{}^j(t) - A_1{}^k(t) + A_1{}^j(t)] \tag{3–3}$$

in case of taking into account an ionospheric layer model. Eq. (3–3) is the testing quantity for method 1. The variation of $\ell DD_{12}^{jk}(t)$ with time is tested. A cycle-slip is detected if a discontinuous change occurs.

Since the double difference is set up from four phase equations of type eq. (2–1), the integer ambiguity in eq. (3–2) is also a combination of four integer ambiguities. This causes the problem of the assignment of the cycle-slips to a satellite and a station. In general the respective satellite can be found by comparing double-differences, e.g. for the satellites SV 6, 9, 11, 12 the differences 9–6, 11–9, 12–11, but the problem of the station is still present. In this context it may be mentioned that the influence of erroneous station coordinates is less important in case of short baselines where the dependency on the stations almost cancels. For long baselines a combination of two close-by receivers at each station could manage the situation.

Advantages and disadvantages of the method:

+ one-frequency receivers may be used;
– data of two stations are necessary therefore presently in kinematic mode not applicable because the station coordinates of the second station must be transferred to the first station (by some kind of data transfer);
– cycle-slips are calculated from a combination of four integer ambiguities;
– satellite and station coordinates are to be known at least approximately;
– compared to the phase the double-difference has much more noise by the multiple differences.

3.2.2. Method with ionospheric residuals

The testing quantity is derived from the phases Φ_{L1} and Φ_{L2} of the two carrier waves L1 and L2 by considering one satellite and one station only. Starting from the initial eq. (2–1), the difference of the two phases reads as

$$\lambda_{L1}\cdot\Phi_{L1}(t) - \lambda_{L2}\cdot\Phi_{L2}(t) = \lambda_{L1}\cdot N_{L1} - \lambda_{L2}\cdot N_{L2} - \frac{A(t)}{f_{L1}^2} + \frac{A(t)}{f_{L2}^2} \quad . \tag{3-4}$$

Note that the main distinction is now made between the frequency dependent quantities shown by the symbol L1 or L2 of the corresponding carrier, e.g. λ_{L1} denotes the wavelength of carrier L1. In order to avoid a too heavy burden of superscripts and subscripts, the satellite and station symbols have been omitted. Especially remarkable is that the distances $\rho_i{}^j(t)$ and the clock errors $\Delta\tau_i{}^j(t)$ totally vanish.

The actual testing quantity is the ionospheric residual which is obtained from eq. (3–4) by dividing the whole equation with λ_{L1}:

$$\Phi_{L1}(t) - \frac{\lambda_{L2}}{\lambda_{L1}}\cdot\Phi_{L2}(t) = N_{L1} - \frac{\lambda_{L2}}{\lambda_{L1}}\cdot N_{L2} - \frac{A(t)}{\lambda_{L1}f_{L1}^2} + \frac{A(t)}{\lambda_{L1}f_{L2}^2} \quad . \tag{3-5}$$

It should be emphasized strongly that this is <u>not</u> the ionospheric-free linear combination L3. This can be seen best if the part on the left-hand side is slightly transformed by eq. (2–3) to $\Phi_{L1}(t) - (f_{L1}/f_{L2})\cdot\Phi_{L2}(t)$, whereas the L3 combination reads $\Phi_{L1}(t) - (f_{L2}/f_{L1})\cdot\Phi_{L2}(t)$. The influence of the ionosphere for the phase differences in eq. (3–5) is

$$\Delta_{iono} = -\frac{A(t)}{\lambda_{L1} f_{L1}^2} + \frac{A(t)}{\lambda_{L1} f_{L2}^2} \qquad (3-6)$$

or similarly

$$\Delta_{Iono} = -\frac{A(t)}{\lambda_{L1} f_{L1}^2} \cdot \left[1 - (f_{L1}/f_{L2})^2\right] \qquad (3-7)$$

which shows the decrease (i.e., from 1 to –0.65) of the ionospheric influence for the phase difference. The 1 within the brackets multiplied by the factor is the influence of a single phase, by the difference of two phases it is reduced to about 65 %. This value results from inserting f_{L1} and f_{L2}, cf. Goad (1986). In addition it should be mentioned that the influence of the ionosphere in the testing quantity in eq. (3–4) can be eliminated by subtracting the corresponding difference of the pseudo-code range measurements.

Advantages and disadvantages of the method:

+ only measurement data are used, no additional information (satellite, station coordinates) is necessary;
+ no problem to assign the cycle-slip to satellite, station and carrier;
+ possible for kinematic applications;
– two-frequency method;
– for certain reasons to be explained later a rough knowledge of the cycle-slip in the amount of 6 – 8 cycles is necessary, this can in general already be achieved by pseudo-code measurements.

The quantity Δ_{iono} in eq. (3–7) only varies with time if A(t) changes. However, the influence is different compared to the method with double-differences. There is neither a satellite nor a station dependency, thus Δ_{iono} may in general even be neglected. This holds true as long as no magnetic storms occur.

Some interesting remarks for this method should be considered. If a jump ΔN of the testing quantity is assumed, this can be written by using eq. (3–5) and eq. (2–3) as

$$\Delta N = \Delta N_{L1} - \frac{f_{L1}}{f_{L2}} \cdot \Delta N_{L2} \qquad (3-8)$$

where ΔN is known from the measurements and ΔN_{L1}, ΔN_{L2} are unknown. Thus for one eq. (3–8) two unknowns (of integer type) appear. This kind of equations is called a diophantic one which can be solved by various combinations of ΔN_{L1}, ΔN_{L2} and by looking for the minimum with respect to the given quantity ΔN.

At first we ask for possibilities where $\Delta N = 0$. Omitting the trivial case $\Delta N_{L1} = \Delta N_{L2} = 0$ we easily find the condition

$$\frac{\Delta N_{L1}}{\Delta N_{L2}} = \frac{f_{L1}}{f_{L2}} = 1.28 \tag{3-9}$$

which can occur for e.g.

$$\begin{array}{lcl} \Delta N_{L1} = 32 & ---> & \Delta N_{L2} = 25 \\ \phantom{\Delta N_{L1} = }\, 64 & & \phantom{\Delta N_{L2} = }\, 50 \\ \phantom{\Delta N_{L1} = }\,\vdots & & \phantom{\Delta N_{L2} = }\,\vdots \end{array}$$

which means that a unique solution is only possible if the cycle-slips are known to at least 25 cycles. This accuracy can certainly be obtained by pseudo-code measurements. All calculations so far included an error free measurement. Now a measurement error for both phases of

$$m_{\Phi} = \pm\, 0.01 \cdot \lambda = \pm\, 0.01 \text{ cycles} \tag{3-10}$$

is assumed. This is not valid in case of a squaring-type receiver where the error for Φ_{L2} is worse than for Φ_{L1} cf. Hofmann-Wellenhof and Lichtenegger (1988), p. 43. The measurement noise will also increase during irregular ionospheric activity. Anyway, the mean error of the variation with time for the residual follows from

$$\Delta N = \Phi_{L1}(t+\Delta t) - \frac{f_{L1}}{f_{L2}} \cdot \Phi(t+\Delta t) - \left[\Phi_{L1}(t) - \frac{f_{L1}}{f_{L2}} \cdot \Phi_{L2}(t)\right] \tag{3-11}$$

and yields with the previous assumption

$$m_{\Delta N} = \pm\, 0.023 \text{ cycles} \quad . \tag{3-12}$$

Taking into account a level of three times the mean error this leads to ≈ 0.07 cycles. Therefore two ΔN, calculated by various combinations of ΔN_{L1}, ΔN_{L2} must differ from each other by more than 0.07 in order to enable a unique assignment.

A few examples follow for a better illustration. The combination $\Delta N_{L1} = -3$ and $\Delta N_{L2} = -3$ yields the residual $\Delta N = 0.84$. Accordingly, $\Delta N_{L1} = +2$ and $\Delta N_{L2} = +1$ yields the residual $\Delta N = 0.72$. For both cases a unique separation is possible since the difference of the two values ΔN is 0.12 and thus greater than the aforementioned limit. As another example consider $\Delta N_{L1} = +9$ and $\Delta N_{L2} = +7$ yielding $\Delta N = 0.04$ which cannot be resolved uniquely from the trivial case $\Delta N_{L1} = \Delta N_{L2} = 0$ yielding $\Delta N = 0.00$.

A systematic investigation shows that for all combinations up to 6 – 8 cycles the minimum difference between the corresponding ΔN is greater than 0.12, thus being uniquely solvable, cf. also the subsequent table containing the results up to 5 cycles.

	−5	−4	−3	−2	−1	0	1	2	3	4	5
−5	1.40	0.12	−1.16	−2.44	−3.72	−5.00	−6.28	−7.56	−8.84	−10.12	−11.40
−4	2.40	1.12	−0.16	−1.44	−2.72	−4.00	−5.28	−6.56	−7.84	− 9.12	−10.40
−3	3.40	2.12	0.84	−0.44	−1.72	−3.00	−4.28	−5.56	−6.84	− 8.12	− 9.40
−2	4.40	3.12	1.84	0.56	−0.72	−2.00	−3.28	−4.56	−5.84	− 7.12	− 8.40
−1	5.40	4.12	2.84	1.56	0.28	−1.00	−2.28	−3.56	−4.84	− 6.12	− 7.40
0	6.40	5.12	3.84	2.56	1.28	0.00	−1.28	−2.56	−3.84	− 5.12	− 6.40
1	7.40	6.12	4.84	3.56	2.28	1.00	−0.28	−1.56	−2.84	− 4.12	− 5.40
2	8.40	7.12	5.84	4.56	3.28	2.00	0.72	−0.56	−1.84	− 3.12	− 4.40
3	9.40	8.12	6.84	5.56	4.28	3.00	1.72	0.44	−0.84	− 2.12	− 3.40
4	10.40	9.12	7.84	6.56	5.28	4.00	2.72	1.44	0.16	− 1.12	− 2.40
5	11.40	10.12	8.84	7.56	6.28	5.00	3.72	2.44	1.16	− 0.12	− 1.40

Table: Residuals ΔN for various combinations of ΔN_{L1} (columns) and ΔN_{L2} (rows).

3.2.3. Method with kinematic mode

Actually, the testing quantity has a different meaning for this method. The principle is a revisiting from time to time to a station with known coordinates and known integer ambiguities. Remondi (1988) speaks of a kinematic method: oe receiver moves and stops and returns to the known point. Therefore the path of this receiver contains several loops (as in case of gravity measurements too). The integer ambiguities are re-computed. As long as they do not change nothing has happened. Thus one could call the integer ambiguities themselves the testing quantities.

Advantages and disadvantages of the method

\+ especially designed for kinematic applications;

− known starting position requested;

− revisiting of the known station.

3.2.4. Method with phase-range combinations

The testing quantity is calculated from a combination of phases and pseudo-code measurements, both corrected for the ionosphere, cf. Beutler et al. (1989), by

$$\lambda \cdot \Phi_i{}^j(t) - R_i{}^j(t) = \lambda \cdot N_i{}^j \qquad (3\text{-}13)$$

where $R_i{}^j(t)$ is the pseudo-code measurement. As mentioned earlier, apart from multipath and measurement noise, the pseudo-code measurement differs from the phase measurement by the integer ambiguity only (and the different sign for the ionospheric term). At present the resolution is the main problem with this method. The wavelength of the P-code amounts to 30 m, the actual resolution is about $\frac{\lambda}{100}$ which yields 30 cm. This corresponds to 1 – 2 cycles of the carrier. This is the reason why this method is presently used for an approximate determination of cycle-slips. In the near future a resolution of $\frac{\lambda}{1000}$ may be expected. This performance is already claimed by a manufacturer. If this becomes accessible then this will be the method!

Advantages and disadvantages of the method

\+ very simple;

\+ possible for kinematic applications;

– present accuracy is not sufficient, necessary accuracy is not yet commercially available.

3.3. Checking the testing quantity for cycle-slips

Given is a time series for the testing quantity. In many cases the data are equidistant, e.g. a value is given every thirty seconds. To be determined are discontinuous jumps in the time series. There are different methods to find a discontinuous behaviour. Three of them are mentioned below.

3.3.1. Polynomial approach

A low degree polynomial is fitted over the time series. This method is used in the Bernese software. Large discrepancies between polynomial and actual value are interpreted as cycle-slip. Disadvantages are:

- it is not applicable for kinematic measurements;
- the shape of the polynomial changes with the number of cycle-slips;
- the statements for the cycle-slips are influenced by the global behaviour of the time series.

3.3.2. Kalman filtering

Using initial observations, a dynamic model is set up which enables the prediction of the subsequent observation value by Kalman filtering, cf. e.g. Landau (1989). From the comparison of the predicted and the observed data a cycle-slip can be detected. The dynamic model is actualized for each data value.

In case of static observations this is closely related to the improvement of a geodetic network by single observations. Advantages and disadvantages are:

+ may be used for kinematic applications;
+ only one measurement is considered; in case of a cycle-slip the correction is performed at once thereby avoiding an influence on subsequent measurement values which is equivalent to a local behaviour.
- set up of an initial condition

3.3.3. Scheme of differences

Why this method is extremely well-suited to be applied can be seen best from a quotation of Zurmühl (1963): "Das Differenzenschema ist ein wichtiges und nützliches Hilfsmittel zur Aufdeckung von Unstimmigkeiten in äquidistanten Zahlenreihen. Auch zur Glättung kann es vorteilhaft verwendet werden, da Fehler in den höheren Differenzen stärker auftreten." (The scheme of differences is an important and useful tool to detect irregularities in equidistant series. Also for smoothing purposes it may be used advantageously because errors appear more strongly in higher differences.) A similar approach for cycle-slip detection is proposed by Hilla (1986).

In principle, differences are a kind of a subtractive filter. The advantage of subtractive filters is the effect of a high-pass filter. This results in damping of low frequencies and eliminating constant parts. High-frequency parts containing the errors are amplified. The subsequent example illustrates the strength of the scheme of differences:

$y(t)$	y^1	y^2	y^3	y^4
0				
	0			
0		0		
	0		ϵ	
0		ϵ		-3ϵ
	ϵ		-2ϵ	
ϵ		$-\epsilon$		3ϵ
	0		ϵ	
ϵ		0		$-\epsilon$
	0		0	
ϵ		0		
	0			
ϵ				

where $y(t)$ is the equidistant signal of the time series, y^1, y^2, y^3 and y^4 the first, second, third and fourth difference. As it can be seen, a jump in the signal is amplified in the differences. Naturally, signal noise in $y(t)$ appears in the differences, but all systematic influences cancel out. Assume small remaining model errors which could be modelled by a polynomial. Those contributions will vanish.

4. Conclusion

One is often asked for the best method to detect and eliminate cycle-slips. We emphasize and recommend the method with ionospheric residuals for setting up the testing quantity and to proceed with the scheme of differences for the detection of the cycle-slips. By the way, this procedure is also very suitable for interpolation. Naturally, two frequency measurements are required.

The method of the kinematic mode will probably become more important since the ambiguity resolution will soon be possible with much less observation epochs, cf. Frei und Beutler (1989).

In the near future, if the resolution accuracy increases, the phase-range combination method should be used for the testing quantity and again the scheme of differences to check it. This presupposes an unchanged P-code signal. The code itself is not to be known.

If all of the methods fail there is still another possibility to detect cycle-slips at least for static applications. The observation data are separated into parts and processed independently leading to ambiguities for each part. In a second step cycle-slips, showing up in significant different ambiguities, are repaired and a new calculation of the whole data set is performed. Obviously, this is applicable only in case of very rare cycle slips.

It would be extremely helpful if cycle-slips could be indicated in all cases by the hardware (for some receiver types this is already done). In case of kinematic applications, this would save unnecessary returns to the station with the known coordinates. Only in case of a cycle-slip a return to the known station would be obligatory. In addition, if cycle-slips are repaired by the receivers internally during field operation, then the whole preprocessing procedure would be superfluous.

Acknowledgment:

The authors are grateful to Dr. Benjamin Remondi for some hints and correction proposals and for being supported by his experience.

5. References

Beutler, G.; W. Gurtner; U. Hugentobler; M. Rothacher; T. Schildknecht; U. Wild (1989): Ionosphere and GPS processing techniques. Paper presented at the 5th International Geodetic Symposium on Satellite Positioning. Las Cruces, New Mexico, March 13-17.

Frei, E.; G. Beutler (1989): Some considerations concerning an adaptive, optimized technique to resolve the initial phase ambiguities for static and kinematic GPS surveying-techniques. Paper presented at the 5th International Geodetic Symposium on Satellite Positioning. Las Cruces, New Mexico, March 13-17.

Gervaise, J.; M. Mayoud; G. Beutler; W. Gurtner (1985): Test of GPS on the CERN-LEP control network. In W.M. Welsch und L.A. Lapine (editors): Proceedings of the Joint Meeting of FIG Study Group 5B and 5C on Inertial, Doppler and GPS measurements for National and Engineering Surveys. Munich, July 1-3. Schriftenreihe der Universität der Bundeswehr München, Vol. 20-2, pp. 337-358.

Goad, C. (1986): Precise positioning with the GPS. CERN Accelerator School, Applied Geodesy for Particle Accelerators, Geneva, Switzerland, April 14-18.

Hilla, S. (1986): Processing cycle slips in nondifferenced phase data from the Macrometer V-1000 receiver. Proceedings of the Fourth International Geodetic Symposium on Satellite Positioning. Texas, Austin, April 28 - May 2, Vol. 1, pp. 647-661.

Hofmann-Wellenhof, B.; H. Lichtenegger (1988): GPS - Von der Theorie zur Praxis. Mitteilungen der geodätischen Institute der Technischen Universität Graz, Folge 62.

Landau, H. (1989): Precise kinematic GPS positioning. Experiences on a land vehicle using TI 4100 receivers and software. Bulletin Geodesique, Vol. 63, No. 1, pp. 85-96.

Mader, G.L.; W.E. Carter; B.C. Douglas; W.B. Krabill (1986): Decimeter precision aircraft positioning using GPS carrier phase measurements. Geodetic Research and Development Laboratory, National Geodetic Survey, NOAA.

Remondi, B.W. (1985): Modeling the GPS carrier phase for geodetic applications. Proceedings of the First International Symposium on Precise Positioning with the Global Positioning System. Rockville, Maryland, April 15-19, Vol. 1, pp. 325-336.

Remondi, B.W. (1988): Kinematic and pseudo-kinematic GPS. Institute of Navigation GPS-88, Colorado Springs, September.

Zurmühl, R. (1963): Praktische Mathematik für Ingenieure und Physiker. Springer Verlag, Berlin, Göttingen, Heidelberg. Vierte verbesserte Auflage.

ON PREPROCESSING OF GPS DATA

Roman Galas*) and Klaus Deichl
Institute for Astronomical and Physical Geodesy
Technical University of Munich, Munich, F.R.G.

INTRODUCTION

From a practical point of view, it is useful to develop a preprocessing software, which could enables one to evaluate all data received immediately after an observational session. The approach presented here make it possible to estimate if observations at a site should be repeated.
The time - consuming process of data transfer from TI data cassettes to floppy disks is skipped by directly collecting the data with a microcomputer. The computer program which was developed allows one to collect the GPS field data for about one hour and forty minutes when the collection rate is 3 seconds.

DATA COLLECTION

In order to simplify collecting data coming from a TI 4100 receiver, a microcomputer, a serial Atari 1040ST, was used. The computer was interfaced to the "PROM" as well to the "RAM" models of TI receivers through RS232 serial ports. Fig.1 briefly explains our simple "on-line" collection of the GPS observations. The strategy presented below is a consequence of the fact that the Atari microcomputer has only one serial port.
Due to this drawback it is not possible to give the receiver commands with the Atari, and during the observational process the operator is obliged to perform appropriate initialization and operating functions with the hand-held Control Display Unit of the receiver. Moreover, when a "RAM" model is used, the TI operational software must be loaded in the usual way prior to connecting it to a computer. However this low-priced microcomputer also has a significant advantage: namely, its operating system enables one to address directly the whole RAM memory of the computer, which afterwards permits the errorless collection of all the coming data every 3 seconds.
On start-up, the developed program TILINK goes at once into a "wait state". This means that the program only tests if the buffer of the

---

*) Now at Sternwarte of the University Bonn.

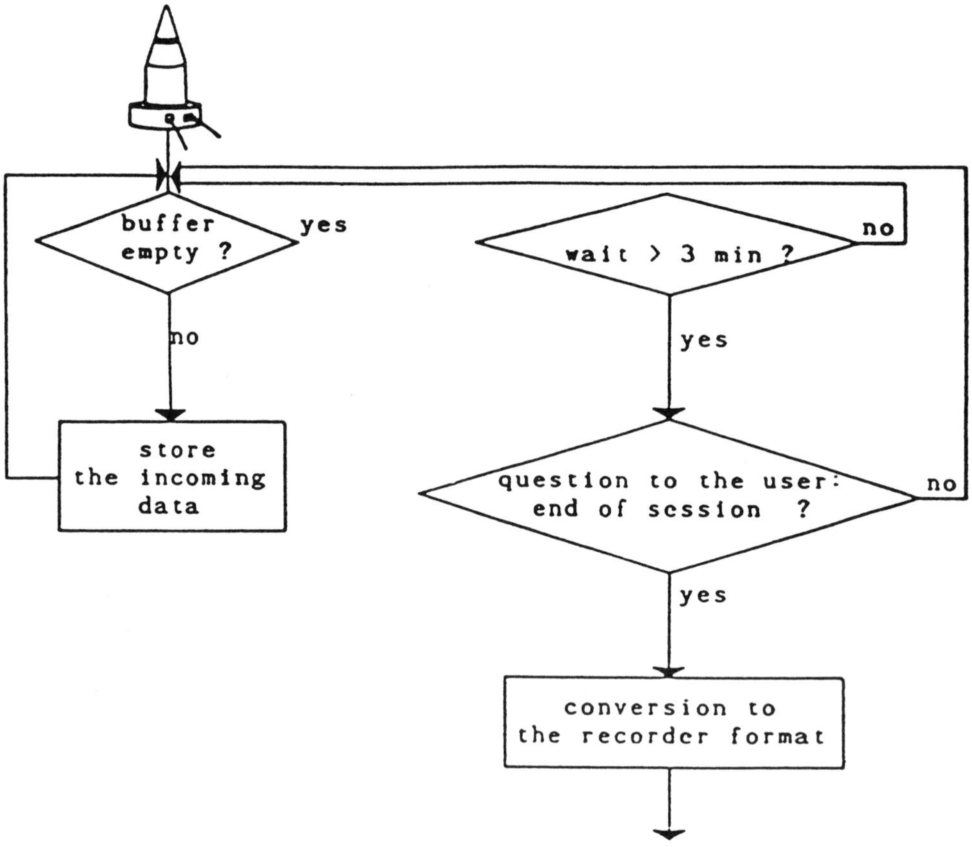

Fig.1 Flow chart of the data collecting.

microcomputer's port is not empty. Using the CDU, the operator now has to command which kind of report will be sent to the microcomputer. As soon as the results of the test are positive, the binary data received are stored in the RAM memory of the microcomputer. When this operation is finished (i.e. the buffer is again empty), the program reverts to the former state. The operator can finish the task of the data transfer from the TI receiver into the main memory of the microcomputer if the RS-232 buffer is empty longer than 3 minutes (this value can be changed on request). After the data collection is completed the information received is converted from the transparent binary text format into the recorder format. The collected observations are examined for validity and saved on archival floppy disks in binary IEEE format in four files, according to the TI reports. During verification, the program checks, among others, the loss of "lock", detects wrong initial epochs, and in Space Vehicle Navigation Data blocks the valididity of the NPRMP map array is confirmed. An additional information file on each floppy disk contains necessary data for later elaboration and processing as well as a summary. These archival floppy disks contain only valid data unchanged by pre-processing.

APPLICATION OF KALMAN FILTERING TO THE DETECTION AND REPAIR OF THE CYCLE SLIPS

The next task of the preprocessing concerns the problem of editing and repairing the cycle slips. Up to the present, many different approaches have been developed. A short discussion of these approaches and a new method were recently presented by P.A.Cross and N.Ahmad (1989).
The concept "one station - one satellite" has been chosen because, as was already mentioned above, it was desired to perform the preprocessing at each individual observation site immediately after an observational session.
The nature of the integer cycle slips suggests that these discontinuities could be verified by approximation and prediction of the accumulated phase observables in both frequencies or of their linear combinations. P.A.Cross and N.Ahmad (1989) have proposed such a method, which also can be applied for one-frequency receivers. In our preprocessing approach, a well-known method originally proposed by Goad (1985) was implemented with a modification concerning the rectification of the integer cycle slips. Goad has proposed to fix cycle slips at a single station by the applying Kalman filtering to the following linear combination (so-called ionospheric residuals)

$$\Delta\phi = \phi_1 - \frac{\nu_1}{\nu_2}\,\phi_2 \qquad (1)$$

where ϕ_1, ϕ_2 are Doppler-corrected phase values and ν_1, ν_2 are frequencies of the carriers. After a cycle slip has been fixed, the size of a "jump" can be determined by the polynomial approximaton to the cycle slip free intervals of the differences (1) and of the phase quantities ϕ_2 . A similar method has been used by G.W.Hein, H.Landau, G.Baustert (1988) and L.Bastos, H.Landau (1988).

It was decided to use only phase observables with any additional infor mation in order to perform preprocessing as quickly as possible. In this case, the analyzed data are L_1 and L_2 cycles corrected only for frequency bias. In these "time - series", discontinuities were detected by discrete Kalman filtering. This approach was performed in the following way (Gelb, 1979):

1. It was accepted after Goad that the state transitions of the signal is a third-order process.
2. From a least-squares, adjustment the initial values for the state update x_k^+ and the propagated error covariance update P_k^+ were derived.

Subindex k denotes the previous time step, and k+1 - the new one. Upper index + refers to the state or covariance prior to inclusion of the new data point, and - after.

3. Calculation of the components of the transition vector ϕ .

4. Error covariance propagation .

$$P^-_{k+1} = \Phi_k \; P^+_k \; \Phi^T_k + Q_k$$

5. State propagation

$$x^-_{k+1} = \Phi_k \; x^+_k$$

6. Design matrix H_{k+1}

7. Kalman Gain Calculation

$$K_{k+1} = P^-_{k+1} \; H^T_{k+1} \; [H_{k+1} \; P^-_{k+1} \; H^T_{k+1} + R_{k+1}]^{-1}$$

8. State update

$$x^+_{k+1} = x^-_{k+1} + K_{k+1} \; [z_{k+1} - H_{k+1} \; x^-_{k+1}]$$

9. Error covariance update

$$P^+_{k+1} = P^-_{k+1} - K_{k+1} \; H_{k+1} \; P^-_{k+1}$$

Where:

z = measurement vector,

Q = covariance matrix of the noise.

If the predicted value z' differs significantly from data one z, then an appearance of a cycle slip(s) in one or both carriers is considered. Before the next cycle slip is fixed, the previous one must be repaired. The corrections ΔN_1 and ΔN_2 could be calculated from

$$\delta_1 = y'_{k+1} - z_{k+1} = \Delta N_1 - \frac{\nu_1}{\nu_2} \Delta N_2$$

where y' denotes a result of an approximation after the jump in $\Delta\phi$. Since there are two unknowns, a second combination of phase observables would be useful. A good accuracy of prediction is needed, especially if there are longer gaps in the observables. Therefore the rate of changes in this second combination must not be too high. It was decided to use the so-called "wide-line" combination. After approximation with a polynomial of the second order before and after the cycle slip, the second value δ_2 was obtained and corrections to both carriers were found. The corrections are displayed on the screen and can therefore be verified. Afterwords, the process of correcting for the cycle slips is continued.

As a result of preprocessing, ASCII files, relatively free from the cycle slips and which contain phase observables are stored on floppy disks. Another piece of preprocessing software produces ASCII files with broadcast ephemeris and pseudo-ranges. The fixing and repairing of the cycle slips, as well the produce mentioned above two more data files is developed for an IBM or compatible computer. All programs described here are written in MS C.

CONCLUSION

The efficiency of the procedure presented above has been checked on the basis of real GPS data. The field data collection can be done with

the use of a 2 mbyte laptop version of Atari ST.
A number of tests with artificially injected cycle slips have also been performed. It was found that the rectification of the cycle slips in the mode "base station - base satellite" works very well when the sampling rate is 3 seconds. It also seems that the proposed field preprocessing could significantly simplify the observational campaign in larger GPS nets.

REFERENCES

Cross, P.A., Ahmad, N. (1988). Field Validation of GPS Phase Measurments. GPS-Techniques Applied to Geodesy and Surveing. E.Groten and R.Strauss (eds.). Lectures Notes in Earth Sciences, Vol.19.

Bastos, L., Landau, H. (1988). Fixing cycle slips in dual - frequency kinematic GPS - applications using Kalman filtering. Manuscripta Geodetica, No.13, pp.:249-256.

Goad, C.C (1988). Precise Positioning with the Global Positioning System. Proceedings of the Third International Symposium: Inertial Technology for Surveying in Geodesy. Banff, Canada, pp.:745-756.

Hein, G.W., Landau, H., Baustert, G. (1988). Terrestrial and Aircraft Differential Kinematic GPS Positioning. GPS-Technics Applied to Geodesy and Surveying. E.Groten and R.Strauss (eds.). Lectures Notes in Earth Sciences, Vol.19.

A COMPARISON OF SEVERAL TECHNIQUES TO REDUCE SIGNAL MULTIPATH FROM THE GLOBAL POSITIONING SYSTEM

Alan G. Evans and Bruce R. Hermann
Naval Surface Warfare Center
Dahlgren, Virginia 22448-5000 U.S.A.

ABSTRACT

Earlier Global Positioning System (GPS) tests and analyses have demonstrated that signal multipath can cause large variations in the pseudorange measurement. The multipath environments are compared by computing the standard deviations from the pseudorange measurements over several hours. These have ranged from 1.3 m in a benign environment to 4.0 and 5.0 m in a highly reflective environment.

This paper presents the results of several tests of procedures to reduce signal multipath. The techniques discussed include the placement of the GPS antenna and the use of absorption material, a ground plane, a choke ring, and antennas with low gain at low elevation angles. The analysis procedure to determine the level of signal multipath uses the difference between ionospherically corrected pseudorange and biased Doppler range values. The multipath characteristics of a number of GPS antennas are analyzed. Day-to-day crosscorrelation confirms the conclusions of the analysis.

INTRODUCTION

For the initial research and development phase of GPS, a number of antennas were intentionally designed to have high gain at low elevation angles. The intended purpose of this design was to receive signals from satellites close to the horizon. There were two reasons for this: first, in the early GPS constellation this was an important factor since there were not many satellites available for measurements. Second, the geometric figure of merit for four satellites, the Geometric Dilution of Precision (GDOP), is generally improved by including satellites at low elevations. However, the consequence of high gain at low elevation angles is that reflected signals are more likely to interfere with the direct signal from the satellite.

A simple procedure to evaluate the signal multipath effects of the GPS pseudorange measurement is to difference the ionospherically corrected pseudorange from its corresponding biased Doppler range (Evans, 1986). This procedure subtracts away system and clock errors and leaves approximately the pseudorange multipath effects plus pseudorange measurement noise. The standard deviation of these differenced measurements can be used as a criteria for signal multipath evaluation of a particular site with a given antenna or, several antennas in specific environments. A further criteria, for fixed site locations, is to compute the crosscorrelation function between consecutive days of these measurements. Since the satellite ground tracks repeat, each day's track being four minutes earlier than the one on the previous day, the day-to-day cross correlation coefficient may be used as an indicator of the amount of multipath present in the data.

This multipath evaluation procedure has been used at a number of sites to evaluate the performance of the commonly used log periodic conical spiral antenna (Evans, 1986). The elements of this antenna transition to a helix near the base for the purpose of enhancing the gain at low elevations. The sites under evaluation ranged from low reflective desert locations to highly reflective laboratory rooftops. The standard deviations for the differenced ranges at these locations spanned the range from 1.3 meters, to over three times this value at sites with high multipath. The 1.3 meter figure is about the noise level of the ionospherically corrected pseudorange measurements due to the receiver alone. Day-to-day comparisons showed that the signal multipath was sufficiently severe to cause the receiver to lose phase lock at the worst locations.

Two types of evaluations are performed in this paper. First, the log periodic spiral antenna is investigated in several configurations designed to reduce its susceptibility to reflected signals. Second, for one particular highly reflective location, several antennas with various grounding techniques are evaluated for their multipath sensitivity. The next section describes these test cases and their corresponding results. The conclusion summarizes the test results.

TEST CASES AND RESULTS

Dual frequency pseudorange and Doppler range measurement data were collected using a Texas Instruments TI-4100 GPS Geodetic Receiver. The receiver resident GEodetic SAtellite Receiver (GESAR) software was used with a 30 second measurement interval. Although a number of satellites were tracked, only results from one satellite (PRN-11) are presented. This satellite was selected due to its favorable ground track. At The Naval Surface Warfare Center (NSWC) this satellite is above 10 degrees elevation for about 7 hours with a maximum elevation angle of 57 degrees. An earlier study (Evans, 1986) also was based upon data from this satellite.

All of the test results below, except for cases 2 and 3 are for antennas at one site. The antennas were mounted on top of a 5 foot pole situated above a laboratory rooftop. This has been shown to be a highly reflective environment. Cases 2 and 3, in which the log periodic spiral antenna was placed directly on the roof, demonstrate the dramatic effect that antenna placement can have on the quality of the GPS pseudorange observations.

In three cases a plot of the difference between the ionospherically corrected pseudorange and biased Doppler data is given. Standard deviations for all cases are determined and presented in Tables 1, 2 and 3. Table 1 presents comparisons of the log periodic spiral antenna in various configurations and locations. Table 2 presents comparisons of the numerical standard deviations for a number of different antennas mounted on top of the 5 foot pole. Table 3 presents results similar to Table 2, but the antenna performance is modified by specially designed choke rings. When evaluating these standard deviations, it should be remembered that the minimum expected value is approximately that of the receiver pseudorange measurement noise, i.e. with no signal multipath. The standard deviation of the ionospherically corrected pseudorange due to receiver measurement noise alone is about 1.3 meters (Evans et al., 1985).

Case 1

The usual method of operation at NSWC is to place the antenna on a small, one foot square metal base plate which is fastened to the five foot high pole. Figures 1a and 1b show the multipath effects for days 283 and 284 (1987) respectively. The figures also demonstrate the day-to-day repeatability of multipath. The numerical standard deviations for this data are given in Table 1.

Case 2

The log periodic spiral antenna, with the one foot square ground plane attached, was placed on the roof. Due to mounting constraints, the top surface of the ground plane was about four inches above the roof. In this case, because signals impinging on the antenna from below are reduced, there is a significant decrease in the standard deviations listed in Table 1.

Case 3

This case is an enhancement of Case 2 above. In order to further minimize the amount of reflected signal hitting the antenna, foam absorption material was placed around the antenna. The foam blocks measured two feet square by one foot high. The antenna was placed in the center of a surrounding square array of eight blocks of this material. This further reduced the standard deviations to about 1.2 times the background level of 1.3 meters. The numerical standard deviations are given in Table 1.

This experiment indicates that the sensitivity of the log periodic spiral antenna to signal multipath is due to the ability of low elevation reflected signals to interfere with the direct satellite signal. This is probably due to the high gain of the antenna at low elevation angles.

Case 4

The success of Case 3 above suggested the use of a ground plane and absorption material when the log periodic spiral antenna is placed on top of the 5 foot pole. The design for this highly reflective environment, called for the antenna to be placed on a four foot diameter circular ground plane. In addition, a six inch high, two foot square block of absorption material was placed tightly around the antenna. Only the top conical portion of the antenna protruded above the absorption material. As can be seen from the pseudorange minus Doppler numerical standard deviation in Table 1, this produced a significant reduction in signal multipath effects when compared with Case 1.

Case 5

This case is the same as Case 4, but without the use of the absorption material. As can be seen by the numerical standard deviation of Table 1, the use of just the ground plane does not significantly reduce the multipath sensitivity. It is necessary to use the absorption material to absorb the low elevation reflected signals. The numerical standard deviations are entered in Table 2 to be compared with other antenna results.

Case 6

Another antenna design tested consisted of a small circularly polarized turnstile antenna atop a 3 foot square ground plane. The center of the ground plane is slightly domed. This configuration was mounted on top of the 5 foot pole. The low standard deviation values are given in Table 2. This antenna was designed to have low gain at elevation angles below 20 degrees. Note that significant reduction of the standard deviations compared with the log periodic spiral plus four foot diameter ground plane of Case 5. Note also that the phase center of the log periodic spiral is physically higher above the ground plane and therefore more susceptible to the reflected GPS signals.

Cases 7 and 8

Several commercial GPS antennas are available without ground planes. Two antennas like this were tested. The first is a drooping turnstile design. The second is noted as a dual frequency Fixed Radiation Pattern Antenna (FRPA-2). Both are constructed to be mounted physically close to (about one inch above) a round plane. These antennas were tested on the two foot diameter circular ground plane supplied with the antennas when borrowed from the National Geodetic Survey. These antennas produced results with significantly less multipath than the log periodic spiral antenna, but more than the antennas of Case 6 above, or Case 9 described below. The standard deviations for these two antennas are also listed in Table 2.

Case 9

The final antenna included was a dual frequency antenna made from microstrip elements and integrated with a radome, ground plane, and preamplifier. The elements are barely 0.3 inches above the ground plane. This is an advantage for minimizing the area susceptible to signals reflected off surfaces below the ground plane. In addition, the gain drops significantly for elevation angles below 20 degrees. Consequently, as seen by the low standard deviation values of 1.55 and 1.59 m given in Table 2, the design features of the microstrip antenna minimize multipath interference effects. Figure 2 is a plot of the range difference data for three consecutive days plotted on top of each other, with successive days shifted four minutes earlier to match satellite geometries.

Cases 10 to 12

The application of a choke ring ground plane to reduce multipath effects on GPS antennas was originally suggested by the Jet Propulsion Laboratory (Young et al., 1988) in Pasadena, California. A choke ring consists of strips of conductor which are concentric about the antenna and connected to a ground plate (Foot, 1986). The tops of the strips are approximately even with the antenna bottom plate.

Test data was collected with the antenna from Case 7 mounted on choke rings built by JPL and NSWC. These are cases 10 and 11, respectively. The two choke rings produce similar results, i.e. range difference standard deviations of about 1.8 m, as given in Table 3. Figure 3 shows two consecutive days data from Case 10 plotted on top of each other in the same manner as Figure 2.

A comparison of this antenna on a 2 foot flat ground plane, whose range difference standard deviation is about 2.4 m (Case 7), shows a significant reduction in the pseudorange multipath effects through the use of a choke ring. A notable improvement was also obtained with the log periodic spiral antenna. However, as seen by the 2.8 m range difference standard deviation (Table 3), the multipath effects on this antenna are still large even with the use of a choke ring. The use of six inch thick absorption ring (Case 4, Table 1), had a more significant multipath reduction effect with a 2.0 m range difference standard deviation.

CONCLUSIONS

The test results of this study indicate that multipath rejection measures need to be designed into the antenna itself. Here, low gain at low elevation angles as with the antennas described in Cases 6 and 9 produced good results. As theory suggests, adding absorption material or a choke ring to a multipath susceptible antenna reduces multipath by modifying the antenna gain, but in practice these may not be desirable alternatives. The low profile microstrip antenna element in Case 9 had the lowest signal multipath of any antenna combination tested at the NSWC site.

The log periodic spiral is one antenna which has the high gain at low elevation angles. Its sensitivity to signal multipath has been demonstrated earlier [1] and also in the first part of this paper. It has been shown above that this sensitivity can be significantly reduced by placing the antenna close to the ground. For situations where it is not practical to place the antenna on or close to the ground, the use of a ground plane and absorption material can significantly reduce the antenna's sensitivity to multipath.

This investigation was not intended to review all possible GPS antenna designs. By its inclusion or exclusion in this paper, the Naval Surface Warfare Center, a U.S. Government agency, does not endorse the use of any particular type of GPS antenna design.

Acknowledgement

The authors wish to thank National Geodetic Survey, Rockville, Maryland and Aero Services, Inc. of Houston, Texas for lending their antennas for signal multipath tests. The authors are also indebted to Trimble Navigation, Ltd. of Sunnyvale, California, Ball Aerospace of Broomfield, Colorado, and the Jet Propulsion Laboratory of Pasadena, California.

Also to be thanked are Joe Patton, Garnett Parker, and Ruth Darnell of the Naval Surface Warfare Center in Dahlgren, Virginia for their assistance. This work was encouraged and funded by the Defense Mapping Agency.

REFERENCES

Evans A. G. (1986). Comparison of GPS Pseudorange and Biased Doppler Range Measurements to Demonstrate Signal Multipath Effects, *Proceedings of the Fourth International Geodetic Symposium on Satellite Positioning*, Austin, Texas, 28 April 1986.

Evans A. G. et al. (1985). Collocation Tests of an Advanced Geodetic Global Positioning System Receiver, *The First International Symposium on Precise Positioning with the Global Positioning System*, Rockville, Md., April 1985.

Young L. E. et al. (1988). GPS Antenna Selection: Preliminary Range and Field Test Results, *Proceedings of ANTEM88*, Winnipeg, Canada, August 9-12, 1988.

Foot, Normal J. (1986). Cylindrical Feedhorns Revisited, *Ham Radio*, February 1986.

Table 1. Numerical standard deviation of the pseudorange minus doppler range data for the log periodic spiral antenna in various configurations.

Case	Description of configuration	Standard deviation first day (meters)	Standard deviation second day (meters)
1	One foot square ground plane atop 5 foot pole	3.95	4.11
2	One foot square ground plane placed directly on roof	2.11	2.07
3	One foot square ground plane placed directly on roof surrounded by absorption material	1.55	1.60
4	Four foot diameter circular ground plane atop 5 foot pole with 6 inches of absorption material around the antenna	2.05	1.99
5	Four foot diameter circular ground plane atop 5 foot pole	3.31	3.34

Table 2. Numerical standard deviation of the pseudorange minus doppler range data for various GPS antennas atop the 5 foot pole.

Case	Antenna	Ground plane	Standard deviation first day (meters)	Standard deviation second day (meters)
5	Spiral	4 foot diameter	3.31	3.34
6	Turnstile	3 foot square	2.06	1.91
7	Drooping Turnstile	2 foot diameter	2.37	2.49
8	FRPA-2	2 foot diameter	2.30	2.65
9	Microstrip	1.6 foot diameter	1.55	1.59

Note the differences in the ground planes. No absorption material was used.

Table 3. Numerical standard deviation of the pseudorange minus doppler range data for choke ring ground planes.

Case	Antenna	Ground plane	Standard deviation first day (meters)	Standard deviation second day (meters)
10	Drooping Turnstile	1.2 foot diameter JPL Choke Ring	1.81	1.80
11	Drooping Turnstile	1.7 foot diameter NSWC Choke Ring	1.57	1.99
12	Spiral	1.7 foot diameter NSWC Choke Ring	2.80	2.81

Figure 1a. Pseudorange minus Doppler range, log periodic spiral antenna, day 283.
Standard deviation: 3.95 meters

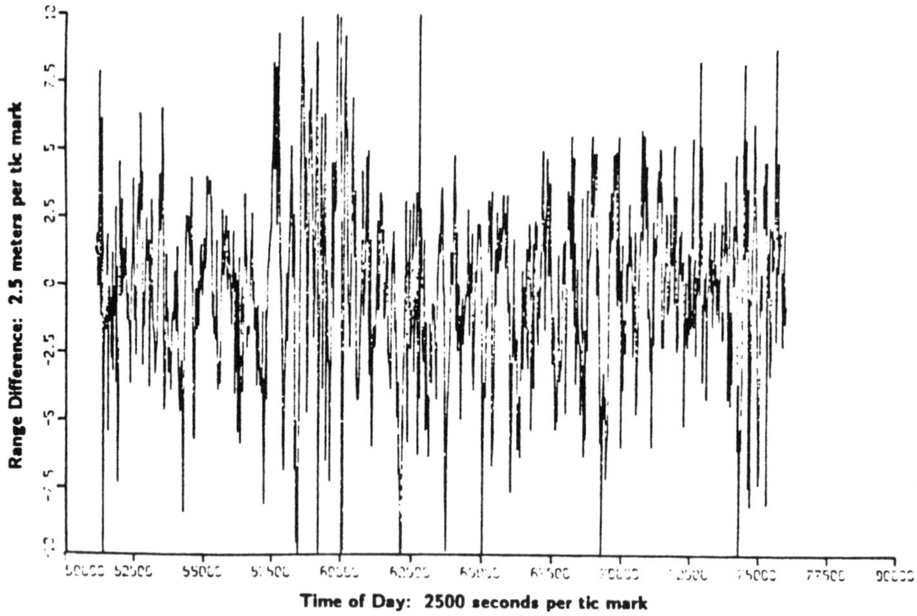

Figure 1b. Pseudorange minus Doppler range, log periodic spiral antenna, day 284.
Standard deviation: 4.15 meters

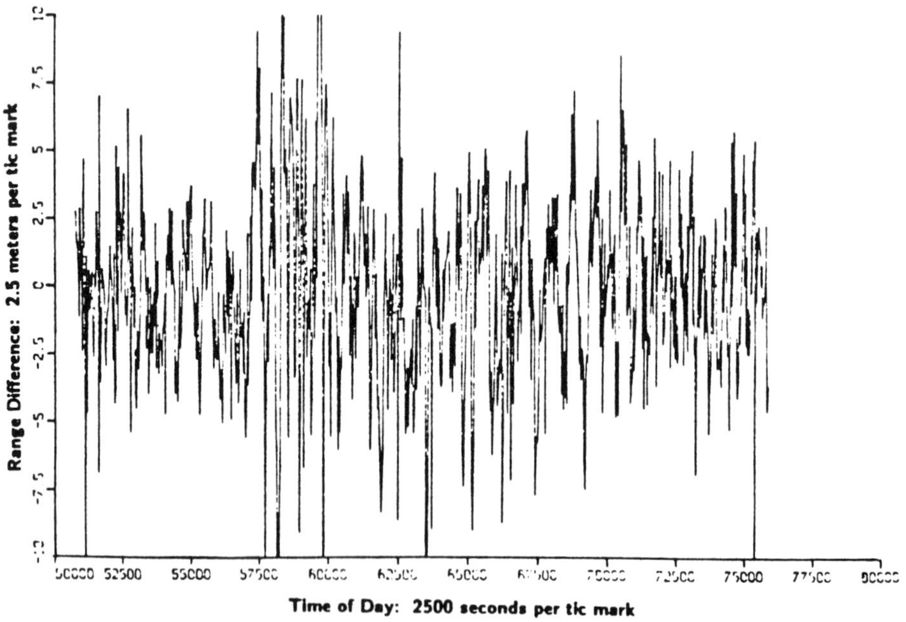

Figure 2. Pseudorange minus Doppler range, microstrip antenna, days 126-128.
Combined standard deviation: 1.56 meters

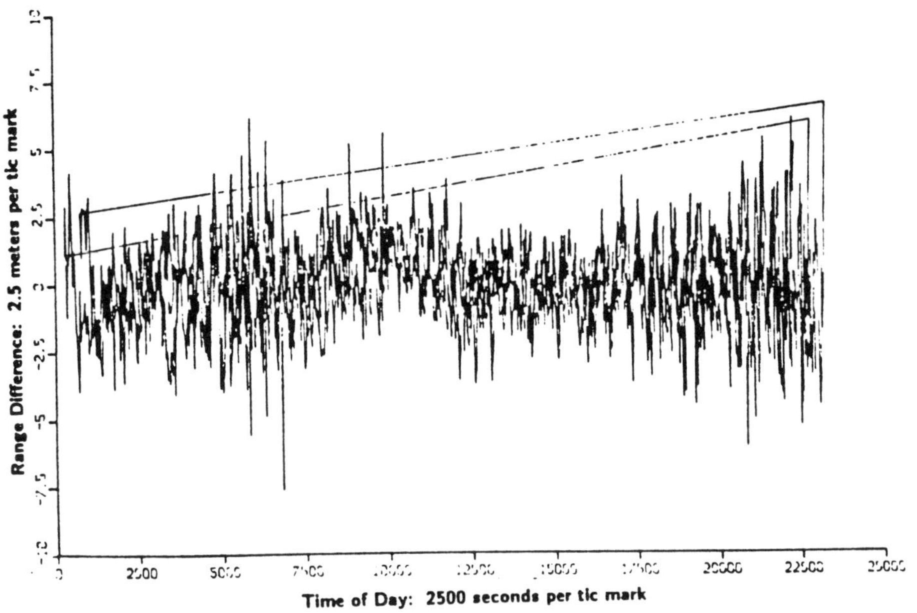

Figure 3. Pseudorange minus Doppler range, drooping turnstile antenna, days 348-349.
Combined standard deviation: 1.80 meters

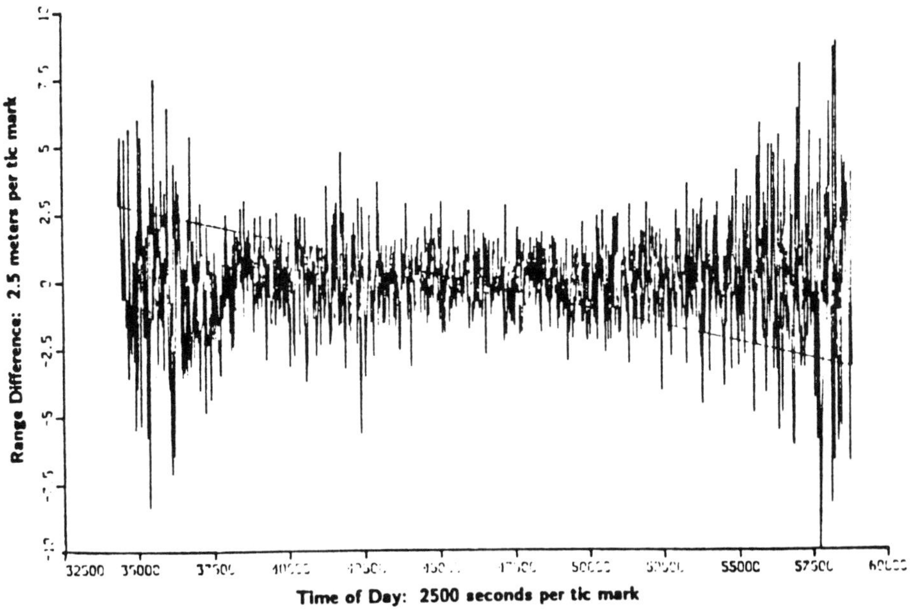

MULTIPATH EFFECTS IN STATIC AND KINEMATIC GPS SURVEYING

Yola Georgiadou and Alfred Kleusberg
Geodetic Research Laboratory
University of New Brunswick, POB 4400
Fredericton, N.B., Canada E3B 5A3

ABSTRACT

We discuss multipath signal variations and multipath induced relative position errors in static and kinematic differential GPS positioning. In the static mode, constraints regarding the location of survey sites (urban surveys) may contaminate the carrier. Two short baselines (400 and 100 m) observed for one hour with the TRIMBLE microstrip antenna were selected to illustrate limitations resulting from multipath in an urban environment. Baseline components convergence in the presence of multipath is two times slower compared to convergence in a clean environment. In kinematic mode, the inability to average out the effect on the moving platform may introduce relative position errors that are significant at the decimeter accuracy level. By treating TI4100 static data collected over a 24 meters baseline as if they were observed on a moving platform, we show that relative position errors of the order of several centimeters may result from multipath interference.

INTRODUCTION

In Georgiadou and Kleusberg (1988), we derived a mathematical model for multipath induced carrier phase errors as a function of the strength of the reflected signal and the satellite-reflector-antenna geometry. Based on TI4100 observations of very short baselines, designed expressly for our multipath studies, we demonstrated that multipath induced carrier phase errors may have amplitudes of the order of centimeters and error periods varying between less than a minute and several hours. Observation-site dependent errors, such as multipath, are acquiring increasing significance as a result of the refined modelling of baseline length dependent errors (e.g. ionospheric delay, orbital errors). This is even more so for urban and engineering surveys where short baselines (less than 2 km) are usually surveyed. In surveys of this type, site-dependent errors resulting from multipath interference may become the main accuracy limiting factor since severe constraints regarding the location of survey monuments may exist. This situation may be remedied in the static case by sufficiently long observation sessions. In kinematic surveys on the other hand, multipath errors will not have the chance to average out, particularly when no more than 4 satellites are being observed (i.e. no redundancy).

In the present paper we illustrate the effect of multipath signal variations based on observations performed in an urban environment in the City of Edmonton. The baselines from the urban survey analysed here are 400 and 100 meters long. Although precautions had been taken to substantially reduce the possibility of multipath interference the microstrip antenna shows some degree of multipath sensitivity in the carrier signal. To show the effect of carrier signal multipath on relative position in a kinematic mode, we analyse static data by treating them as if they were observed on a moving platform. These data have been described and analysed in static mode in Georgiadou and Kleusberg (1988).

DATA AND RESULTS

Multipath effects in the GPS carrier phase measurements are similar from day to day because of repeated satellite-antenna-reflector geometry (ibid). In the adjustment of GPS measurements, a part of these errors will be absorbed by the adjustment residuals and/or the estimated parameters. The division line varies depending on the length of the observation interval. As more observations are collected, the cyclic pattern of multipath cannot be absorbed anymore by the parameters and the errors are mapped into the residuals. This process leads to slow and non-uniform convergence of the parameter estimates in the sequential adjustment. For short observation intervals, the error is almost completely absorbed in the parameters, a special case being a kinematic survey where the observation interval is a single observation epoch.

Consequently, similar patterns in the least squares adjustment residuals (static case) or the estimated trajectory errors (kinematic case) of the moving platform from daily repeated GPS surveys indicate presence of multipath.

Static case

In 1987 the City of Edmonton, Alberta initiated a research project to examine the suitability of GPS in an urban environment (Rapatz et al., 1987). The Edmonton test network (1km x 1km x 1m) was observed with TRIMBLE 4000SX single frequency GPS receivers featuring the TRIMBLE 4000SX microstrip antenna equipped with a 0.8 m^2 ground plane. Observation sessions were one hour long with a sampling rate of 15 seconds.

Figure 1 shows the pattern of TRIMBLE double difference adjustment residuals from carrier phase observations of a 400 m baseline. The graph is an overlay of residual time series obtained for days 292 and 294, 1987. The residuals for day 294 are shifted by twice 236 seconds along the time axis to allow for their direct comparison for repeated geometry after two sidereal days (see e.g. Evans, 1986). The y-labels on the left hand side of the graph apply to the upper curve. The y-labels on the right side apply to the lower curve.

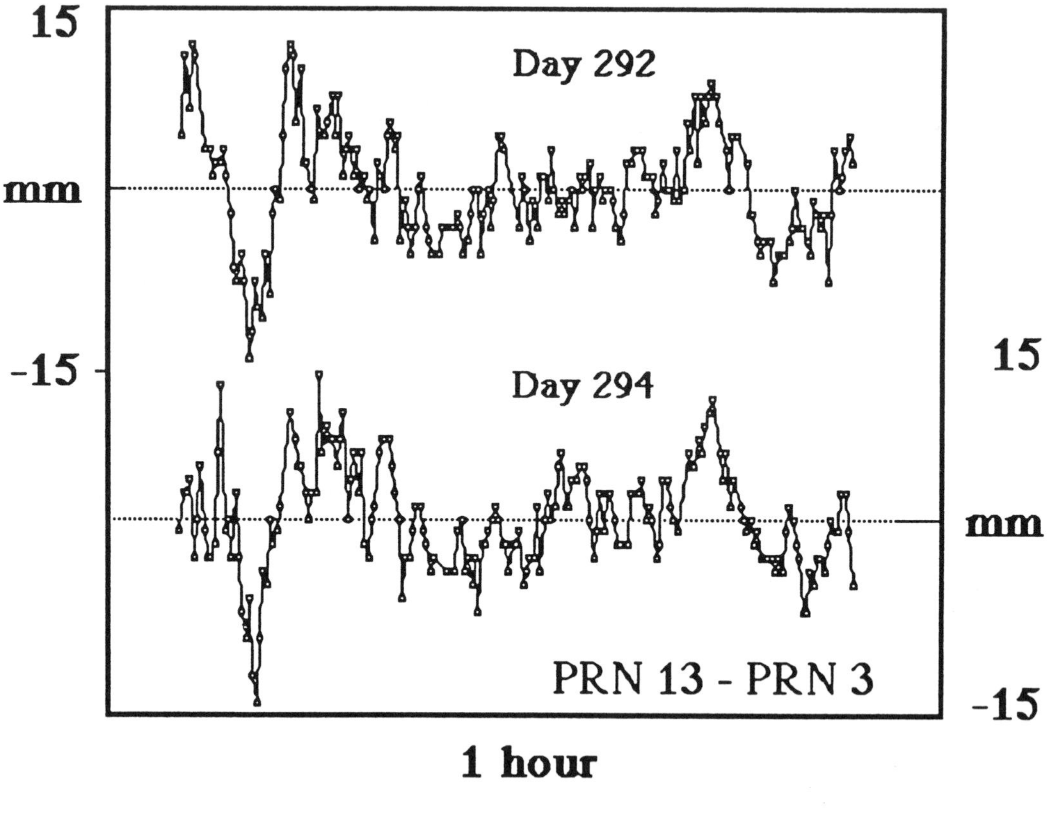

Fig. 1.TRIMBLE adjustment residuals for Days 292 (upper curve) and 294 (lower curve)

The pattern of systematic variations is practically identical for both days with a peak-to-peak variation of 2.5 centimeters. The random noise is much lower.

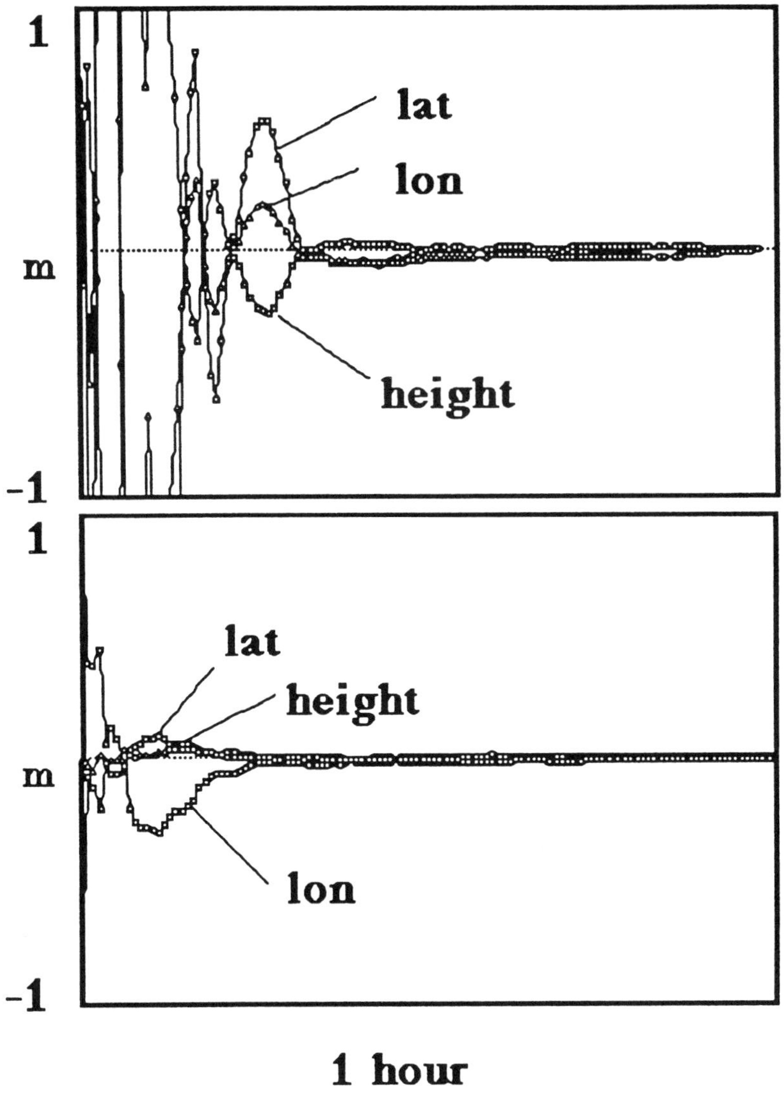

Fig. 2. TRIMBLE solution convergence with (upper curve) and without (lower curve) multipath interference

The cyclic pattern of the adjustment resuduals is accompanied by a cyclic and slow convergence of the baseline components to their final values in the sequential adjustment. Another important consequence is slow convergence of estimated carrier phase ambiguities meaning that their integer nature can be exploited after long observation sessions only.

Figure 2 shows two cases of convergence of baseline components. The first refers to the 400 m baseline where multipath interference was obvious from the inspection of the residuals (see Figure 1). After 20 minutes of observations, the baseline components are within one decimetre of their final values. The second plot shows the solution convergence for a 100 m baseline of the same network from observations without significant multipath contamination. The baseline components converge to within one decimetre of their final values after 10 minutes only. Thus, in a signal reflecting environment observation sessions may have to be twice as long to achieve reasonable results.

Kinematic case

Results shown here are computed from a 1 hr subset of TI4100 data observed on a 24 m baseline on two consecutive days (May 7 and 8, 1987). The sampling rate was 30 seconds. No conducting material was present within 15 metres of one antenna. The other antenna was placed on an aluminium box on a trailer to generate reflected signals. For more details, also regarding the satellite geometry during the observation session, we refer to Georgiadou and Kleusberg (1988). Double difference ambiguities were resolved from the static solution, based on the 2.5 hrs data span and were considered known for the kinematic test. Relative positions of the antenna on the trailer with respect to the antenna on the tripod were estimated using the L1/L2 ionospheric free carrier signal combination for each observation epoch.

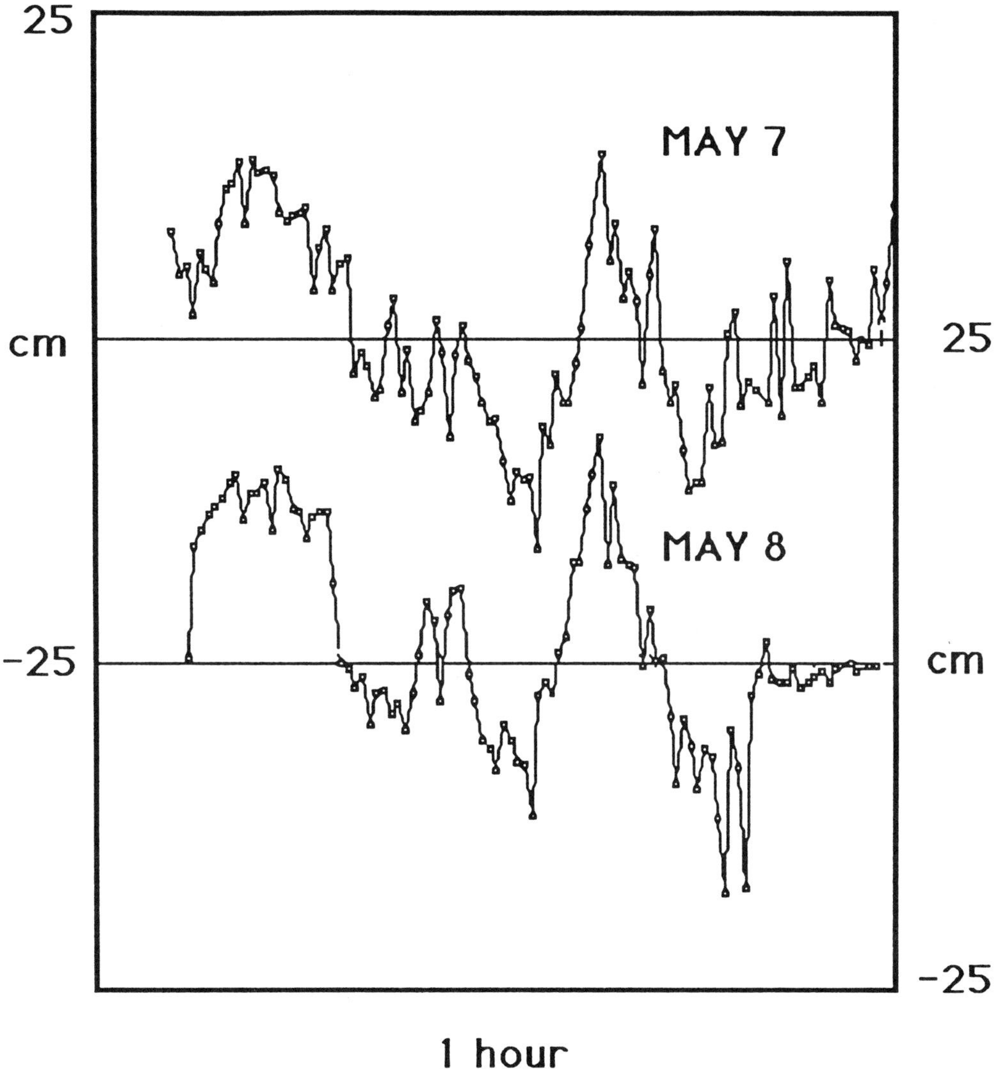

Fig.3. Height error due to multipath in kinematic mode

Figure 3 shows the relative height of the "moving receiver" as a function of time for two consecutive days. The y-labels on the left hand side of the graph apply to the upper curve. The y-labels on the right side apply to the lower curve. The resulting "trajectories" are shifted by 236 seconds to allow for direct comparison. The high correlation for this time lag suggests strong multipath interference resulting in errors larger than 10 cm in the height component.

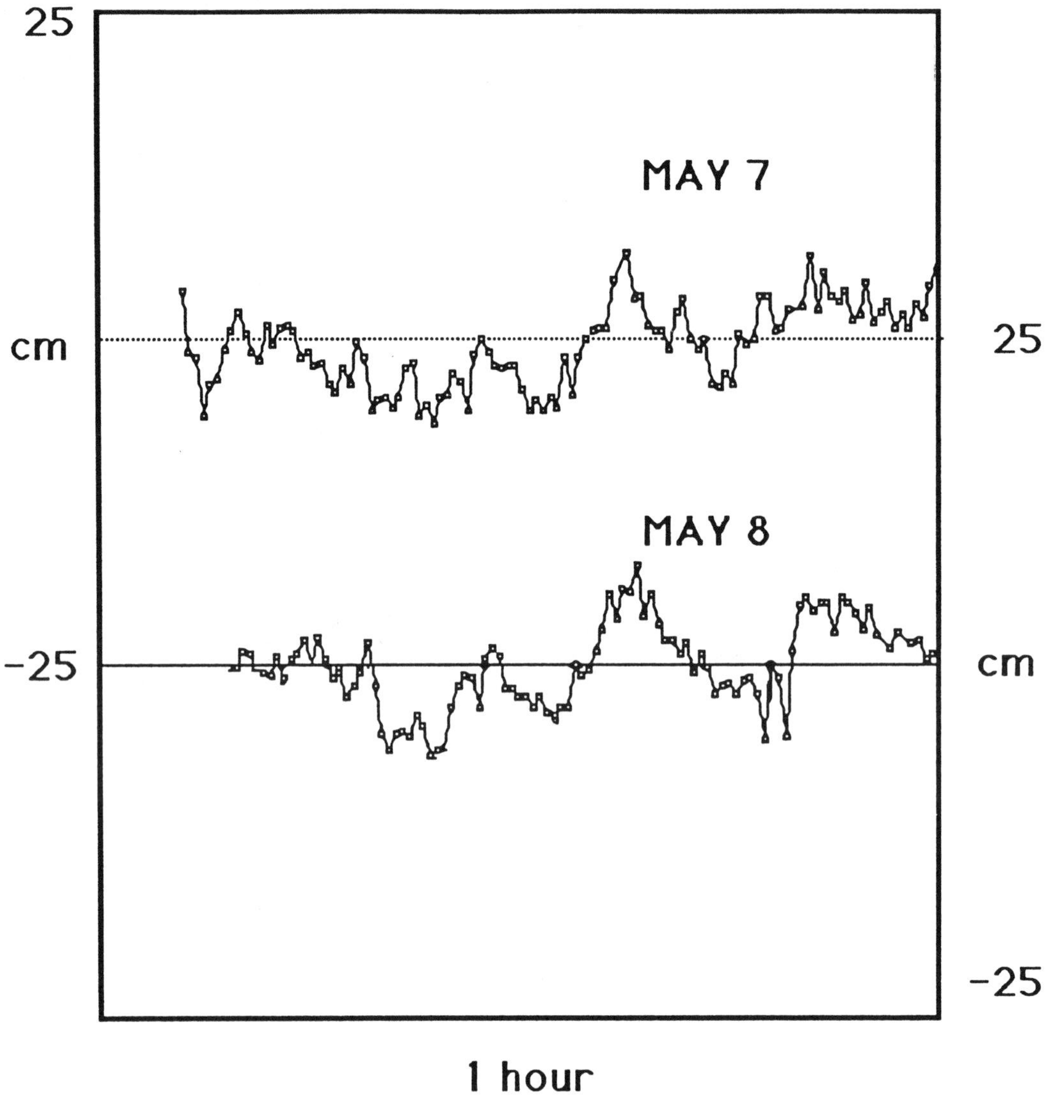

Fig.4. Distance error due to multipath in kinematic mode

Figure 4 shows the "changing" distance of the "moving receiver" as a function of time for two consecutive days. The resulting "trajectories" are again shifted by 236 seconds to allow for direct comparison. Day-to-day correlation is evident again with distance errors of the order of 5 cm.

CONCLUSIONS

Microstrip antennas may exhibit some degree of multipath contamination in carrier phase observations. Compared to multipath free observations, the estimated baseline components converge two times slower. If possible, sites should be carefully selected to avoid conducting environment. If this is not possible in an urban environment, longer than usual observation sessions have to be planned. In a kinematic mode, on the other hand, multipath reduction through redundancy may be provided by the simultaneous tracking of more than four satellites. Otherwise, estimated relative positions may be significantly biased by signal multipath effects.

Acknowledgement

Assistance from and discussions with UNB colleagues are appreciated. This work has been partially funded through a Natural Sciences and Engineering Research Council (NSERC) strategic grant entitled "Application of Differential GPS Positioning".

REFERENCES

Evans AG (1986): *Comparison of GPS pseudo range and biased Doppler range measurements to demonstrate signal multipath effects*. Proc.4th Intern.Symp.Sat.Pos., Austin,Tx, 28 April - 2 May, 573-587

Georgiadou Y, Kleusberg A (1988): *On carrier signal multipath effects in relative GPS positioning* Manuscripta Geodaetica, Vol. 13, 172-179

Rapatz PJV, Craymer MR, Kleusberg A, Langley RB, Quek SH, Tranquilla J and Wells DE (1987): *Procedures and specifications for urban GPS surveys*. Department of Surveying Engineering, Technical Report No. 131, University of New Brunswick, Fredericton, N.B.

GPS-OBSERVATIONS IN A LOCAL NETWORK COVERING BIG HEIGHT DIFFERENCES

Günter Klein, Gerd Boedecker
Bayerische Kommission für die Internationale Erdmessung
Bayerische Akademie der Wissenschaften, München, Fed. Rep. of Germany

INTRODUCTION

In the framework of an investigation on integrated GPS-INS observations for positioning and gravity vector recovery (sponsored by the *Deutsche Forschungsgemeinschaft*) a test network has been established in Upper Bavaria in the Alpine foreland and into the Alps including 29 stations along GPS-INS traverses (see figure 1).

Fig. 1. Traverses of GPS-INS test network "Upper Bavaria"

For logistic reasons one of the three TI 4100 GPS receivers available was located permanently on top of the Laber mountain some 1,000 m above the average station altitude. Height differences to Laber mountain and different climate of stations in the different valleys led to problems in modelling tropospheric refraction.

Because of limited availability of the GPS-receivers, the observation period at each station was only 50 minutes. Taking into account considerable refraction effects and noise from multipath etc., ambiguity determination required special attention.

The observations have been evaluated using the Bernese Software, rel. 2. Observing four satellites simultaneously, as is the usual procedure with TI 4100 receivers, long series are needed in order to recover ambiguities as parameters unambiguously.

AMBIGUITY DETERMINATION FROM SHORT OBSERVATION SERIES

For the reliable determination of ambiguity parameters from adjustment, a 50 minutes period is about the limit in our production environment. Reduction of observation periods to a few minutes requires other methods like that of Wübbena (1988). In either case it is necessary to check the ambiguities found, e. g. by the study of the stochastic behaviour of residues.

Wübbena's Method

The ambiguity may be determined independently of the geometry and tropospheric refraction from a combination of PRN-code and carrier beat phase; this also accounts for the ionospheric refraction. Because of big noise in PRN-code, a combination with so-called *wide lane* (L5)-phase is preferred, which refers to the difference of L1 and L2, yielding 86 cm wavelength.

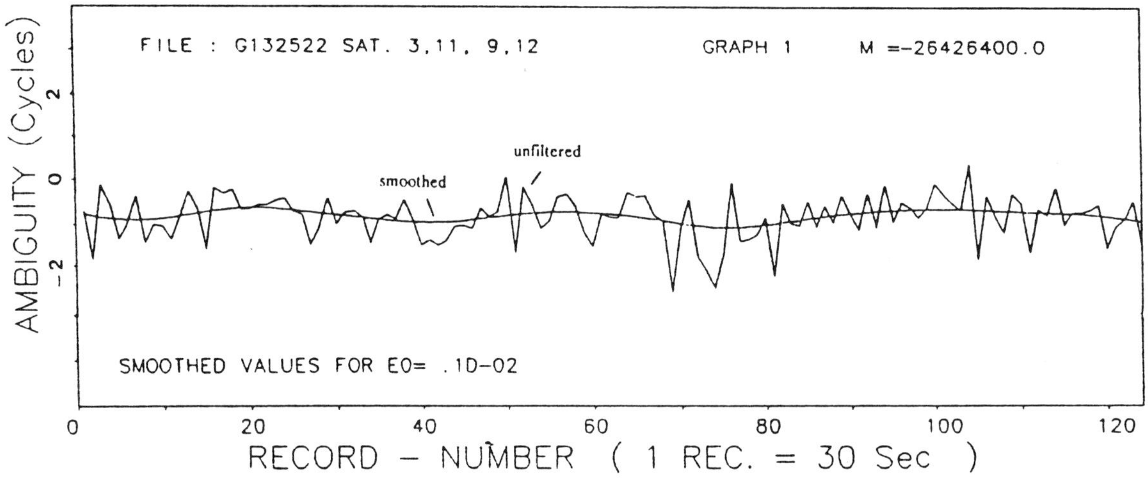

Fig. 2. Wide lane ambiguities ΔN

Figure 2 shows wide lane ambiguities ΔN every 30 seconds, computed from (Wübbena 1988):

$$\Delta N = (F1 - F2)\, T_{\Sigma} - ((\varphi 1 - \varphi 2) + t_r (F1 - F2)) \qquad (1)$$

where

$F1, F2$	carrier frequencies
$T_\Sigma =$	$(T1 + T2)/2 + 0.0625\,(T1 - T2)$
$T1, T2$	transmit epoch of PRN-code signals
$\varphi 1, \varphi 2$	carrier beat phases
t_r	receive epoch

Because of ionospheric effects the computed PRN-code signal transmit times will not be equal, since they are referring to the identical receiver epoch. Depending on multipath effects and noise, a final value ΔN has to be found from a more or less sophisticated estimation procedure. Analogously, L5 ambiguities can be computed from single differences and these be reduced for wide lane ambiguities, thus avoiding ambiguity determination from adjustment.

Ambiguity as Parameters

The Bernese software offers three options with respect to ambiguity determination:

- rounding of real ambiguity value from adjustment to nearest integer
- in the vicinity of the original real ambiguities different combinations of integer ambiguities are tried to find the set with minimum residue standard deviation
- search only in the range of real ambiguity ± 3 ambiguity r.m.s.

The last sigma dependent ambiguity search can provide proper values for longer observations series, for short observation series, however, the ambiguity r.m.s. estimated typically is too small and therefore the method fails. Hence, the time consuming second option is recommended for short observation series. In our network, in 90 % out of 56 single difference files with observation duration of 50 -60 minutes option 2 permitted to determine unambiguous ambiguities.

Problems, Checks, Suggestions

Because of local extension of our network, the ionospheric effect vanishes for single differences and L1 and L2 single differences can be evaluated separately, disregarding of whether L2 and L1 ambiguities had been determined via wide lane or parameter. If the coordinate differences between L1 and L2 solutions as also the formal errors are small, the evaluation appears finished. In rare cases, however, the solution is still in error. Figure 3 shows an example where L1 and L2 coordinates agree quite well. Inspection of the residue autocovariance function with time reveals big values and only the second solution yields satisfactory results (figure 3). It may also happen that residual errors are small, but L1 and L2 coordinates differ quite much. Inspection of residual autocovariance function with time again will lead eventually to good results.

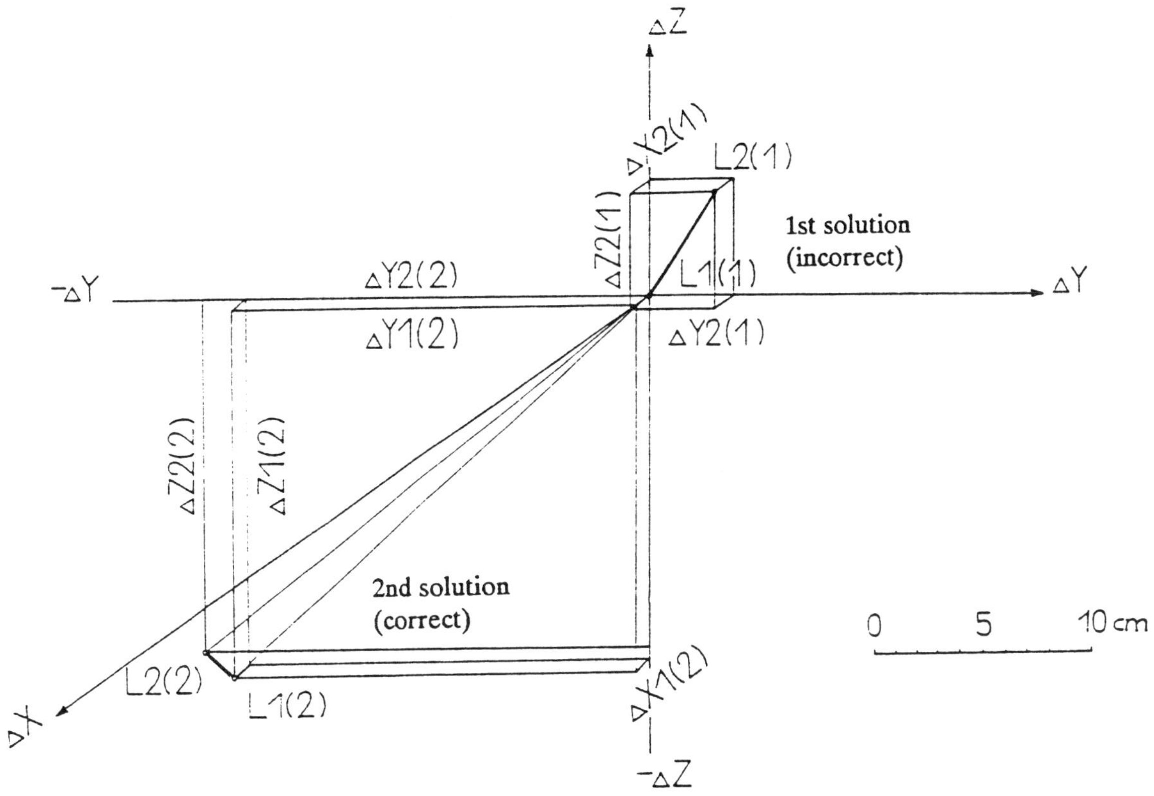

Fig. 3. Coordinates before and after test by covariance

If more satellites can be observed simultaneously, the observation duration can be shortened.

Another improvement in the Bernese software would be the extension of ambiguity search by a few more cycles, the range may be defined by a multiple of ambiguity standard deviation.

TROPOSPHERIC REFRACTION IN A LOCAL NETWORK COVERING BIG HEIGHT DIFFERENCES

Regarding tropospheric refraction, five evaluation computations have been made:

- Saastamoinen model using a global model with sea level values (Saastamoinen 1973)

$t_0 = 18^{\circ}$ $\quad p_0 = 1\,013.2$ mbar $\quad H_0 = 50\ \%$

and extrapolated values as below

$t = t_0 - 0.0065 \cdot h$ — temperature

$p = p_0 (1 - 2.26 \cdot 10^{-5} h)^{5.225}$ — air pressure

$H = H_0 \exp. [-6.396 \cdot 10^{-4} h]$ — humidity in %

- Saastamoinen model using locally observed meteorological data
- Modified Hopfield model (Remondi 1984) using global values as above
- Modified Hopfield model using locally observed meteorogical data
- For comparison without any tropospheric correction

Because Saastamoinen (Saastamoinen 1973) and Hopfield (Hopfield 1971) models both employ essentially identical physical principles, namely exponential air pressure and linear temperature decrease with height, the implication of global models yields virtually identical coordinates, the application of the same local model gives little differences less than 12 mm.

Differences arising from the use of local vs. global meteorological data, however, very much affect the height and to some extent the horizontal coordinates. The errors in height are eight times as much as in plane coordinates.

If no tropospheric reductions are applied, heights are by 90 cm and positions by 10 cm in error.

Proposals from literature for improved tropospheric models on the basis of observed meteorological data are difficult to apply if the stations are located in different valleys. A model appears necessary that fits the local observations but is common for higher altitudes.

REFERENCES

Hopfield, M. S. (1971). Tropospheric effect on electromagnetically measured range: Prediction from surface weather data, *Radio Sci.* **6**

Remondi, B. (1984). *Using the Global Positioning System (GPS) Phase Observable for Relative Geodesy: Modelling, Processing and Results,* Ph. D. thesis, The University of Texas, Austin

Saastamoinen, I. (1973). Contribution to the Theory of Atmospheric Refraction, Part I,II, *Bull. Geod.* **106, 107**

Wübbena, G. (1988). GPS Carrier Phases and Clock Modelling, *Proceedings of the International GPS-Workshop April 10 - 13, 1988*, Darmstadt

INTERIM STATUS REPORT ON DÖNAV

Günter Seeber
Institut für Erdmessung
D-3000 Hannover 1, Nienburger Str. 5
Federal Republic of Germany

PROJECT DÖNAV

The German-Austrian NAVSTAR Campaign (DÖNAV) is a joint project of the following scientists and institutes

Geodätisches Institut, Universität Bonn (A. Müller)
Alfred-Wegener-Institut, Bremerhaven (H.W. Schenke)
Amt für Militärisches Geowesen, Euskirchen (F.J. Lohmar)
Institut für Angewandte Geodäsie, Frankfurt (H.Seeger, W.Schlüter)
Institut für Weltraumforschung, Graz (K.Rinner, P. Pesec)
Geodätisches Institut, Universität Hannover (H. Pelzer)
Institut für Erdmessung, Universität Hannover (G.Seeber, G.Wübbena)
Geodätisches Institut Kopenhagen (F. Madsen)
Institut für Astronomische und Physikalische Geodäsie, Technische Universität München (R.Sigl, K. Deichl)
Institut für Astronomische und Physikalische Geodäsie, Universität der Bundeswehr München (G.Hein, H. Landau)
Institut für Geodäsie, Universität der Bundeswehr, München (A. Schödlbauer, H. Glasmacher, K. Krack).

Main objectives of the campaign are to

- create a set of dual frequency data on a large number of simultaneously occupied stations for various investigations

- create a set of 3D-coordinates on first order stations of the national networks for various comparisons.

Some DÖNAV key factors are

11 days of observations, November 4-14, 1986
12 TI 4100 Dual frequency P-Code receivers
51 Stations
7 reference stations with permanent observations (*part time)
Graz* Laser, EDOC, DÖDOC
Zimmerwald Laser, EDOC, DÖDOC

Wettzell Laser, VLBI, EDOC, DÖDOC
Kloppenheim DÖDOC
Hohenbunstorf* Edoc, DÖDOC
Kootwijk Laser
Onsala VLBI

6 mobile receivers roving in groups from the south to the north with two overlapping points kept fixed from day to day.

A first report on DÖNAV was given during the IUGG General Assembly at Vancouver in 1987. A further report with maps and observation schedule can be found in the Proceedings of the Darmstadt International GPS-Workshop 1988. The purpose of the status-report is to give an indication on the final results which will be published in in the near future as a joint publication of all participating groups.

COMPUTATIONS

The complete data-set was processed in five different processing centers with three different software packages. One package, the Bernese software, was used in three different versions and modes. The key factors of the five different solutions are listed in the sequel. More details of the individual packages are published elsewhere.

Bernese Version 3 (computation H. Glasmacher, München)

- Broadcast Orbits
- Standard Atmosphere (Saastamoinen)
- Ionospheric corrected Phase
- Ambiguities solved during preprocessing
- Orbit improvement
  - 3 arcs with 4, 4 and 3 days length
  - 6 Keplerian parameters and 1 radiation pressure scale bias
- all correlations considered

Bernese Version 2 (computation P. Pesec, Graz)

- Broadcast Orbits
- Standard Weather - Saastamoinen
- L3 observations
- different fixed reference stations from session to session
- all session solutions combined to a complete solution via 7-parameter transformations

Bernese Version 2 (modified) (computation A. Müller, Bonn)

- Precise Ephemeris (NSWC)
- L3 Phase combination (Ionospheric corrected observation)
- no ortbit improvement
- no tropospheric bias parameter
- common adjustment of all observations

GEONAP (author and computation G. Wübbena, Hannover)

- Broadcast Orbits
- tropospheric correction with mean weather and Hopfield-Black Model
- Ambiguity solution through code/carrier combination and geometry (appr. 90% solved)
- Orbit improvement with 6 Keplerian elements/session
- 1 tropospheric scale bias/station/session
- datum introduced as obervation
- session solutions with subsequent network adjustment

TOPAS (author and computation H. Landau, München)

- Broadcast Orbits
- Tropospheric correction with standard weather and modified Hopfield Model
- Dynamic orbit improvement
  - two arcs with 5 rsp. 6 days length
- 1 tropospheric scale bias/station/session
- Ambiguity solution through code/carrier combination and geometry

The detailed comparison of results will give valuable insights into the reliability of GPS network computation. In order to make a rigorous comparison of individual solutions possible, the group agreed upon the definition of a preliminary "DÖNAV Datum" in such a way, that 7 components within the whole network were held fixed. This unambiguous datum definition was

- 3 coordinates Wettzell from a laser solution
- 3 dimensional vector Wettzell-Onsala from a VLBI solution
- the length of the radius vector to Kootwijk from a laser solution.

In the meanwhile it turned out that most probabely for other international projects like EUREF the coordinates from the Terrestrial System of the International Earth Rotation Service (ITRF) will be used as "fiducial sites" for datum definition. In order to be compatible a new DÖNAV solution, starting from the ITRF coordinates on all Laser/VLBI stations (five sites) is being computed in the processing centers. Possibly these solutions, which are in the best way "geocentric" will be considered to form the final DÖNAV solution. The pertinent decision of the group will be made once the computations are finished.

PRELIMINARY RESULTS

The accuracy prospects for the final solution can be reflected by one typical result from the computation in the preliminary DÖNAV datum. In this solution the Laser/VLBI -coordinates of Wettzell and Onsala as well as the "Laser-height" of station Kootwijk were introduced as highly precise "observations" into the solution. The adjustment was done with the GEONAP software.

Figure 1 shows the relative error ellipses between adjacent stations They are nearly all considerably below 5 cm for interstation distances of 100-150 km. These values will certainly decrease when five to six datum stations are introduced as "fiducial points". The accuracy of the "fiducial stations", coming from a global solution of Laser und VLBI observations, however, is also in the order of ± 5cm and thus corresponds to the GPS results.

Since only 7 from 15 coordinates of the permanent stations have been used for datum definition, it is possible to compare the GPS network results with distances computed from the laser solution. This is done in Table 1. The original results are compared without any transformation. Station Zimmerwald (ZMW) and Graz (GRZ) did not contribute to the datum and provide an independent check. Station Kootwijk is considered to give the weakest laser solution.

Line	Laser (m)	GPS (m)	Difference (m)
WTZ-GRZ	30 2087,019	30 2087,035	,016
WTZ-ZMW	47 5895,298	47 5895,331	,033
WTZ-KTW	60 2457,351	60 2457,283	-,068
GRZ-ZMW	61 0851,157	61 0851,142	-,015
GRZ-KTW	89 9498,744	89 9498,700	-,044
ZMW-KTW	60 1438,683	60 1438,791	-,108

Tabl.1: Comparison between interstation distances from Laser and GPS solution.

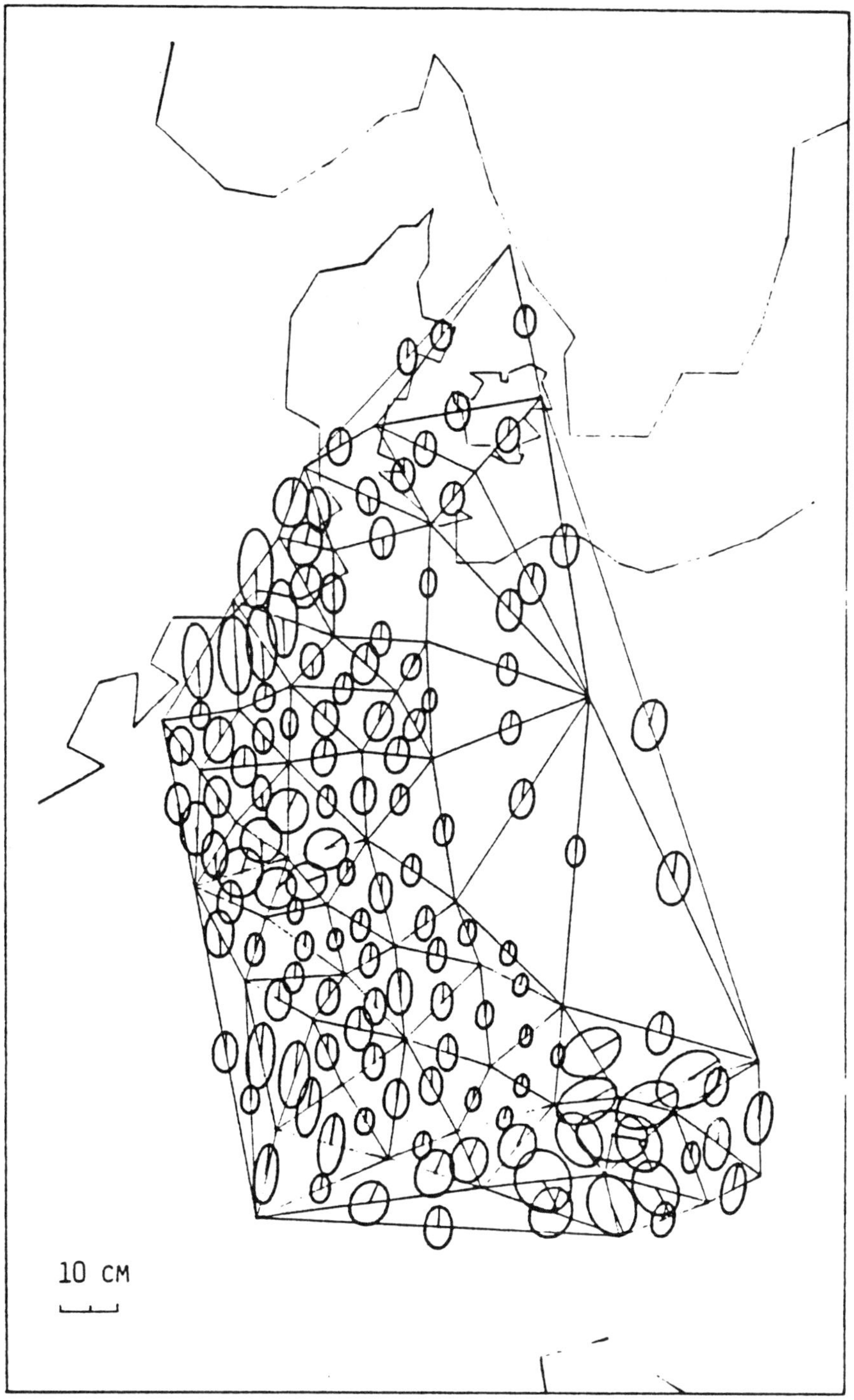

Fig. 1: Relative error ellipses in one typical DÖNAV solution (GEONAP)

When distances to Kootwijk are not regarded, the GPS derived and Laser-derived distances agree on the $5 \cdot 10^{-8}$ level.

FURTHER PLANS

The final computations in the definite DÖNAV datum and the comparisons with various data sets will be finished before the end of 1989. A joint publication of the results will come out in the series B of the "Deutsche Geodätische Kommission" in 1990.

REPEATABLE MEASUREMENTS OF BASELINE LENGTH BY GLOBAL POSITIONING SYSTEM IN CENTRAL JAPAN

Fumiaki Kimata, Yoshio Sumino, Masaru Nakamura, Rikio Miyajima, Takashi Okuda, *Mikio Satomura, *Yuji Sasaki, **Yoshimi Sasaki
Research Center for Seismology and Volcanology, Nagoya University
Nagoya 464-01, Japan
*Faculty of Liberal Arts, Shizuoka University, Shizuoka 422, Japan
**Faculty of Education, Gifu University, Gifu 501-11, Japan

INTRODUCTION

An accurate determination of baseline length is of basic importance in the interpretation of crustal motions in earth's interior for the study of earthquake prediction. A complete understanding of the crustal motion of a particular region requires a dense set of geodetic measurements in space and time. During the past decade, Global positioning System (GPS) at the United State National Geodetic Service has been developed chiefly for instantaneous positioning and dynamic navigation. This system has developed further with the capability of measuring baseline length on the order of 1 to 2 ppm accuracy for distances of several ten to several hundred kilometers in length, by using phase interferometric method with the GPS carrier wave length (e.g,, Hothem and Fronczek, 1983).

In 1987, delivery of accurate and proven GPS survey system was made to Geographical Survey Institute of Japan where it was used for experimental study of geodetic and crustal motion monitoring (Tsuji et al., 1988). Recently, several GPS receivers were delivered to our institute for earthquake prediction project in the Tokai area of Central Japan (Kimata et al., 1988). In this paper, we reported the preliminary results of repeatable measurements of baseline length in the territory of Central Japan under poor satellite geometry and coverage currently available, and some discussions were made on the obtained results.

MEASUREMENTS AND DATA PROCESSING

During the summer of 1989, five GPS receivers were used to measure five-station network in the southern part of Central Japan with baselines ranging from 23 km to 172 km. The sites occupied by the GPS receivers are shown in Figure 1 and the approximate distances between the sites are found in Table 1.

Table 1. Approximate distances between stations.

Station	INUYAMA	GIFU	TAKAYAMA	SHIZUOKA
NAGOYA	23	40	111	135
INUYAMA		29	88	135
GIFU			85	164
TAKAYAMA				173

Distances are in kilometers.

Measurements were made using TRIMBLE 4000 series dual frequency type GPS receivers (Allison et al., 1988). The principle observing session on each day consisted of four satellite, which included SVs 3, 8, 11, and 14, continuing for about 3 hours (Figure 2). These observations used a newly revised software version called TRIMVEC 89-03 which provided a smooth transition between observed sessions. During the observed sessions, a few cycle slips were found in the phase data. In the post-processing part of data analysis, TRIMVEC data processing program was used. The triple difference technique was used in the processing of the phase data, since this technique is a very useful for cycle

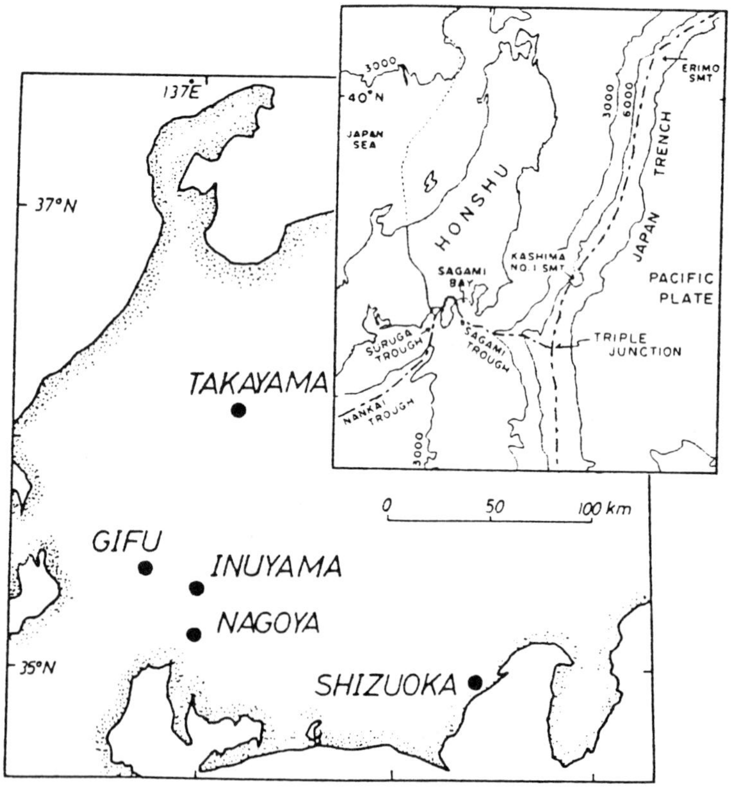

Figure 1. Location of GPS measurement sites in Central Japan.

slip detection and repair. The influence of cycle slip error was greatly reduced in the triple difference technique compared to the double difference technique (Carroll and Kyle, 1988).

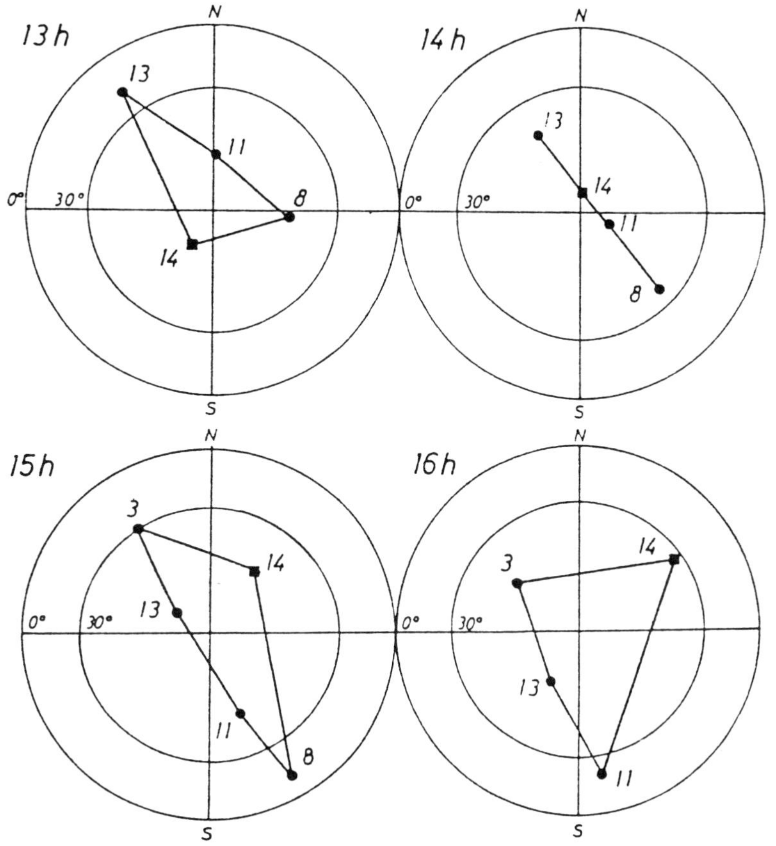

Figure 2. Polar plots of satellite orbits for NAGOYA, Central Japan, June 15, 1989, from 13:00 to 16:00.

TEMPORAL VARIATION OF BASELINE LENGTH

The results for baseline lengths and each component of baseline vector during June 2 - 30, 1989 are shown in Figure 3. As shown in this figure, the baselines to north-south direction such as Nagoya-Takayama (111 km) and Gifu-Takayama (85 km) were found to fluctuate as little as 5 cm a day. However, the baselines to east-west direction such as Nagoya-Shizuoka (135 km) and Takayama-Shizuoka (173 km) showed a large fluctuation with time, and the change in daily value sometimes amounted 10 cm or more. In this figure, it was found that the fluctuation of the east-west component of the baseline vector was very large when compared to the fluctuation of the north-south component.

Figure 4 shows the repeatability for the determination of baseline length and the component of baseline vector. In this figure, the repeatability of baseline determination is expressed by the standard deviation (SD) of the monthly mean values, given by the maximum allowable error at a 95 % confidence level. As shown in Figure 4, SD of repeatability in the baselines to north-south direction such as Nagoya-Takayama (111 km), Inuyama-Takayama (88 km), and Gifu-Takayama (85 km) became within 0.5 ppm during one month. However, SD of repeatability in the baseline to east-west direction such as Inuyama-Gifu (29 km) became more than 1 ppm during one month. These results may be due to the geometry of satellite constellation, since the satellites aligned mostly to north-south direction in the sky as shown in Figure 2.

In conclusion, by using the triple difference technique the baseline lengths in the range of about 100 km in length were determined within 5 cm (0.5 ppm) without tropospheric correction for north-south direction, which was equivalent to the alignment of satellite constellation. Further data on the variation of baseline length with time are certainly

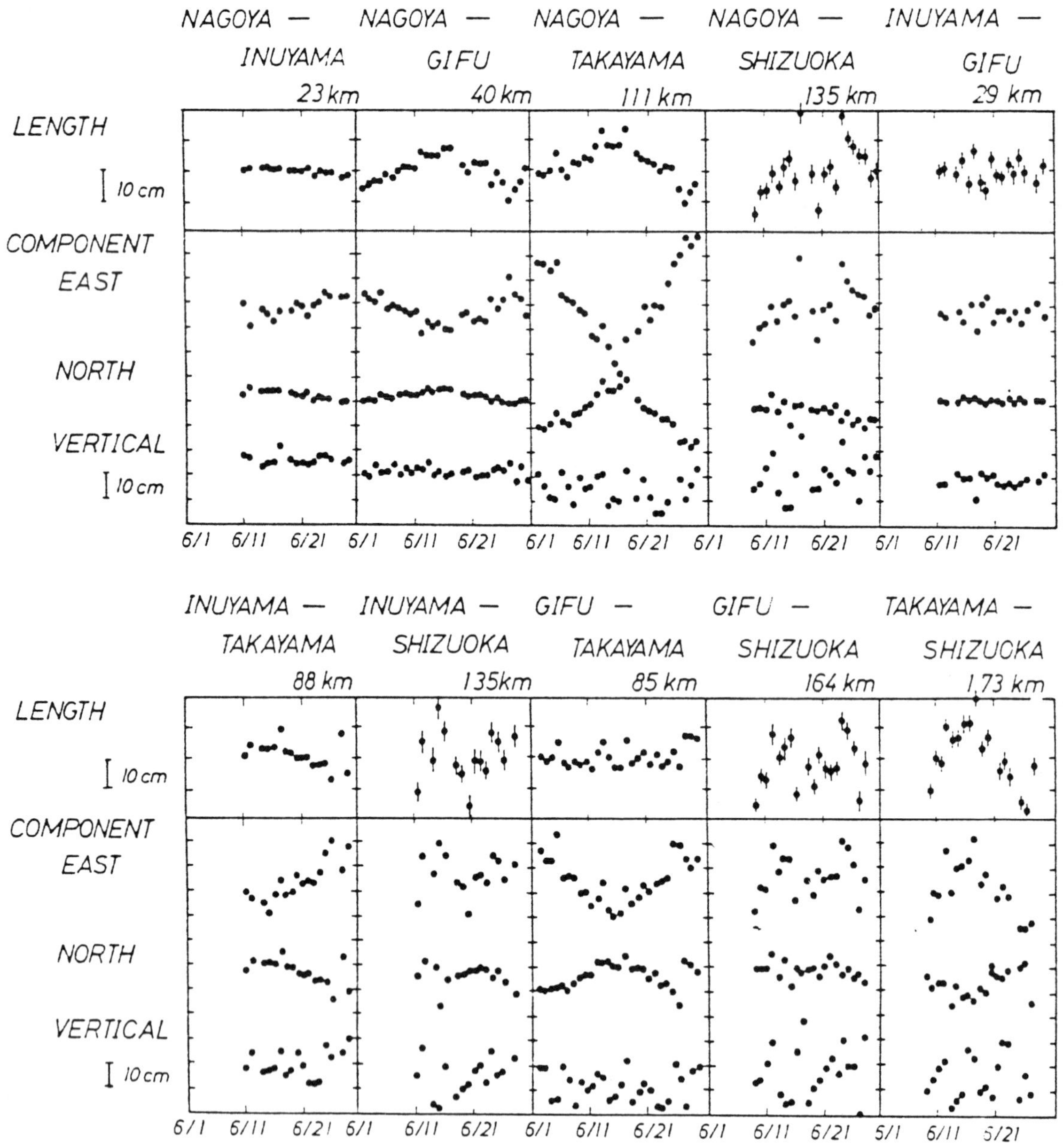

Figure 3. Baseline lengths and baseline vectors observed in June, 1989.

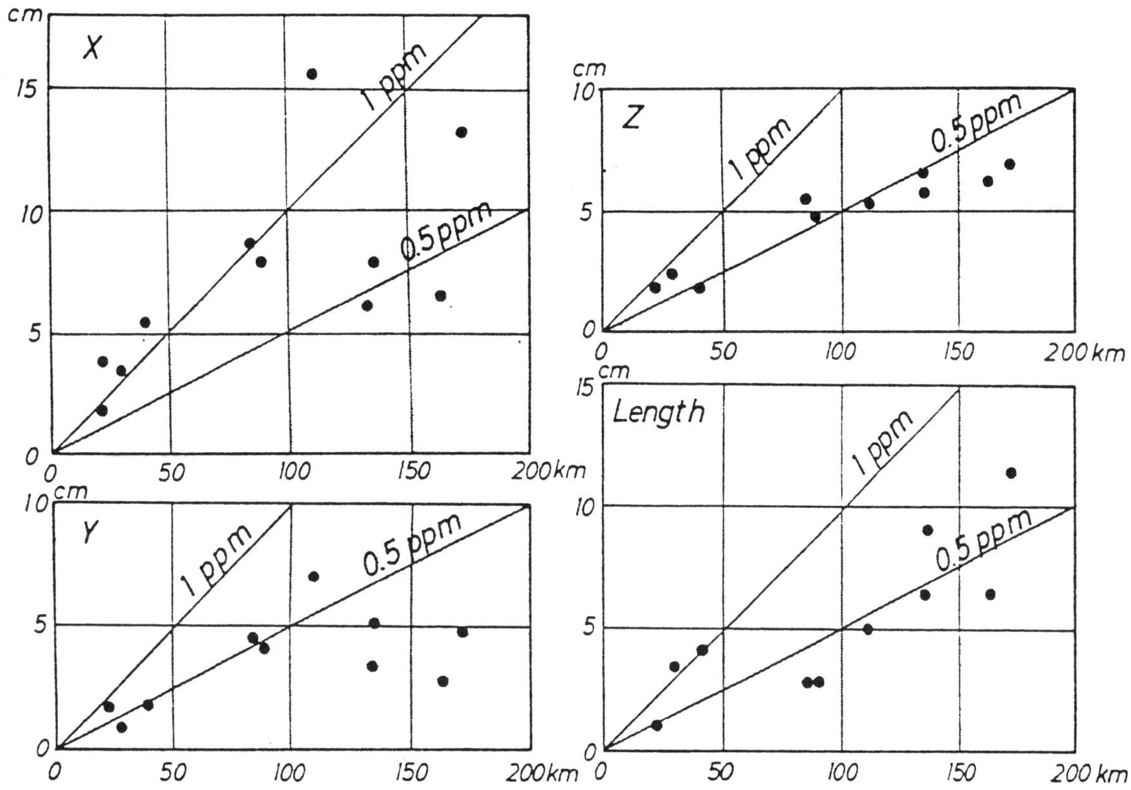

Figure 4. Standard deviation (SD) of repeatability for east-west (X), north-south (Y), vertical (Z) components, and baseline length vs. baseline length.

necessary in order to give an acceptable interpretation of the present results and to find regional tectonic motions. A complete satellite constellation coverage in the early stage of 1990 will give us more accurate data sets of baseline length in the Central Japan.

REFERENCES

Allison, T., R. Eschenbach, R. Hyatt, and B. Westfall (1988). C/A-code dual frequency surveying receiver - Architecture and performance, Positioning Location and Navigation Symposium, IEEE Aerospace and Electronics System Society, Florida, Decembar.

Carrol, L. E. and D. W. Kyle (1988). Triple difference vs. double difference - Is double difference always the best ?, EOS Trans. Geophys. Union, 69, 324.

Hothem, L. D. and C. J. Fronczek (1983). Report on test and demonstration of MACROMETER model V-1000 interferometric surveyor, Federal Geodetic Control Committee :FGCC-IS-83-2,Rockville, Maryland.

Kimata, F., M. Nakamura, T. Okuda, R. Miyajima, M. Yamada, Y. Sumino, M. Satomura, Y. Sasaki, and Y. Sasaki (1988). Repeatability in baseline length under the constellation of three GPS satellite in Central Japan, Zisin, 41, 615-618.

SOME FIRST EXPERIENCES WITH THE WM102 GPS RECEIVER

Jadwiga Piechocinska and Lars E. Sjöberg
The Royal Institut of Technology
Department of Geodesy
S-100 44 STOCKHOLM, SWEDEN

Since January 1989 our department is gaining experience the L2 frequency updated Wild-Magnavox receiver (WM102). During spring 1989 some GPS campaigns were carried out. The practical experiences of these campaigns are reported. The overall experience of the receivers is that they are both reliable and easy to operate.The preliminary results for baselines with different lengths (long, medium and very short) are generally satisfactory, but significant PoPS software problems occured.

1. BACKGROUND

Since 1987 the Department of Geodesy has three WM101 GPS receivers at its disposal. In January 1989 the recevers were updated to WM102 to take also the L2 frequency. The equipment has been used together with the PoPS software in some campaigns, such as the Jozefoslaw - Borowiec Travers in Poland (6-10 March, 1989; Zdunek,1989) and a regional deformation network in the Ethiopian Rift Valley (20 April-12 May,1989).

2. WM102 RECEIVER

The WM102 is a eight-channel dual frequency receiver. Seven channels provide the L1 signal processing and the eighth one the L2 frequency. The instrument tracks up to six satellites simultaneously. It uses both the C/A- and P-code. The instrument is a robust, easily transportable GPS receiver. According to the manufacturer the unit can operate in the temperature range -25°C to +50°C. We have not been able to check these limits, but the units have been operating between -5°C in Sweden to +45°C and under rain period in Ethiopia without any problems. The robustness of the receiver was proved under transport on bad roads in the Ethiopian Rift Valley.The overall experience of the receiver and antenna unit is that they are both reliable and easy to operate.

3. POPS SOFTWARE

PoPS is the name of a post-processing software package to process both the WM101 and WM102 data. It runs in a PC environment and works under DOS. It is an interactive, menu-driven system (Fig.1). The computation component of PoPS is able to process a network of up to 10 stations. The updated version V3.02 of PoPS arrived only at the beginning of June

1989. A most well-come new feature is a possibility to provide up to eight-hour observation time in one session (to compare with three hour limit in previous versions). Other features are:

- evaluating of the option "AutoPoPS", allowing the software to run in a time-saving automatic mode,
- more flexible use of log-files,
- using different computation strategies for lines of different lenghts (L1,L2,L3 or L3+L5 observables).

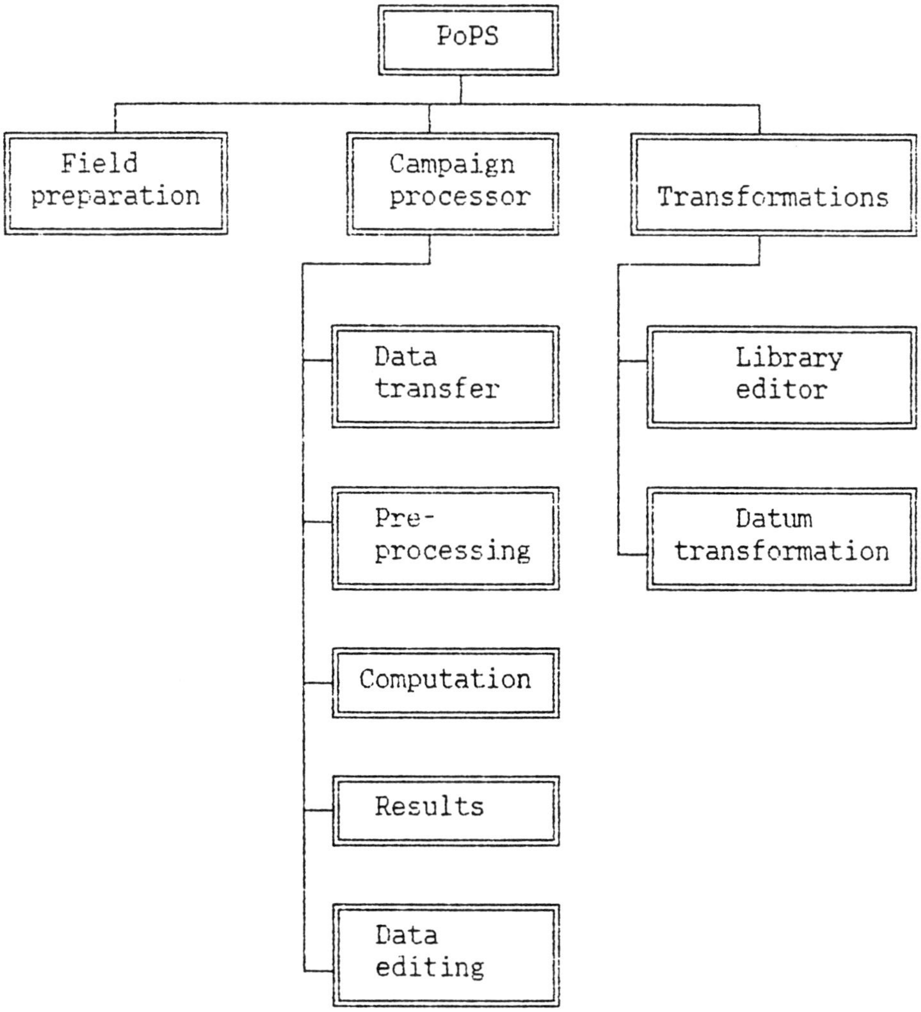

Figure 1. The PoPS structure

4. PRELIMINARY RESULTS

Some preliminary results were obtained for the Polish campaign concerning the long baseline 01JOZ - 12BOR, which was observed during four days and the test network at Borowiec and for the Ethiopian campaign concerning the regional deformation network. The following options was used in computation component:

- Saastamoinen tropospheric model,

- standard atmospheric model for the meteorological information,
- 15° elevation cut-off angle,
- broadcast ephemerides,
- single-layer model for the ionosphere for L1, L2 observables and no model at all for ionosphere-free linear combination L3 observables,
- take into account all correlations from phase to double-differences.

4.1. THE 01JOZ - 12BOR BASELINE

The baseline was measured during four observation sessions. Tables 1,2,3 contain the results of least squares adjustments of carrier phase double difference L3, L1, L2 observables."HEIGHT DIFF" means ellipsoidal height differences between stations 01JOZ and 12BOR. The r.m.s. of double differences in L3 adjustments is about 2 times less than in L1 and about 3 times less than in L2. Repetibility on L3 is the best (less than 0.2 ppm), too. The WM102 receiver provides more sofisticated tracking of L1 carrier than of L2, and this results in difference of precisions between L1 and L2 observable adjustments.

4.2. THE TEST NETWORK AT BOROWIEC

The test network at Borowiec was measured during the fifth observation session with four WM102 receivers. Tables 4 and 5 contain results of different adjustments of the network. In all computations the coordinates of the station 12BOR was fixed. As was expected from the theoretical point of view, processing of L3 for very short baselines deteriorates results. For the baselines 12BOR - B1 and 12BOR -B2 there

SESSION	IONOSPHERE-FREE LINEAR COMBINATION OF L1, L2 (L3)							
	DD RMS	BASELINE LENGTH	RMS	DIFF *	PPM *	HEIGHT DIFF	RMS	DIFF *
DAY 2	.061	271463.208	.021	.017	.06	26.282	.014	.044
DAY 3	.071	.276	.026	-.051	.19	.360	.018	-.122
DAY 4	.065	.221	.020	.004	.01	.177	.013	.061
DAY 5	.072	.233	.023	-.008	.03	.136	.015	.102
ALL DAYS	.072	271463.225	.012			26.238	.008	

Table 1. Slope distances and height differences in L3 adjustments. Units: meters, *) Relative to all days

SESSION	L1 OBSERVABLES							
	DD RMS	BASELINE LENGTH	RMS	DIFF *	PPM *	HEIGHT DIFF	RMS	DIFF *
DAY 2	.092	271462.722	.032	-.120	.04	26.421	.020	-.186
DAY 3	.100	.400	.036	.202	.07	.353	.025	-.118
DAY 4	.140	.247	.042	.355	1.3	.205	.029	.030
DAY 5	.161	271463.238	.051	-.636	2.3	25.907	.034	.328
ALL DAYS	.158	271462.602	.026			26.235	.017	

Table 2. Slope distances and height differences in L1 adjustments.
Units: meters, *) Relative to all days

SESSION	L2 OBSERVABLES							
	DD RMS	BASELINE LENGTH	RMS	DIFF *	PPM *	HEIGHT DIFF	RMS	DIFF *
DAY 2	.132	271462.527	.046	-.181	.67	26.282	.029	-.289
DAY 3	.155	461.985	.056	.361	1.3	.266	.039	-.111
DAY 4	.225	461.774	.068	.572	2.1	.142	.046	.013
DAY 5	.262	463.392	.082	1.046	3.8	25.679	.055	.476
ALL DAYS	.248	271462.346	.040			26.155	.027	

Table 3. Slope distances and height differences in L2 adjustments.
Units: meters, *) Relative to all days

are no essential differences between various computation strategies. Concerning the baseline 12BOR -B4, it seems to be some problem with the L2 measurements, which influences L2, L1+L2 and L3 adjustments. Under the data screening 26 cycle slips on L2 were detected and 39 L2 observations were rejected. It could be caused by using too long an antenna cable (100m) without proper signal amplification at this station, which results in a noisy L2 signal.

COMPUT. STRATEGY	RMS of DD	BASELINES					
		12BOR - B1		12BOR - B2		12BOR - B4	
		LENGTH [m]	RMS [m]	LENGTH [m]	RMS [m]	LENGTH [m]	RMS [m]
L1	.005	211.451	.0004	209.592	.0004	211.484	.0004
L2	.026	.454	.0017	.596	.0018	.281	.0092
L1+L2	.024	.452	.0011	.594	.0012	.454	.0015
L3	.043	.444	.0042	.590	.0142	.786	.0174

Table 4. Slope distances at the test network.

COMPUT. STRATEGY	RMS of DD	BASELINES					
		12BOR - B1		12BOR - B2		12BOR - B4	
		HEIGHT [m]	RMS [m]	HEIGHT [m]	RMS [m]	HEIGHT [m]	RMS [m]
L1	.005	1.364	.001	0.704	.001	3.112	.001
L2	.026	1.376	.004	0.713	.004	3.207	.006
L1+L2	.024	1.372	.002	0.708	.003	3.125	.003
L3	.043	1.361	.009	0.685	.010	2.968	.010

Table 5. Height differences at the test network.

COMPUT. STRATEGY	BASELINE: METEHARA - HURUTA							
	DD RMS [m]	BASELINE LENGTH [m]	RMS [m]	DIFF * [m]	PPM *	HEIGHT DIFF [m]	RMS [m]	DIFF * [m]
L1	.107	98842.304	.021	.856	8.7	1463.275	.016	-.221
L2	.173	842.796	.037	1.348	13.6	.195	.028	-.301
L5+L3	.022	841.448	.001			.496	.003	

Table 6. Slope distances and height differences in the different adjustments, *) Relative to L5+L3

4.3 THE ETHIOPIAN CAMPAIGN

The deformation network around Metehara is situated near the equator (≈9°lat.) and contains 6 points. The baseline lengths are about 100km. Height differences between the net points vary very strongly from 200m to 2000m. The net was measured during six observation sessions with three WM102 receivers.Table 6 contains the preliminary results of different adjustments of a medium long baseline METEHARA - HURUTA in the network. In all computations the coordinates of the station METEHARA was fixed.

5. CONCLUSIONS AND FINAL REMARKS

Our WM102 receivers have proved to be reliable and easy to operate. The computing results are generally satisfactory and well within the 5mm + 1ppm of accuracy as prescribed by Wild Magnavox Survey Co. However the preliminary computations revealed some major problems with the PoPS software V3.02.

6. REFERENCES

Zdunek, R. at all (1989). First results of the measurements with GPS the traverse Jozefoslaw - Borowiec in March 1989; the 18th Intercosmos Symposium, 12-17 June, 1989;Cracow

Wells, D. (1986). Guide to GPS positioning; Canadian GPS Associates

Scherrer, R. (1987). The WM GPS primer; Wild Heerbrugg

MONITORING DISPLACEMENTS BY GPS : A CALIBRATION TEST.

B. Coulon & Y. Caristan
CEA/RCP, B.P. 12, 91680 Bruyères-le-Châtel, France

ABSTRACT

In April 1988, the position of three stations located near Paris were monitored with single frequency GPS SERCEL receivers. The three base lengths were 30.030 km, 28.168 km and 2.715 km. One station was equipped with an antenna mounted on a stage and free to move in three directions. The displacements were known with a precision of one millimeter through the stage micrometers. Repeated measurements were done, first without displacements, then by steps varying from 1 to 50 cm. A priori coordinates were obtained from the pseudo-range data. Large differences between results obtained with or without satellite 8 are probably due to satellite 8 clock problems.

Baseline vector was determine by phase inversion. Comparaisons between GPS and stage measurements show that the inversion result quality is function of two parameters : the r.m.s. of the residuals after inversion and the number of data taken into account during the inversion. When the displacements are computed from GPS baseline vector components, the scatter reach 10 ppm. Repeatability baseline measurements have a scatter of 1-3 ppm on the baseline lengh and reach 5 ppm on the baseline vector components. However, the r.m.s. on the repeatability measurements is less than one ppm on the baseline length and varies up to 2 ppm on the baseline vector components.

INTRODUCTION

The GPS system can be used for monitoring displacements. Many survey projects are based on the GPS measurements in active sismic zone (Prescott et al., 1989 ; Bock et al., 1989 ; Kato et al., 1989) or man-made structures survey (Bock et al., 1989 ; Collins, 1986). In the case of 24 hours surveying, specific problems arise such as data storage and data processing (Bock, 1989). If the aera is isolated (i.e. without fiducial point) the a priori coordinates must be computed from GPS data.

A GPS campaign was done near Paris to calibrate GPS measurements against known antenna displacements. The purpose of this work was :

1. To improve our software.

2. To determine which statistical parameters best qualify the inversion results.

2. To test the single frequency receivers as baselines measuring equipment.

GPS CAMPAIGN

During April 1988, three sites (B, M, and P) located in Paris (fig. 1) were equipped with single frequency GPS SERCEL receivers. The baseline lengths were 2.7 km **(MB)**, 28.1 km **(PM)** and 30.0 km **(PB)**.

Site B antenna was mounted on a stage allowing three dimensional displacements with a precision of one millimeter. The stage in site B was horizontal. Its axes were not oriented with respect to WGS84 axes. Thus only the displacement modules and the height variations were studied.

Two constellations of four satellites were tracked (fig. 2). The first constellation corresponds to sessions numbered with even numbers and the second constellation corresponds to sessions numbered with odd numbers. The displacement module and height variation of point B with regard to point M are computed as follows :

Let (X_i, Y_i, Z_i) be the components of the vector MB computed with GPS data from the session i and (X_j, Y_j, Z_j) from the session j.

The GPS displacement module $D_{GPS}(i,j)$ between session i and j is equal to :

$$D_{GPS}(i,j) = ((X_j-X_i)^2 + (Y_j-Y_i)^2 + (Z_j-Z_i)^2)^{1/2} \qquad (1)$$

The GPS height variation $H_{GPS}(i,j)$ between session i and j is equal to :

$$H_{GPS}(i,j) = h_{GPS}(j) - h_{GPS}(j) \qquad (2)$$

where $h_{GPS}(i)$ is the ellipsoidal height difference between station B and M obtained from the MB vector components, X, Y, Z.

The stage displacement module $D_{stage}(i,j)$ between session i and j is equal to :

$$D_{stage}(i,j) =((x_j-x_i)^2 + (y_j-y_i)^2 + (z_j-z_i)^2)^{1/2} \qquad (3)$$

where (x_i, y_i, z_i) denote the coordinates of point B in the stage reference frame for the session i and (x_j, y_j, z_j) for the session j.

The stage was horizontal, therefore the stage height variation $H_{stage}(i,j)$ between session i and j is equal to :

$$H_{stage}(i,j) = z_j - z_i \qquad (4)$$

Because only elevation differences were computed and these differences were small, the orthometric (stage measurements) and ellipsoidal (GPS measurements) height variations were assumed identical.

During the campaign, surface meteorological measurements were done at points P and B. At point M, the surface meteorological data were the same as at point B.

There were 53 sessions for base MB, 41 sessions for base PB. Some problems arised at differents points, therefore only 37 sessions for base MB and 30 sessions for base PB were processed.

A comparaison between GPS and stage measurements allowed to test the capacities of single frequency GPS SERCEL receivers and our software for gtound displacement monitoring and to determine which statistical parameters best qualify the solutions.

PSEUDO-RANGE DATA PROCESSING

The a priori coordinates used in the phase data processing were computed from the pseudo-range data.

Data processing was done with the broadcast ephemerides and the broadcast ionospheric model as well as tropospheric model (Hopfield, 1969) based on surface meteorological data.

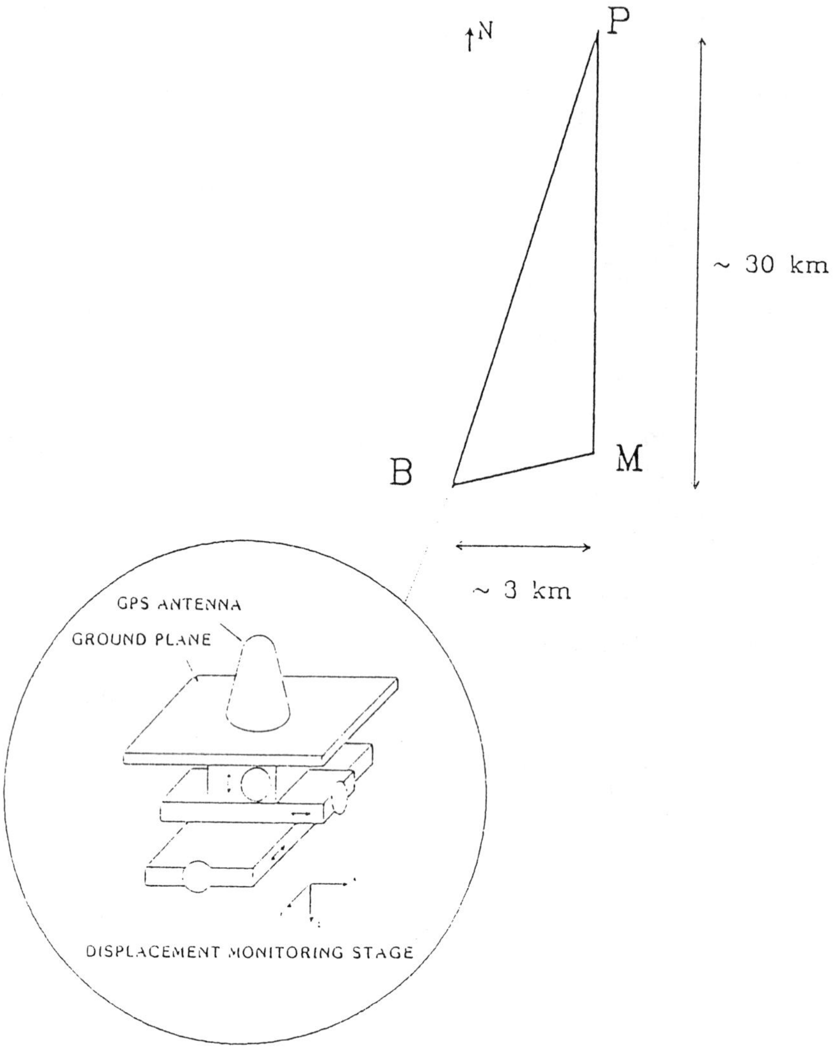

Fig. 1. Points B, M and P. Relative localisation.

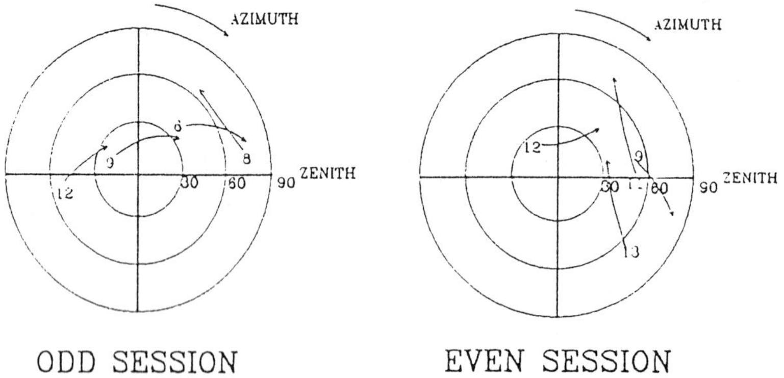

Fig. 2. Satellite constellations.

Table 1. WGS84 coordinates of the 3 sites computed from the pseudo-range data and without satellite 8.

Site	X (m)	Y (m)	Z (m)
B	4223220 +/- 5	162749 +/- 3	4761090 +/- 5
M	4222285 +/- 6	165197 +/- 5	4761803 +/- 4
P	4201744 +/- 4	172821 +/- 3	4779505 +/- 4

The a priori coordinates were taken equal to the weighted mean on all sessions. Satellite 8 was taken into account in the odd sessions had clock problems, and therefore the solution was degraded (fig. 3 and fig. 4). This satellite was then removed for the pseudo-range processing. Results are resumed on table 1.

PHASE DATA PROCESSING

The phase data processing software used the triple difference algorithm to edit cycle slips and the double difference algorithm to compute the equations to inverse (Remondi, 1984 ; Daniel, 1986). The least squares inversion algorithm was used to estimate the baseline vector, and no atmospheric correction was done. Satellite 8 was taken into account in the processing because a large part of its clock problems was removed with the double difference algorithm.

The least square inversion is a statistical processing, and two parameters are used to qualify the solution :

1. **N** which is equal to the number of data taken into account for the inversion.
2. σ which is equal to the root mean square (r.m.s.) of the residuals after inversion.

These parameters are specific of the GPS measurements. Let D be the difference between stage measurement and GPS measurement. Then :

$$D = | D_{GPS} - D_{stage} | \qquad (5)$$

Comparaison between stage measurements and GPS measurements allowed to define lower limit for **N** and upper limit for σ such that D remains small enough.

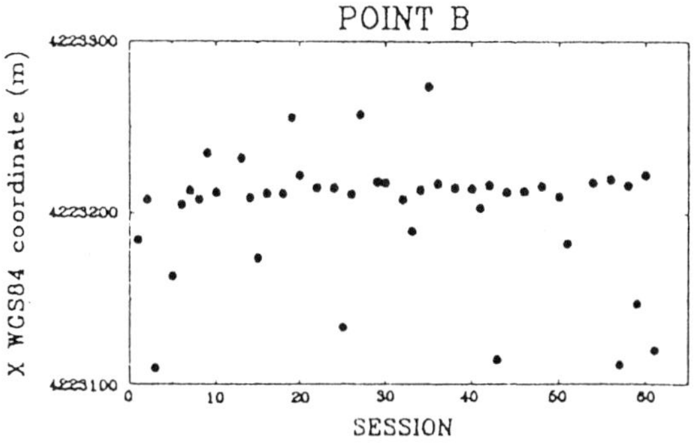

Fig. 3. X WGS84 component values of the point B computed from pseudo-range data with satellite 8.

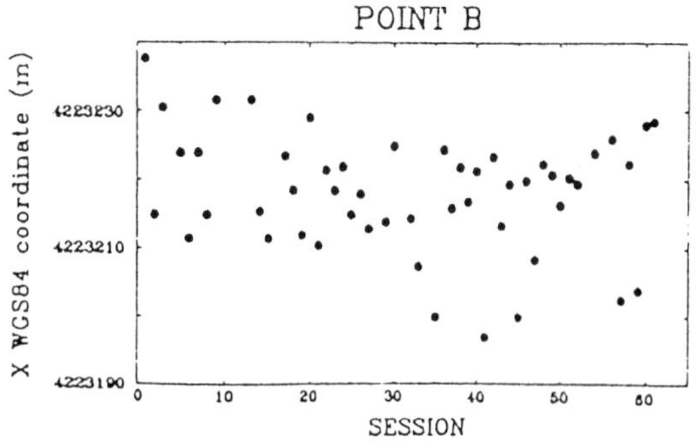

Fig. 4. X WGS84 component values of the point B computed from pseudo-range data without satellite 8.

RESULTS

Figures 5 and 6 show D as a function of N and σ for baselines MB and PB.

Note that when σ and N are both small or when σ and N are both large, D is very scattered.

From figures 5 and 6, the best agreement between GPS measurements and stage measurements is obtained when the criteria, given on table 2, on the inversion quality parameters are observed.

Stage and GPS measurements for displacement modules and height variations are compared on figures 7 for MB baseline and on figure 8 for PB baseline. The scatter is larger on the 30 km baseline (PB) than on the 3 km baseline (MB). It is also lager on height variations than on displacement modules. This is due to the influence on atmospheric delay getting larger for the 30 km baseline and which has a lager the vertical component (Beutler et al., 1989). The receiver located in M had problems. This explains why results obtained on the 3 Km baseline were of poor quality. The scatter varies up to 10 ppm.

During the campaign, repeatabilty measurements were done. They corresponded to the sessions where the stage coordinates of the antennea located in site B were the same. Figure 9 shows the MB baseline repeatability and the figure 10 shows the PB baseline repeatability. On these figures, results were selected according to the criteria of table 2. They are summarized in table 3.

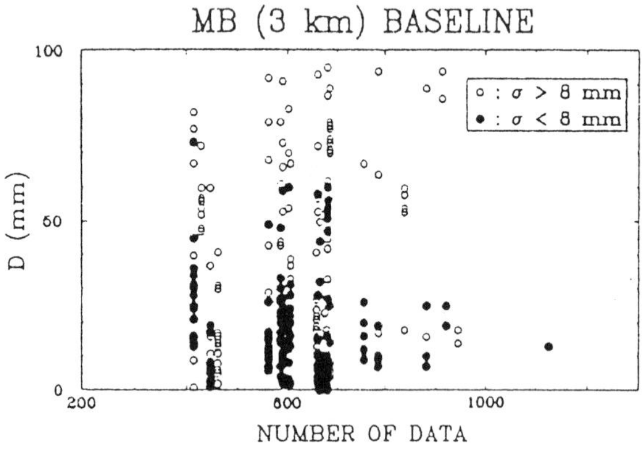

Fig. 5. D (= | D_{GPS} - D_{stage} |) values on the MB baseline as a function of N and σ.

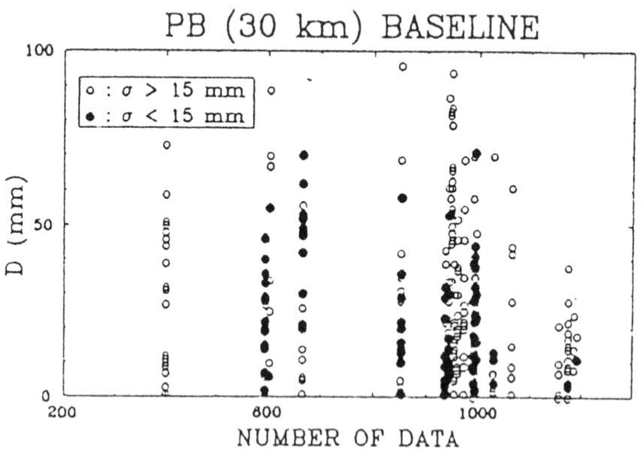

Fig. 6. D (= | D_{GPS} - D_{stage} |) values on the PB baseline as a function of N and σ.

Table 2. Criteria on the inversion quality parameters.

Baseline	σ	N
MB (3 km)	< 8 mm	> 600
PB (30km)	< 15 mm	> 700

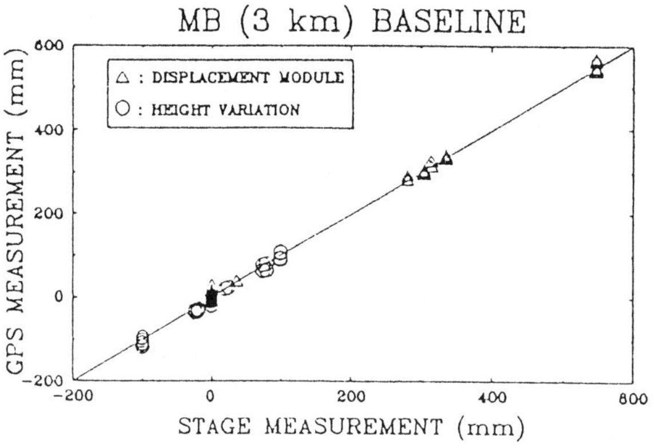

Fig. 7. GPS measurements versus stage measurements when the criteria are observed on the MB baseline.

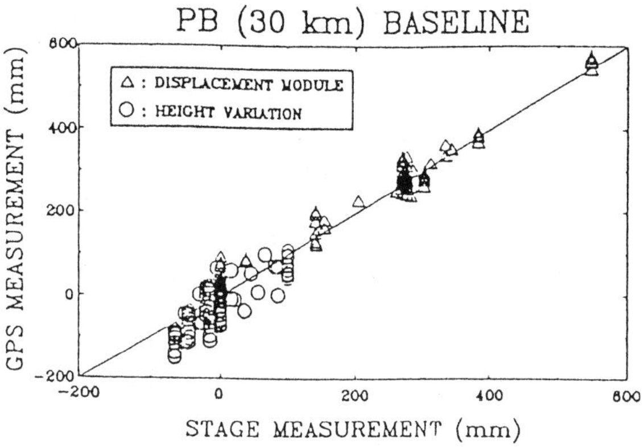

Fig. 8. GPS measurements versus stage measurements when the criteria are observed on the PB baseline.

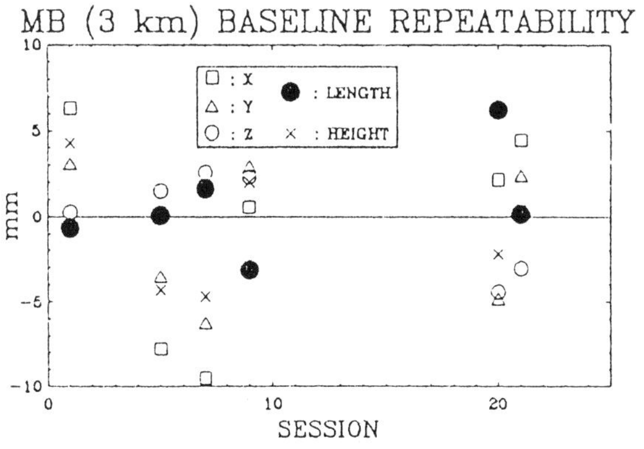

Fig. 9. MB baseline repeatability results when the criteria are observed.

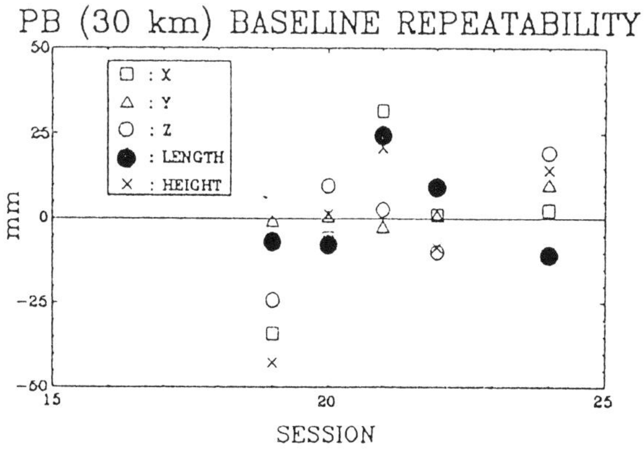

Fig. 10. PB baseline repeatability results when the criteria are observed.

Table 3 Repeatabilty measurements. These values are computed from the GPS measurements plotted in figures 8 and 9. The mean and the r.m.s. are computed with the GPS results weighting by their r.m.s. getting of the inversion.

Component	MB (3 km) baseline	PB (30 km) baseline
X (r.m.s.)	936.074 (.006) m	21476.290 (.013) m
Y (r.m.s.)	-2447.106 (.003) m	-10072.250 (.003) m
Z (r.m.s.)	-712.202 (.002) m	-18415.526 (.015) m
Length (r.m.s.)	2715.105 (.002) m	30030.196 (.013) m
Ellipsoidal height (r.m.s.)	21.529 (.003) m	53.040 (.022) m

DISCUSSION

Tracking problems arised in the sites, hence, the data processing was often done with only 3 satellites instead of 4 and **N** was smaller than expected. An other source of trouble was the calibration data. The SERCEL receivers used during this campaign must be calibrated before each measurement sessions. This calibration values correspond to the electronic delays between the five channels of the SERCEL receiver. They are necessary to fixe the ambiguities. Because of recording problems at point M, some calibration data were not correct. Therefore after data processing the residuals r.m.s. σ increased and the MB (3 km) baseline results were degraded.

Comparaisons between GPS measurements and stage measurements give a scatter which varies up to 10 ppm. The repeatability measurements show that baseline vector components were scatered up to 5 ppm. However, the scatter of the baseline length varies up to 3 ppm on MB (3 km) baseline and 0.9 ppm on the PB baseline.

Therefore, to monitore displacements with GPS and with a good accuracy, two variables can be used :

1. GPS baseline length. In this case, to determine the displacement vector, three fixed stations and the mobile station must be activated together.
2. GPS baseline vector components. The accuracy is not good as in the previous case, but two stations only are needed.

Finally, when possible, repeated sessions will improve accuracy as $\sqrt{n}$ where n is the number of sessions.

CONCLUSION

Three sites (B, M and P) were occupied with single frequency SERCEL GPS receivers. The antenna at point B was mounted on a stage allowing three dimensional displacements with a precision of one millimeter. Comparaisons between GPS measurements and stage measurements gave criteria on two parameters which qualify the inversion results :

1. **N** which is the number of the data taken into account during the inversion it should be greater than 700. But this number will be easy to reach when the full GPS constellation is operational.
2. σ which is the residual r.m.s.. This parameter should be below 8 mm.

With these criteria the GPS baseline vector components have a scatter which varis up to 10 ppm. The GPS baseline repeatabilty measurements show that the baseline vector components are more scaterred (up to 5 ppm) than the baseline length (up to 3 ppm).

The accuracy of the GPS displacement measurements can be improve by repeated measurements or by using only the baseline length . In the last case, four points must be implemented to obtain the displacement module and its direction.

Acknowledgements

We gratefully acknowledge Dr. P. Briole and Dr. J.C. Ruegg of the Institut de Physique du Globe de Paris who ensured the measurements at point P.

REFERENCES

Beutler G., Bauersima I., Botton S., Gurtner W., Rothacher M. and Schildknecht T (1989). Accuracy and biases in the geodetic application of the Global Positioning System. Manuscipta Geodaetica 14 : 28-35.

Bock Y. and Shimada S. (1989). Continuously monitoring networks for deformation measurements. Paper presented at general meeting of the International Association of Geodesy. Edinburgh, Scotland. August 3-12.

Colins J. (1986). Measuring platform subsidence using GPS satellite surveying. Offshore Technology Conference. Houston, Texas, May 5-6.

Daniel L. (1986). Evaluation du positionnement en temps réel par le recepteur GPS TR5S SERCEL et traitement de la phase. EPSHOM, France.

Hopfield H. S. (1969). Two-quartic tropospheric refraction on the doppler shift of a satellite signal. Journal of Geophysical Research, vol. 74, n° 18, August 20.

Kato T. and Japenese Universities' Consortium for GPS Research. (1989). Use of GPS for the prediction of M=7 earthquake at the Sagami area, South Kanto, Japan. Poster presented at general meeting of the International Association of Geodesy. Edinburgh, Scotland. August 3-12.

Prescott W. H., Davis J. L. and Svarc J. L. (1989). Global Positioning System measurements for crustal deformation : precision and accuracy. Science, vol. 244, 16 une 1989.

Remondi B. W. (1984). Using the Global Positioning System (GPS) phase observable for relative geodesy : modeling, processing and results. Ph. D. dissertation, CRS-84-2, The University of Texas at Austin, National Geodetic Information Center, NOAA, Rockville.

SOME RESULTS IN THE PRELIMINARY DATA ANALYSES OF THE FIXED-POINT GPS BASELINE DETERMINATION NETWORK IN CENTRAL JAPAN

Yukio Fujinawa*, Seiichi Shimada*, Tokuo Kishii*,
Shoji Sekiguchi*, Takao Eguchi*, Yoshimitsu Okada*,
Shingo Watada**

* National Research Center for Disaster Prevention,
Tennodai 3-1, Tsukuba-shi, Ibaraki 305 JAPAN

** Earthquake Research Institute, University of Tokyo,
Yayoi-cho, Bunkyo-ku, Tokyo, 113 JAPAN

ABSTRACT

We present preliminary results of data analyses of GPS fixed point geodetic measurements of 70 - 260 km baselines in the Kanto-Tokai district, central of Japan. The system was introduced for the purpose of earthquake prediction researches in the district. Firstly, the analysis employs the determination of initial positions of ten sites by means of the result of the experiment using one VLBI point and one SLR point. Next, one of the station was fixed to determine other receiver locations and the wet term tropospheric delays using the orbital relaxation technique, based on the dual frequency carrier phase and surface meteorological data.

The measurements repeatability for baseline vector components was 5.4cm + (0.1ppm of BL, Baseline Length) in the case taking an average of three principal axes of error ellipsoid in comparison with that of 1.9cm + (0.00ppm of BL) for the best direction under the present satellite configuration over the district, during the period 97 - 128 DOY of 1988.

A long term crustal deformation of the central Japan was investigated by calculating the positions of each stations at two period, April, 1988 and March, 1989. The result was in general agreement to that anticipated from the motion of the plates colliding in the area. A precursory crustal deformation prior to the 1989 volcanic eruption off east of the Izu peninsula was observed. The deformation amounted some 14cm of the baseline of length about 10km in about a week before the eruption.

The estimated path delays for wet troposphere ranged from -10 to +150cm in comparison with the calibration result of 4 - 20cm based on the vertical profile of meteorological data. The day-to-day variation of the path delay, however, does correlate scarcely to that of calibration and with the correction of satellite location amounting order of 50m. Those errors are thought to be caused from the potential systematic error of inaccurate satellite ephemerides. The wet tropospheric delays on highland was discussed using the estimated value by the GPS analysis software taking account of the meteorological characterisctics over the sites. A modified formula of the wet tropospheric delays was proposed and was shown to be in better agreement to the calibration data than those due to original one by Saastamoinen (1972).

INTRODUCTION

In the earthquake prediction research, monitoring the crustal deformation seems to be one of most important and most fruitful, since recent experiences and studies on the precursory crustal motions indicated that there are occasionally precursory slips or vertical movements or creep rate changes before moderate or large earthquakes. By continuously monitoring the crustal deformation with high accuracy in the earthquake-prone area, we may be able to detect how the crustal motion proceeds in each of the earthquake cycle and to find how the short-term and imminent precursory fault slip occurs. These knowledge will enable us to make a more confident quantitative earthquake prediction.

The Global Positioning System (GPS) offers us possibility to make continuous geodetic measurement with much higher accuracy in comparison with the conventional geodetic techniques. NRCDP has established an extensive crustal monitoring on-line network in central Japan (Okada, 1984), the network is composed of about 90 observation stations with seismometer and/or ground-tilt meters and a central data acquisition and analyses computer. In addition, we started in 1987 to compliment this network by establishing the fixed-point GPS network (Shimada et al., 1988) to measure the surface crustal motion in central Japan.

Here we will present preliminary results of the data analyses obtained by the network. Repeatability of baselines was calculated for data during about a month in 1989. Because of the deviated constellation of the satellites the error ellipsoide has an elongated configuration leading to the repeatability depending on a direction in the geocentric coordinate system.

We will show two examples of analyses of crustal deformations. A long term crustal deformation in central Japan over past about one year was determined by fixing one of the points. Displacements of several centimeters were obtained, and those deformations will be discussed by comparing the values of plates motions and local geophysical events of past years. A submarine volcanic eruption occurred off east of the Izu peninsula which situated just between one permanent station and another temporary station. A process of a clear precursory crustal deformation amounting about 14cm in about 10 days was observed.

The wet tropospheric path delays was studied with regards to geodetic baseline measurements using the GPS system. The wet delays at zenith obtained by the software (SONAP, provided by Aero Service Division, Western Atlas International, Inc.) of simultaneous estimation of geodetic parameters were compared with the calibration value using the radio sonde data. Evaluation of the wet tropospheric path delays τ_z^{ω} based on the actual meteorological conditions over central Japan, especially taking account of the deviated distribution from the assumed profiles of previous authors in the near surface layer, led to a modified formula of τ_z^{ω} to that of Saastamoinen (1972). The procedures to infer the wet-term delays at zenith on the highland are proposed, and the resultant correction coefficient versus the altitude is presented.

Observation

In the first stage of establishing the GPS network we selected ten fixed observation sites in central Japan as is shown in Figure 1. The region has been regarded as one of the earthquake prone area and designated an extensive observation region, and is covered by dense crustal monitoring network of NRCDP and of other organizations. Furhter the area is the very place where four plates, the Eulasian, Philippines and Pacific or North-American plate interact. With this choice of coverage of the GPS network we intend to find long-term intra- and inter crustal deformations caused by the interaction between those plates as well as the precursory crustal movement of earthquakes and volcanic eruptions.

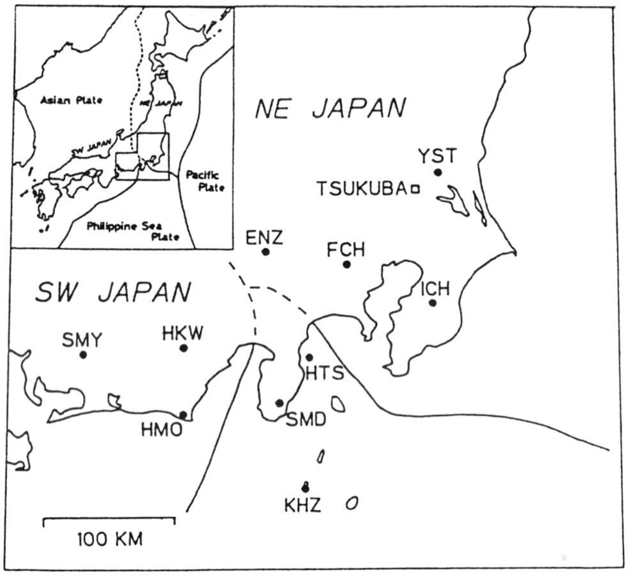

Fig.1. Locations of the observation stations in the Kanto-Tokai district

Most of the stations have originally 100 meter depth borehole for the seismic observation. At the stations YST, FCH, ICH, ENZ and HMO, the GPS antennas are settled at the top of the borehole. The boreholes are thought to be stable even when surface soil is deformed at the time of freezing or precipitation. At HTS, SMD and HKW stations, however, antennas are set at the top of the observation pier, which has a depth about 2m so as to reduce the effect of the soil freezing. At KHZ and SMY stations, antennas are mounted on the roof of the observation hut because of obstacles around the sites.

For the fixed point measurement, the GPS receiver of MINI-MAC 2816 AT is adopted, which can track both L1 and L2 frequencies. Temperature, barometric pressure and humidity are measured at each station in order to correct the excess pass delays in the troposphere. Observed data are telemetered to NRCDP in Tsukuba using public telephone lines. The data analyzing system in Tsukuba is composed of three networked computers (32 bit micro-computers, COMPAQ 386). One of which is used for automated and remote controlled measurement and data acquisition, while the other two computers for the simultaneous calculation of baseline lengths of the network. Details of the observation system are described in Shimada et al. (1988).

Data are analysed by a central processor using the software 'SONAP' provided by the Aero-Service (1988). The software provides a single-session network analysis using two bands data for about 200 min. with simultaneous determinations of the satellite orbital parameters, atmospheric data parameters, phase ambiguities and least-square adjustment of station positions.

Error distribution in positioning

Positioning accuracy using GPS data depends on the satellite configurations. The accuracy can be discussed quantitatively using the error ellipsoid determined by the variance-covariance matrix of the least-squares solution. For the case of very short baselines the observed error distributions have been shown to be in agreement with the calculated one (Kuroishi et al., 1988). In the case of intermediate and long baselines the error distribution may be contaminated by the effects of the multipath and atmospheres.

Use was made of the baseline vectors solution for data from April 6 through May 8, 1988. The baseline length ranges from about 70km to about 230km. The observation was made during a session of about two and half hours every day under the same satellites configuration by advancing the start time by four minutes every day. The satellite configuration during a session is shown in Figure 2, indicating that the visible satellites SV6, 9, 8, and 11 constitute a poor geometry with lower altitudes and roving only in the NW-quadrant.

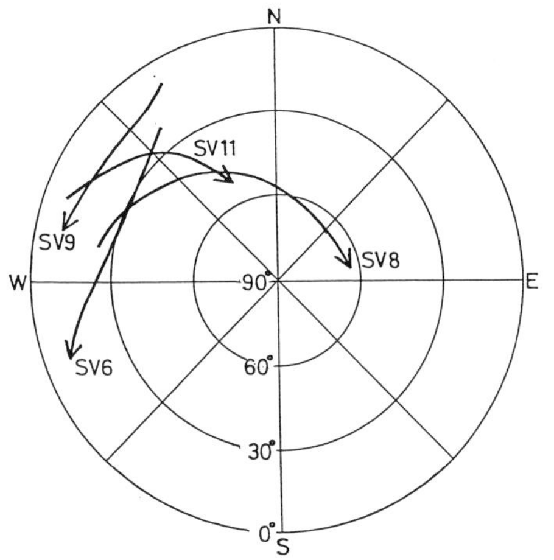

Fig.2. Sky plots of satellite visibility during a tracking session of the observation in 1988.

In Figure 3 is shown a distribution of deviation of all normalized baseline vectors from their mean baselines amount 28. The solutions are projected on the XY, YZ, ZX plane of WGS84 reference frame by normalizing the magnitude of the deviation with its baseline length. Total number of data is 372 and the data are simultaneous solutions of baseline vectors and integer bias. Figure 3 is the projection of the 2σ

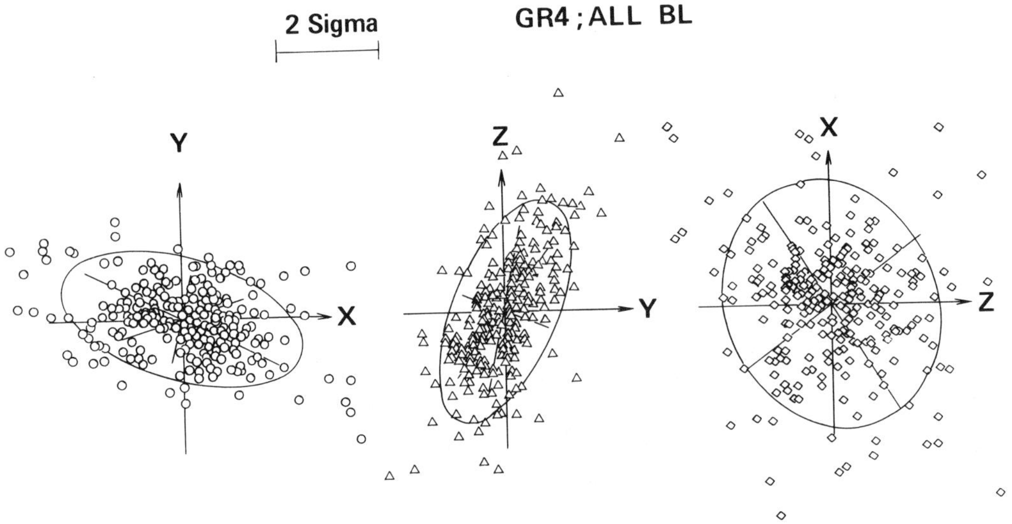

Fig.3. Distribution of the baseline vectors observed by the network. Vectors are those for 28 baselines observed from 97 to 129 day of year in 1988. Total number of the vectors plotted is 372. Ellipsoid is that for 2σ deviation, i.e. the ellipsoid with scales of twice that of the error ellipsoid, and derived from the vectors normalized by the baseline length of the solution vectors in each baseline.

ellipsoid whose lengths of axes are twice those of the error ellipsoid, and the indicated scale of '2 Sigma' in upper part is the radius of the cube whose volume is the same with that of the 2σ ellipsoid. The observed error distribution is qualitatively in good agreement with the calculated one depending only on the satellite-receiver configuration (Watada, 1989). A detailed comparison of the predicted and observed error ellipsoids will help us to find the origin of various errors such as inappropriate reference frame, bias misfixing and inaccurate mapping functions adopted.

Repeatability

One of the most important factors to achieve high accuracy is to use good fiducial station coverages. In our case we could use two fiducial points, one at Tsukuba VLBI point of the Geographical Survey Institute (GSI), and another at Shimosato SLR of the Hydrographic Department (HD). Solution based on these two fiducial points will be denoted as Version 3, solutions obtained by use of the Tsukuba NNSS precise ephemerides point as Version 2, and Version 1 concerns the result in which the conditions of calculation is the same as Version 2 except erroneously treat Tsukuba in the frame of WGS 72 instead of WGS 84.

We can assess the accuracy of the solutions of GPS measurements by taking the repeatability of determined vectors. Here we choose the scale of the error ellipsoid as a measure of the repeatability. The configuration of the ellipsoid is solely dependent on the constellation of the satellites over the network, and is shown to be reduced by several times in the case of the constellation over north America (Watada, 1988).

First we take an average of three axes of the error ellipsoid and plot versus the baseline length ranging from some 40 km to 260 km in Figure 4. Each of three pairs of figures compares the result of Version 1 and 2, and that for Version 2 and Version 3. We can see remarkably high accuracy was obtained in Version 3 indicating how the choice of accurate fiducial points are important in order to achieve high accuracy in baseline determination. Further the repeatability is shown to be 5.4cm + 0.1ppm of the horizontal baseline length. If we take the repeatability for the shortest direction of the error ellipsoid the repeatabilty is found to be almost irrelevant to the baseline length, i.e., 1.9cm + 0.00ppm, and has a value of 0.1ppm for 200km length baseline.

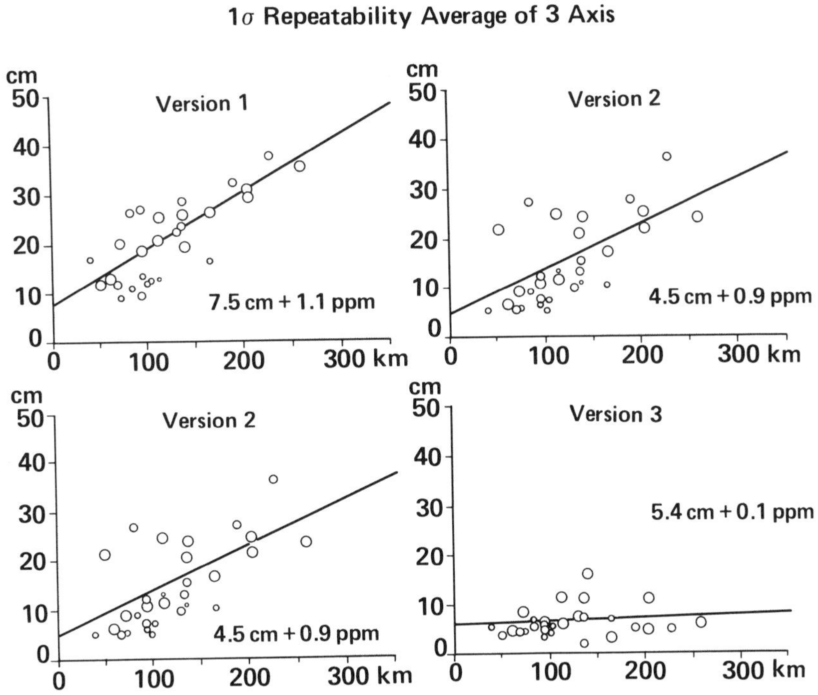

Fig.4. Relation between baseline lengths and the average of three principals axes of the error ellipsoid for each baseline. Two pairs of horizontal graphs (Version 1/Version 2, Version 2/Version 3) shows increase of accuracy for the solution corresponding larger number version.

Crustal Deformation

In addition to the data during April - May, 1988, we obtained the GPS solution for a period of March, 1989. The conditions of analysis are the same for two periods. Several days of March, 1989 did not yield convergent solutions, which seems to be caused from the insufficient inonospheric correction. The adopted criterion for good solution is the value of the normalized chi-square $\chi^2<1.8$ in the least-squares geodetic solutions. The estimated wet term tropospheric delays at zenith ranges some -50cm to +50cm.

Mean position of each site was calculated, weighting according to the reciprocal of the value χ^2. Figure 5 shows the horizontal components of the difference vectors between those two periods of apart about a year. Many places resulted in displacement of several centimeters. The largest value of about 10cm at HTS off east coast of the Izu peninsula might be due to the crustal deformation which is related to the 1988 seismic swarm activity south of HTS. Relative displacements between HKW, HMO and SMD are quantitatively in agreement with those obtained from the geodetic data of period 1955/1957 ~ 1882/1899 (Nakane and Fujii, 1981), but the magnitudes are several times larger. A very little displacement at YST is in agreement with the geodetic measurements of recent five years conducted by GSI. A large scatter of data of order 2 ~ 3cm caused mainly by the one-sided satellites constellation does not lead us discuss more in detail quantitatively.

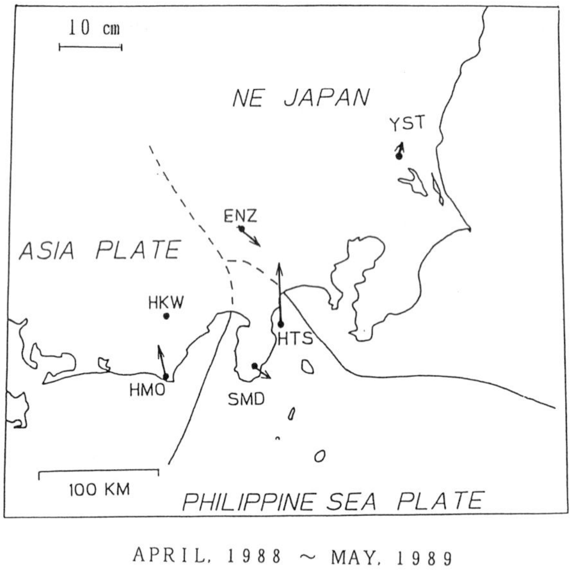

Fig.5. A long term crustal changes during about a year (1988~1989) in the south central Japan.

East off Izu Peninsula in our network, there have been frequently occurring seismic swarms in the area since 1978, and the upheaval of about 30cm in the area of 40km × 20km is found in the east coast of Izu Peninsula, in the west of the swarm epicentral area. We installed a GPS receiver at ITO site just near the upheaval area in January, 1989, in addition to the HTS site as is shown in Figure 6. At ITO site we cannot determine the position of the site because of obstacles in west and south directions, until a new satellite PRN#14 has become available at the midst of May. The last seismic swarm east off Izu Peninsula began around May 20, 1989, and was most active during July 4 and 10. After the peak of the activity there occurred continuous tremors on July 11 and 12, and a submarine volcanic eruption occurred on July 13. The volcano was named Teishi Volcano. Figure 6 shows the locations of GPS sites, the swarm region, and the submarine volcanic eruption.

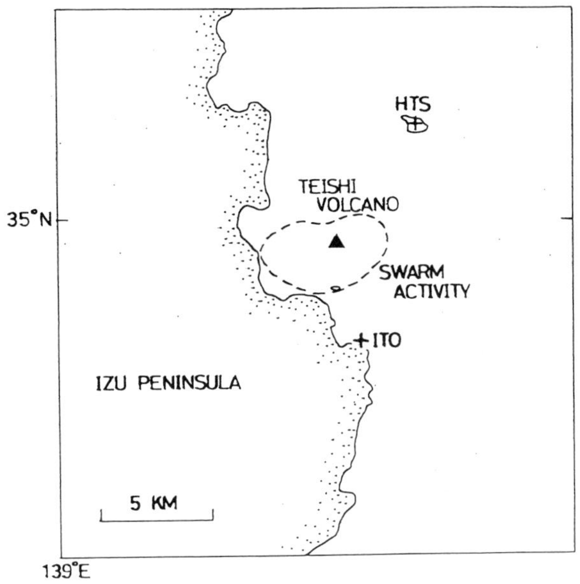

Fig.6. Locations of the fixed-point GPS sites, HTS and ITO, the seismic swarm region during May and July, 1989. The Teishi submarine volcano erupted on July 13, 1989.

We analyzed data tracked every day at HTS and ITO sites in the period of swarm and eruption activities (Shimada et al., 1989). Length of data is about 4 hours under the condition of every 30 second sampling and four satellites tracking. Sky plots of the satellites used are shown in Figure 7. Figure 8 shows time variations of the ITO site position in reference to HTS, as well as HTS-ITO baseline length. We could not obtain geodetic solution for four days of the peak swarm activity in July, because the broadcast ephemerides of the satellite are not available.

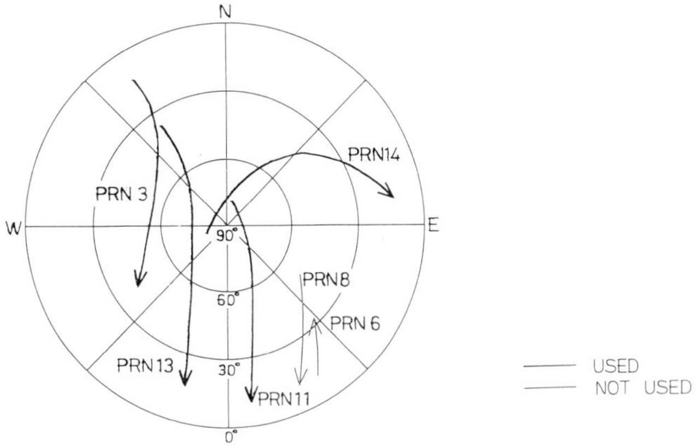

Fig.7. Sky plots of the satellites in the tracking period for each day. Broad lines concern satellites used for the analyses and narrow lines for not used. Satellites PRN #3, #11 and #13 are on the same orbit track and PRN #14 on another. Arrow indicates the direction of satellites movement.

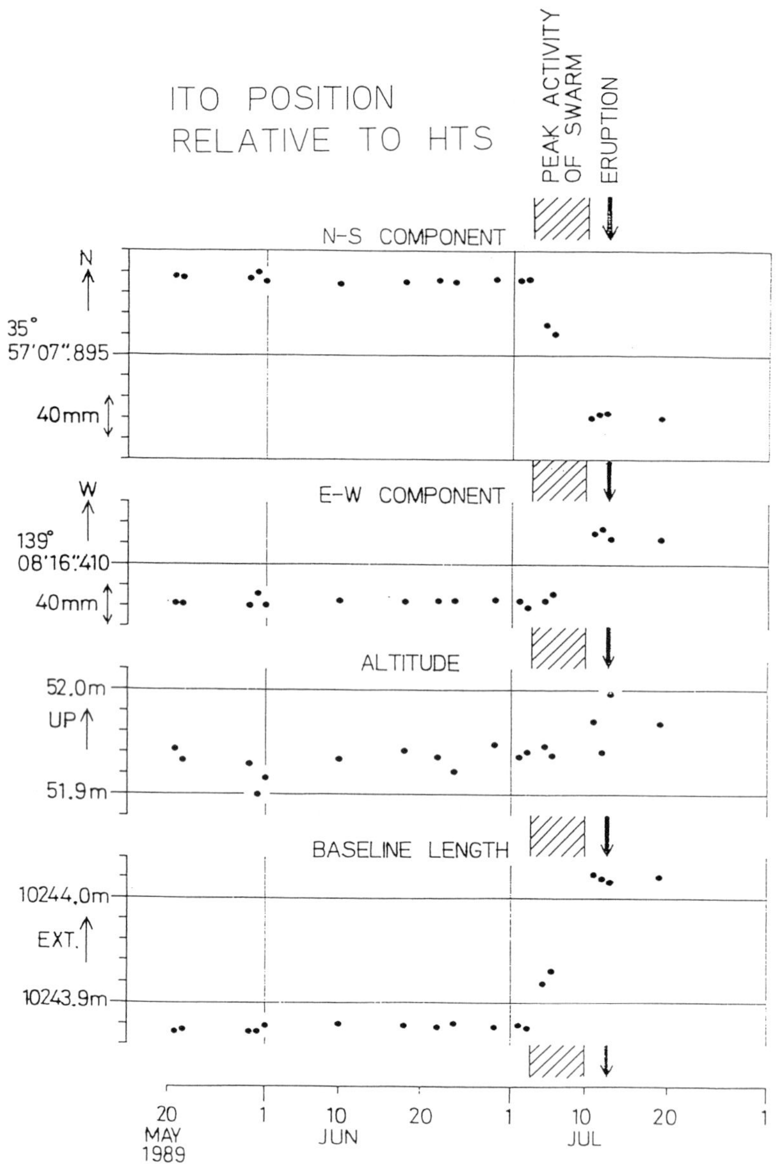

Fig.8. Time variations of the position of ITO site in reference to that of HTS site (Shimada et al., 1989).

We could see obvious crustal deformations in the period of the peak swarm activity (indicated in the figure). ITO site moved 13cm southward and 6cm westward in reference to HTS site, and the baseline length between HTS and ITO extended 14cm. The position of ITO site is seemed to have moved slightly upward, although the amount of upward movement is equal or less than the repeatability. Except the period of the swarm peak activity, there does not seem appreciable crustal movements indicating a dominant crustal deformation ended a few days before the continuous volcanic tremors and the volcanic eruption.

In the east part of Izu Peninsula, there are a borehole tiltmeter of NRCDP at ITO site, a borehole volumetric strainmeter of JMA in Higashi-

Izu site about 28km southwest of swarm epicentral area. And, moreover, EDP was measured everyday between HTS and Kusumi site about 3km west of ITO site by Earthquake Research Institute, University of Tokyo. Their data show that the obvious crustal movements were detected in the period of the peak swarm activity in agreement with the present result.

The success of detection of precise evolution of crustal deformation accompanied with volcanic and seismic swarm activity east off Izu Peninsula by the fixed-point GPS measurement demonstrates for the first time the effectiveness and usefulness of this technique for the observation of precursory crustal movements for the prediction of earthquakes and volcanic eruptions.

The wet term zenith delay

In the software we used, the tropospheric zenith delay can be estimated in the course of the data analyses using GPS positioning software. Since the dry term delay can be estimated quite accurately by the use of the atmospheric pressure at the ground surface, we can infer the wet-term delay, which is shown in Figure 9 by the solid line through open square.

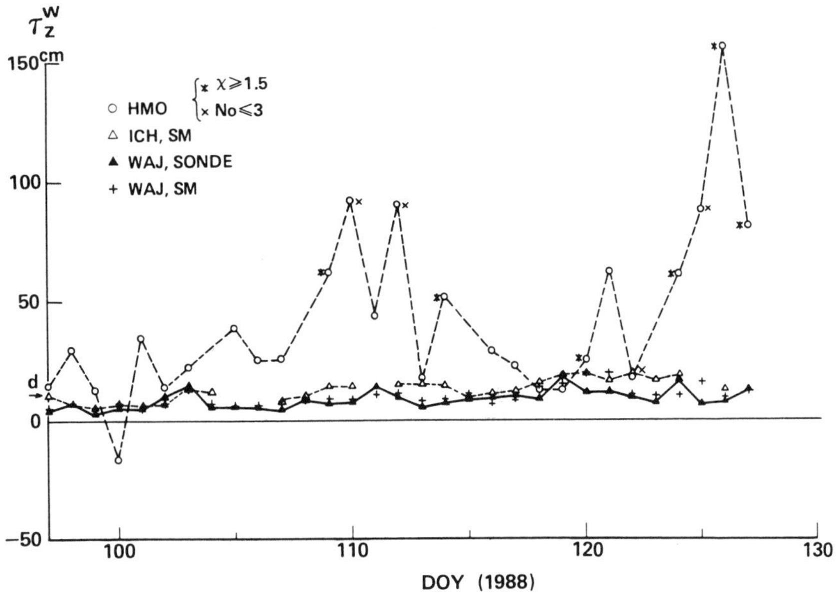

Fig.9. Wet tropospheric patch delays at zenith τ_z^{ω} estimated by several methods during the period between DOY 97 and DOY 128, 1988. The value at HMO(○) was obtained by the GPS data analysis software, assuming constant during a tracking session lasting about 2.5 hr. Data attached by a symbol (*) have normalized χ^2-value larger than 1.5 suggesting a large systematic error and those attached by a sumbol (×) correspond to observations which could use data only at three stations, the number being nearly half compared with other ordinary observations. Also are shown the wet term zenith delay (Δ) based on surface meteorological data at ICH using the formula of Saastamoinen (1972), the delay (▲) based on the radio sonde data of JMA at Wajima, north of the NRCDP network, and the delay (+) on the near surface (h = 14m) meteorological data at Wajima using the formula of Saastamoinen (1972).

The dotted line in the Figure is the value calculated making use of the vertical atmospheric profile of the radio sonde of Japan Meteorologicall Agency (JMA).

Although the site Wajima is situated a little north to our network, a comparison of wet-term zenith dealys at Wajima, Sendai (northeast of the network) and Shionomisaki (west of the network) suggest that the differences are within about 20%. On the other hand the estimated value is about several times larger compared to the calibration data and sometimes show negative values (Fujinawa et al., 1989, Shimada et al., 1989). The difference could be largely attributed to inappropriate ephemerides as can be suggested from the large value of the satellite initial position correction (Shimada et al., 1989) obtained in the geodetic and orbital parameter determination procedures. Use of more exact ephemerides and better satellite constellation will enable us to improve the estimation of tropospheric wet term delays.

The wet-term zenith delay in the troposphere is usually estimated by the formula of Saastamoinen (1972) using surface meteorological data. However the temperature and water vapor distribution profiles are a little bit different from those assumed by Saastamoinen, especially at around the ground surface where the water vapor density is usually most dense. Figure 10 shows monthly average temperature profile over Tateno in central Japan, showing fairly good fittness to the constant lapse rate profile at especially higher layer than 1km, but beneath the level there are appreciable deviations. It is suggested that we should use the linear extrapolated surface temperature T_1^* rather than the observed surface value T_1 to represent the temperature profile by two parameters.

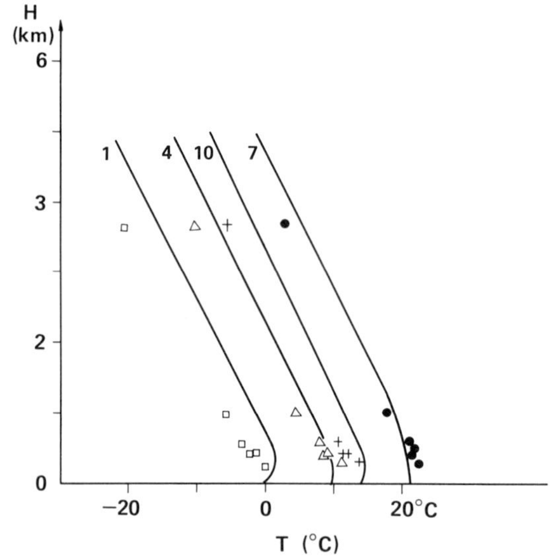

Fig.10. The monthly mean atmospheric temperature distribution T over Tateno for four seasons and atmospheric temperature (symbol, +) at several places of highland in central Japan. Highland is seen to be colder by 5°C, 1°C, 2°C in January, April, October, respectively, but warmer by 1.5°C in July in average. The dependence of the difference to the altitude is not clear.

In Figure 11 we show the water vapor pressure e versus the two kinds of normalized temperature, one is using the T_1 and the other T_1*. The figure demonstrates a quite good fit to the long-linear relation indicating the exponential decay of the water vapor in the normalized temperatures. Moreover the upper limit of the normalized temperature is nearly constant with the value less than unity in the case of the normalization using T_1* , in contrary to the value larger than unity and with a significant scatter in the case of the normalization using T_1. These points are important in the evaluation of the integral to get the zenith delay calculation.

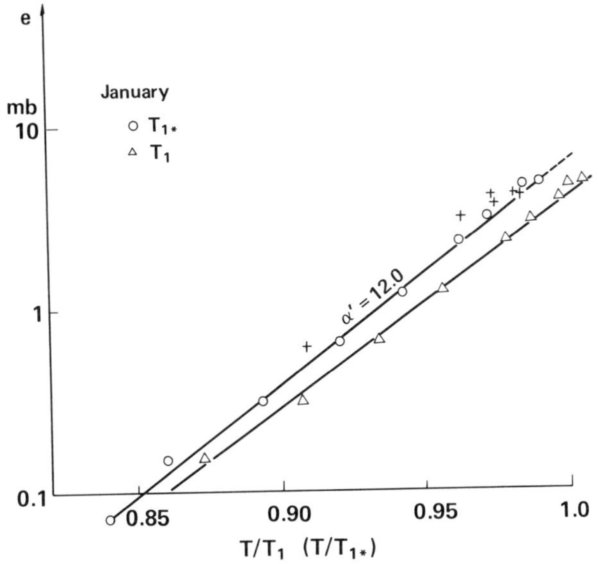

Fig.11. The monthly mean tropospheric water vapor pressure distribution e(mb) over Tateno versus the normalised temperature. Two cases of the normalization are shown; the surface temperature T_1 is used for one case (△) and the extrapolated temperature T_{1*} at the ground surface extended from the constant-lapse rate distribution is for second case (○). The case using T_{1*} can be seen to fit the data comparatively better. It is convenient and appropriate for the evaluation of the wet path delays that the upper limit of the normalized temperature T/T_{1*} is less than unity and nearly constant (≒ 0.99) for all seasons. The data with symbol (+) indicates the values on highland in central Japan (Kofu: 270m, Nagano: 418m, Iida: 480m, Matsumoto: 610m, Karuizawa: 999m, Mt. Fuji: 3772m).

By taking account of these properties of the profiles the wet term zenith delay was calculated following the procedures adopted by Saastamoinen (1972). A modified formula is $\tau_z^{\omega} = 191 e_1*/T_1*$ (cm), where $e_1* = 1.3e$, with e_1 being the water vapor pressure (mb) on the ground surface and T_1* being the extrapolated temperature (K) on the ground surface from the constant-lapse rate distribution prevailing in the dominant region of troposphere. Figure 12 shows the wet term zenith delay calculated both by the original Saastamoinen's formula and by the present one, versus the value determined from the profile data of radio sonde. The figure indicates that the modified formula gives better agreement with the calibration values especially in a smaller value of the wet term delay. Data analyses at other two sites around central Japan resulted in the same conclusion.

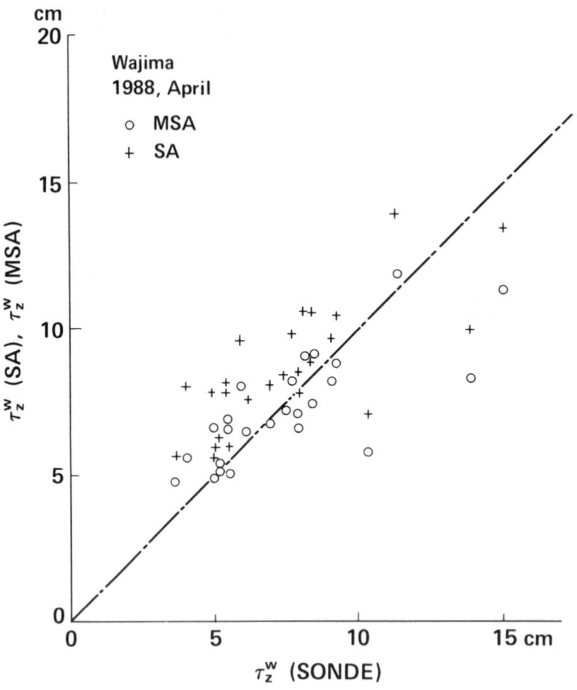

Fig.12. The wet tropospheric path delays at zenith τ_z^{ω} (SA) based on the formula of Saastamoinen (1972) and that τ_z^{ω} (MSA) based on the present formula versus the calibration values τ_z^{ω} (SONDE) based on the radio sonde data of JMA at Wajima in the central Japan during April, 1988. In the range τ_z^{ω} (SONDE) $\lesssim$ 10cm, the present formula gives better agreement with the calibration data. But in the range τ_z^{ω} (SONDE) $\gtrsim$ 15cm worse agreement is obtained, which would be caused by unsteady climate conditions.

Meteorological conditions on highland with respect to those at near the mean sea water level was studied (Fujinawa et al., 1989), and revealed that the difference of atmospheric temperature and water vapor pressure at the same altitude depends usually on season and hours of day. We should take into account the possibility that the surface meteorological data are not good parameters to represent the profile over the highland. So that it is recommended that in order to estimate the wet term excess path delay, at least as a first approximation, in the case that we have no profile data, we should calculate the delays contributed from the altitude of the highland to the tropopause assuming the same vertical distributions on low land. The lower limit of the vertical integration in the evaluation of the excess path delay following Saastamoinen (1972) is transformed to that of the normalized temperature. Figure 13 shows a ratio between the zenith wet term tropospheric delays τ_z^{ω} (h = H) at the altitude h = H and that at the ground surface of the mean sea level. Atmospheric vertical profiles used are those for central Japan.

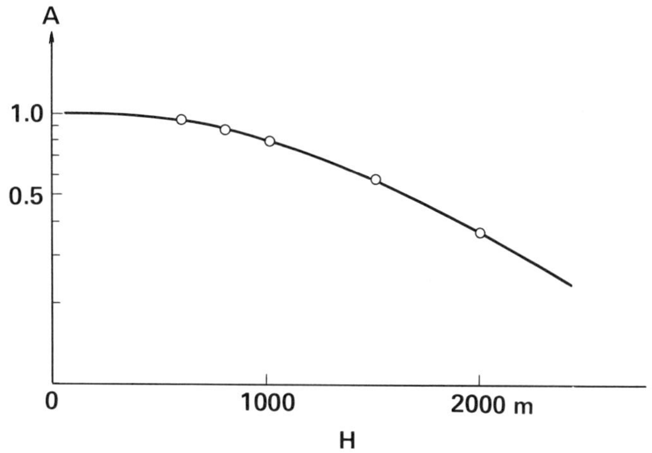

Fig.13. The wet tropospheric path delays τ_z^{ω} (h = H) at the altitude H in relation to the value τ_z^{ω} (h = 0) at the ground surface of the mean sea level. Starting from the estimated value τ_z^{ω} (h = H), it is suggested to infer the correct value by using the GPS positioning software.

Conclusion

We have constructed the fixed-point GPS network in central Japan for the purpose of observation of precursory crustal movements of earthquakes and volcanic eruptions and for studies of intra- and inter plate motions. Preliminary analyses of data during April and May in 1989 showed that it is essential to attain high accuracy that we can use the fiducial points whose coordinates are precisely known and the precise orbital parameters as well as a uniform satellite constellation. A careful check of data enabled us to attain the repeatability of order 0.1ppm for the baseline of several hundred kilometers, with the error in the best direction under the present satellite constellation over Japan being almost irrelevant to the baseline length.

The fixed-point GPS network has shown that it is possible to observe the evolution of crustal deformations occurring in the plate boundaries in a long term. The displacements at many sites of our network have proved to be grossly in agreement with the results of ordinary geodetic measurements. Under a better condition of satellite constellation, it will be possible to observe the evolution of the crustal movement occurring at plate boundaries. The system has also proved its ability to follow the short term evolution of the crustal deformation associated with the volcanic eruption occurred east off Izu peninsula central Japan on July 13, 1989.

The wet tropospheric path delays were studied with regards to geodetic baseline measurements for about 30 days data during April and May, 1988. The wet zenith delays simultaneously estimated with geodetic parameters by the software were found significantly larger than those calculated using the radio sonde data. Evaluation of the wet tropospheric path delays τ_z^{ω} based on the meteorological measurements over central Japan, especially taking account of the deviated distribution in the near surface layer, led to a modified formula to that by Saastamoinen (1972). The procedures to infer the wet-term zenith delays on the highland are proposed, and the resultant correction coefficient versus the altitude is presented.

References

Aero Service (1988). Simultaneous Orbit Network Adjustment Package, Aero Service Division, Western Atlas International.

Fujinawa, Y., Shimada, S., Kishii, T., Sekiguchi, S., Eguchi, T., Okada, Y. (1989), Estimation of tropospheric path delays using the fixed-point GPS simultaneous baseline determination network of NRCDP(1), Rep. Nat. Res. Centr. Disaster Prevention, 44, in press.

Kuroishi, Y., Murakami, M. and Tsuji, H. (1988). A simulation study on GPS double difference positioning. J. Geod. Soc. Japan, 34, 137-149 (in Japanese).

Mogi, K. (1981). Earthquake prediction program in Japan, in Earthquake Prediction - An International Review, Maurice Ewing Series, 4, 635-666.

Nakane, K. and Fujii, K. (1981). Horizontal crustal movement in KANTO-TOKAI district (III), J. Geod. Soc. Japan, 27, 140-147.

Okada, Y. (1984). First results from Japanese network for earthquake prediction, Nature 312, 500-501.

Saastamoinen, J. (1972). Atmospheric correction for the troposphere and stratosphere in radio ranging of satellites, The use of artificial satellites for geodesy, Geophys. Monogr. Ser., 15, Amer. Geophys. Union, Washington D.C., 247-251.

Shimada, S., Sekiguchi, S., Eguchi, T., Okada, Y. and Fujinawa, Y. (1988). Simultaneous baseline determination network in Kanto-Tokai area, central Japan, Proc. Japanese Symp. on GPS, Jan. 27-29, 1988, 266-270.

Shimada, S., Sekiguchi, S., Eguchi, T., Okada, Y., Fujinawa, Y. (1989). Preliminary results of the observation by fixed-point GPS simultaneous baseline determination network in Kanto-Tokai district, J. Geod. Soc. Japan, 35, 85-95.

Shimada, S., Fujinawa, Y., Sekiguchi, S., Eguchi, T., Okada, Y. (1989). GPS detection of a volcanic fracture opening in Japan, submitted to Nature.

Watada, S. (1989). Relashionship between error distribution in positioning and satellites configuration in GPS technique, Master Thesis, University of Tokyo.

THE RESOLUTION OF MEAN SEA LEVEL ANOMALIES ALONG THE NEW SOUTH WALES COASTLINE BY GPS

Roderick.T.MacLeod
SAGEM Australasia Pty Ltd
2/122 Euston Road, Alexandria, Sydney, N.S.W. Australia.

A.H.W. Kearsley & C.Rizos
School of Surveying, University of New South Wales
Kensington, Sydney, N.S.W. Australia

INTRODUCTION

The Public Works Department (PWD), in the state of New South Wales, Australia, is responsible for the operation of the state's coastal tide monitoring system and the information obtained therefrom is used in the determination of Mean Sea Level. Due to intensified development along the coast, information such as flood heights and Mean High Water (which are used to define flood prone lands, cadastral boundaries and in extensive numerical modelling of rivers) must be accurately related to a geodetic levelling datum such as the Australian Height Datum (AHD). Anomalies were found in determinations of Mean High Water and Mean Sea Level (MSL) from ocean gauges and that derived from spirit levelling between tide gauges along the entire New South Wales coast (see Figure 1).

In order to investigate the existing coastal levelling network and its relationship to MSL, an independent levelling traverse using the new satellite technology of the Global Positioning System (GPS) was undertaken. In May 1987, a 40 station GPS survey using Trimble 4000SX single frequency receivers was carried out from Lakes Entrance in Victoria to Caloundra in Queensland. The exact station positions were constrained by the location of the tide gauges. Generally the nearest bench mark (or where possible a state permanent mark) related to each tide gauge was used as the GPS station.

To obtain the most accurate results in GPS, dual frequency receivers would be required in order to minimize the effect of the ionosphere on the heighting results. Unfortunately in May 1987 the receivers available in Australia were almost exclusively of the single frequency variety. Therefore a total of five single frequency Trimble 4000SX receivers were used.The survey took 17 days to complete.

In September 1987 the availability of four Texas Instruments TI4100 dual frequency receivers allowed a survey to be carried out that connected all the existing substantial GPS surveys in NSW together. Those connected included the PWD coastal project, a WM101 network and the Segment network which includes the Very Long Base Interferometry (VLBI) stations at Parkes, Fleurs and Tidbinbilla and the total network is shown in Figure 2. The accuracy of the combined network was expected to be greater than the single frequency coastal network. In addition the connection into the VLBI network would provide very accurate global coordinates to the network.

POST PROCESSING

Generally, a GPS campaign will set out to determine a set of coordinates for those stations within a network, and possibly corrections to satellite orbits. A number of receivers will be deployed at sites simultaneously tracking carrier phase data from a number of satellites over a period of time. This is commonly referred to as a session. The receivers are deployed over several sessions until all sites in the network have been occupied.

TRIMBLE Session Processing - TRIMVEC Baseline Solutions

After the completion of the Trimble 4000SX survey in May 1987, the data was reduced using the Trimble commercial baseline software called TRIMVEC (TRIMBLE, 1987). All possible baseline combinations within each session were calculated and checked for processing consistency using a set of geometric tests similar to those used in LETHABY et al (1987). Closures of the polygons formed on each day would show any outliers, which could then be analysed and reprocessed. The baselines observed in more than one session were then compared in all 3 dimensions to show the system's and software's overall consistency. These inter-session comparisons are shown in Figure 4 with the worst result being 4.2 ppm.

TRIMBLE Session Processing - BATCH-PHASER Multi-Station Solutions

While commercial "baseline" software such as TRIMVEC treat mathematical correlations rigorously when processing a single baseline, they neglect stochastic correlation between baselines measured within a session (MACLEOD & RIZOS, 1988). A multi-station reduction package which took into account these stochastic correlations was needed to check the validity of the TRIMVEC solutions.

The PWD used the multi-station software package called BATCH-PHASER. The original program was written by GOAD (1985) but has been substantially modified by the South Australian Department of Lands, (JONES et al, 1987). Processing of the data was initially carried out at the South Australian Lands Department in July 1988. The precise ephemeris obtained from the Defense Mapping Agency (DMA) in the USA was included for the first time. The session results were combined into a network adjustment using the mainframe version of NEWGAN (ALLMAN, 1988). Repeated interstation vector comparisons are compared to the TRIMVEC comparisons and shown in Figure 3.

While both software packages showed similar consistency in their repeated baseline comparisons, the BATCH-PHASER solutions differed from the TRIMVEC solutions by as much as 4 ppm. Although differences in the mathematical models would account for some variations ie. multi-station verses multi-baseline, these would not be expected to create baseline differences of this magnitude. Further investigations have lead to the belief that TRIMVEC omits one of the keplerian elements (i) that will effect the solution over extended observation sessions (TALBOT, 1989). Therefore the BATCH-PHASER solutions using the precise ephemerides have been adopted.

TI4100 Session Processing - NOVAS Baseline Solutions

Here the results from the TI4100 dual frequency network were obtained. The TI4100 campaign was reduced in September 1987 by the NSW Land Information Centre using the broadcast ephemeris and the baseline software NOVAS (WANLESS & LACHAPELLE, 1987). Like the TRIMVEC processing, all possible baseline combinations were used in the network solution. Again geometric tests were carried out and baselines measured more than once were compared (Figure 4). The results show very good consistency over long distances. This is ofcourse largely due to the elimination of ionospheric effects by the use of dual frequency observations.

NETWORK ADJUSTMENT

The two networks were then combined into one adjustment. One of the major problems in this project was that the combined solution included both multi-station and multi-baseline sessions, using different receivers, ephemeris and software. JONES (1988) discussed this problem, noting that unlike existing terrestrial methods, GPS users do not have a large amount of past experience to call upon, when trying to determine the expected accuracies of various GPS solutions. One must take into account such things as the receiver type, the observables, the validity of the mathematical model and ephemeris used.

The method used to scale the VCV's of the NOVAS baseline solution was to add an external variance to the diagonal elements of the baseline's internal VCV. This external variance is a function of the baseline length and reflects the external influences such as tropospheric and orbital errors on the baseline solution.The use of a external variance factor in ppm of the baseline's length instead of a constant is supported by research on GPS networks in South

Australia (JONES, 1988) and overseas (BOCK et al, 1985).

The BATCH-PHASER single frequency Trimble network processed with precise ephemeris required another method of scaling the multi-station VCV of each session. An external ppm factor would be difficult to implement, so a scalar was introduced.

The final solution assumed each mathematical model was equivalent and scale the two network solutions accordingly and produced an a posteriori variance factor of 1.07. The minimally constrained adjustment held the WGS84 coordinates of Tidbinbilla fixed. These coordinates have been defined via GPS to Orroral NLRS, Mirror 7 which has a Centre for Space Research (CSR) LAGEOS LONG ARC 8511 adjusted values and is considered to be better than 5 metres absolute, ie. its position on the earth's surface is known to within 5 metres in the WGS84 datum (LID, 1989).

ORTHOMETRIC HEIGHTS FROM GPS

In engineering studies the problem with satellite positioning systems such as GPS is that they are based on an earth centred coordinate system then transformed onto a reference ellipsoid which can be either global or regional. The ellipsoid heights derived are not the same as the orthometric heights required. The orthometric heights are related to an equipotential gravitational surface, the geoid.

The difference between the orthometric and ellipsoid heights at a point is the geoid-ellipsoid separation N and is shown in Figure 5 with the relationship being

$$\mathbf{H = h - N}$$

Because the systematic errors in both h and N at points A and B will be similar, it is preferable when seeking the highest precisions to adopt a differential approach, ie. to use line elements dh and dN. For this reason, we adopt the approach

$$\mathbf{dH_{AB} = dh_{AB} - dN_{AB}} \qquad (1)$$

to propagate orthometric height differences through the tide gauge network. If the precision in dH is to match that of dh, then the desired precision of dN should also at least match that of dh.

Determination of the Geoid Height Differences (dN)

There are various methods to determine N and dN eg Geometric methods, but many are unsuitable for independent, high precise determinations. Therefore, dN was calculated by Stokes integral, using the ring integration technique

(RINT). This technique has been shown, in theory, to be capable of producing a relative error in dN (δdN) of 2-3ppm of the line length (KEARSLEY, 1985;1986). Tests have confirmed this precision to be possible, particularly in areas were the geopotential model recovers the geoid signal well (4-6ppm) (KEARSLEY, 1988; HOLLOWAY, 1988). This is providing certain conditions for the accuracy of the gravity field are met, which would result in the mean gravity anomaly for a 10km * 10km block to be known to 3mGal. While this is achievable for continental Australia, the lack of data off the south east coast of Australia means that the precision of dN in these coastal regions (the southern third of the network) is probably degraded.

The gravity data set was provided by the Australian Bureau of Mineral Resources, and has been edited to remove duplicated and obviously erroneous data. The gravity is converted to IGSN 71 values and then referred to both OSU81 (RAPP, 1981) and OSU86 (RAPP & CRUZ, 1987), the geopotential models used to provide the medium and long wavelength components of the geoid signal. The calculations were carried out at the School of Surveying at the University of NSW, using a suite of programs on the University's VAX mainframe computer. The details of the software and the RINT technique can be found in HOLLOWAY (1988).

RESULTS

The differential GPS ellipsoid heights have been calculated, subtracted from the terrestrially derived orthometric height and compared to the gravimetrically derived dN values defined in (1). These terrestrial values are available in the form of the differences in the AHD values.

Figure 6 shows how the integration of the gravity field data using OSU86E into the solution improves the comparisons between the dN derived by gravimetry and that calculated by subtracting the terrestrially derived orthometric heights (AHD) from the GPS ellipsoid heights. Comparisons of the orthometric heights at the bench marks derived from GPS and gravity integration and existing AHD values show the best agreement at between a cap size of 0.2 to 0.5 degrees. The network has been broken up into two regions, being north and south of Crescent Head. This is because there is sufficient gravity data offshore only to the north, and this is reflected in the results.

The GPS/gravimetry approach using the combined solution seems to produce orthometric heights that agree with the terrestrial levelling data to third order levelling specifications. There is however a systematic trend in heights generated if only TRIMVEC or NOVAS baseline solutions are used. This could be due to a bias in the broadcast ephemeris itself or the way it is used in the respective post-processing packages, since the bias seems to be latitude dependent. It could also be a reflection on the fact that the stochastic correlations are not fully taken into account.

Finally the combined GPS solution using OSU86E was compared to the calculated AHD heights of MSL at each tide gauge. The GPS/gravimetry derived orthometric heights used an integration of the spherical cap to a radius of 0.4 degrees (about 40kms). The orthometric heights derived at the GPS stations have been transferred via spirit levelling connections to the tide gauges. The results are biased by the errors in the GPS height (h) and geoidal height (N) determinations at the origin of the network at Tidbinbilla (MACLEOD et al, 1988). However, the GPS/gravimetry heights have been adjusted so that the GPS/gravimetry RL at Fort Dension equals the AHD RL (ie. the systematic error in the GPS/gravimetric solution has been removed). Figure 7 shows the apparent correlation between the two sets of data, and indicates that the terrestrial levelling and AHD are in fact reliable, since the AHD values fit within the variations of the GPS/gravimetric results.

CONCLUSIONS

The Public Works Department of NSW through its tidal analysis and numerical modelling, had encountered a number of levelling anomalies along the coast. The results from GPS/gravimetry have been added to new tide gauge data and has provided an evaluation of the existing levelling network that connects the gauges together. Amendments to the original tide gauge analysis due to this GPS survey have located a levelling error of about 0.1 metres near Eden. There remains the dip in MSL between Newcastle and Yamba and this is now thought to be due to the Sea Surface Topography. These revised relationships between the river datums now allows the Department's Engineers to relate their river studies accurately.

REFERENCES

Allman. J.S., "NEWGAN - Preliminary Users Manual.", June 1988

Bock. Y., Abbot. R.I., Counselman. C.C., Gourevitch. S.A., King. R.W., "ESTABLISHMENT OF THREE DIMENSIONAL GEODETIC CONTROL BY INTERFEROMETRY WITH THE GLOBAL POSITIONING SYSTEM.", Journal of Geophysical Research, Vol 90, No B9, August 1985, pp 7689 -7703.

Goad. C.C., "PRECISE RELATIVE POSITION DETERMINATION USING GLOBAL POSITIONING SYSTEM CARRIER PHASE MEASUREMENTS IN A NONDIFFERENCE MODE.", Proceedings of the First International Symposium on Precise Positioning with the Global Positioning System, Vol 1, National Geodetic Survey, Rockville, Maryland, pp 347 - 356, April 1985.

Grant. D., "COMBINATION OF TERRESTRIAL AND GPS DATA IN NEW ZEALAND EARTH DEFORMATION STUDIES.", PhD Thesis, Unisurv

Report S-32, School of Surveying, University of New South Wales, Kensington, NSW, 1989.
Holloway. R., "THE INTEGRATION OF GPS HEIGHTS INTO THE AUSTRALIAN HEIGHT DATUM (AHD).", Unisurv Report S-33, School of Surveying, University of New South Wales, Kensington, NSW, 1988.
Jones. A.C., "POPULATION VARIANCE FACTORS FOR MULTI-STATION GPS DATA SETS.", Proceedings of the International Symposium on Global Positioning Systems, Brisbane, October 1988.
Kearsley. A.H.W., "TOWARD THE OPTIMUM EVALUATION OF THE INNER ZONE CONTRIBUTION TO GEOIDAL HEIGHTS.", Aust. J. Geod. Photogram. Surv., No 42, pp 75 -98, 1985.
Kearsley. A.H.W., "DATA REQUIREMENTS FOR DETERMINING PRECISE RELATIVE GEOID HEIGHTS FROM GRAVIMETRY.", J. Geophys. Res., Vol 91, No B9, pp 9193 - 9201, 1986.
Kearsley. A.H.W., "TESTS ON THE RECOVERY OF PRECISE GEOID HEIGHT DIFFERENCES FROM GRAVIMETRY.", J. Geophys. Res., Vol 93, B6, pp 6559 - 6574, 1988.
Land Information Division (LID) of the NSW Department of Lands, "WGS 84 POSITIONS RESULTING FROM COMBINED GOVERNMENT DEPARTMENTS GPS SURVEY OF SEPTEMBER 1987 AND GPS CAMPAIGNS UNDERTAKEN BY THE CENTRAL MAPPING AUTHORITY.", Letters to the Public Works Department of NSW, January 1989.
Lethaby. J., Zahra. C., Bosloper. C., "THE NSW REGION 4 GPS NETWORK: PROCESSING STRATEGIES AND PRELIMINARY RESULTS." R.M.I.T. Centenary GPS Conference, Melbourne, August 1987.
Macleod. R.T., "THE RESOLUTION OF MEAN SEA LEVEL ANOMALIES ALONG THE NEW SOUTH WALES COASTLINE BY GPS - THE PLANNING AND LOGISTICS OF A LARGE SCALE GPS SURVEY.", Proceedings of the 30th Australian Survey Congress, Institution of Surveyors, Sydney, April 1988.
Macleod. R.T., Rizos. C., "CORRELATION IN GPS SURVEYING - A CAUSE FOR CONCERN ?", Proceedings of the International Symposium on Global Positioning Systems, Brisbane, October 1988.
Rapp. D.T., "THE EARTHS GRAVITY FIELD TO DEGREE AND ORDER 180 USING SEASAT ALTIMETER DATA, TERRESTRIAL GRAVITY DATA AND OTHER DATA.", Report NO 322, Dept Geod. Sci. Surv., The Ohio State University, Columbus, Ohio, 1981.
Rapp. R.H., Cruz. J.Y., "SPHERICAL HARMONIC EXPANSIONS OF THE EARTH'S GRAVITIONAL POTENTIAL TO DEGREE 360 USING 30' MEAN ANOMALIES.", Report No 334, Department of Geodetic Science and Surveying, Ohio State University, Columbus, Ohio, USA, 1986.
Talbot. N., personal communication, March 1989.
Wanless. B., Lachapelle. G., "NOVAS - AN AUTOMATED PROGRAM R THE PRECISE REDUCTION OF GPS STATIC CARRIER PHASE OBSERVATIONS." Paper submitted to Manuscripta Geodaetica October, 1987, accepted December 1987.

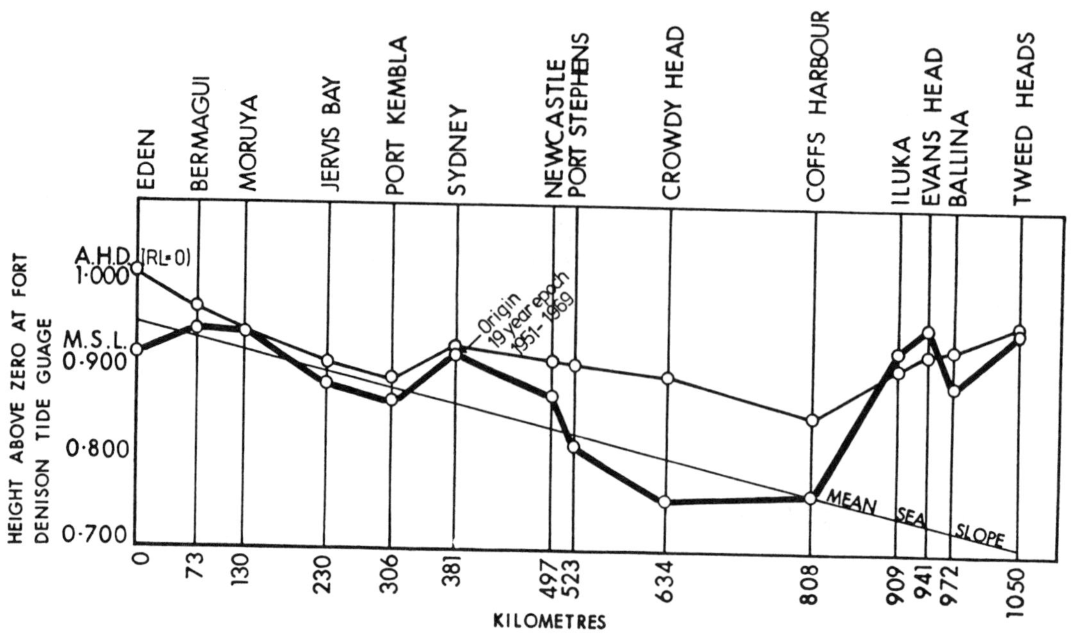

Figure 1: Comparison of M. S. L. to A. H. D. along the N. S. W. coast

FIGURE 3

TRIMBLE REPEATED BASELINES

TRIMVEC & PHASER RESULTS

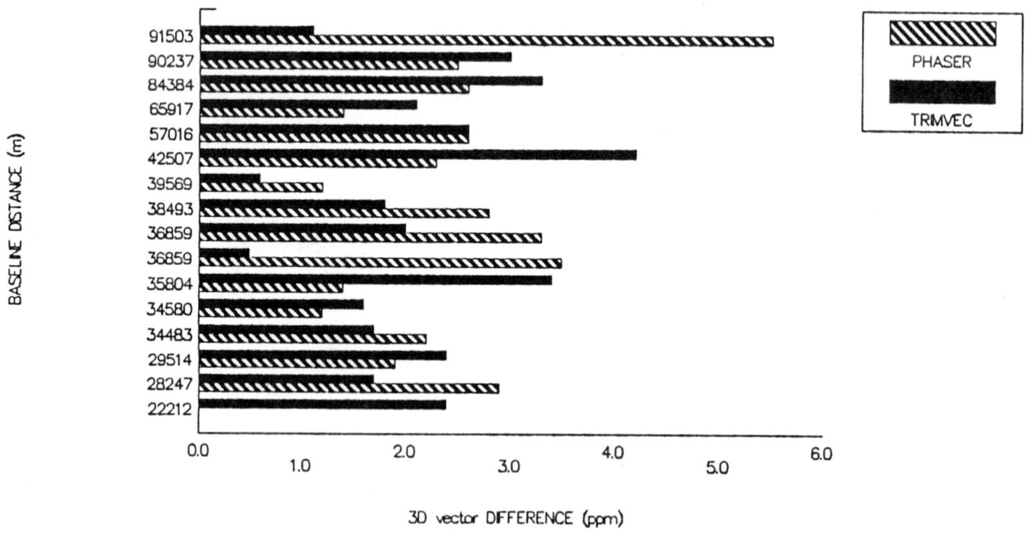

Figure 2: G. P. S. Network for the Analysis of Tide Gauges.

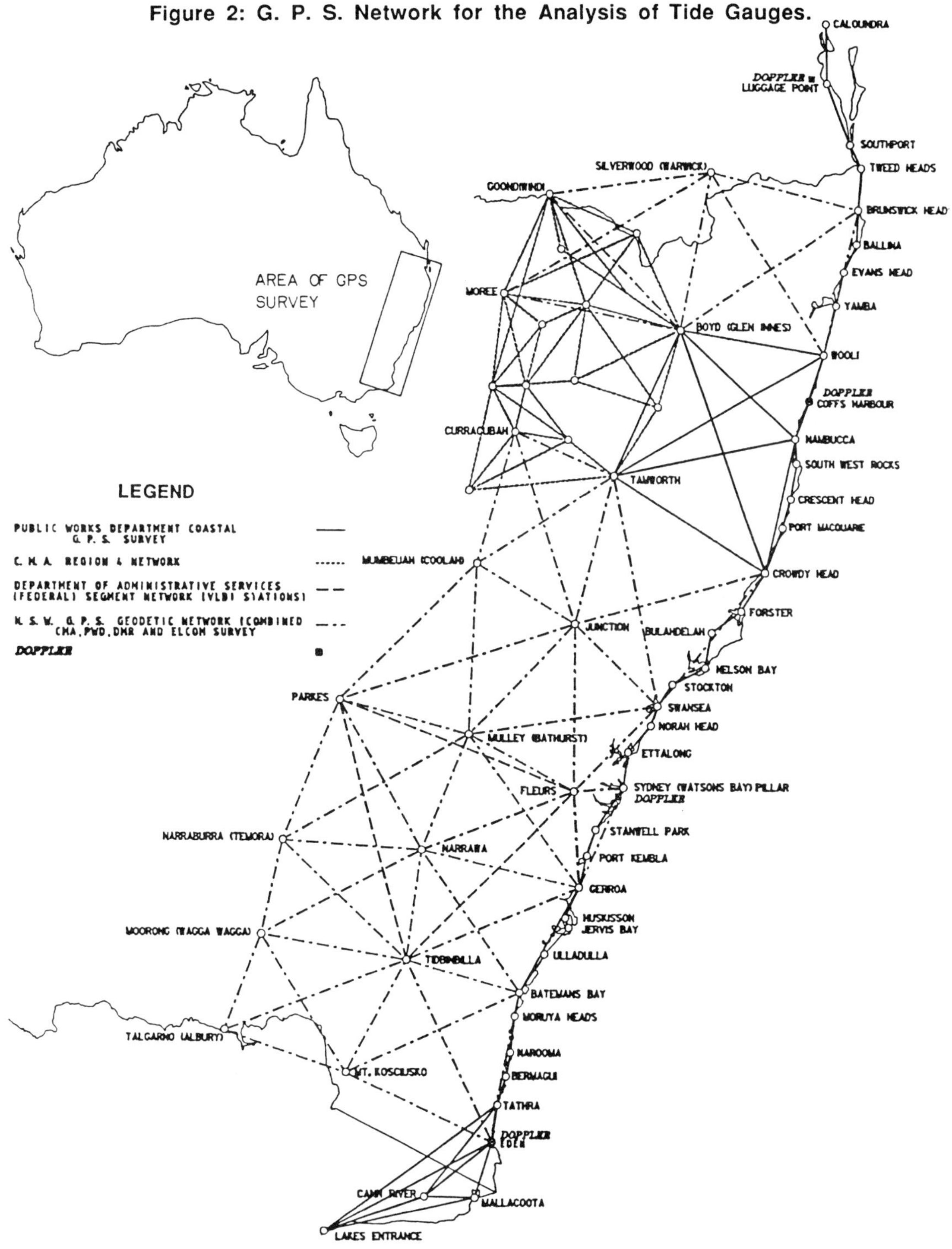

FIGURE 4

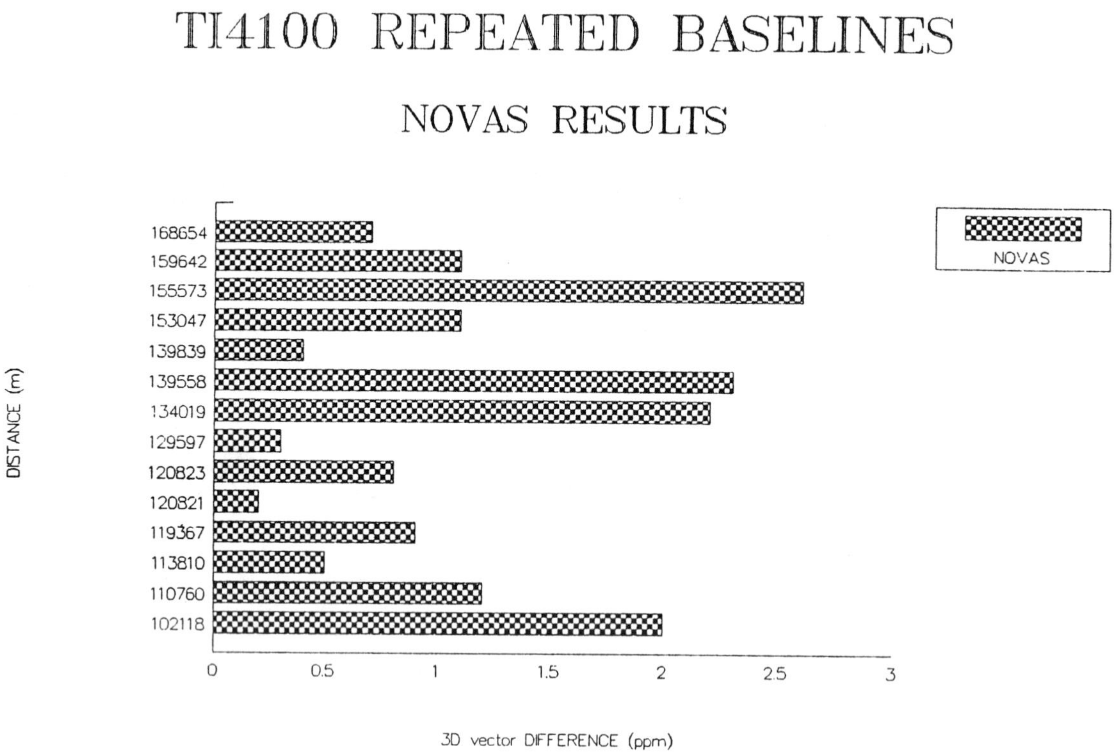

Figure 5: Orthometric and ellipsoidal heights

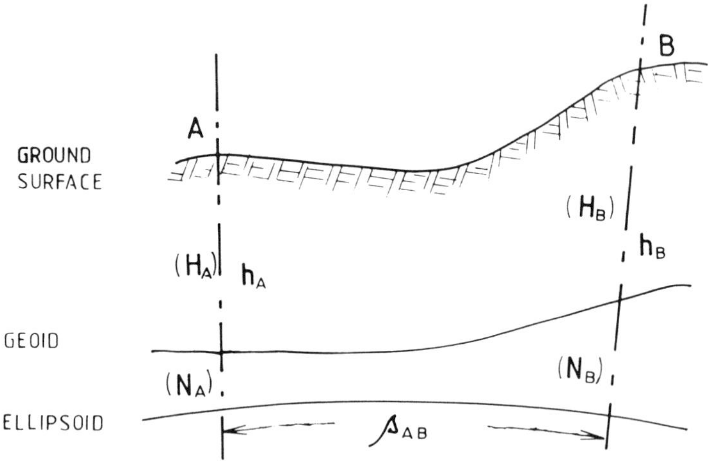

FIGURE 6

AGREEMENT BETWEEN TERRESTRIAL AND GPS/GRAVITY DERIVED ORTHOMETRIC HEIGHTS

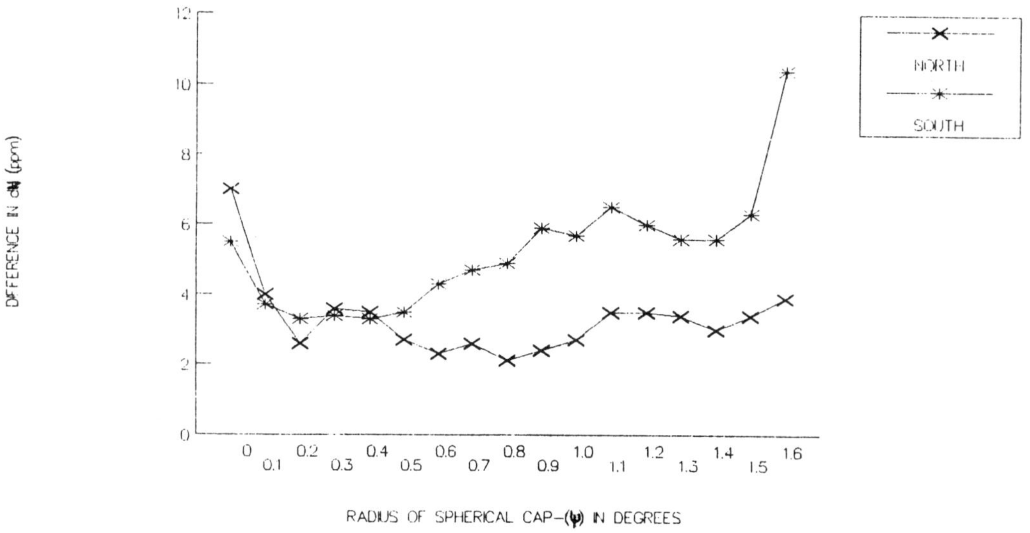

FIGURE 7

HEIGHT OF MSL PHASER/OSU86 and AHD

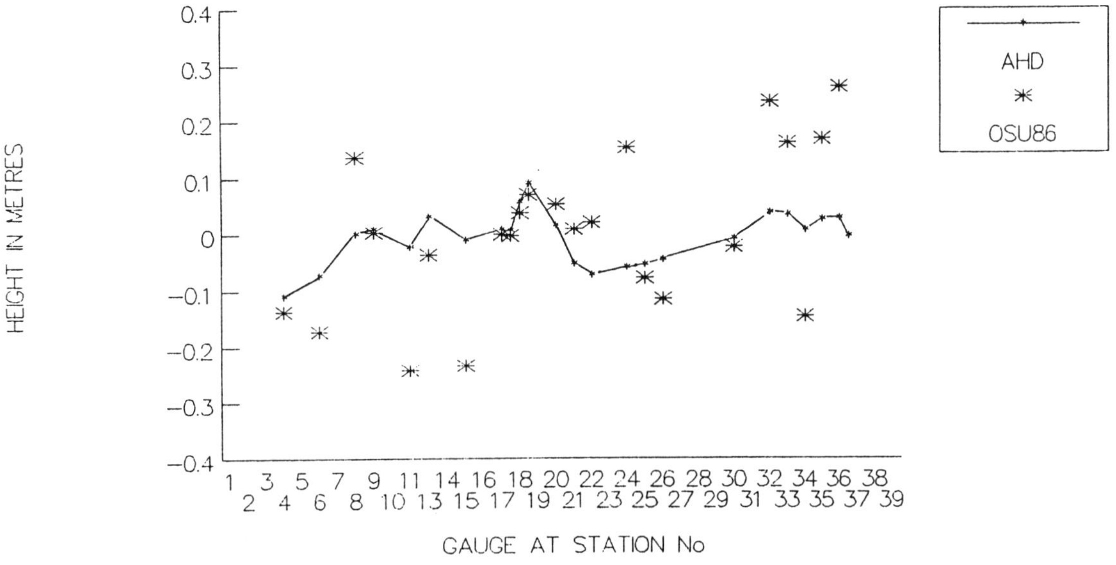

HIGH ACCURACY GLOBAL POSITIONING SYSTEM ORBIT DETERMINATION: PROGRESS AND PROSPECTS

Stephen M. Lichten
Jet Propulsion Laboratory
California Institute of Technology
Pasadena, California 91109 USA

INTRODUCTION

The Global Positioning System (GPS), by the mid-1990s, will include 21 navigation satellites launched by the U.S. Air Force. About one-third of the constellation is presently in orbit; however, most of these satellites are of a developmental Block I design and eventually will be replaced by the operational Block II satellites. The GPS constellation is designed to evenly distribute the satellites with circular orbits in three different planes at an altitude of about 20,000 km so that ground users can track typically six to eight GPS at any time from most locations.

For the civilian user community, GPS can provide precise positioning capability for numerous scientific applications. Vigorous research efforts have been underway for more than five years now to optimize and improve GPS orbit determination and relative positioning techniques. Over the past five years, GPS orbit determination capabilities have improved by about an order of magnitude, from 5-10 m to better than 1 m, even with fairly limited ground tracking networks and the available subset of the full constellation. In and of itself, this is an unusually accurate orbit capability for a satellite in high-earth orbit, but is even more significant when seen from the perspective of scientific research into earth crustal motions, oceanography, and other disciplines which have traditionally relied on space-based observations, either using observations involving satellites or observations of fixed celestial objects such as quasars. With m-level GPS orbits, a number of different groups have reported ground baseline measurements with accuracy and precision of about 0.5 cm +/– several parts in 10^9 relative to baseline length over distances ranging from less than 100 km up to 6000 km. With the pattern of increasingly sophisticated GPS parameter estimation techniques and improved orbit determination set over the past few years, we can expect to continue to improve to better than 1 part in 10^9 using the full GPS constellation by the time it is operational in the mid-1990s. For geodetic applications, GPS provides a unique and versatile ground-based system. In addition to being small, portable, and relatively inexpensive (with costs expected to decrease in the future when GPS receivers are produced in larger quantities), GPS technology combines some of the desirable aspects of both very long baseline interferometry (VLBI) and satellite laser ranging (SLR). When used either alone or in combination with SLR and/or VLBI, GPS will be an extremely precise yet flexible geodetic technique. For earth orbiter missions where high-accuracy orbits are needed to calibrate orbiting scientific instruments, there is virtually no technique other than differential GPS tracking, using flight and ground GPS receivers together, which has the potential for cm-level low-earth orbit accuracy needed for future missions such as the Earth Observing System (EOS) [Yunck et al. 1989]. The TOPEX/POSEIDON oceanography satellite, to be launched in 1992, will carry a GPS flight receiver designed to demonstrate altimetric orbit accuracy of about 10 cm (Wu et al. 1989).

In this paper, the effect of GPS orbit accuracy on ground baseline errors will be reviewed. Some of the techniques which have matured over the past five years for estimation of precise GPS orbits will be discussed, including linearization of the problem, orbit modeling, estimation of phase biases, and treatment of atmospheric-induced delays.

Of particular importance is the calibration of the GPS reference frame using VLBI or other external techniques. Efforts by different groups to produce precise GPS ephemerides routinely in an operational sense will be discussed and compared, and the use of such precise ephemerides for estimation of ground baselines will also be discussed. With concentrated efforts centered on specific data sets, sub-meter GPS orbits have been demonstrated from the current limited constellation and a regional ground tracking network. Presumably, estimation strategies to achieve sub-meter GPS orbit accuracy can eventually be streamlined and adopted for routine GPS processing. GPS orbits should improve to the decimeter level by the early to mid-1990s, when the 21-satellite constellation is operating in conjunction with a worldwide ground tracking network equipped with advanced receivers.

GPS ORBIT ERROR EFFECT ON GROUND BASELINE ESTIMATES

When differential GPS positioning techniques are used, in the limit of very closely spaced GPS antennas, the error of the baseline vector for their relative positions is independent of GPS orbit error. On the other hand, for widely separated GPS antennas, the baseline error is essentially proportional to the baseline length. In general, the baseline error dependence on GPS orbit error can be expressed as

$$\sigma^2(baseline) \approx A^2 + B^2L^2 \tag{1}$$

The orbit-independent contribution to the baseline error is A. This will include local site tie errors, certain tropospheric effects, unmodeled antenna phase center variations (including multipath), relative station geometry, and measurement noise. B represents orbit-related errors which generally grow larger with baseline length. This generally includes orbit modeling, fiducial station coordinate errors, and tropospheric effects on GPS orbit accuracy (Fig. 1).

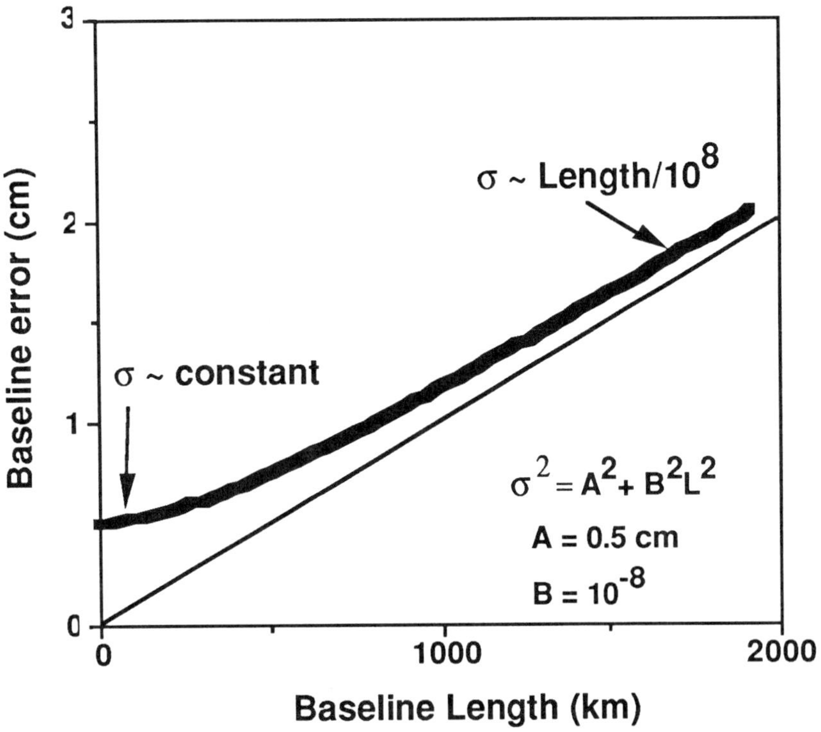

Fig. 1. Expected variation of baseline error with baseline length for 1 part in 10^8 accuracy and orbit independent error of 0.5 cm.

Sometimes an error affects both A and B and the distinction may be somewhat vague. For example, troposphere mismodeling can be a serious source of GPS orbit error (B); however, even with perfect orbits, troposphere mismodeling also contributes to baseline error (A). Errors in local GPS antenna site ties usually result in a contribution to A, but at fiducial sites, these errors will also distort the orbits (B). B can be considered to be largely a geometric effect (Beutler et al. 1987):

$$B \approx \frac{\beta \sigma_{orb}}{R} \tag{2}$$

where R is the receiver-GPS distance (typically about 23000 km for a ground receiver), and β is a geometrical factor. From recent results (Lichten and Bertiger 1989) indicative of simultaneous 1-m orbit accuracy and about 1 part in 10^8 baseline accuracy, β is probably about 0.2, at least for the regional networks used in conjunction with a constellation of seven developmental GPS satellites.

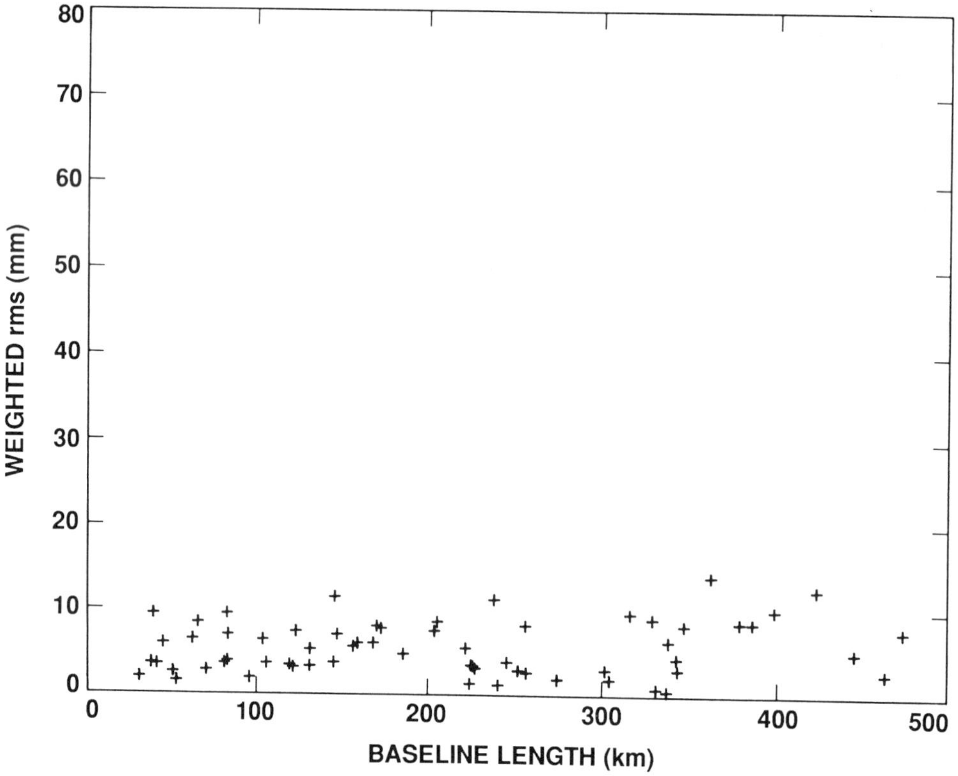

Fig. 2. East component daily baseline repeatability in California from Bock (1989) with bias-fixed carrier phase. The rms scatter is about 0.5 cm with little dependence on baseline length, showing that the advanced estimation techniques (Dong and Bock 1989) used have minimized orbit-related errors which grow with baseline length.

The form of eqn. (1) implies that the variation of baseline error with baseline length will have two different regimes. This can be seen in Fig. 1. For short baselines (less than 500 km for the case illustrated in Fig. 1), we expect sub-cm accuracy and relatively little variation with baseline length. For long (greater than 1000 km) baselines, we expect baseline accuracy to behave like a linear function of baseline length. Observed GPS baseline daily repeatability and agreement with VLBI measurements does, in fact, follow these patterns. Fig. 2 shows baseline precision (daily repeatability) in California (Bock

1989) up to 500 km. There is only a slight, if any, linear trend to the data up to about 300 km length for these baselines, which have been bias fixed.

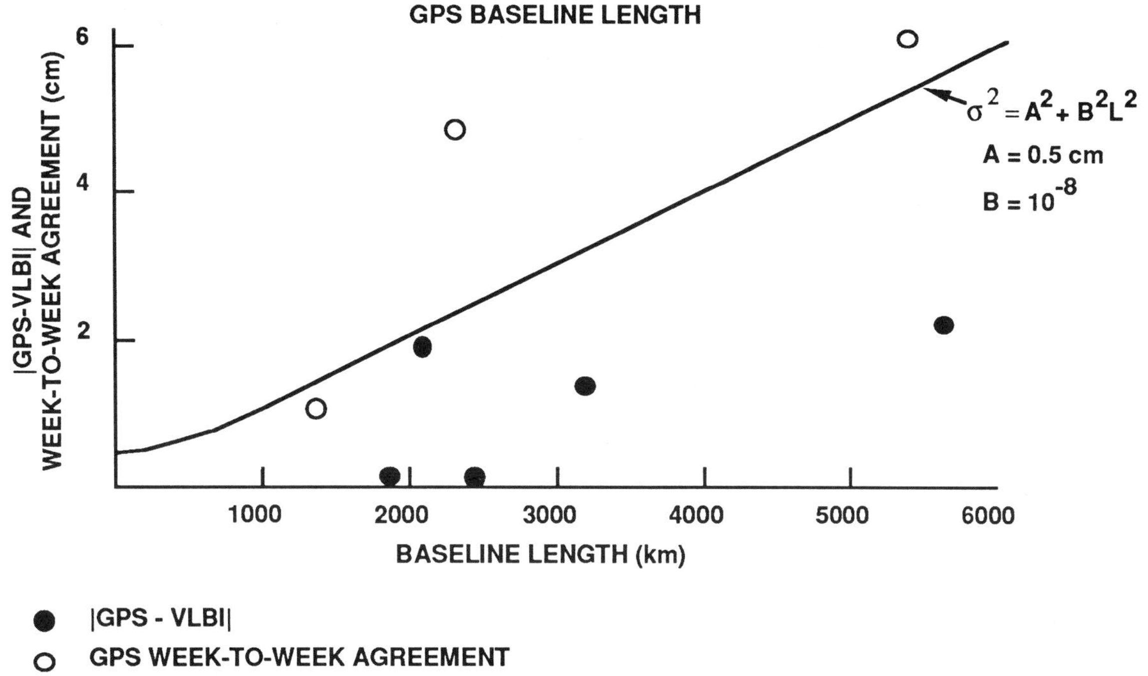

Fig. 3. GPS long baseline accuracy (agreement with VLBI from Lichten et al. 1989) and precision (week-to-week agreement from Schutz et al. 1989) of several parts in 10^9 has been demonstrated for a number of baselines longer than 1500 km, including some intercontinental baselines.

Fig. 3, which is based on data from Lichten et al. (1989) and Schutz et al. (1989), shows baseline length accuracy (agreement with VLBI) and precision (weekly repeatability) for baselines up to 6000 km in length, showing several parts in 10^9 in most cases. Table 1 summarizes these results showing also horizontal and vertical baseline components. Horizontal vector baseline components are generally slightly less precise than the scalar length measurements, but in most cases are at the 1 part in 10^8 level. Note that the baselines shown here are all either between a non-fiducial site and the *nearest* fiducial (fixed) site or between two non-fiducials which are *closer* than the distance to the nearest fiducial. Occasionally results have been presented in the literature where performance has been exaggerated by using the size of the overall tracking network (which can be several thousand km) as a reference instead of the distance to the nearest fiducial (which can be a few hundred km). True long baseline performance can only be determined with ground stations estimated thousands of km away from the nearest fiducial.

Table 1. GPS long baseline precision and accuracy.

Baseline	ΔEast (cm)	ΔNorth (cm)	ΔVertical (cm)	ΔLength (cm)	Length (km)	Ref.
Haystack/ Onsala	+6.4	+3.2	+19.2	-2.3	5601	(a)
FtDavis/ Kokee	-2.6	+0.9	-8.9	+6.2	5409	(b)
FtDavis/ Haystack	-0.01	+3.2	+5.1	+0.7	3136	(c)
FtDavis/ Richmond	+0.5	+3.4	+0.6	+0.02	2363	(c)
Canberra/ Blk Birch	+6.6	+2.2	-4.7	+4.8	2255	(b)
Richmond/ Haystack	-1.3	-1.9	-2.4	-1.9	2046	(c)
Hatcreek/ FtDavis	-1.3	-1.7	-1.1	-0.2	1933	(c)
Onsala/ Tromso	-2.2	+1.0	-3.3	-0.9	1406	(b)

Baseline results displayed in local geodetic coordinates.

(a) This paper, 1988 Casa Uno experiment, GPS-VLBI agreement, multi-day arc orbits.

(b) Schutz et al. (1989), 1988 Casa Uno experiment, week-to-week baseline repeatability, multi-day arc orbits.

(c) Lichten et al. (1989) November 1985 data, GPS-VLBI agreement, multi-day arc orbits.

DIRECT MEASURES OF GPS ORBIT ACCURACY AND PRECISION

Intercomparisons of baseline estimates can provide a strong indication of orbit accuracy. However, as shown above, ground baselines are affected by a variety of errors, some of which are independent of orbits. Most ground baselines measured with GPS have been relatively short (where baseline length is defined as the distance to the nearest fiducial site or to a closer non-fiducial site). Therefore, at least for the more accurate determinations, baseline repeatability over several hundred km distances measured in regional networks, particularly where carrier phase bias have been fixed to integers (as described below in discussion on carrier phase ambiguity resolution), does not show a strong dependence of baseline length (Fig. 2). The flatness of this baseline repeatability versus baseline length plot shows that the orbits here have been effectively removed to a large extent as a major error source by advanced GPS data processing methods, particularly network ambiguity resolution as described by Dong and Bock (1989). The longer baselines (Fig. 3 and Table 1) have been rather sparsely sampled and have limited geometry. The presence of the geometrical factor β in eq. (2) adds more uncertainty in assessments of GPS orbit accuracy from baseline intercomparisons.

A more direct method for measuring GPS orbit accuracy and precision consists of comparing GPS orbits directly during periods of overlap for two separate solution arcs. Early direct orbit comparisons of this sort in 1985 (Swift 1985; Abbot et al. 1985) indicated orbit precisions of about 10 m. Since that period, increasingly sophisticated orbit determination techniques have improved GPS orbit determination capabilities by about an order of magnitude.

GPS orbit determination at present can usually be categorized as either a specialized effort for a specific data set, or as part of a routine orbit production activity. The latter includes the estimation of the broadcast ephemeris by the Department of Defence, the precise ephemeris determined by the Naval Surface Warfare Center (NSWC) and Defense Mapping Agency (DMA), and the precise orbits routinely determined at the University of Texas. The National Geodetic Survey (NGS) is also becoming involved in routine GPS orbit determination. Specialized orbit determination efforts are usually concentrated on specific experiments, such as has been done at the Massachusetts Institute of Technology (MIT), University of Berne, Jet Propulsion Laboratory (JPL), University FAF (Munich), University of Nottingham, Canadian Geodetic Survey, and numerous other centers of scientific activity where a variety of softwares are now being used for geodetic GPS-based studies. For routine orbit production, usually global tracking networks are used if available, such as the 10 worldwide sites used by the NSWC and the CIGNET stations used by the University of Texas. Routine orbit production also generally involves multi-day arc orbit solutions, in contrast to specific geodetic occupations where shorter (single-day) arcs have been used with smaller networks. For the more specialized, regional experiments, analysts frequently can spend more time tuning the orbits to the region of interest with specialized techniques. Routine orbit production, however, often precludes the time and effort which would be required to do this on a global scale, week after week. A worthwhile goal for the GPS geodetic community is to eventually arrive at an orbit strategy which combines certain desirable aspects of both types of orbit determination activity.

To accurately model the GPS orbits, the equations of motion are numerically integrated. The initial epoch states may be obtained from a variety of sources, including the broadcast ephemeris. Since the broadcast ephemeris does not integrate well over intervals longer than a few hours, frequently a least-squares fit is made to broadcast elements from several days. Using the nominal GPS trajectories, precise models for the earth's motion and receiver locations enable computation of station-satellite ranges. Observation residuals are then formed, and provided that the nominal models are within the linear regime, linear partials of the observed minus computed ranges with respect to various parameters are calculated. These partials are then combined with the measurements and a priori information to determine a least-squares solution for the orbits and other estimated parameters. For geodetic work with GPS, models should include GPS solar radiation pressure, earth orientation, antenna and transmitter phase center offsets and variations, and corrections for general relativity.

Orbit comparisons done at overlaps from weekly solutions by the NSWC have recently shown consistency at the 2-3 m level (Gouldman et al. 1989). The NSWC uses a number of advanced orbit estimation strategies in its weekly solutions. These include 2nd order Gauss-Markov process noise models for satellite and station clocks, and 1st order Gauss-Markov process noise models for solar radiation pressure coefficients. Regional baselines have been determined by the National Geodetic Survey (NGS) with a constrained partial adjustment of the NSWC precise orbits. The results of this procedure (Strange and Zurflueh 1988; Strange et al. 1989) show several cm agreement with VLBI and daily baseline repeatability. Similar orbit comparisons from the University of Texas (Schutz et al. 1989) show week-to-week orbit agreement of 2-4 m. The Texas orbits rely primarily on doubly differenced carrier phase, in contrast to the NSWC's use of pseudorange with more elaborate clock modeling, but extra degrees of freedom were allowed for GPS force parameters over the arcs of 1-week or longer to absorb unmodeled effects. Double differenced data types (or estimation of white noise clock parameters, which is equivalent) are commonly used in regional GPS experiments where sites have substantial common visibility of the GPS satellites. Long baselines (> 1000 km) determined with the University of Texas orbits show precision of a few parts in 10^8. Both the Texas and NSWC orbits were determined with multi-day arc fits.

Additional orbit tuning with regional networks has resulted in sub-meter orbit accuracy (Lichten and Bertiger 1989). Fig. 4a shows an orbit intercomparison in which GPS orbits from one week are predicted into the next week and an rms is computed from the two independent GPS orbit solutions. Fig. 4b shows a similar comparison with two interleaved arcs without any common data. The estimation procedure followed by Lichten and Bertiger (1989) used as initial conditions globally determined orbits from either the broadcast ephemeris or from the NSWC precise ephemeris; data from a denser North American network was used for further orbit improvement and resulted in GPS orbits tuned to the North American region. However, the rms differences showed in Fig. 4 were computed while the satellites were *not* over North America. Baselines of 2000—3136 km length computed with these orbits agree with VLBI to several parts in 10^9 in length, 1 part in 10^8 horizontally, and better than 2 parts in 10^8 in the vertical. Additional refinements to the orbit estimation here included solar radiation pressure process noise modeling (similar to the NSWC strategy), troposphere delay process noise modeling, simultaneous processing of pseudorange and carrier phase data types (which effectively constrains the clocks and bias parameters, strengthening in particular the down-track orbital components and the east baseline components) and multi-day arc orbit fits.

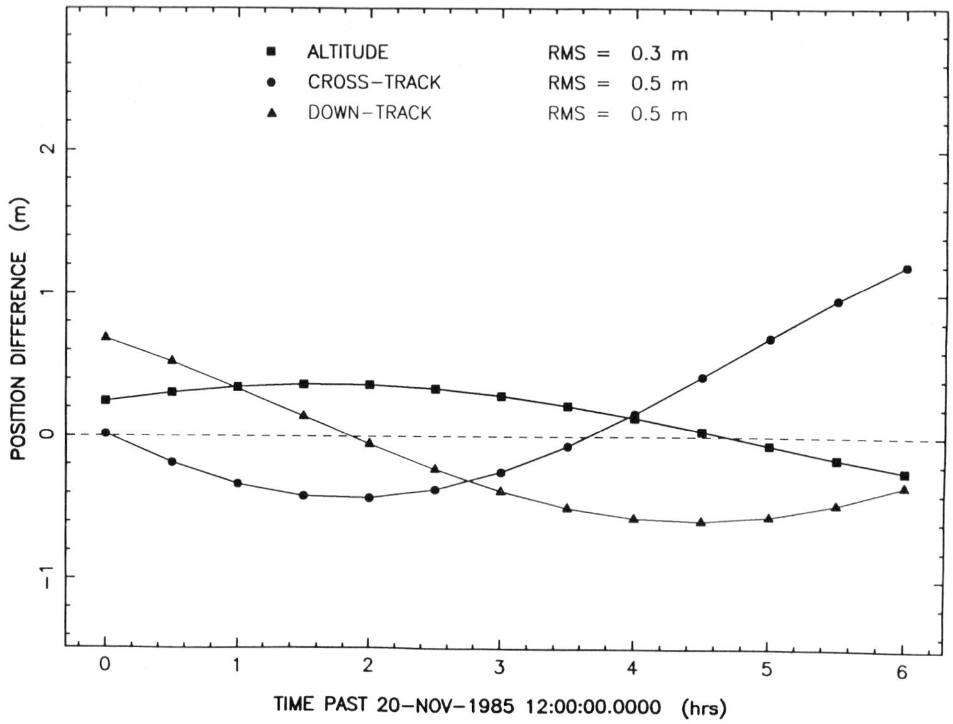

Fig. 4a. Difference in GPS 8 orbit when 1-week arc solution is predicted ahead into the next week and compared to independent solution from a separate 1-week arc. The two arcs have no common data points (see also Fig. 6). Rms shown here for GPS 8 is computed over a 6-hr interval, but remained below 1 m even when computed over a 72-hr interval.

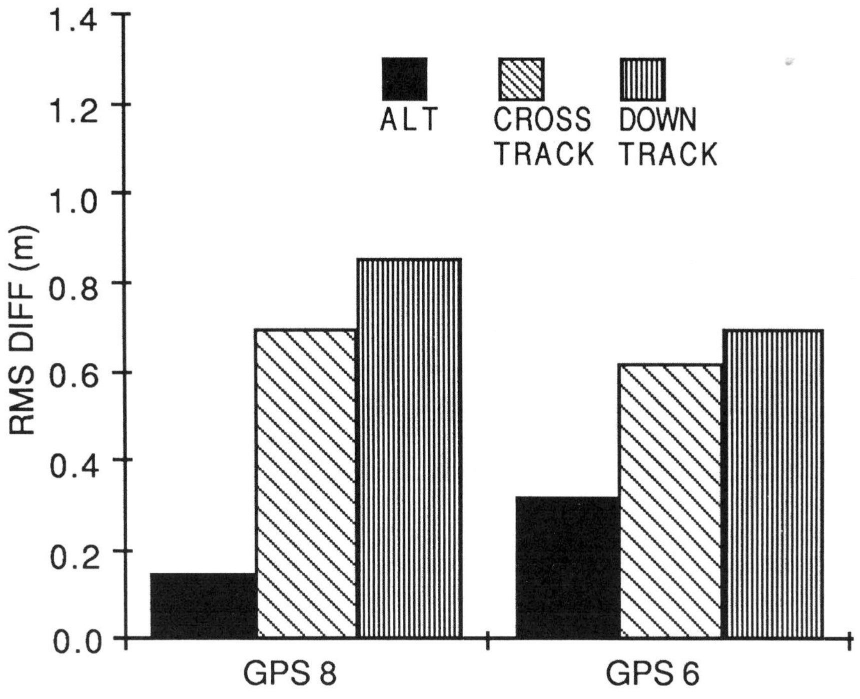

Fig. 4b. Rms for orbits computed from data collected on 1985 November 13,15 and 18 differenced with orbits computed from data collected on November 14, 16 and 19 (multi-day arc orbit estimation arcs). The rms was computed over a 24 hour period (November 17) which was not used for either set of orbits. Rms differences for the other five satellites with shorter tracking arcs, were 0.5-1.5 m.

HIGH-ACCURACY GPS ORBIT DETERMINATION TECHNIQUES

Dynamic Modeling for GPS Satellites

Dynamic modeling for GPS satellites is fairly straightforward for data arcs of 1 day or less. Typically a gravity field truncated to degree and order 8 from the GEM-L2 MERIT standard model (Melbourne et al. 1983) is used along with the ROCK4 GPS solar radiation model (Schutz et al. 1989; Lichten and Bertiger 1989). Scale factors close to unity are often estimated for the solar radiation pressure coefficients, which include a Y-bias imbalance factor. However for arcs longer than several days, GPS orbit determination requires additional degrees of freedom to absorb unmodeled accelerations. These accelerations have been estimated with dynamic process noise filtering (Lichten and Bertiger 1989; Swift 1987), with additional drag and solar pressure parameters (Schutz et al. 1989; Powell and Gaposchkin 1988), or with resonant trigonometric force parameters (Colombo 1989). There have been indications that force mismodeling may be correlated with the GPS eclipse season, when solar pressure effects are more difficult to model. If this is true, then the new Block II GPS satellites may be even more complicated to follow over long arcs since their shapes are somewhat more intricate than those of the Block I satellites.

The results of Lichten and Bertiger (1989) and of Schutz et al. (1989) suggest that GPS orbit accuracy may be currently limited to some extent by accuracy of the force models for the satellites. When GPS orbit determination arcs are extended beyond 1-2 weeks, the benefits from additional averaging down of random errors is offset by growing errors from dynamic mismodeling, particularly during GPS eclipse seasons, when solar radiation effects reach a maximum. Fig. 5 shows daily baseline repeatability for a 729 km baseline in California estimated at JPL from the 1988 Casa Uno experiment. Note the significant improvement in precision when the process noise force model is used for the GPS solar radiation pressure coefficients. Four of the seven satellites tracked during this experiment were in eclipse. The level of process noise used allowed two solar pressure coefficients to vary by a few percent per day, but the Y-bias parameters were estimated as constants.

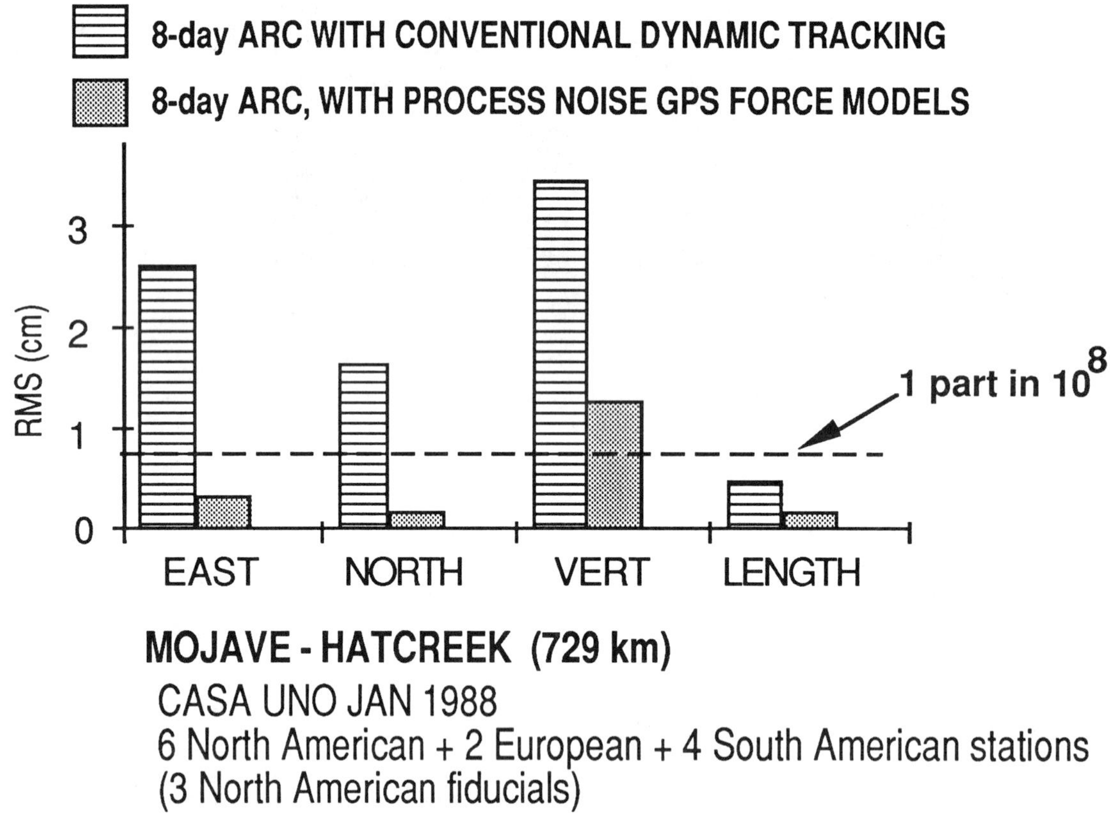

Fig. 5. Improvement in baseline repeatability resulting from process noise filtering for estimated GPS solar radiation pressure coefficients. The process noise model allows for only a few percent variation in solar pressure coefficients per day.

Effect of Troposphere Errors on GPS Orbits

The wet troposphere is variable and can be difficult to model (see Tralli et al. 1988). Since the measured range between a receiver and a GPS satellite includes a delay from the troposphere, this can be a serious error source in GPS orbit determination. Lichten and Border (1987) and Lichten and Bertiger (1989) have demonstrated improvements in GPS orbits when zenith troposphere parameters are estimated with a random walk process noise model. Fig. 6 compares predicted and estimated GPS orbits for one of the satellites for constant and process noise troposphere zenith delay estimation. The orbits improve by nearly 40% with the stochastic troposphere model. Fig. 7 compares GPS-determined stochastic zenith troposphere delays at Haystack during the 1988 Casa Uno campaign with the sum of wet delays from water vapor radiometers (WVRs) and dry delays from surface measurements. Sub-cm agreement between these techniques shows the power and sensitivity of GPS data for tropospheric delay estimation. Even more significant is the ability of the GPS technique to track large and rapid fluctuations in the troposphere (about 2-3 cm at zenith in 2 hours on January 21, 1988). With zenith tropospheric path delays as variable as this, clearly estimation of a single constant zenith scale factor for a daily observing session will result in systematic errors. Fig. 8 shows an improvement by more than a factor of 2 in GPS and VLBI agreement for a 2000 km baseline when stochastic troposphere parameters are estimated for orbit determination. Baselines would be affected by tropospheric mismodeling even if the orbits were perfect, so the contrast in Fig. 8 (a factor of more than 2) is somewhat greater than in Fig. 6 due to the fact that the baseline errors include both the orbit and non-orbit troposphere-related effects.

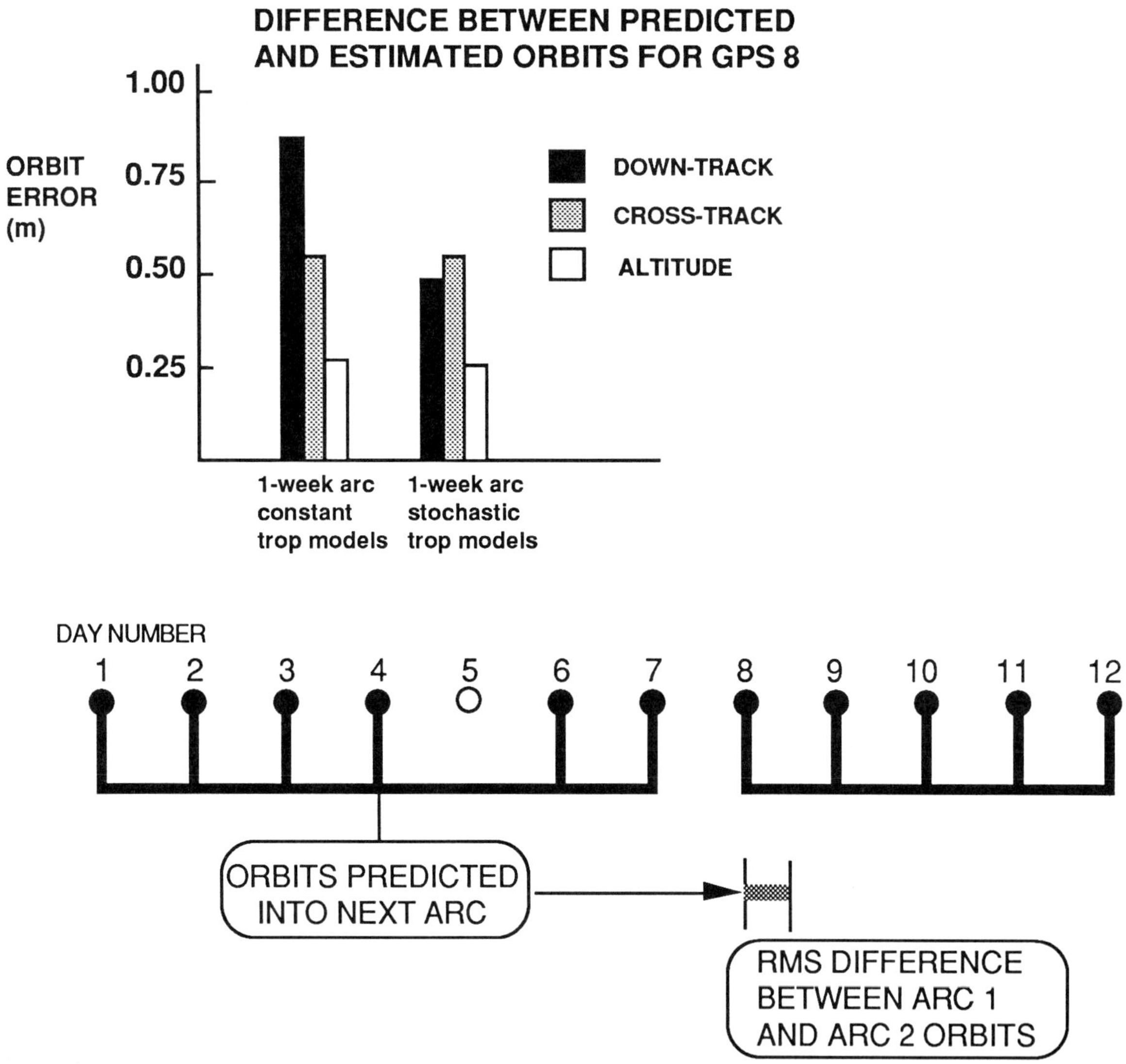

Fig. 6. Improved orbit accuracy for GPS 8 with stochastic (random walk) zenith troposphere delay models (November 1985 experimental data). Similar improvements were seen for the other six GPS satellites tracked.

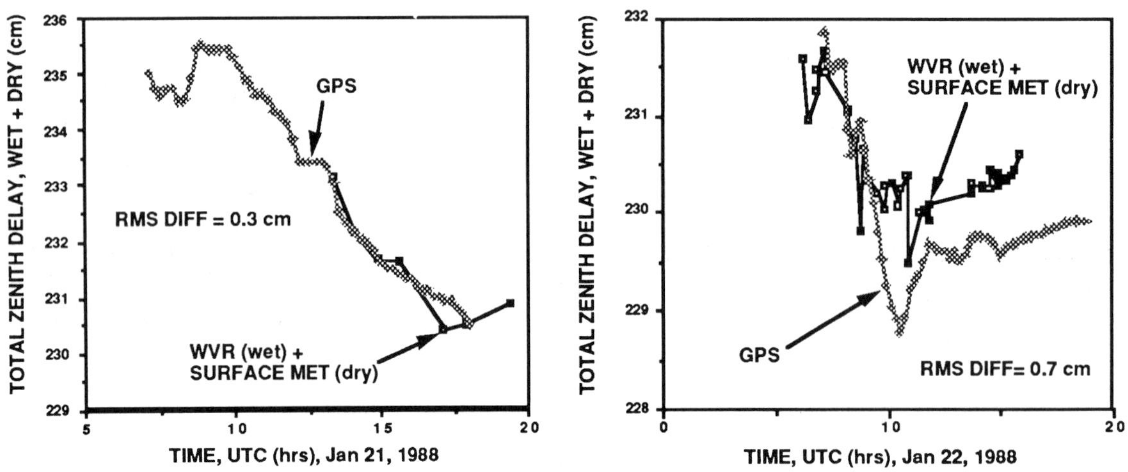

Fig. 7. Comparison of GPS estimates for the total (wet + dry) zenith troposphere delay from process noise filtering with the sum of wet delay from a WVR and dry delay from surface data. Results from two days during the Casa Uno experiment are shown.

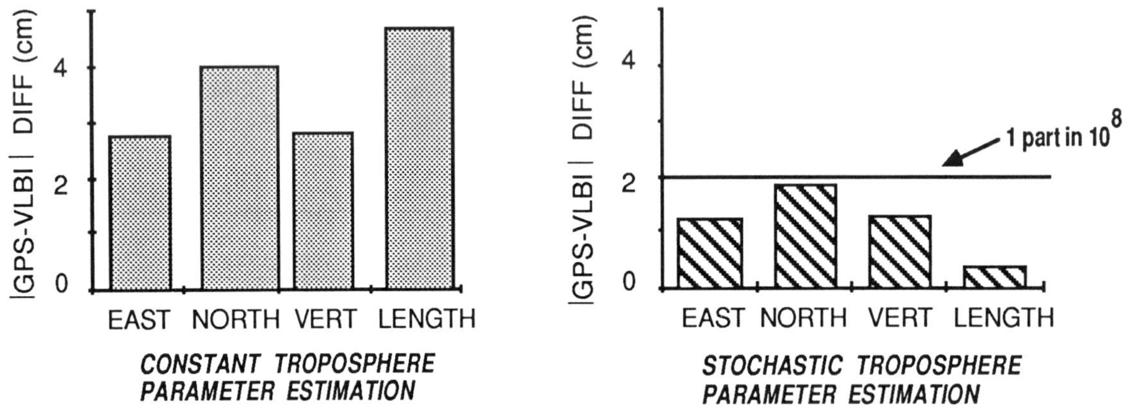

Fig. 8. Comparison of GPS-VLBI agreement for the Hatcreek, CA — Fort Davis 1933 km baseline estimated from a 2-week arc in November 1985 with two different troposphere parameter estimation schemes. The stochastic model was a random walk.

Carrier Phase Ambiguity Resolution and Use of Pseudorange

The GPS carrier phase (dual-frequency, ionosphere-corrected) is precise to millimeters over several-minute averaging intervals with modern receivers. Short-term tropospheric fluctuations, however, probably limit this precision to about 0.5 cm. The range which is derived from the GPS carrier phase is ambiguous to an integer multiple of the wavelength (the L1 and L2 wavelengths are about 20 cm). Thus continuously tracked carrier phase provides a precise record of *range change*. The integer range ambiguity must either be estimated as one of the parameters to be determined, or it can be fixed to an integer value using ambiguity resolution techniques (Dong and Bock 1988, 1989; Blewitt 1989). For each pair of stations and satellites in common view, there is a unique combination of doubly differenced carrier phase biases. Counselman and Abbott (1989) have demonstrated that single-day arc bias fixing can improve the accuracy of the GPS orbits over the region which has been bias fixed. This in turn can result in more biases fixed over longer baselines. Dong and Bock (1989) suggest that multi-day session GPS orbits can improve bias fixing, based on post-fit statistics. There is little data available on multi-day arc bias fixing at this time, but it is logical that improving the orbits through longer arcs would enable more biases to be fixed. Multi-day session bias fixing is presently a rather computationally intensive procedure, but future streamlined approaches may make it easier. It is clear from the references above, that for single-session data processing, bias-fixing can substantially enhance GPS orbits over the region being bias fixed.

Bias fixed carrier phase is equivalent in information content to *carrier range*, which would be produced by a receiver whose pseudorange had sub-cm data noise. It is interesting to compare this ideal data type (which is presently only attainable by bias fixing over an entire tracking network) to various combinations of ambiguous carrier phase + noisier pseudorange. Figs. 9 and 10 show expected levels of orbit accuracy for one GPS satellite and expected baseline accuracy for a 1314 km baseline with various different combinations of carrier phase and pseudorange. The scenario used in this study corresponds to nine well-spaced ground sites in a 5000 km North American network which, because of the wide spacing of the ground sites, would be very difficult to completely bias fix. Figs. 9 and 10 shows that use of pseudorange along with ambiguous carrier phase can produce comparable accuracy to bias fixed phase. In a network with a wide range of baseline lengths, even relatively noisy pseudorange can be expected to substantially improve results when there are baselines over which ambiguity resolution is impossible. The case shown for phase + 100 cm pseudorange might be indicative of expected future performance with advanced receivers in the codeless mode with anti-spoofing (see discussion below in Future Prospects section).

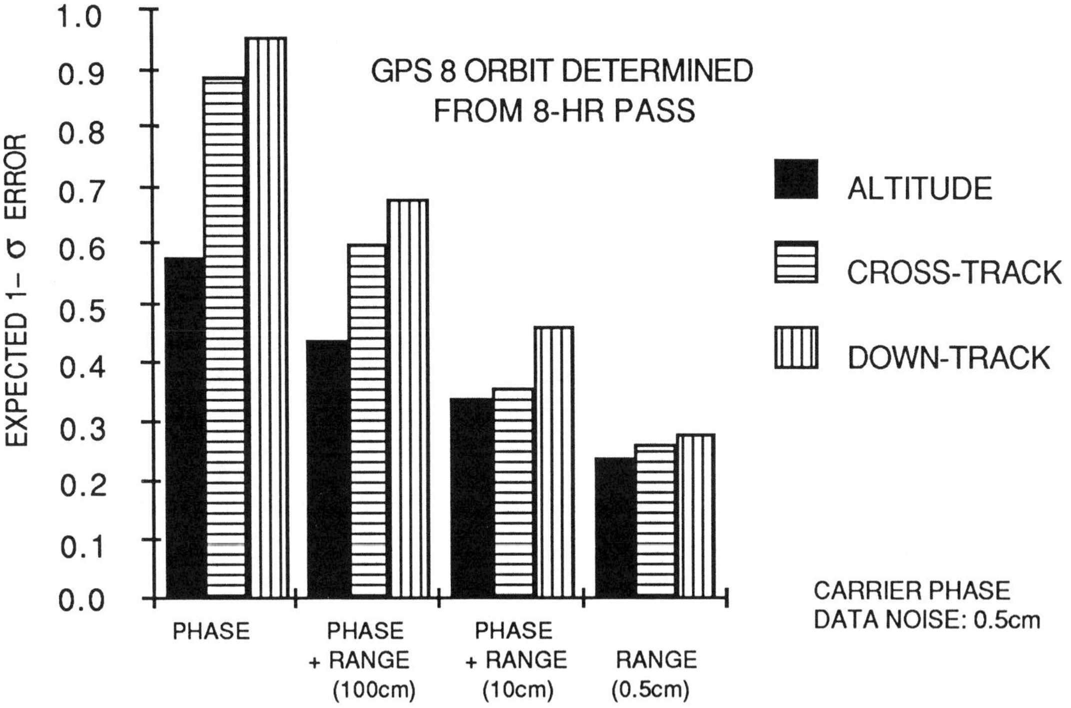

Fig. 9. Predicted accuracy from covariance analysis for North America tracking only with 6-GPS Block I constellation, assuming 3 fiducial sites with 2 cm coordinate errors each.

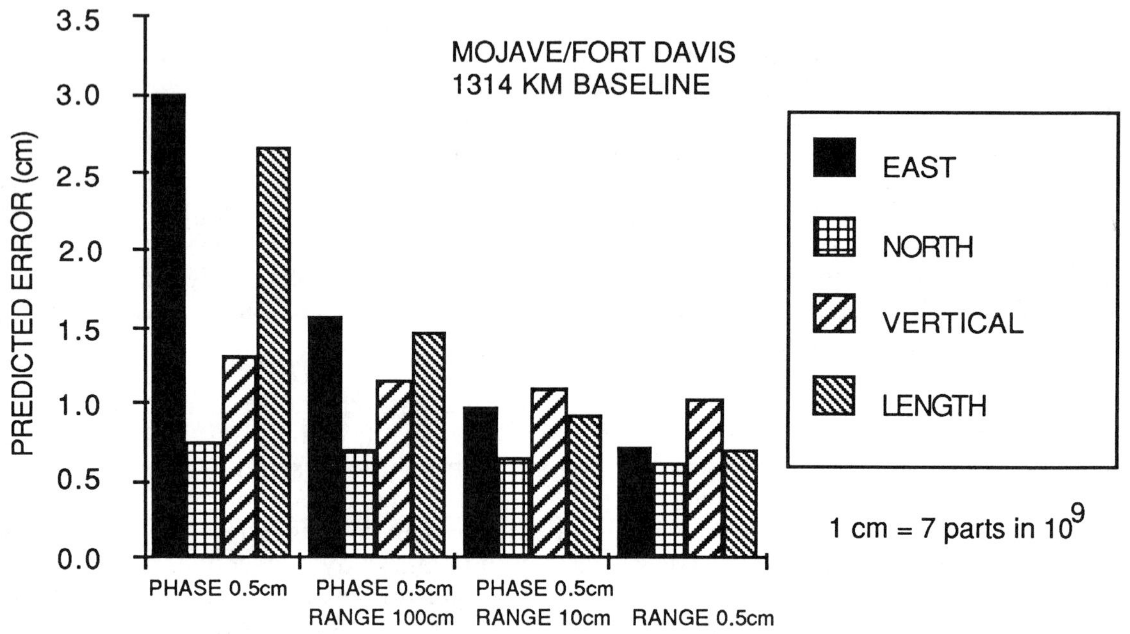

Fig. 10. Comparison (corresponding to Fig. 9) of expected baseline errors for combinations of GPS phase + pseudorange and for bias fixed GPS phase (0.5 cm range).

Global GPS Tracking Networks

A worldwide ground tracking network would be expected to produce more globally accurate GPS orbits than would regional networks. Covariances analyses by Wu et al. (1988) predicted improved GPS orbit accuracy when data from the South Pacific is added to a tracking network with just North American sites. However, regionally determined orbits might be tuned for the baselines of interest. In practice, data from a worldwide network are combined with regional data. The orbits would improve both from global tracking data as well as from the denser regional coverage, which could include a number of bias fixed baselines. Fig. 11, from Kornreich et al. (1989), shows formal orbit errors for GPS 8 from experimental data processed from the 1988 Casa Uno experiment . There is lowering of orbit formal errors as the tracking network expands geographically. Fig. 12 shows baseline repeatability for the three tracking networks. The baselines range from about 50 to 1000 km and are all in South America, somewhat outside the North American fiducial network. The baseline repeatability improves particularly for the longer baselines as the tracking network becomes more global. Since the baselines studied here are all in South America, well outside the U.S. fiducial network, the improvement in repeatability is probably due to orbit improvement resulting from the larger tracking network.

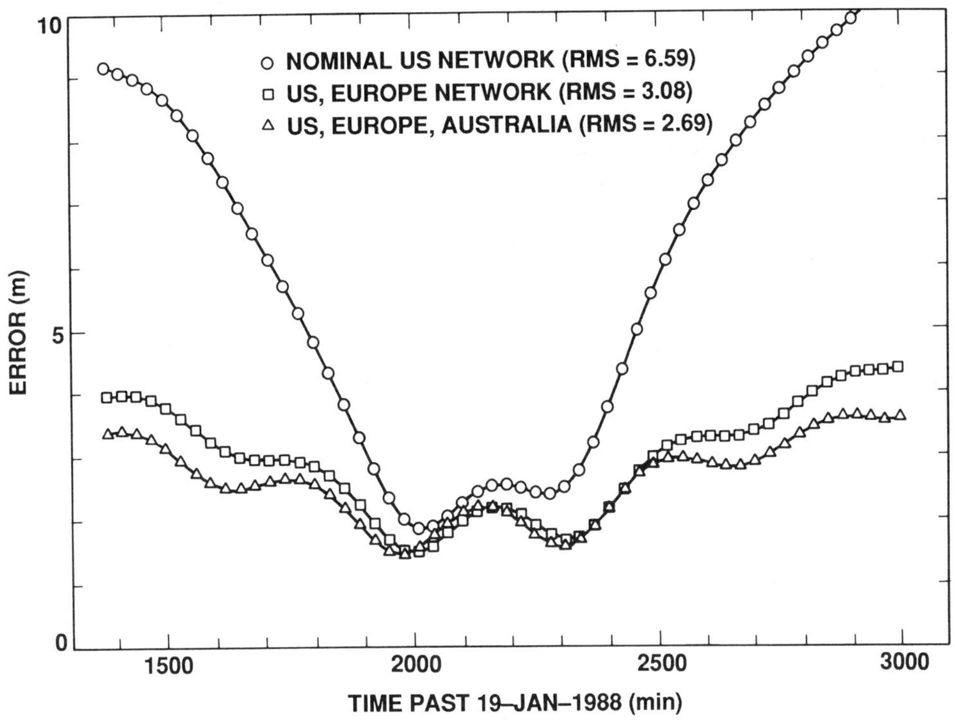

Fig. 11. Formal GPS 8 orbit errors show improved orbit accuracy expected for a more global tracking network. The formal errors are for a single 24-hr data pass during the 1988 CASA UNO experiment. From Kornreich et al. (1989).

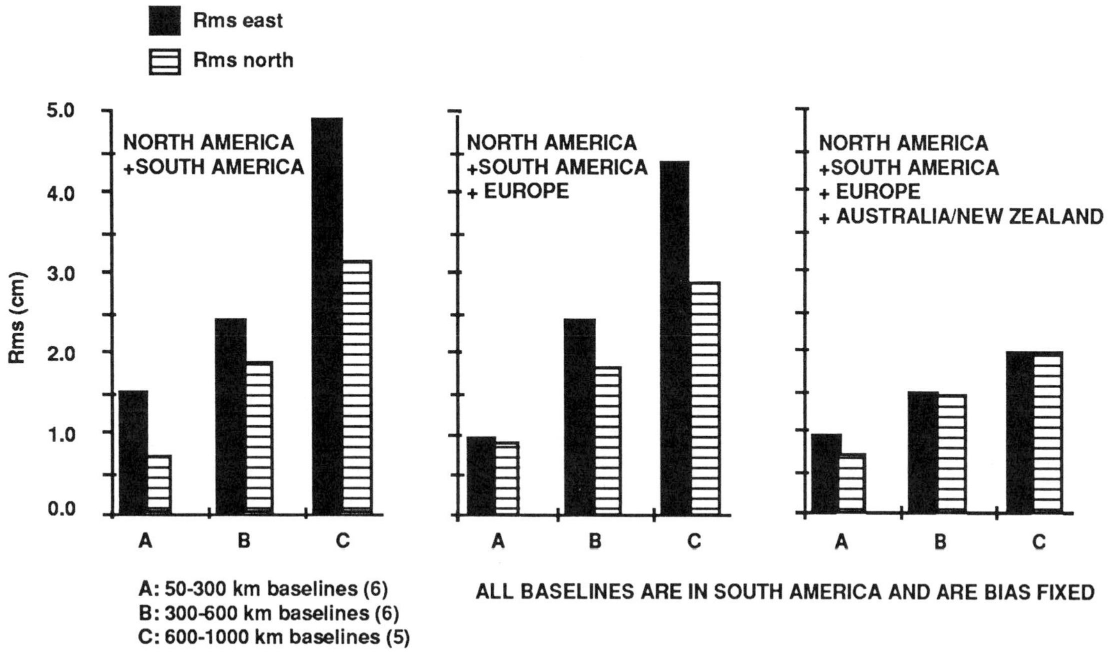

Fig. 12. Daily baseline repeatability for South American baselines with three tracking networks, showing improved precision with the more worldwide network. Rms values plotted are averaged over the five or six baselines in each of the three ranges of baseline length. From Kornreich et al. (1989).

GPS ORBIT ACCURACY: FUTURE PROSPECTS

By the mid-1990s, we expect that the 21-satellite GPS constellation will be complete. With a global tracking network of ten ground stations, GPS orbits should be routinely determined to 10-20 cm with just 12 hrs or less of tracking. This is illustrated in Fig. 13, which is based on a covariance analysis at JPL and shows, for comparison purposes, recent sub-meter orbit repeatability achieved with a regional North American network. The covariance analysis for the mid-1990s assumes that the ground receivers will be able to track up to 8 satellites, will use carrier phase, and will either operate in a code mode with high-precision (5 cm) pseudorange, or operate in a codeless mode with substantially noisier pseudorange (200 cm). This kind of pseudorange precision with both code and codeless operation has already been demonstrated in the field recently with new, advanced receiver and antenna designs such as the Rogue (Meehan et al. 1987; Blewitt et al. 1989), which can also track up to 8 satellites. A fiducial ground network consisting of the three NASA deep space tracking sites (which are presently being equipped with advanced Rogue receivers) was postulated to have 3 cm accuracy by the mid-1990s; judging from Table 1, even *current* performance with GPS may not far from this target. Seven additional receivers were assumed in the study to be deployed worldwide, evenly spaced.

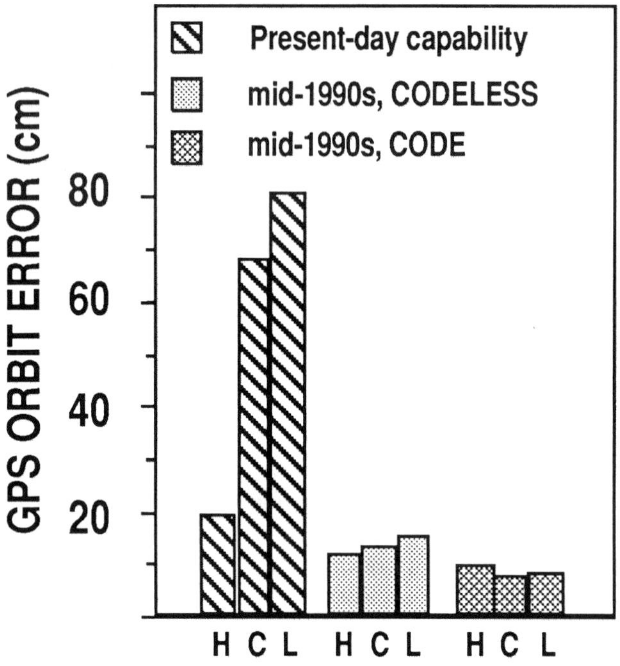

Fig. 13. Present-day GPS orbit precision (as judged by orbit repeatability from Lichten and Bertiger, 1989) achieved with a regional network, compared to expected GPS orbit accuracy in the mid-1990s using carrier phase and either precise (code) pseudorange or degraded (codeless) pseudorange with ground receivers distributed worldwide in a network consisting 10 sites. Orbit errors are shown for altitude (H), cross-track (C), and down-track (L). Present-day repeatability is based on multi-day orbit arcs, while the decimeter accuracy predicted in the 1990s is based on only 12-hr arcs.

Issues potentially affecting the accuracy of the orbits include: a) multipath degradation of the pseudorange; b) GPS force modeling; and c) military restrictions on the GPS signals. The effect of multipath can be minimized through antenna design to be consistent with the assumptions of this analysis (Meehan et al. 1987; Blewitt et al. 1989). The problems with

GPS force modeling will probably be small with short arcs of 12 hrs or less, but extra degrees of freedom could be added to account for additional GPS accelerations. The military restrictions could include selective availability, which is similar in effect to a highly variable clock, and anti-spoofing, which makes the code pseudorange unavailable without a classified key. The selective availability will not affect receivers used in the differential mode where double differencing or white noise clock estimation is performed as long as the receiver can maintain lock on the GPS signal. A properly designed receiver should be able to maintain lock and track continuously. At least two of the GPS satellites launched earlier in 1989 have been operating with selective availability and a number of different receiver types have been tracking them without difficulty. The anti-spoofing is potentially more serious, but a receiver like the Rogue receiver can still operate in a "codeless" mode, producing dual-band carrier phase, a linear combination of P1 and P2, and the C/A code. The information available from these data types in the codeless mode is comparable (or slightly better) than what is currently obtained with the P-code by older commercially available receivers such as the TI-4100. We expect, however, that the anti-spoofing will be activated only intermittently and that at most times, the full data strength will be available. Nonetheless, Fig. 13 shows that the penalty for tracking GPS in a codeless mode is not great.

Fig. 13 compares present-day GPS repeatability (which is a measure of precision) based on a regional network in North America only, with potential GPS global orbit accuracy for the next decade provided that a worldwide tracking network operates regularly. Note that no bias fixing is postulated in the covariance analysis from which the mid-1990s orbit errors in Fig. 13 were computed, since 10 widely spaced sites were assumed to be located worldwide. Regional GPS experimenters could use the global GPS orbits as initial values and (depending on the baseline lengths) further refine the orbits using the denser data from their regions of interest. These further refinements would include bias fixing, inclusion of additional data, and perhaps additional models. At some sites, a more complex tropospheric estimation or calibration scheme might be required. Current trends seem to favor wide dissemination of high-accuracy orbit determination software either generated in-house or purchased from other institutions. To reach the ultimate levels of accuracy, local regional refined GPS orbits will probably be generated regularly. The worldwide GPS scientific community needs to consider the formation of the organization which will establish common standards and be responsible for globally determined sub-meter GPS orbits which could serve as a central data base to initialize locally generated regional orbit and baseline solutions. Appropriate reference frames and models definitions will need to be agreed upon so that GPS data can be combined with VLBI and SLR data.

SUMMARY

Since the mid-1980s, GPS orbit determination accuracy has improved about an order of magnitude to better than 1 m. Two approaches have been used for GPS orbit estimation. The first relies on widely spaced, worldwide tracking sites, while the other emphasizes denser, regional networks, which can tune the orbits to a particular region. The more global tracking approach tends towards longer (multi-day) orbit arcs and more sophisticated GPS force modeling. The shorter regional experiments emphasize ambiguity resolution, first over short baselines, and eventually to longer baselines as the orbits improve as a result of the previous bias fixed lines. Both approaches can benefit from stochastic tropospheric delay estimation. Baselines determined with GPS have reached 1 part in 10^8 accuracy in the horizontal, and several parts in 10^9 for determination of baseline length. Vertical component accuracy is believed to be several times worse. For very short baselines, there is presently a non-orbital error component which is independent of baseline length amounting to about 0.5 cm. Baselines as long as 6000 km can be estimated with GPS, and a full worldwide capability is expected in a few years after more GPS satellites are launched.

There are several key components to further GPS orbit improvement. One is enhanced modeling for GPS forces and tropospheric delays. The need for complicated GPS force models may diminish as the constellation fills out and shorter observation sessions are needed. The troposphere will remain as a serious error source. Another is the development of advanced GPS receivers and antennas. Sub-decimeter pseudorange will enable significant orbit improvement over a wide range of baseline scales, particularly over longer distances where bias fixing is impractical. Multipath reduction for pseudorange will be necessary to exploit improved receiver performance, requiring improved antenna designs. Although the GPS signals may be restricted from time to time, appropriate receiver design utilizing codeless techniques will enable minimal loss of precision. As more GPS satellites are launched, we expect to rely more on global tracking networks for GPS orbit determination. Preliminary results already show that significant orbit improvements will result with global tracking. As global tracking becomes routine and GPS baselines begin to span longer intercontinental distances, it has become clear that a central international organization is needed to oversee the operation of a worldwide network and to define the reference frame and standards which will facilitate intercomparisons between different institutions using GPS as well as between GPS and other space geodetic techniques such as SLR and VLBI.

Acknowledgement. The work described in this paper was carried out by the Jet Propulsion Laboratory, California Institute of Technology, under contract with the National Aeronautics and Space Administration.

The author thanks B. Schutz, M. S. Kornreich, Y. Bock, G. Beutler, W. Strange, and W. Gouldman for providing assistance and unpublished materials which were used in this paper.

REFERENCES

R. I. Abbott, Y. Bock, C. C. Counselman III, R. W. King, S. A. Gourevitch, and B. J. Rosen, Interferometric Determination of GPS Satellite Orbits, *Proceedings First International Symposium on Precise Positioning with GPS-1985,* (ed. C. Goad), Vol. I, pp. 63-72, National Geodetic Information Center, NOAA, Rockville, MD.

G. Beutler, I. Bauersima, W. Gurtner, M. Rothacher, and T. Schildknecht, Evaluation of the 1984 Alaska Positioning System Campaign with the Bernese GPS Software, *Journal of Geophysical Research,* Vol. 92, pp. 1295-1303, Feb. 10, 1987.

G. Blewitt, Carrier Phase Ambiguity Resolution for the Global Positioning System Applied to Geodetic Baselines up to 2000 km, *Journal of Geophysical Research*, Vol. 94, pp. 10187-10203, Aug. 10, 1989.

G. Blewitt, T. P. Yunck, S. M. Lichten, W. I. Bertiger, and S. C. Wu, GPS Geodesy: A Status Report, in *High Precision Navigation*, ed. K. Linkwitz and U. Hangleiter (Proceedings of the International Workshop on High Precision Navigation, May 1988, Stuttgart, Germany), Springer-Verlag, Berlin, 1989 (pp. 74-85).

Y. Bock, 1989 (private communication).

O. Colombo, The Dynamics of Global Positioning System Orbits and the Determination of Precise Ephemerides, *Journal of Geophysical Research,* Vol. 94, pp. 9167-9182, July 10, 1989.

C. C. Counselman III, and R. I. Abbot, Method of Resolving Radio Ambiguity in Satellite Orbit Determination, *Journal of Geophysical Research,* Vol. 94, pp. 7058-7064, June 10, 1989.

D. Dong, and Y. Bock, GPS Network Analysis: Ambiguity Resolution, *EOS*, vol. 69, p. 325, April 19, 1988.

D. Dong and Y. Bock, Global Positioning System Network Analysis with Phase Ambiguity Resolution Applied to Crustal Deformation Studies in California, *Journal of Geophysical Research*, Vol. 94, pp. 3949-3966, April 10, 1989.

M. W. Gouldman, B. R. Hermann, and D. L. Weedon, Evaluation of GPS Production Ephemeris and Clock Quality, Proceedings for the 5th International Geodetic Symposium on Satellite Positioning, pp. 210-222, March 13-17, 1989, Las Cruces, NM.

M. S. Kornreich, T. Dixon, and J. T. Freymueller, The Effect of Tracking Network Configuration on GPS Baseline Estimates for the CASA UNO Experiment, *Geophysical Research Letters*, 1989 (submitted).

S. M. Lichten and W. I. Bertiger, Demonstration of Sub-Meter GPS Orbit Determination and 1.5 Parts in 10^8 Three-Dimensional Baseline Accuracy, *Bulletin Geodesique*, Vol. 63, pp. 167-189, 1989.

S.M. Lichten and J.S. Border, Strategies For High Precision GPS Orbit Determination, *Journal of Geophysical Research,* Vol. 92, pp. 12751-12762, Nov. 10, 1987.

S. M. Lichten, U. J. Lindqwister, and W. I. Bertiger, The Effect of Fiducial Network Strategy on High-Accuracy GPS Orbit and Baseline Determination, Proceedings for the 5th International Geodetic Symposium on Satellite Positioning, pp. 516-525, March 13-17, 1989, Las Cruces, NM.

T. K. Meehan et al., ROGUE: A New High Accuracy, Digital GPS Receiver, IUGG Conference, Vancouver, Canada, Aug. 1987.

W. Melbourne, R. Anderle, M. Feissel, R. King, D. McCarthy, D. Smith, B. Tapley, and R. Vicente, Project MERIT Standards, U.S. Naval Observatory Circular No. 167, 1983.

G. E. Powell and E. M. Gaposchkin, Modelling of Non-Gravitational Effects on GPS Satellites, Proceedings of the AIAA/AAS Astrodynamics Conference, paper AIAA-88-4291-CP, August 15-17, 1988, Minneapolis, Minnesota (AIAA, Washington, D. C.).

B. E. Schutz, B. D. Tapley, C. S. Ho, H. J. Rim, and P. A. M. Abusali, GPS Orbit Determination: Experiments and Results, Proceedings for the 5th International Geodetic Symposium on Satellite Positioning, pp. 201-210, March 13-17, 1989, Las Cruces, NM.

W. E. Strange, M. D. Abell, and E. G. Zurflueh, The Establishment of a Primary GPS/VLBI Strain Monitoring Network in the Eastern United States, *EOS*, vol 70, p. 308, April 11, 1989.

W. E. Strange and E. G. Zurflueh, Initial Results and Evaluation of a GPS Crustal Motion Network for the Eastern Two Thirds of the United States, Chapman Conference on GPS Measurements for Geodynamics, September 19-22, 1988, Fort Lauderdale, Florida (American Geophysical Union).

E. R. Swift, NSWC's GPS Orbit/Clock Determination System, *Proceedings First International Symposium on Precise Positioning with GPS-1985,* (ed. C. Goad), Vol. I, pp. 51-62, National Geodetic Information Center, NOAA, Rockville, MD.

E. R. Swift, Mathematical Description of the GPS Multisatellite Filter/Smoother, Naval Surface Warfare Center, NSWC TR 87-187, Dahlgren, Virginia, Oct. 1987.

D. M. Tralli, T. H. Dixon, and S. A. Stephens, The Effect of Wet Tropospheric Path Delays on Estimation of Geodetic Baselines in the Gulf of California Using the Global Positioning System, *Journal of Geophysical Research,* Vol. 93, pp. 6465-6557, June 10, 1988.

S. C. Wu, T. P. Yunck, and G. A. Hajj, Toward Decimeter TOPEX Orbit Determination Using GPS, AAS/AIAA Astrodynamics Specialist Conference, August 7-10, 1989, paper AAS 89-359, Stowe, Vermont.

T. P. Yunck, S. C. Wu, S. M. Lichten, W. I. Bertiger, U. J. Lindqwister, and G. Blewitt, Toward Centimeter Orbit Determination and Millimeter Geodesy With GPS, Proceedings for the 5th International Geodetic Symposium on Satellite Positioning, pp. 272-281, March 13-17, 1989, Las Cruces, NM.

THE DEVELOPMENT OF A GLOBAL GPS TRACKING SYSTEM IN SUPPORT OF SPACE AND GROUND-BASED GPS PROGRAMS

R.E. Neilan, W.G. Melbourne
Jet Propulsion Laboratory,California Institute of Technology, Pasadena, CA 91109 USA
G.L. Mader
National Geodetic Survey, Rockville MD 20852 USA

INTRODUCTION

A GPS Global Tracking System for the 1990's is called for to maintain accurate satellite ephemerides and terrestrial reference frames which support a broad class of geodetic, geophysical, oceanographic and engineering applications. The rationale for an international and civilian-operated GPS tracking system is quite evident in terms of its diverse applications and significant benefits to the scientific community. A description follows for a proposed organizational structure, physical configuration, operational characteristics and user interfaces, which together constitute a highly reliable internationally sponsored GPS Global Tracking System.

GPS Tracking Networks

The first precise post-fit ephemerides from a global GPS tracking network were made available to the civilian community in 1983, by the Naval Surface Warfare Center (NSWC) and the Defense Mapping Agency (DMA). These ephemerides were initially estimated using data from a network of five stations and demonstrated a consistency at that time between weekly solutions of 10-20 meters. This network now includes ten sites world-wide located at Colorado Springs, Ascension Island, Diego Garcia, Kwajalein, Hawaii, Australia, Argentina, England, Bahrain, and Ecuador. Orbital position comparisons show much improvement over the last six years with current weekly solutions indicating agreement at the 2-3 m level [Gouldman, et al., 1989]. Data from this network is also used for the daily determinations of the Broadcast Ephemerides.

In 1984 various universities and institutions began to develop in-house GPS orbit determination capabilities with the goal of obtaining higher levels of precision and accuracy. These efforts spawned the evolution of the fiducial concept [Thornton, Fanselow and Renzetti, 1986]. The fiducial technique places three to four receivers at sites having well-defined coordinates in a consistent reference frame with positions determined by methods of VLBI or SLR, and other GPS receivers at new locations. These Fiducial sites are held fixed while the coordinates of the new sites, including orbital states and other parameters are estimated simultaneously. The March '85 High Precision Baseline Test was the first interagency experiment coordinated by JPL designed to test the fiducial concept [Davidson, et al., 1985]. This test, and subsequent experiments, successfully proved the robustness of the fiducial concept within the United States using as fiducials the well known reference sites of Haystack, Richmond and Ft. Davis [Beutler, et al., 1986]. GPS receivers located at other non-fiducial sites such as Owens Valley Radio Observatory

(OVRO), Hatcreek and Mojave, (all VLBI sites) enable independent comparison between GPS and VLBI solutions. The approach has proven highly successful for these regions with good satellite observing geometries; it has since yielded baseline accuracies on regional scales of several millimeters in the horizontal components when ephemeris information based upon multi-day arcs is used and when the region of geodetic interest is embedded within or near the fiducial network [Blewitt, 1988, Dong and Bock, 1988]. On the other hand, baseline accuracies in many regions of the world of tectonic interest, including third world areas and across oceanic basins where fiducial sites will be sparse, are not likely to fare as well with this approach.

In 1986, the U.S. National Geodetic Survey NGS began to concentrate on acquiring GPS data on a continuous basis at fiducial locations to support the increasing use of GPS receivers by the surveying, engineering and geodetic communities. The Cooperative International Geodetic Network (CIGNET) was thus formed in late 1986 and early 1987. This is a cooperative venture of several international geodetic agencies interested in accessing raw GPS tracking data in support of geodetic operations. The sites are concentrated in the Northern hemisphere, located in the U.S, Canada and Europe. This network is currently a mix of GPS instrumentation (both P-code and P-codeless), antenna types and methods of data retrieval. The Mini-Mac receivers at the U.S. fiducial locations have a telephone access for data retrieval, and all other data from the CIGNET sites are sent by mail to the NGS on a weekly or bi-weekly basis. Data are available on request on magnetic tape medium [Chin 1989], and all data are archived by the NGS. The participants of this cooperative network are listed in table 1.

Table 1. The locations and sponsors of CIGNET stations are listed along with the types of GPS receivers installed and the initial date that the station began to record data. (Some of the earlier stations have had changes in instrumentation since the first date, replacing TI-4100 with codeless Mini-Mac receivers).

STATION	DATE	SPONSOR
WESTFORD^{m}, MA	10/86	U.S. NATIONAL GEODETIC SURVEY (NGS)
MOJAVE^{m}, CA	11/86	UNIV. OF TEXAS, NGS, U.S. GEOLOGICAL SURVEY (USGS)
YELLOWKNIFE^{t}	1/87	GEODETIC SURVEY OF CANADA
WETTZELL^{m}, FRG	11/87	INSTITUT FUR ANGEWANTE GEODASIE (IFAG)
TRÖMSO^{r,t}, NORWAY	12/87	STATENS KARTVERK
ONSALA^{t}, SWEDEN	12/87	ONSALA SPACE OBSERVATORY, SWEDEN STATENS KARTVERK, NORWAY
RICHMOND^{m}, FL	2/88	NGS, U.S. NAVAL OBSERVATORY
KOKEE^{t}, HI	4/88	NGS, NASA/GSFC CRUSTAL DYNAMICS PROJECT
TSUKUBA^{m},	7/88	GEOGRAPHICAL SURVEY INSTITUTE, JAPAN MINISTRY OF CONSTRUCTION (CRL)

^{t} TI-4100 ^{m} MINI-MAC ^{r} ROGUE SNR-8

GPS tracking stations are also being installed at the three NASA/JPL Deep Space Communication Complexes (DSCC) which make up the Deep Space Network (DSN). These three sites are located at Goldstone, California; Canberra, Australia; and Madrid,

Spain. GPS Rogue receivers and ancillary equipment are being installed as part of the DSN Media Calibration Subsystem. This subsystem of the DSN is initially intended to provide highly accurate, near real-time ionospheric calibration data for deep space navigation of spacecraft that typically transmit at a single frequency. The ionospheric calibrations and maps are derived from dual-frequency GPS observations of up to eight satellites. GPS tracking support for the Topex/Poseidon GPS Precise Orbit Determination Program requires at least a six-station global configuration of precise P-code type receivers [Melbourne and Davis, 1987]. Thus, the three DSN stations will be complemented by three additional stations currently being installed through cooperative agreements with the French and Japanese space agencies. ISAS, Institute of Space and Astronautical Science, will be installing a Rogue GPS receiver at Usuda Deep Space Station in Japan. The French space agency, Centre National d'Etudes Spatiales (CNES) is proceeding with plans to install a similar receiver in Africa. Proposed locations for a South American station are being reviewed at JPL, and candidate locations are Santiago, Buenos Aires or Sao Paulo. The six station network will be operational by June 1991 (figure 1).

Table 2. Listed in the following table are the sponsors, dates, locations and instrumentation in support of the TOPEX/Poseidon Precise Orbit Determination Program. A combination of these sponsors with those listed in table 1 constitute the beginning of the global GPS Tracking Network for the 1990's. Because of national interests the network is likely to contain additional sites. Figure 1 shows station locations, including other potential tracking sites.

STATION	DATE	SPONSOR
GOLDSTONE[r],CA	12/89	NASA/JPL
CANBERRA[r], AUSTRALIA	1/90	NASA/JPL
MADRID[r], SPAIN	2/90	NASA/JPL
USUDA[r], JAPAN	4/90	INSTITUTE OF SPACE AND ASTRONAUTICAL SCIENCE
HARTEBEESTHOEK[r], AFRICA	12/90	CENTRE NATIONAL D'ETUDES SPATIALES
SOUTH AMERICA[r]	2/91	to be determined

[r] ROGUE SNR-8

Another type of GPS tracking network is created by the temporary deployment of tracking sites in support of specific GPS experiments, mostly for geodynamical investigations. A crucial component that has contributed to the success of these campaigns is the configuration, reliability and data quality of GPS stations used in the ad hoc tracking network. The CASA Uno '88 campaign is a good demonstration of the scientific community's need for a global GPS tracking network. CASA Uno '88 was an international GPS experiment involving 13 countries, 26 agencies and over 40 GPS receivers coordinated by JPL in January 1988. The scientific study in this experiment was the tectonically interesting area of Central and South America, however the GPS satellite constellation at that time offered poor satellite visibility and geometry over this area. As a result, JPL implemented a number of GPS tracking sites in the South Pacific, augmenting the existing CIGNET stations in order to achieve higher precision orbits. Results from the analysis of the GPS data show a factor of 2 improvement in the formal

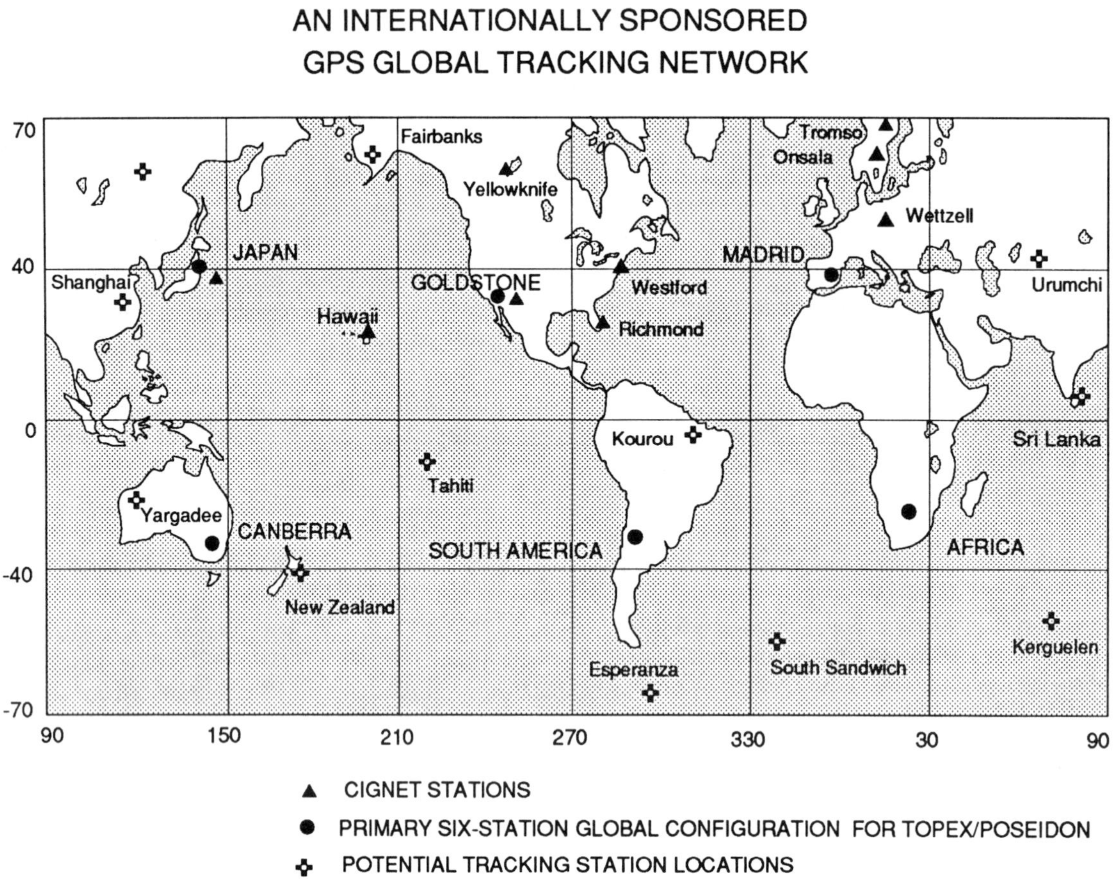

Figure 1. GPS Global Tracking Network station locations, existing, planned and potential.

satellite orbit errors and a marked improvement in daily baseline repeatability for the South American baselines, particularly for baselines between 600-1000 km where the average rms repeatability for horizontal components of the bias-fixed baselines improved from 4 cm to 2 cm [Kornreich Wolf, et al., 1989, Lichten 1989]. These kinds of results demonstrate the benefits of an extended tracking network and illustrate the utility of a truly global system in support of the international scientific and geodetic communities. The relevance of temporary tracking networks will diminish as the permanent global network is established.

GPS IN THE 1990'S

The 1990's will bring a host of planned flight and ground observation programs which reflect an increasing urgency in acquiring timely information about fundamental physical and biological processes in the troposphere and stratosphere, in the oceans, and in the solid Earth. GPS will play an important role in some of these observation programs through its proven ability to provide centimeter accuracy in differential positioning. These Earth observations programs are international in origin involving all of the major space agencies, geodetic agencies and numerous other organizations that will be dependently linked through the requirement for a global tracking system.

Ground-based Programs

Low cost, high accuracy GPS-based geodetic systems for regional deformation studies are proliferating dramatically. Their costs are two orders of magnitude below those of VLBI and SLR systems and yet they are providing comparable accuracies. During the early 1990's high density continuously operating and remotely monitored arrays of GPS receivers will be in place in key regions of tectonic interest for measuring strain distribution rates with spatial resolutions of a few kilometers up to several hundred kilometers and temporal resolutions of a few hours over periods of years. GPS arrays on ocean platforms for seafloor geodesy [Speiss, 1987] and sea level monitoring will be extensively used. Geophysical and geodetic campaigns involving arrays of portable GPS receivers will be mounted in increasing number. A recommendation of the NASA-sponsored Coolfont conference on Solid Earth Sciences is the establishment of FLINN (Fiducial Laboratory for an International Natural Science Network), a 200-station, international network of geophysical GPS sites that will continuously monitor earth strain, deformation and related geopotential variations. The challenging goals of 1 mm accuracy for regional geodesy and 1 cm accuracy for global geodesy were also proposed by the scientific community at the Coolfont conference [Bosworth, 1989].

The GPS ephemerides and reference frame information for the ground programs are currently provided through regional networks tracking the GPS satellites and collocated at VLBI sites. With the full deployment of the Block II constellation in the early 1990's, a promising alternative approach to GPS ephemeris and reference frame maintenance will be the use of the global tracking network to provide a global framework based upon the dynamical consistency of the satellite ephemerides and key reference frame controls.

Space Missions

Flight missions planning to use GPS for high accuracy differential positioning have grown considerably since the first navigation demonstration on Landsat 4 in 1982. The NASA/CNES-sponsored TOPEX/Poseidon altimetric mission for mapping dynamic ocean topography, planned for 1992, will be the first to use GPS as a flight experiment for precision orbit determination. The objectives of TOPEX/Poseidon are to obtain global ocean circulation information, both average and variable, over a nominal three year lifetime. In addition to the Monarch GPS receiver aboard TOPEX/Poseidon [Carson, et al., 1988], this flight experiment will include a global network of six or more concurrently operating GPS ground tracking stations referred to above. Solutions will be expressed in a terrestrial reference frame that uses a priori defined locations of key fiducial sites where GPS trackers are collocated at VLBI and/or SLR sites. Differential positioning of TOPEX/Poseidon at 5-10 cm accuracies and significant gravity recovery up to wavenumbers of about 25 are expected from this flight experiment.

NASA's proposed Gravity Probe B, for 1995, will use GPS and a ground tracking network for navigation and for gravity recovery up to wavenumber 60 [Everitt, et al., 1989]. NASA's proposed Eos-A and Eos-B polar platforms, for 1997 and 1999, will use GPS and a ground network for centimeter-level satellite positioning and geodesy, attitude determination, ionospheric tomographic studies, and temperature profiling of the atmosphere [Yunck and Melbourne, 1990]. ESA's planned Aristoteles gravity gradiometry mission, scheduled for 1995, may use GPS for navigation and to complement the gravity information obtained from gradiometry.

BENEFITS OF A GPS GLOBAL TRACKING SYSTEM

In order to achieve the goals for both gound-based and space programs, it is necessary to establish a highly reliable, continuously operational, near real-time GPS Global Tracking System that will support missions like those listed above. The System will provide a worldwide service designed to meet the different levels of accuracy and precision required. Covariance studies of this system predict that ephemeris accuracies of a few parts in 10^9 could be obtained from a network of roughly a dozen globally well-distributed tracking sites. Accuracies at the 1-5 mm level would result for the horizontal components of ground baselines from regional scales up to continental scales. The aggregate benefit in cost savings to the users of this System will be enormous. Many of the independent tracking and computational efforts would not be required on an operational basis provided the System met its functional and performance goals. Also, the System will provide standards for all users as well as a conventional terrestrial reference frame for GPS-based geodesy. Its tracking and ephemeris products will facilitate an enhanced time resolution of short term variability in crustal deformation using continuously monitored regional arrays. It will provide valuable earth orientation information to the International Earth Rotation Service (IERS). Although GPS and SLR depend on VLBI for maintaining the inertial reference system and for eliminating possible longitude/right ascension drifts inherent in dynamical systems, the introduction of GPS into Earth orientation monitoring will fundamentally change the mix of these three techniques for this application.

A low Earth orbiter (LEO), such as TOPEX/Poseidon, bearing a high accuracy GPS receiver that is concurrently tracking would provide valuable additional information for ground programs through increased geometrical strength in the aggregate observational scenario. The LEO mediates the transfer of tracking information between ground stations by simultaneously observing GPS satellites that are not co-visible from the ground stations, thereby expanding the range of tracking intervals over which concurrent observations (and hence clock cancellation) can be made. The rapid parallax resulting from its relatively high angular motion leads to shortened resolution times for the ground stations. Also, the LEO to ground station baseline has on average a large "vertical" component and is corrupted by tropospheric errors at only one end of the baseline. This will result in improved accuracy of the vertical components of baselines within ground arrays by a factor of two in certain cases, and it may reduce the sensitivity of the vertical components of the ground baselines to mis-modeled tropospheric delay. The LEO also enhances the accuracies of the global solutions for the GPS ephemerides.

The principal benefits that the majority of the international community will accrue through the global tracking system are:

1. Cost savings through reduced operations and computation tasks;
2. Standardization of the Terrestrial Reference Frame for GPS applications;
3. Standardization of data management and information systems;
4. Timely tracking and ephemeris products through a world-wide distribution system;
5. Increased accuracy, particularly for regions with weak fiducial constraints
6. Shortened resolution time for a regional array to achieve a given accuracy level;
7. Complementary Earth orientation information for the IERS;
8. Enhanced accuracy and temporal resolution from concurrent tracking by LEO's.

Essentially, the formation of an internationally sponsored network with internationally developed standards of performance, operation and response is called for in the 1990's.

During NASA's Crustal Dynamics Project Principal Investigators Meeting held in Munich, FRG in October 1988, a meeting was held establishing a working group within the Commission VIII International Coordination of Space Techniques for Geodesy and Geodynamics (CSTG) GPS Subcommission that would begin to address the formulation of an internationally sponsored tracking network that would insure global coverage. The main thrust of the working group is to address specifications on the operational aspects of the GPS tracking stations and to contribute to the evolution of an international tracking network.

ORGANIZATION OF A GLOBAL TRACKING SYSTEM

The GPS Global Tracking System that is envisioned will be internationally sponsored and monitored to insure operational reliability, data quality and technical integrity for all users or applications of the system. The organization is modelled after the coordinating and computational organizations used in the IAU/IUGG-sponsored MERIT Project (Monitoring Earth Rotation and Intercomparison of Techniques), and it would operate in a manner similar to the current IERS (International Earth Rotation Service) for Earth orientation. The IERS will be a primary recipient of the data products provided by the System. The System will be functionally divided into four major components (figure 2):

1. International Governing Board
2. Operational Network Center and possibly regional centers
3. Global network of GPS tracking stations
4. Computational Centers providing products to users

The International Governing Board

The International Governing Board would set policy and maintain standards of operation. It would ensure that internationally equitable arrangements are maintained in supporting and operating the System, and that the appropriate interfaces are maintained between the System and its users.

The Operational Network Center

The principal functions of the Operational Network Center are to coordinate implementation, operations, data collection from the tracking stations (such as GPS, meteorological, water vapor radiometer, clock, real-time navigation message, etc.), and data dissemination to the Computational Centers; to conduct real or near real-time monitoring of the tracking network performance, to maintain configuration control and standards, including facilities control and security, and to coordinate system development, sustaining engineering and maintenance. The Operational Network Center is responsive to the computational centers and maintains operational interfaces with the IERS.

GPS Tracking Stations

The GPS tracking stations and local network communication facilities would be operated by the cooperating host countries and agencies, in compliance with policies set by the Governing Board and under the direction and coordination of the Network Center. Three

ORGANIZATION FOR AN INTERNATIONALLY SPONSORED GPS GLOBAL TRACKING SYSTEM

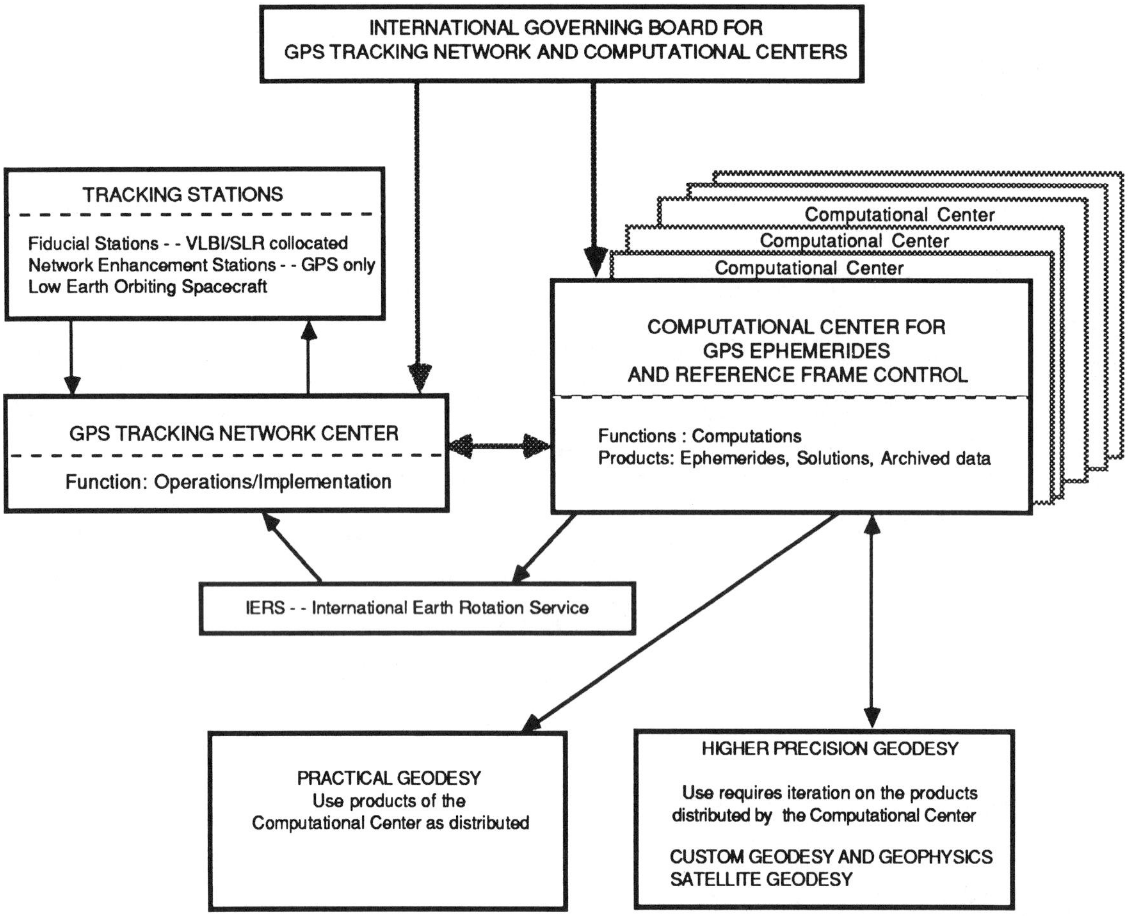

Figure 2. Simplified organization structure of the GPS Tracking System.

types of stations are envisioned. The primary Fiducial stations are collocated with VLBI and SLR and have a priori baseline vectors available, a priori relative velocities, ties to the inertial reference frame, existing hydrogen maser frequency standard, existing water vapor radiometers and many other operational advantages. The second type are Network Enhancement Stations with GPS instrumentation (non-fiducial or possibly secondary VLBI stations) that provide a homogeneous, global distribution of sites, enhance mutual visibility of the fiducial stations, provide increased geometrical strength of the network, and allow for redundancy and system reliability. The third type is Low Earth Orbiting Spacecraft which provide increased geometrical strength, enhanced tracking observations between ground stations, shortened resolution times for ground stations, and improved accuracy of baseline vertical components.

Computational Centers

The functions of the Computational Centers would be data editing, parameter estimation, tracking data archiving and distribution, off-line monitoring of network performance, generating tracking data predicts, and system analysis and research. Its major products would be ephemerides (0.01 ppm - 20 cm or better), tracking station locations and velocities, a geocentric reference frame, Earth orientation information, global solutions including full information and covariance matrices, and the edited tracking and calibration data. These functions include transmitting the raw data to other regional computational centers.

System Users

The users of the System naturally fall into two categories: Practical Users and Custom Users. Practical Users, using the ephemeris products and station coordinates as produced by a central or regional Computational Center, would benefit from significant cost-savings as well as accuracy and stability. The practical user group would include applications such as continuously operating arrays, reference frame ties between pre-existing networks, site stability studies, monitoring volcanic uplift, deformation studies near plate boundaries, pre- and post-seismic strain monitoring, engineering applications, seafloor geodesy, dynamic positioning, precision navigation and mapping. Practical Users would take the products of the Computational Center as they stand; their accuracy or functional requirements do not compel them to make any adjustments to the products.

Custom Users have accuracy requirements as stringent as those of the System itself and they may have other functional requirements. Custom geodesy and geophysics include continuous arrays for millimeter accuracy monitoring of crustal strain, plate tectonics, VLBG (Very Long Baseline GPS), development of estimation strategies, ionospheric studies, etc. Custom satellite geodesy users include TOPEX/Poseidon orbit determination, IERS, Eos, Space Station, Gravity Probe-B, and the Space shuttle. Custom Users might further process the tracking data or the ephemeris solutions and information matrices to refine the accuracy and time resolution of solutions from their own networks. For example, the IERS will use these products to further strengthen its Earth orientation monitoring function.

GPS GROUND STATIONS -- DESCRIPTIONS AND LOCATIONS

The products available from the Computational Center are generated from continuous tracking of both carrier phase and high accuracy pseudorange at a set of globally distributed tracking stations. Currently, the primary limiting error source for baselines longer than about 100 km is the uncertainty in the position of the GPS satellites. Figure 3 shows the improvement in GPS satellite position errors as the number of ground tracking stations increases [Melbourne, et al., 1988]. Covariance analyses indicate that a primary six station network, referred to above and highlighted in figure 1, is the minimum ground station configuration for global orbit determination and for determining the position of earth orbiting spacecraft. These stations are closely enough spaced to have ample satellite mutual visibility for eliminating clock errors. Ideally, particularly for precise geodynamical applications, about a dozen stations would complete the international network, if the additional stations are uniformly distributed. There are a number of assumptions that should be noted with regards to the covariance analysis results depicted in figure 3, particularly with regard to future ground station implementation. The analysis assumed that the ground stations were instrumented with high accuracy P-code receivers

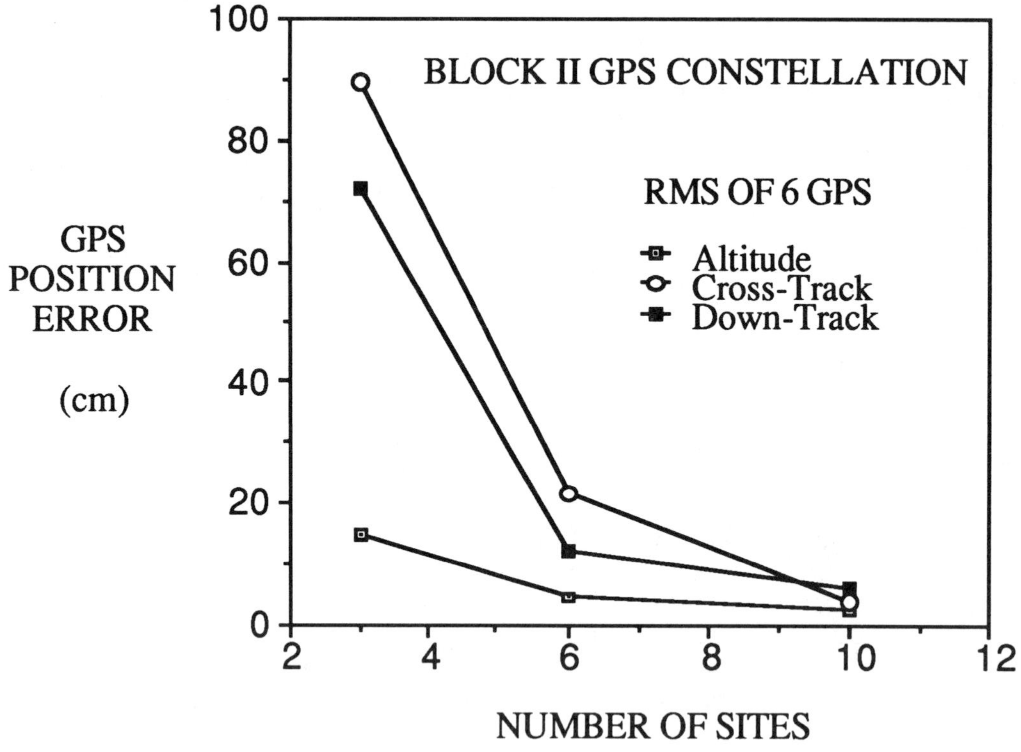

Figure 3. Tracking performance improvement with an increased number of GPS Ground Tracking Stations instrumented with precision P-code GPS receivers. RMS of 6 Block II satellites averaged over 1 orbit.

demonstrating low noise pseudorange, high suppression of environmental multipath, and a 10° elevation cutoff.

The Rogue GPS Receiver being implemented for the Topex/Poseidon mission is a third generation, advanced GPS receiver developed at JPL [Thomas, 1988]. The Rogue is a dual frequency receiver which simultaneously tracks up to eight GPS satellites using advanced digital signal processing techniques. Output data is simultaneous C/A-code and P-code (P1 and P2 group delay) pseudorange observables as well as carrier phase observables on L1 and L2. The Rogue is also unique in that it can operate in both a code and 'non-code' mode. The non-code mode employs a cross-correlation method that automatically activates for one or more satellites if P-code encryption or Anti-Spoofing (AS) is enabled or detected. The data that are collected in the codeless mode are C/A code, L1 carrier phase, differential P2-P1 group delay, and differential carrier phase L1-L2 for ionospheric corrections. This allows the L2 carrier to be recovered instead of its second harmonic which most codeless receivers recover. This hybrid operation allows versatile and continuous use of the dual frequency GPS satellite transmissions for ionospheric calibration and time transfer. Although there is no guarantee, information from the Department of Defense suggests that AS will infrequently be enabled; perhaps two weeks per year for testing purposes and possibly with advance notice to users of the system (barring of course any national crisis situation). The Rogue production units are being manufactured by Allen Osborne Associates, Incorporated, Westlake Village, CA.

Why has NASA been so interested in developing a GPS receiver that can acquire precise P-code pseudorange? The Rogue receiver, operating in its code mode with an advanced antenna/backplane and sited in a low multipath environment, has demonstrated an accuracy of 10 cm on smoothed P-code pseudorange over 2 minute averages and better than 5 cm on 30 minute averages. A number of reasons and applications favor P-code tracking receivers that provide both precise pseudorange and carrier phase:

1. Accurate pseudorange with carrier phase can improve orbit determination capability by a factor of 2-3 in a global tracking network [Lichten, 1987]
2. Accurate dual band pseudorange permits absolute determination of ionospheric delay and nanosecond accuracy time transfer,
3. Accurate P-code strengthens network solutions and allows bias fixing on long geodetic baselines, and also aids in rapid editing of the carrier phase data,
4. Accurate P-code allows dynamic applications of GPS, such as aircraft attitude determination, seafloor geodesy, and other kinematic platform positioning requirements
5. P-code tracking receivers operate satisfactorily, i.e., recover carrier phase to the requisite accuracy, under harsher signal conditions than codeless receivers, during episodes of large and scintillating ionospheric conditions [Srinivasan, et al., 1989]

Collocation of the GPS tracking stations at VLBI sites is preferable for operational and logistical reasons, including use of their ultra-stable frequency standards and timing. The network would use conventional standards such as the nutation series and the constant of precession. For user convenience, a current tectonic plate motion model could be included as part of the reference system. Earth orientation information from an Earth monitoring system such as the IERS would be incorporated to maintain alignment with the Conventional Terrestrial System; Polar motion and length of day (LOD) variability would be detected through the diurnal rotation of the Earth and provided to the IERS. One should use the a priori site location information provided from VLBI and SLR systems wherever practical to minimize potential systematic errors in the GPS system. But this a priori information need not dominate the solution set for station locations as it does in the regional fiducial approach for ephemeris control because of the dynamical information derived from the global network. In its extreme form, the only non-GPS observational information that the Global Tracking System would need to maintain a consistent reference system is the longitude of a single GPS tracking station and UT1 to tie down the degree of freedom associated with an unobservable rotation of the combined GPS constellation and Earth system; SLR uses a similar approach.

COMMUNICATION, DATA FLOW AND ACCESS

Details of performance standards, configuration control, user interfaces, maintenance and sparing philosophies will be guided by the participating host agencies implementing the GPS ground stations and by the GPS Global Network Working group. The DSN plan for handling the TOPEX/Poseidon GPS data, from both ground and flight receivers, is exemplary of the type of network envisioned, and may well become the skeletal structure of the network. International space agencies, having demonstrated that continuous operations and configuration control can result in a high level of confidence and data return, serve as both suitable partners in the endeavor and practical models for implementing permanent GPS stations in areas where an active space or geodetic agency is nonexistent. The geodetic agencies of the world also play a very strong role in implementation and use of the system, particularly due to the impact that GPS technology is having on conventional geodetic techniques.

The current plan for TOPEX/Poseidon support is for the data from the sites to be transmitted daily over existing high speed (9.6 Kbps) communication lines or via satellite links to the navigation computer (NAV) at the Network Operations Control Center (NOCC) located at JPL in Pasadena, California. Data are monitored both at the sites and at the NOCC for continuous quality control and for rapid response in the event of a system failure.

The navigation computer could handle data from up to ten ground stations, plus the TOPEX/Poseidon GPS flight receiver, the Monarch produced by Motorola. The expected quantity of data for the primary six stations plus the GPS flight receiver is about 101 Mbytes daily (5 Mbytes a day from each ground stations at a 10 second data rate, however 60% of the data is generated by the GPS flight receiver). As the data flows into the navigation computer, the daily archive, including the initial files from preprocessing, editing and conditioning, is about 325 Mbytes. A twice weekly archive of 755 Mbytes is planned. The operational plan is to archive the raw, conditioned and final data output (ephemerides, etc.). Plans are developing to enable primary participating agencies access to the data from the network, and those agencies would in turn disseminate the data to the users they support.

This system will be in place by June 1991, and in the interim it is desirable to enable access to the GPS data acquired by the new sites as they are installed and are proven viable. This will allow access to the raw network data on a weekly basis as the overall System is developing.

CONCLUSION

With the full deployment of the Block II constellation in the early 1990's, a promising alternative approach to GPS ephemeris production and reference frame maintenance will be the use of the GPS Global Tracking System. This System will provide a global framework based upon the dynamical consistency of the satellite ephemerides and key reference frame controls. Implementation of the System will result in an aggregate benefit in cost savings and enhanced accuracies, which will be enormous for most users. Many of the independent tracking and computational efforts would not be required on an operational basis provided the System met its functional and performance goals. The System can provide standards for all users as well as a conventional terrestrial reference frame for GPS-based geodesy.

The realization of such an Internationally sponsored system will take a great deal of cooperation, dedicated effort and resources, similar to the MERIT project. The development and evolution of the system will not occur overnight, but a realistic goal is a primary, real-time network functioning by January 1992.

REFERENCES

Beutler, G., Gurtner, W., Rothacher, M., and Schildknecht, I. (1986). Evaluation of the March 1985 High Precision Baseline (HPBL) Test: Fiducial Point Concept Versus Free Network Solutions, EOS Trans. AGU, Vol. 67, p. 911.

Blewitt, G. (1989). Carrier Phase Ambiguity Resolution for The Global Positioning System Applied to Geodetic Baselines up to 2000 KM, J. Geophys. Res., in press.

Bosworth, J. (1990). Solid Earth Sciences Program Plan (NASA document, in preparation).

Carson, L., Davis, R., Geier, G., Hailey, L., Huth, G., Munson, T. (1988). Design and Predicted Performance of the GPS Demonstration Receiver for the NASA TOPEX Satellite, Proc.. IEEE PLANS 88, Orlando, FL

Chin, M., editor, (1989). GPS Bulletin, CSTG GPS Subcommission, NGS, Vol 2 No. 5, Rockville MD

Davidson, J.M., Thornton, C.L., Vegos, C.J., Young, L.E., Yunck, T.P. (1985). The March 1985 Demonstration of the Fiducial Network Concept for GPS Geodesy: a Preliminary Report, Proceedings, First International Symposium on Precise Positioning with the Global Positioning System, Rockville MD.

Dong, D., and Bock, Y. (1989). GPS Network Analysis with Phase Ambiguity Resolution Applied to Crustal Deformation Studies in California, J. Geophys. Res., in press.

Everitt, C., Breakwell, J., Tapley, M., DeBra, D., Parkinson, B., Smith, D., Colombo, O., Pavlis, E., Tapely, B., Nerem, R., Yuan, D., and Melbourne, W. (1989).Gravity Probe B As A Geodesy Mission and Its Implications for TOPEX, CSTG Bull. No. 11.

Gouldman, M.W., Hermann, B., and Weedon, D. (1989). Evaluation of GPS Production Ephemeris and Clock Quality, Proceedings 5th International Geodetic Symposium on Satellite Positioning, pp 210-222, Las Cruces NM.

Kornreich Wolf, S., Dixon, T.H., Freymueller, J.T. (1989). The Effect of Tracking Network Configuration on GPS Baseline Estimates for the CASA Uno Experiment, submitted to Geophys. Res.

Lichten, S., and Border, J. (1987). Strategies for High-Precision Global Positioning System Orbit Determination, J. of Geophys.. Res., Vol. 92, No. B12, p. 12751-12762.

Lichten, S. (1989). High Accuracy Global Positioning System Orbit Determination: Progress and Prospectives, International Association of Geodesy, General Meeting Edinburgh Scotland (Springer-Verlag, in press)

Melbourne, W., and Davis, E.S. (1987). GPS-Based Precision Orbit Determination: A TOPEX Flight Experiment, AAS/AIAA Astrodynamics Specialist Conference, Paper AAS-87-430, Kalispell, Montana.

Melbourne,W., Blewitt, G., Lichten, S., Malla, R., Neilan, R., Schutz, B., Wu, S. (1988). Establishing a Global GPS Tracking System for Fiducial Control and Ephemeris Production, Paper No. 621A-02, AGU Meeting, May 1988. See also EOS Trans. AGU, Vol. 69, No. 16, p. 323.

Spiess, F. (1987). Seafloor Geodesy by the Year 2000, EOS Trans. AGU, Vol. 68, No. 44, p. 1241.

Srinivasan, J., Meehan, T., and Young, L. (1989). Code and Codeless Ionospheric Measurements with NASA's Rogue GPS Receiver. Proc. ION GPS-89, Institute of Navigation, pp 451-453.

Thomas, J. (1988). Functional Description of Signal Processing in the Rogue GPS Receiver, JPL Pub. 88-15, Jet Propulsion Laboratory.

Yunck, T., Melbourne, W., and Thornton, C. (1985). GPS-Based Satellite Tracking System for Precise Positioning, IEEE Trans. Geosci. and Remote Sens., GE-23(4), pp. 450-457.

Yunck,T. and Melbourne, W. (1990). Geoscience from GPS Tracking by Earth Satellites. International Association of Geodesy, General Meeting Edinburgh Scotland (Springer-Verlag, in press)

GPS ORBITS AND BASELINE EXPERIMENTS: MINI-MAC/TI COMPARISONS

B.E. Schutz, P.A.M. Abusali, C.S. Ho, and B.D. Tapley
Center for Space Research, University of Texas at Austin
Austin, Texas 78712-1085 USA

ABSTRACT

Major changes in the Cooperative International GPS Network (CIGNET) tracking hardware took place in 1989. Prior to 1989, the CIGNET solely consisted of TI-4100 receivers except for the Mini-Mac receiver at Tsukuba. In January, 1989, the National Geodetic Survey (NGS) installed Mini-Mac 2816AT receivers at Mojave, Westford and Richmond which, with the Tsukuba receiver, brought the network to an almost equal number of TI and Mini-Mac receivers. Simultaneous operations of TI and Mini-Mac receivers were conducted by the NGS at the three U.S. sites for a period of several weeks. Using this simultaneous data set in addition to the other CIGNET data, an investigation has been conducted into the compatibility of the all-TI and the mixed receiver configuration. The preliminary results have verified the existence of a 1 ms time tag discrepancy in the Mini-Mac data reported by others, however, a 2.5 cm unresolved discrepancy was found on the short TI to Mini-Mac baseline at Mojave. When the discrepancy in GPS time was corrected, several meter RMS orbit differences were found; however, the 1400 km test baseline between Onsala and Tromso demonstrated less than two parts in 10^8 agreement between the two cases. The results also demonstrated the existence of a change in Tromso antenna position from the location previously reported.

INTRODUCTION

Beginning in 1987, the National Geodetic Survey has been active in the development of the CIGNET. The site locations as of November 1988 are illustrated in Fig. 1 (CSTG Bulletin, Vol. 1, No. 3, 1988). All site locations, except Tsukuba, consisted of TI-4100 receivers using the CORE software developed by the Applied Research Laboratory of the University of Texas (UT).

In January, 1989, Mini-Mac 2816AT receivers were installed at Mojave, Westford and Richmond. For several weeks after this installation, the TI receivers were operated simultaneously with the Mini-Mac. In addition, NGS performed local surveys to determine the position vector of the Mini-Mac antenna with respect to other geodetic markers in the vicinity of very long baseline interferometry (VLBI) systems.

Results obtained from mixed receiver types have been reported by Gurtner et al. (1989), Rocken et al. (1989) and Chin private communication, (1989). In the former case, a 1 ms time tag discrepancy with GPS time was observed when TI and Mini-Mac data were mixed. The source of the discrepancy was traced to the Mini-Mac software; however, the discrepancy is not apparent when only Mini-Mac data were used.

Because of the observed problems with mixing receiver types, the purpose of this study was to investigate the compatibility between TI and Mini-Mac in the use of CIGNET data for GPS orbit and baseline determinations, especially for longer baselines than those used in the previous studies. It was believed that this study was essential for the understanding of results obtained using CIGNET data after the TI receivers were removed and to aid in establishing historical links to prior campaigns using only TI receivers. The results presented in this paper are preliminary, and further detailed studies are planned.

DATA

The simultaneous operation of Mini-Mac and TI receivers at Mojave, Westford and Richmond began on January 30, 1989, in GPS Week 473 using Mini-Mac receiver software Version 1.49. The TI operations continued for approximately two weeks at Westford and Richmond after the Mini-Mac installation. At Mojave, the UT receiver operated until June. At all three sites, the displacement between the TI and Mini-Mac antennas was about 5 m or less.

For the preliminary analysis reported in this paper, only data from the first three days of simultaneous TI and Mini-Mac operation were used (January 30 to February 1, 1989). Since prior work had suggested that the TI data from Richmond would not be usable, these data were initially excluded from the analysis. However, further investigation showed the existence of a significant amount of usable TI data from Richmond. These data were used in some experiments, along with TI data from Kauai, Yellowknife, Wettzell, Onsala and Tromso.

PROCEDURE

The TI and Mini-Mac data at the three U.S. sites were used in two separate, but complementary analyses. First, the TI and Mini-Mac data were used for short baseline tests (<5 m) in which the TI to Mini-Mac vector baseline was determined at each of the three sites. The purpose of this test was to assess the reported time tag discrepancy and phase center of the respective antennas. Because of the very short baselines, L_1 only solutions were obtained, as well as ionospherically corrected solutions. For these analyses the orbits were fixed to a nominal ephemeris obtained from either pseudo-range or double-difference phase data. All baseline solutions were obtained with phase data.

Second, the GPS orbits were determined from double-difference phase data using a) only TI data and b) Mini-Mac data from Mojave, Westford and Richmond plus TI data from the other sites. In these cases, the typical estimated parameter set consisted of the epoch position and velocity for each satellite in the three-day arc, a scale parameter for the ROCK4 radiation pressure (Fliegel et al., 1985), a parameter for y-bias force, three-hour troposphere zenith delay, appropriate biases and selected antenna coordinates. No *a priori* constraints were used on any estimated parameters. The modeling of the GPS force and kinematic characteristics and the site coordinates are given by Schutz et al. (1989a). In all cases, Tromso was treated as a site with unknown coordinates.

SHORT BASELINE TESTS

Analyses were performed on the short baseline between the TI antenna and the Mini-Mac antenna at each of the three U.S. sites. These analyses verified the existence of the 1 ms time tag correction required for the Mini-Mac data. In addition, a more detailed analysis was performed with the short baseline estimates at Mojave.

The orbit used for this analysis was derived from double difference phase data using only TI instruments (Case B in the next section). This orbit was fixed and the coordinates of the Mojave Mini-Mac were estimated with respect to the TI. The TI antenna at Mojave was the FRPA-2 and further information on the relative positions is given in the CSTG GPS Bulletin for May-June, 1989. The results are shown in Table 1 in which the Mini-Mac position vector is given with respect to the FRPA antenna used on the TI-4100. The Mini-Mac coordinates were obtained from the average of three individual solutions, each comprising a complete day of double differences between the TI and the Mini-Mac. This result was compared with the vector difference of positions for the respective antennas obtained on previous surveys. The resulting 2.5 cm vertical discrepancy is under investigation. No significant time tag adjustment to the Mini-Mac was obtained after application of the 1 ms discrepancy. Furthermore, the estimate was obtained with and without bias-fixing; however, both cases gave the same result to within a few millimeters.

ORBIT AND LONG BASELINE TESTS

Several cases were used for the orbit and long baseline tests. These cases were:

A. TI only data from Mojave, Westford, Richmond, Kauai, Wettzell, Onsala and Tromso,
B. Same as Case A except Richmond TI data were excluded,
C. Mini-Mac data from Mojave and Westford, no Richmond data and TI data from other sites.

The orbits that resulted from Case B and Case C are compared in Fig. 2 for PRN-9 and a tabular summary for all satellites is given in Table 2. For comparison, the tabular summary for the differences between Case A and Case B are shown in Table 3, showing the influence of Richmond TI data. Comparison of Table 2 and 3 indicates agreement between the three cases at the few meter level.

Since Tromso cannot yet be linked directly into a VLBI coordinate system (the mobile VLBI occupied a site at Tromso in July, 1989), Tromso has been treated as a site with unknown coordinates. The results obtained from the three cases considered are shown in Table 4, where the coordinates are given with respect to Onsala. The table also contains the number of double-difference measurements and the RMS of the global set of measurement residuals. All three cases give Tromso coordinates that agree to within 1.6 cm in (x, y, z) and 1.5 cm in the 1400 km baseline length, thus suggesting internal consistency at the level of a part in 10^8.

Comparisons of the Onsala to Tromso baseline with previously determined results are shown in Table 5. The previous results are given by Schutz et al. (1989a; 1989b). While the previous results agreed with each other within about 8.5 cm in all components and in the baselines, the Week 473 results disagreed with the others by about 81 cm in height.

This discrepancy prompted further investigation to identify the source. Engen private communication, (1989) indicated that a new pillar for the GPS antenna had been constructed at Tromso in November, 1988. The results from earlier weeks given in Table 5 were obtained with data collected prior to November, 1988, whereas the Week 473 results of this paper used data recorded after the pillar was installed. Detailed measurements of the pillar by the Norwegian Mapping Agency showed a change in height of about 81 cm, thereby justifying the conclusion that the Tromso discrepancy was caused by a change in the antenna height. This result is of further interest since the initial GPS result was obtained without knowledge of the antenna height change, thereby providing a "blind test." An additional test of the GPS determination of Tromso will be possible when the VLBI result is available at the end of 1989.

SUMMARY AND CONCLUSION

Data collected for a period of three days using the TI-4100 and the Mini-Mac 2816AT receivers collocated at Mojave and Westford were processed along with data from other CIGNET sites. Three-day orbit comparisons indicate that the impact of replacing TI-4100 with Mini-Mac on the orbits was less than four meters RMS. However, the effect on the Onsala to Tromso baseline determination was insignificant. A previously reported Mini-Mac one millisecond time tag discrepancy from GPS time was verified in the process of determining short (~5 m) baselines between the TI-4100 and the Mini-Mac antennas collocated at Mojave and Westford, however, an unresolved 2.5 cm discrepancy exists in the Mojave result. The baseline between Tromso and Onsala (~1400 km) was determined with a repeatability between three cases of 1 part in 10^8. An 81 cm change in antenna height at Tromso (due to new construction) was detected without any prior knowledge of the change. While this test can be regarded as a "blind test," an essential component of the test will be provided by the VLBI determination. Based on the analysis of three days of data, it can be tentatively concluded that TI-4100 and Mini-Mac 2816AT CIGNET receivers are sufficiently compatible for orbit determination to support a part in 10^8 baseline determinations. However, orbit differences at the few meter level require further investigation.

Acknowledgements. The data were preprocessed by D. Carroll, L. Burton and H. Rim at the Center for Space Research. The information provided by G. Mader and M. Chin at NGS for this study is gratefully acknowledged. This research has been supported by NASA.

REFERENCES

Gurtner, W.G., Beutler, G., and Rothacher, M. (1989). Combining GPS observations made with different receiver types, *Proceedings of the Fifth International Geodetic Symposium on Satellite Positioning,* Las Cruces, New Mexico.

Fliegel, H.F., Feess, W.A., Layton, W.C., and Rhodus, N.W. (1985). The GPS radiation force model, *Proceedings of the First International Symposium on Precise Positioning with Global Positioning System,* NOAA, Rockville, Maryland.

Schutz, B.E., Tapley, B.D., Ho, C.S., Rim, H.J., and Abusali, P.A.M. (1989a). GPS orbit determination: Experiments and results, *Proceedings of the Fifth International Geodetic Symposium on Satellite Positioning,* Las Cruces, New Mexico.
Schutz, B.E., Ho, C.S., and Bevis, M. (1989b). Analysis of southwest pacific campaign data: July 1988, *Proceedings of the Fifth International Geodetic Symposium on Satellite Positioning,* Las Cruces, New Mexico.
Rocken, C. and Meertens, C.M. (1989). GPS antenna and receiver tests: multipath reduction and mixed receiver baselines, *Proceedings of the Fifth International Geodetic Symposium on Satellite Positioning,* Las Cruces, New Mexico.

Table 1. Mojave short baseline test (TI to Mini-Mac).

	Components and length (m)			
	x	y	z	l
L_1 Average of three days*	−1.197	−2.478	−4.180	−5.005
RMS of three days	0.001	0.001	0.001	0.006
Independent solution**	−1.181	−2.466	−4.196	5.008

*The three days in Week 473 were days 30, 31 and 32
**The independent solution was obtained by TI to TI and Mini-Mac to Mini-Mac local surveys at different times (*GPS Bulletin,* Vol. 2, No. 3, May-June 1989)

Difference (m) between the three-day and the independent solution

North	East	Up
0.0013	−0.0054	0.0239

Table 2. Summary of orbit differences: EPH (C)-EPH (B).

	PRN-3	PRN-6	PRN-8	PRN-9	PRN-11	PRN-12	PRN-13
				Mean (m)			
Radial	−0.026	0.007	−0.010	0.017	0.027	0.038	0.008
Along-Track	3.276	2.089	1.534	1.280	1.843	1.510	1.213
Cross-Track	−0.015	−0.013	−0.026	−0.027	−0.038	−0.003	0.006
				RMS (m)			
Radial	0.957	0.922	0.890	0.298	0.165	0.497	0.251
Along-Track	3.935	2.794	2.354	1.427	2.003	2.001	1.341
Cross-Track	1.580	0.768	2.475	1.005	1.631	0.982	0.966

Table 3. Summary of orbit differences: EPH (B)-EPH (A).

	PRN-3	PRN-6	PRN-8	PRN-9	PRN-11	PRN-12	PRN-13
				Mean (m)			
Radial	0.028	0.013	0.024	−0.006	0.015	−0.010	0.014
Along-Track	2.811	5.387	1.619	1.429	−0.577	−2.873	0.441
Cross-Track	−0.003	−0.024	−0.012	−0.023	−0.009	−0.017	0.001
				RMS (m)			
Radial	0.519	0.902	0.083	0.317	0.297	0.727	0.280
Along-Track	3.037	5.725	1.758	1.574	0.862	3.250	0.769
Cross-Track	0.553	1.679	0.965	0.734	0.376	0.884	0.346

Table 4. GPS week 473 results.

	DD	No. of	Onsala to Tromso baseline (m)			
Case	RMS(m)	DD	Δx	Δy	Δz	l
A	0.021	25353	−1267718.165	9692.552	608406.658	1406187.026
B	0.019	19734	−1267718.149	9692.552	608406.661	1406187.014
C	0.019	19371	−1267718.154	9692.568	608406.647	1406187.011

DD = double differenced phase

Table 5. Onsala to Tromso baseline comparison with other results.

	Components and length (m)			
	Δz	Δy	Δz	l
Week 473 Case B	−1267718.149	9692.552	608406.661	1406187.014
Week 419 1 week soln	−1267718.396	9692.560	608405.885	1406186.900
Week 420 1 week soln	−1267718.417	9692.533	608405.862	1406186.909
Week 444 1 week soln	−1267718.414	9692.512	608405.882	1406186.915
Week 444 3 day soln w/o S. Pacific	−1267718.480	9692.566	608405.899	1406186.983

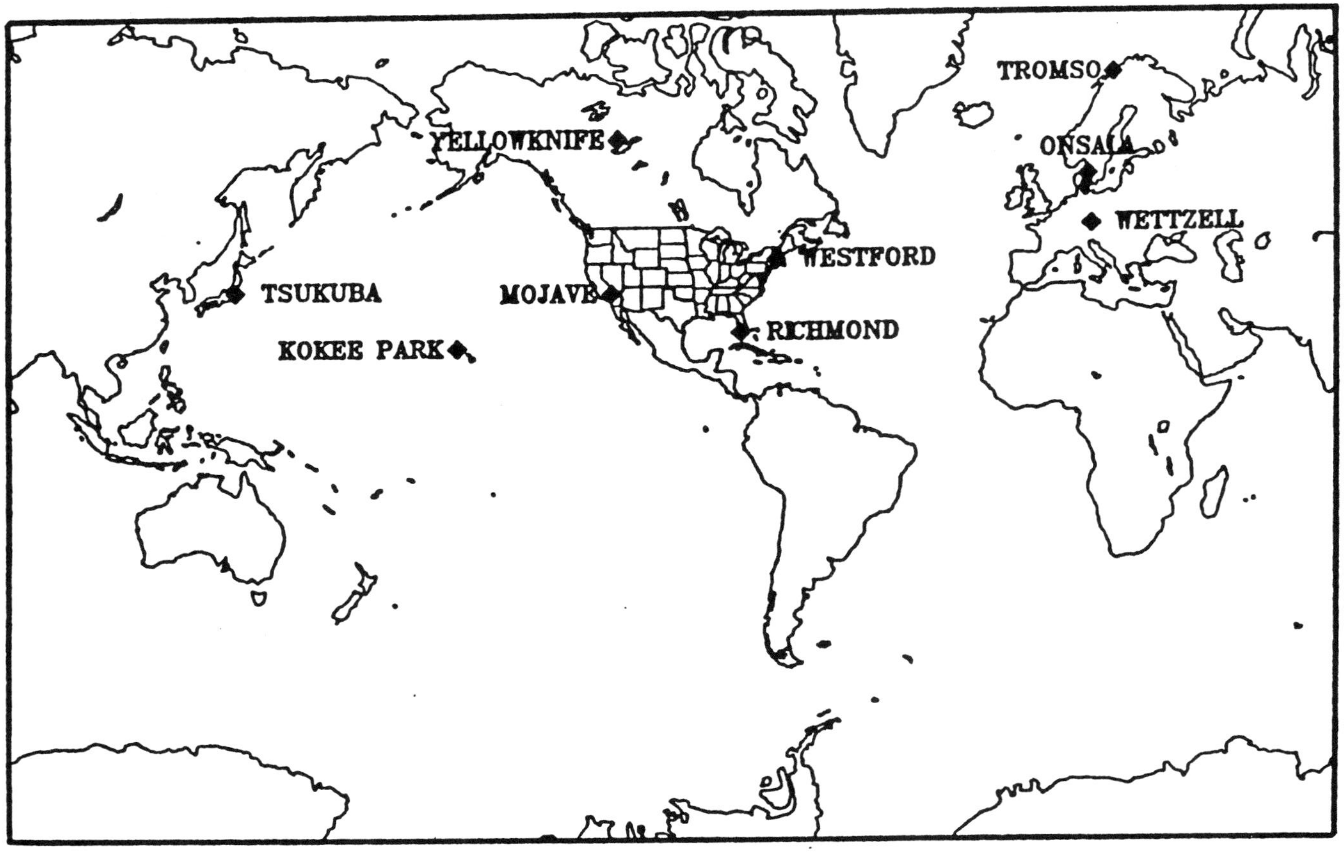

Figure 1. CIGNET Tracking Sites [CSTG GPS Bulletin]

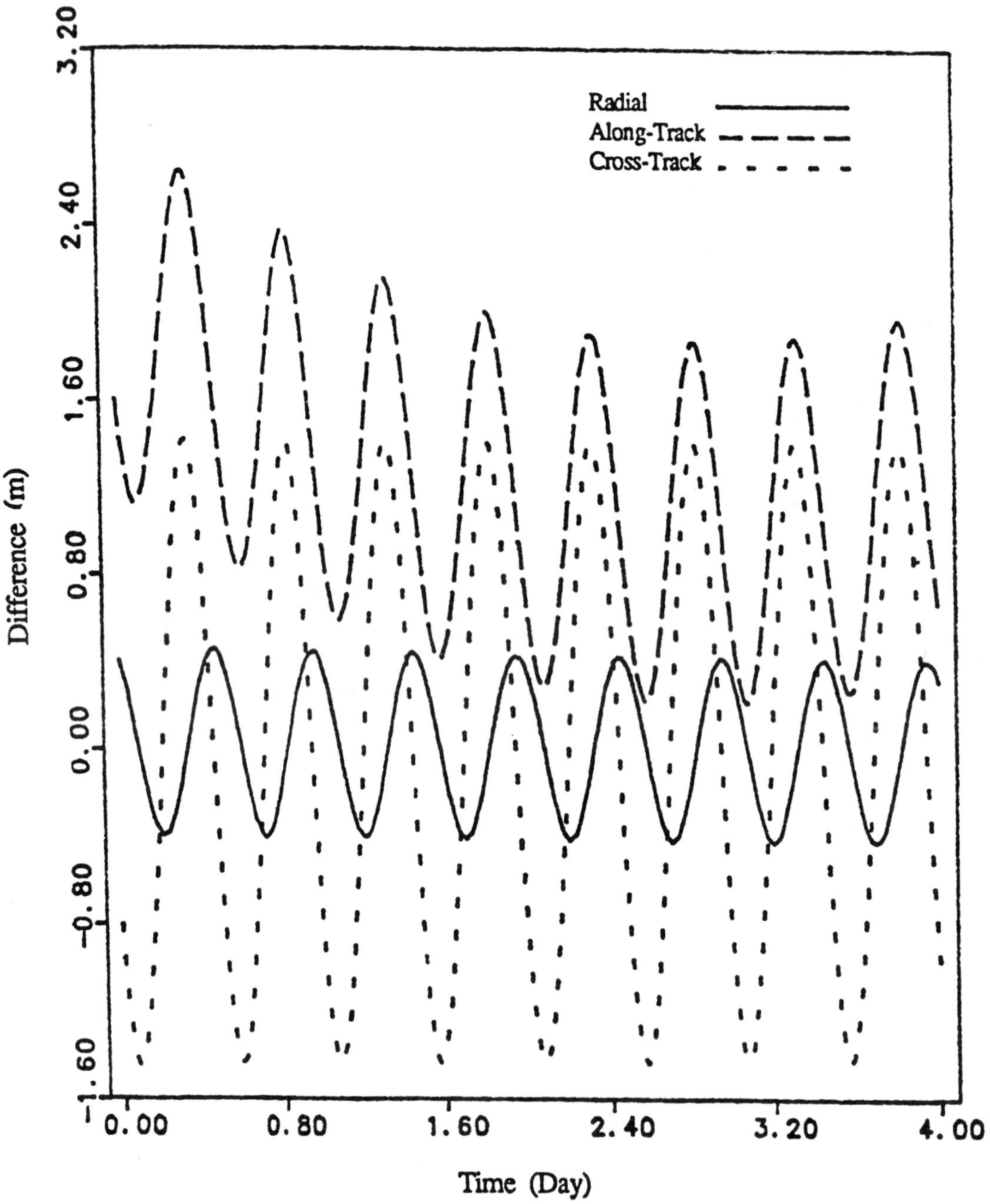

Figure 2. Comparison of solution trajectories in J2000 system for PRN-9
(Case C - Case B) Epoch - Jan.29,1989 00:00:00

ORBIT DETERMINATION FOR GPS SATELLITES

V Ashkenazi, C Hill, T Moore, and S Whalley
Institute of Engineering Surveying and Space Geodesy,
University of Nottingham

ABSTRACT

Conventional GPS processing techniques use the broadcast ephemeris to obtain relative station coordinates. Errors in the broadcast ephemeris can cause baseline errors of the order of 1 ppm. For high order geodetic positioning accuracies, it is therefore necessary to determine precise satellite orbits.

Predicted orbits can be generated by numerically integrating a model of the forces acting on the satellite. Carrier phase data from a number of tracking stations can then be used to improve this predicted orbit. At Nottingham, the phase data is double differenced, and used in a simultaneous least squares adjustment to produce corrections to selected station coordinates and to the predicted orbit.

INTRODUCTION

To compute the position of a point on the earth (or a position difference) from GPS pseudorange or phase observations, the 3-d coordinates of the satellites, at every observational epoch are required. In practice, this orbital information is expressed as the satellite ephemerides (Ashkenazi and Moore, 1986). With the exception of the 'codeless' receivers the majority of GPS receivers have access to the broadcast ephemeris contained in the Navigation Message. For many applications, including navigation and low-accuracy positioning, the broadcast ephemeris is of sufficient accuracy, around 10 m to 25 m for each of the three satellite coordinates. These orbital accuracies have an almost one-to-one correspondence with the precision of the absolute coordinates of points determined by using these satellites. In relative positioning, the orbital errors tend to cancel out for short baselines. However, relative positioning accuracies of better than 0.1 parts-per-million (10^{-7} of the corresponding baseline length), can only be achieved if satellite orbits of a corresponding high accuracy are either made available or computed as part of the solution.

A package of computer programs for the determination of satellite orbits has been developed at the University of Nottingham over the last eight years. This paper describes the principles and models adopted at Nottingham for GPS Orbit determination, and presents a sample of representative results obtained using the US Standard GPS Data Set.

ORBIT DETERMINATION

The cartesian position of a GPS satellite, in its orbit, can be predicted by using a model which represents all the known forces acting upon the space craft. The acceleration vector of the satellite obtained from this force model can be numerically integrated from initial position and velocity (state) vectors, once to obtain velocity and twice to give position, with respect to time. The numerical integration procedure adopted at Nottingham consists of a 4th order Runge - Kutta (single-step) starting procedure, followed by an 8th order Adams-Bashforth (predictor - corrector) scheme (Moore, 1986). This predicted orbit is subsequently improved by introducing observations from a network of tracking stations, and determining small corrections to the initial state vectors and force model components.

The forces acting upon the GPS satellites can be divided into the gravitational forces and the surface forces. The gravitational forces include the gravitational attractions of the earth, moon, sun, the planets and the effects of the solid earth and ocean tides. The earth's gravitational attraction can easily be modelled for the high altitude GPS satellites by using the first few terms of a geopotential model (such as GEM-TI or WGS 84). Tests have shown that an expansion up to degree and order 8 is suitable for predicting orbital arcs of a few days. When using orbital arcs of this length, the effect of the planetary attractions can also be ignored (Landau and Hein, 1986).

The principle surface force acting upon the satellites is due to the solar radiation pressure. This force is one of the most difficult to model due to the complex shape of the GPS satellites. The ROCK IV model (Fliegel et al, 1985) has been used to represent the surface of the satellite by describing different specularity and reflectivity coefficients for different parts of the satellite. However, for short orbital arcs (three or four days) a simpler model can be adopted by resolving the acceleration caused by the solar radiation, into two components. One of these is in the direction of the sun to the satellite and the other lies along the solar panel axis of the satellite (y bias). The high altitude of the GPS satellite avoids the need to model the effects of atmospheric drag.

The result is a relatively simple force model with the principle uncertainties being due to the solar radiation pressure model. This can be improved by solving for scaling coefficients to the direct solar radiation pressure, and the y bias acceleration for each satellite in the second stage of the orbit determination process. This involves the processing of GPS carrier phase data from a network (either global or regional) of receivers. A least squares adjustment procedure is used to determine small corrections to the satellite initial state vectors, the solar radiation pressure model coefficients, the integer phase ambiguities and the coordinates of the receivers. It is however, necessary to hold fixed the coordinates of some of these tracking receivers to already known WGS 84 coordinates. Clearly, to ensure that the computed ephemeris is not degraded, very accurate coordinates must be known for these reference or 'fiducial' stations. The required accuracies (of the order of 10^{-8} or 0.01 parts per million) are only achievable by either the VLBI (Very Long Baseline Interferometry) or SLR (Satellite Laser Ranging) techniques. As a result GPS receivers are now routinely operated at a number of SLR and VLBI sites and the data from these fiducial networks are available for orbit determination.

GPS Data Set

A number of tests have been carried out on the Nottingham GPS Orbit determination software using data from the "Standard GPS Data Set", distributed by the CSTG GPS subcommission. The data were observed between the 3rd and 7th January 1987, mainly from sites in California but also sites elsewhere in North America. A subset of the data has been used, from the sites shown in fig. 1. All the observations were taken using T1-4100 GPS receivers, recording data primarily from satellites (PRN nos.) 3, 6, 9, 11 and 13.

The data set is supplied in both raw (full-rate 30-second measurement interval) and compressed (120 second measurement interval) formats. The compressed format data has been corrected for cycle slips. Since the total span of data is more important than the total number of observations, the compressed, slip-free data set has been used in the following tests.

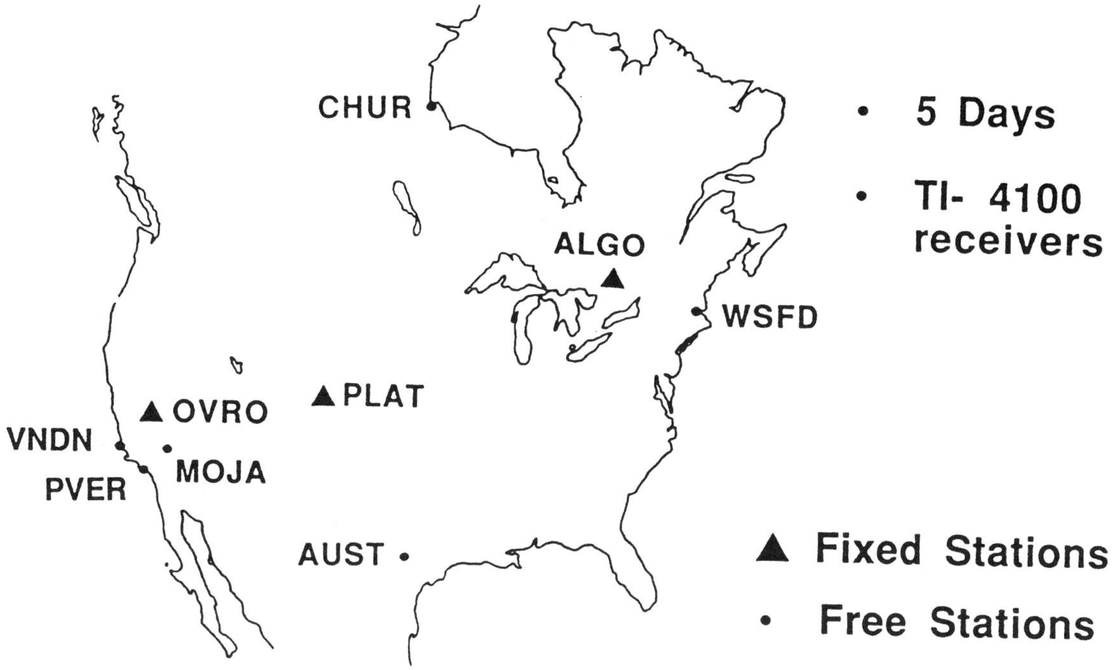

Fig. 1. U.S. GPS Network - January 1987

Processing Strategy

The orbit integration satellite force model included the WGS 84 geopotential model up to degree and order 8, with the initial state vectors obtained from the NSWC precise ephemeris. For the simultaneous least squares adjustment, the coordinates of Platteville (PLAT), Owens Valley (OVRO) and Algonquin (ALGO) were held fixed to the SV4 coordinate set, developed at the Massachussetts Institute of Technology.

The one-day solutions included estimates for the

Non fiducial receiver coordinates,

Initial double difference integer phase ambiguities,

Satellite state vectors $(x, y, z, \dot{x}, \dot{y}, \dot{z})$.

The two-day solution included estimates for the

Non fiducial receiver coordinates,

Initial double difference integer phase ambiguities (new set each day),

Satellite state vectors $(x, y, z, \dot{x}, \dot{y}, \dot{z})$,

Direct solar radiation pressure scaling coefficients.

RESULTS

A comparison has been made between the baseline components (east, north and height) estimated in the single-day GPS solutions, and the VLBI determined coordinates (SV4). These results are shown in fig. 2.

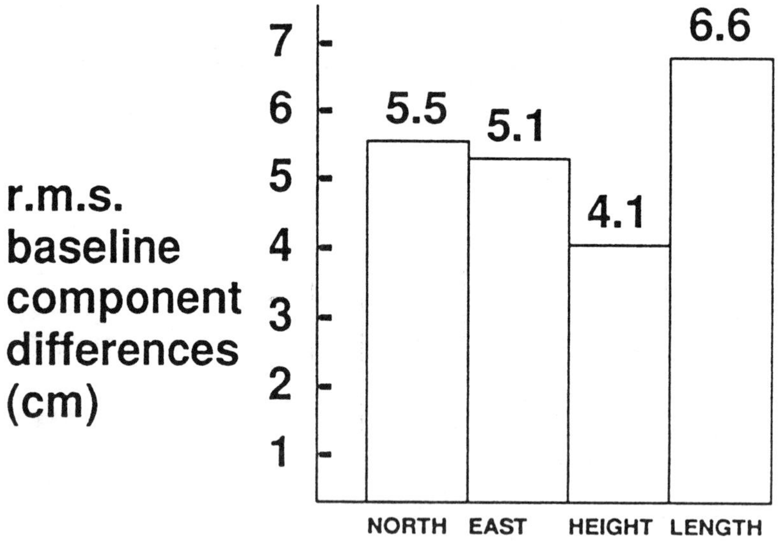

Fig. 2 - Single-day GPS solution compared with VLBI values

The figure shows the root-mean-square (rms) baseline component differences between the two coordinate sets for 7 baselines. The agreement can be seen to be better than six centimetres in all three components.

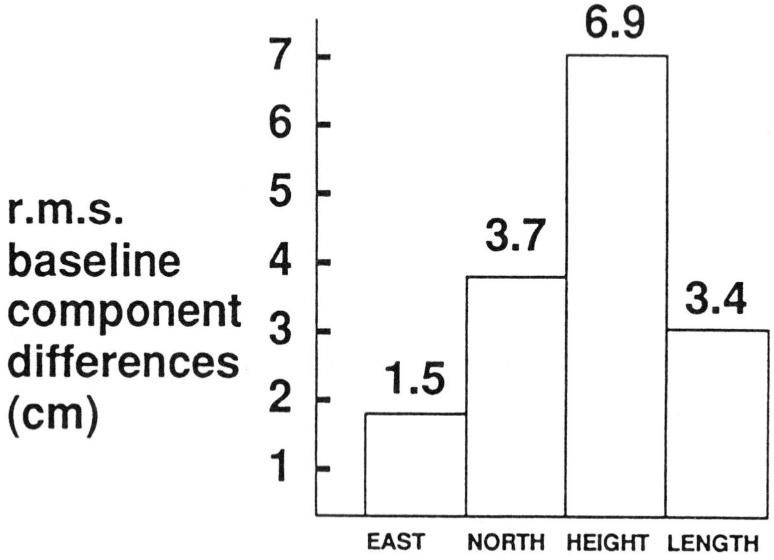

Fig. 3. Two day GPS solution compared with VBLI values

In fig. 3, the rms baseline component differences between the two-day GPS solution and the VLBI solution are shown. In this case there is a much better agreement in the east and north baseline components, giving a much lower rms baseline difference of 3.4 cm. The errors in the individual baselines from this solution are shown in table 1. The baseline differences vary from a few parts in 10^8 up to the largest difference of 1.5 parts in 10^7, which occurs on the shortest baseline. Based on the assumption that the baseline error is proportional to the orbital error (Landau and Hein, 1986), this would correspond to an approximate orbital error of around 3 metres.

	length (km)	difference (cm)	ppm
ALGO - MOJA	3407	4.0	0.012
- PVER	3611	6.2	0.017
- VNDN	3727	2.4	0.006
PLAT - MOJA	1196	3.3	0.028
- PVER	1407	4.6	0.033
- VNDN	1533	1.3	0.008
OVRO - MOJA	245	0.7	0.029
- PVER	387	5.1	0.132
- VNDN	363	-0.6	0.017
MOJA - PVER	224	1.7	0.076
- VNDN	351	-1.4	0.040
PVER - VNDN	223	3.5	0.157
	rms =	3.4	

Table 1 Two-day GPS baseline differences from corresponding VLBI baselines

Orbit Comparison with the NSWC Precise Ephemeris

Fig. 4 gives a comparison between the NSWC precise ephemeris and the orbit of satellite PRN 11, determined from a three-day solution (IESSG orbit). This solution estimated the same parameters as the two-day solution, but included an estimate for the y bias acceleration for each satellite.

During the observation periods, the agreement between the two orbits is generally better than five metres, but this degrades up to fifteen metres when the satellite passes away from the tracking sites. Clearly, it is impossible to distinguish between errors in the precise ephemeris or the IESSG ephemeris. In addition, differences between the two coordinate reference frames could introduce systematic bases between the two solutions.

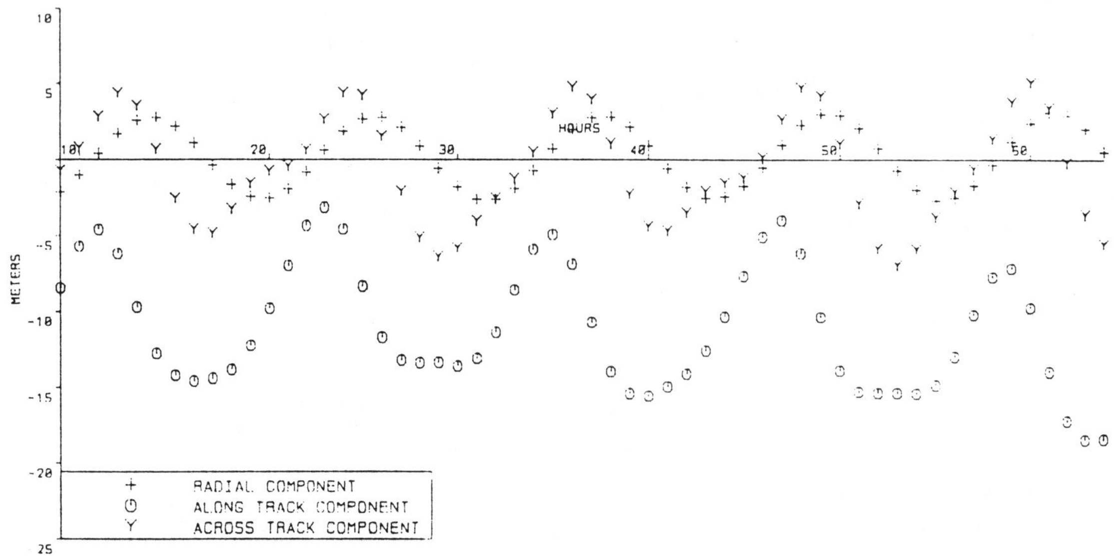

Fig. 4 Orbit Comparison with precise ephemeris

Orbital Repeatability

A comparison has been made between two independent, two-day arcs, which were computed from four consecutive days of data. The predicted satellite state vectors for the first two-day arc were integrated as far as the end of the observation period at the start of the second arc. This is represented in the schematic diagram, fig. 5.

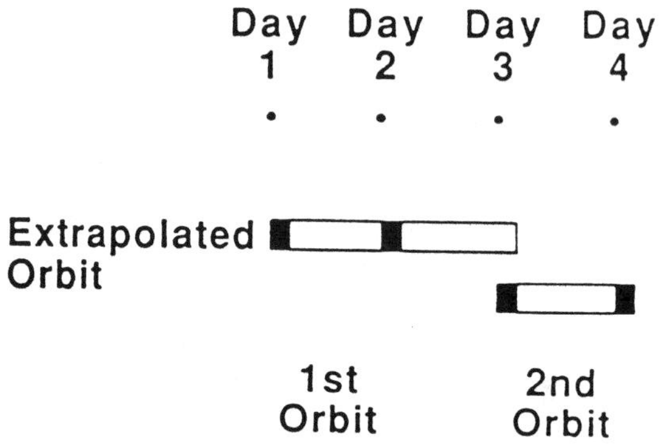

Fig. 5. Orbit Repeatability test

The four-hour overlap period of the two orbits was used to compute the differences between the along track, across track and radial components of the orbits and these are shown in fig. 6.

This process of extrapolating the orbit will magnify the effect of any errors in the satellite force model or in the initial state vector and it therefore provides a stringent test of orbit repeatability. Nevertheless for these two-day orbits, the average differences of the radial, along track and across track components are only 1.8, 5.9 and 6.3 metres respectively.

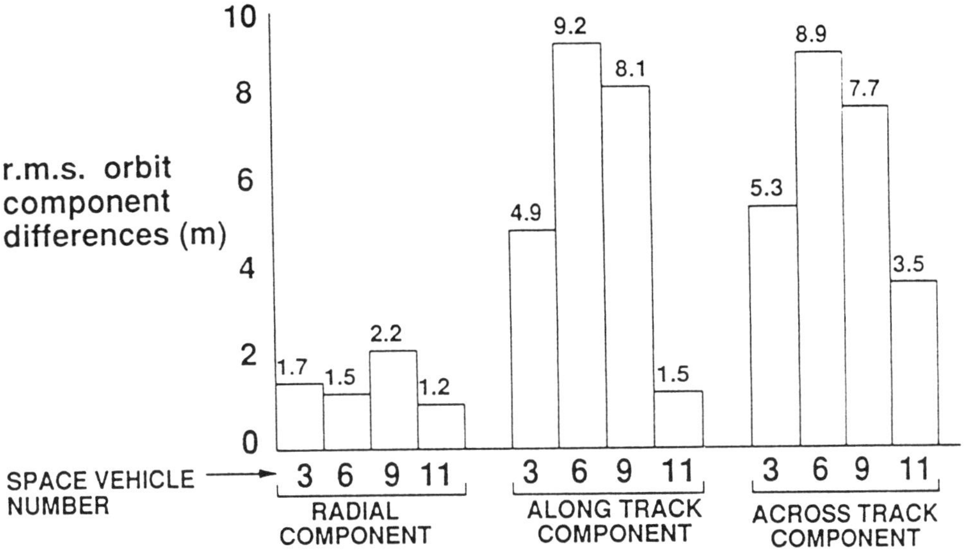

Fig. 6. Orbit repeatability test results

Conclusions

The discrepancies between the GPS and VLBI baseline components demonstrate a precision of between one and two parts in 10^7 or better. The implication of this is that orbital errors have also been reduced to the order of 1 or 2 parts in 10^7, ie 2-4 metres. The comparison with the precise ephemeris indicates that this level of accuracy is limited to the region over the tracking sites, with periodic degradation occurring outside this region. These results were obtained by using only three or four hours of data each day. The accuracy is expected to improve significantly with the use of the longer period of data which will become available as the satellite constellation improves.

Acknowledgments

Research on GPS high precision orbit determination at Nottingham University is sponsored by Shell UK Exploration and Production. Acknowledgements are also due to Dr Y Bock and Dr S Lichten for their help and advice.

REFERENCES

Ashkenazi, V., and Moore, T., 1986. *"The Navigation of Navigation Satellites", Journal of Navigation, Vol 3, No 3.*

Fliegel, H.F et al. 1985. *"The GPS Radiation Force Model". Proc. Ist International Symposium on Precise Positioning with the* Global Positioning System, Rockville, Maryland

Landau, H., and Hein, G., 1986. *"Preliminary Results of a Feasibility Study for a European Tracking Network". Proc. 4th International Geodetic Symposium on Satellite Positioning, Austin, Texas*
Moore, T. 1986. *"Satellite Laser Ranging and the determination of Earth Rotation Parameters". PhD Thesis, Nottingham University.*

HIGH PRECISION GPS POSITIONING BY FIDUCIAL TECHNIQUES

V Ashkenazi, T Moore, G Ffoulkes-Jones, S Whalley and M Aquino
Institute of Engineering Surveying and Space Geodesy, University of Nottingham

ABSTRACT

GPS routinely achieves relative coordinate accuracies of 1 ppm. For networks up to 100 km in size, this represents an error of 10 cm. However, for larger networks incorporating baselines of up to several thousand kilometres, an error of 1 ppm corresponds to a relative error of the order of 1 metre. To obtain centimetric relative positioning accuracies in such cases, an improvement of one or two orders of magnitude is required. This may be achieved by including several highly accurate 'fiducial' stations, based at SLR and VLBI sites in the network. The accuracies of these known stations are propagated into the rest of the network through the improved satellite ephemerides, which are solved for as part of the global least squares adjustment.

INTRODUCTION

One of the major sources of inaccuracy in GPS positioning is due to the orbital errors of the broadcast ephemeris, whose accuracy is restricted to around 20 m. Typically, most users of GPS will only have access to the broadcast ephemeris (and in a limited number of cases, the precise ephemeris), and these satellite coordinates are fixed throughout the processing of GPS data. As a result the orbital errors are transferred to the unknown parameters, and in particular the receiver coordinates. Fortunately, in relative positioning these orbital errors do tend to cancel out for short baselines. However, for high precision positioning over very long distances (up to several thousand kilometres) the orbital errors must be reduced.

These problems are substantially reduced by using an orbit improvement technique in combination with the 'fiducial' network approach. Various methods of orbit improvement are available, but these can only improve the relative positioning accuracy of GPS if observations are included from a network of GPS receivers at known reference or 'fiducial' sites.

In practice the relative positions of these receivers need to be known to around 10^{-8} (or 0.01 parts per million of the length of the baselines between the sites). This accuracy is only achievable by either the VLBI (Very Long Baseline Interferometry) or SLR (Satellite Laser Ranging) Techniques. Ideally, the fiducial stations should be several thousand kilometres apart in order to form a geometrically stable regional or continental tracking network. The scale and orientation of this network is controlled by constraining the coordinates of the fiducial stations to their known values and allowing the satellite orbits to adjust. The improved compatibility between the fixed ground stations and the satellite orbits results in an improvement in the relative positioning accuracy of the unknown GPS stations.

Two distinctly different techniques have been developed at Nottingham University to improve the GPS satellite orbits. Firstly, a rigorous orbit integration procedure has been adopted, which is discussed in the following section. Secondly, a method of orbit relaxation for the broadcast ephemeris has been developed and is described in the 3rd section. The results of comparisons between these two techniques are given in the 4th section, and the paper is concluded in the final section.

ORBIT INTEGRATION

At Nottingham the orbits of the GPS satellites have been predicted using the integration procedures of the in-house developed SODAPOP Package (Satellite Orbit Determination and Analysis Package Of Programs). SODAPOP was initially developed for orbit determination and the processing of Satellite Laser Ranging data and has been adapted for the GPS satellites (Moore and Ashkenazi, 1987). In principle, orbit determination consists of modelling all the forces which act on a satellite and using this acceleration model in a double-integration procedure to calculate the position of the satellite as a function of time. Further details of the Nottingham orbit integration technique are given in a paper presented at IAG, Edinburgh (Ashkenazi, Hill, Moore and Whalley, 1989). This predicted orbit is subsequently improved by introducing observations from a network of tracking stations, and by determining small corrections to the orbit and force model parameters. This process can be repeated and the orbit determination further refined, by using an iterative procedure. To start the process the approximate position and velocity vectors of the satellites must be known at some initial epoch. As well as the integrated orbit, SODAPOP produces the partial derivatives required to solve for the orbital unknowns in the subsequent least squares adjustment.

The GPS phase observables are double differenced and processed in a least squares procedure solving for the following:-

corrections to the initial satellite state vector, $(dX, dY, dZ, d\dot{X}, d\dot{Y}, d\dot{Z})$

receiver coordinates of the non-fiducial sites,

initial integer phase ambiguities,

solar radiation pressure scale factor,

"Y - bias" acceleration,

The Nottingham software can either process single or multi-arc orbits. However, for the comparisons made in this paper only single arc solutions are considered.

The current GPS satellite force model used during the orbit integration includes contributions for the following:

Earth Gravitational attraction
GEM-T1 or WGS 84 (up to degree and order 8)

Lunar, Solar and planetary attractions
JPL DE200

Solid earth tides
simplified Wahr

Ocean tides
GEM-T1 or Schwiderski

Direct solar radiation pressure and the GPS satellite "Y-bias" acceleration

ORBIT RELAXATION

The technique of "orbit relaxation" is conceptually a much simpler procedure than that of orbit integration. Here the orbit is allowed to relax from the broadcast ephemeris and move slightly, resulting in a closer fit to the observations. Corrections to 5 of the keplerian elements, (Ashkenazi and Moore, 1986) which describe the mean orbit of the satellite are solved for as part of the least squares adjustment.

A single-arc and a multi-arc approach to orbit relaxation has been developed at Nottingham. Full details of orbit relaxation are given in an MPhil Thesis currently under preparation at the University of Nottingham (Aquino, M., 1990). The technique as yet has only been applied to a single day solution, and the single/multi-arc refers to the number of arcs into which the satellite pass in split. The single arc method uses only one set of broadcast ephemeris parameters per satellite and solves for only one set of corrections to the keplerian elements. However, in the multi-arc solution the broadcast ephemeris is updated hourly as new values become available, and a set of corrections for each hourly arc is computed.

In order to fully release the orbits the following six keplerian elements are released:-

M_0 - Mean anomaly

e - Eccentricity

$\sqrt{a}$ - $\sqrt{}$ of semi-major axis

Ω_0 - right Ascension

i_0 - inclination

ω - argument of perigee

However, for short periods of data the argument of perigee (ω) is poorly determined and closely correlated to the mean anomaly. As a result only five of the keplerian elements are released and the argument of perigee is constrained to its broadcast ephemeris value.

Initially it may appear that some of the other keplerian elements are also poorly determined, and yet the accuracy of the unknown receiver coordinates improve. Orbit relaxation has the effect of absorbing the errors in the satellite orbits and removing their effects from the parameters of importance, such as the station locations. As a result, this method is not intended to be used for orbit prediction, unlike orbit integration.

Orbit relaxation has several advantages over a rigorous integration technique for high precision GPS. As no complex integration of satellite orbits is required, it is a much simpler and quicker technique, and is better suited for use on micro-computers. In fact the only difference to the conventional GPS processing algorithms is the addition of extra unknown parameters, five for every satellite.

Clearly, over longer periods of data, orbit relaxation will not be as accurate as the full orbit integration procedure, but where speed and ease of use are important considerations, it should be regarded as a more compact and efficient option.

COMPARISON OF TECHNIQUES

The two techniques described above have been compared using data from a North American GPS continental network, observed between the 3rd and 7th of January 1987. This data is generally referred to as the "standard data set" of the International Association of Geodesy Special Study Group 1.104 (Agnew et al, 1987, 1988). The observation schedule consisted of up to 14 TI-4100 GPS receivers recording for 7 hours per day, for a period of 5 consecutive days. The network observed was designed in three levels. At the local level there were 9 sites in California with inter-site baselines ranging from 40 km to 140 km. Surrounding these there were another three sites making up the Californian regional section, with baselines up to 400 km. The network is completed with 5 continentally separated sites across the US, with separations of up to 4500 km.

In the following comparisons, a subset of the data has been used, which essentially involved only the longer baselines formed by the regional and continental receiver sites.

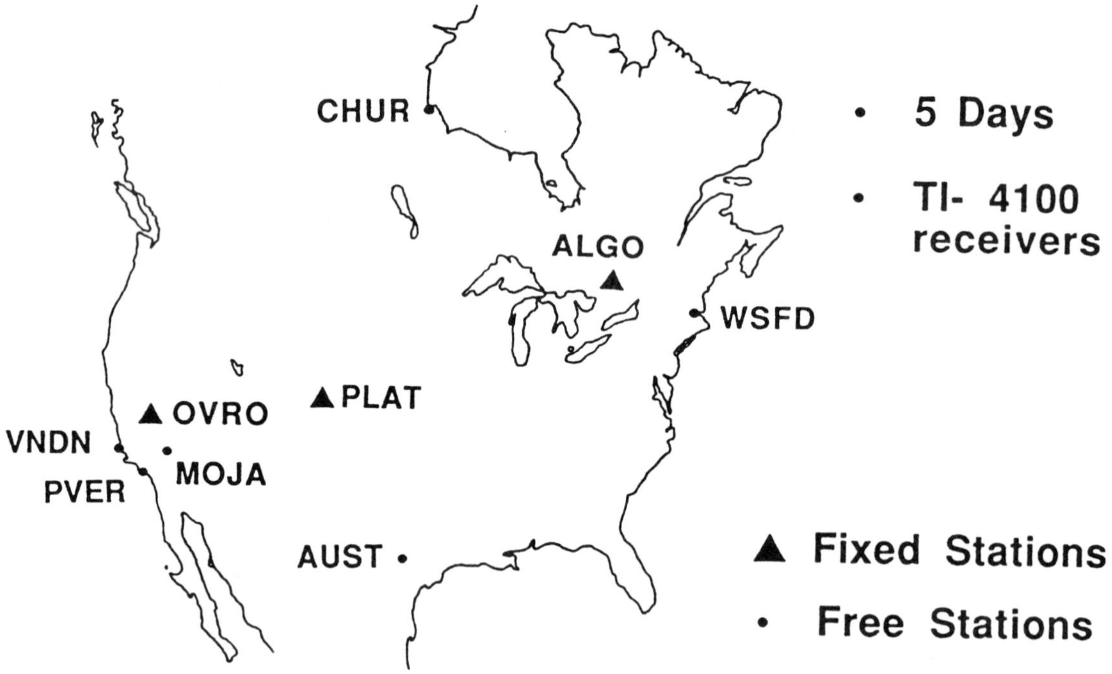

Fig. 1. US GPS Fiducial Network - January 1987.

This data set has been processed with each of the three orbit improvement techniques described above, namely orbit integration and single and multi-arc orbit relaxation.

For the results given in this section, the ground 'truth' is taken from the VLBI solution, and the errors quoted are with reference to this 'truth'.

Tables 1, 2 and 3 show differences between the VLBI and GPS measured baseline lengths for the three techniques of orbit improvement employed. Each solution is based only on data from the first day of observations. As would be expected the orbit integration results show an improvement over orbit relaxation, but this is only marginal. The orbit relaxation results are encouraging, due to the simplicity of the method and the easily attainable 0.1 ppm of baseline length for the majority of the lines measured. Clearly, a given coordinate error at a single station will represent a variable fractional error (ppm), depending on the length of the baseline over which it is compared. Consequently, if

Baseline	Length (km)	VLBI-GPS (cm)	ppm
ALGO - VNDN	3727.6	-7.5	0.020
- WSFD	642.5	7.4	0.115
PLAT - VNDN	1533.3	-6.7	0.044
- WSFD	2752.9	7.1	0.026
OVRO - VNDN	363.4	-7.0	0.193
- WSFD	3929.4	7.1	0.018
VNDN - WSFD	4228.9	-0.3	0.001
		rms = 6.6	

Table 1. Orbit Integration Baseline Differences

Baseline	Length (km)	VLBI-GPS (cm)	ppm
ALGO - VNDN	3727.6	-10.1	0.027
- WSFD	642.5	-5.7	0.089
PLAT - VNDN	1533.3	-11.5	0.075
- WSFD	2752.9	-4.7	0.017
OVRO - VNDN	363.4	-12.4	0.342
- WSFD	3929.4	-3.3	0.008
VNDN - WSFD	4228.9	-12.3	0.029
		rms = 9.3	

Table 2. Single Arc Orbit Relaxation Baseline Differences

Baseline	Length (km)	VLBI-GPS (cm)	ppm
ALGO - VNDN	3727.6	11.1	0.030
- WSFD	642.5	-4.6	0.072
PLAT - VNDN	1533.3	9.9	0.065
- WSFD	2752.9	-8.3	0.030
OVRO - VNDN	363.4	7.8	0.213
- WSFD	3929.4	-6.0	0.015
VNDN - WSFD	4228.9	5.5	0.013
		rms = 7.9	

Table 3. Multi Arc Orbit Relaxation Baseline Differences

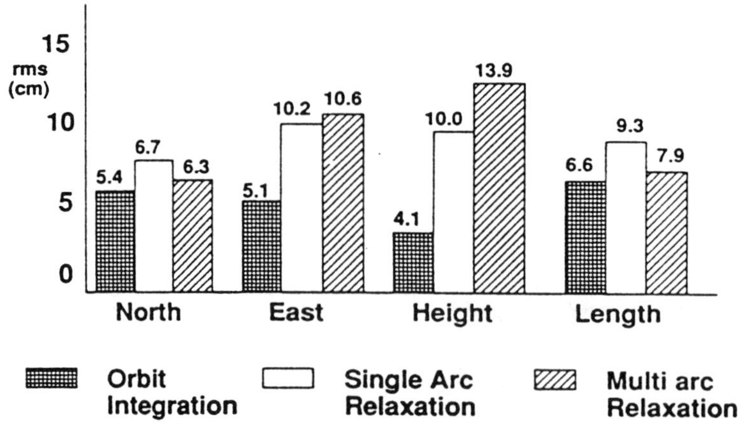

Fig 2. Root Mean Square Baseline Component Differences (VLBI-GPS) for single day solutions

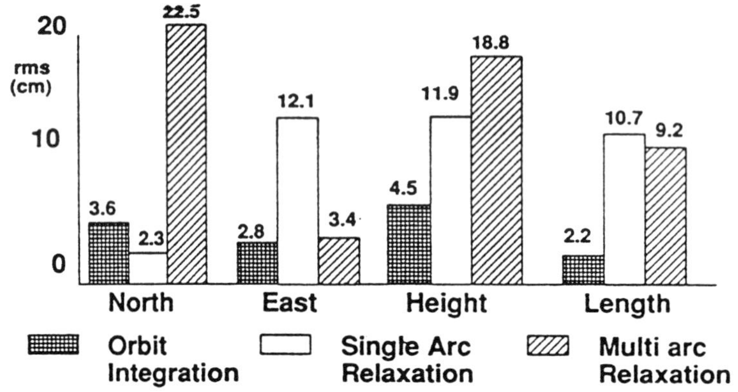

Fig 3. Root Mean Square Baseline Component Differences (VLBI-GPS) for the average of 3 single day solutions

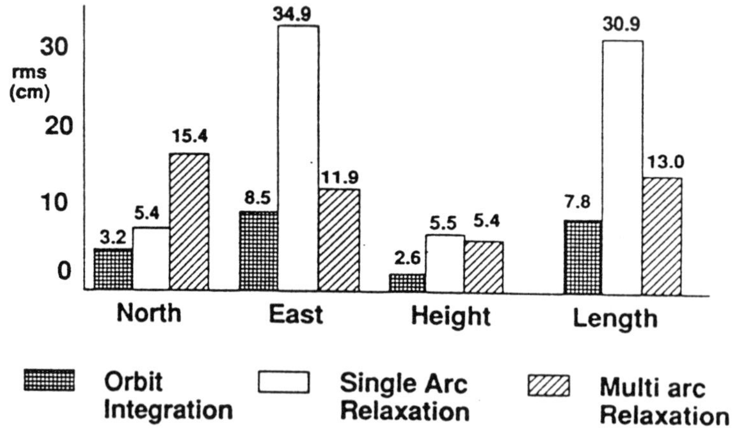

Fig 4. Day to day Root Mean Square Baseline Component Repeatability for 3 single day solutions

parts-per-million' results are quoted with respect to distant fiducial sites, they may appear artificially low. They should therefore be treated with caution.

The accuracies of the individual vector components for the above mentioned single day solutions are shown in a histogram in fig 2. The values shown are the root mean square coordinate differences between the VLBI and GPS solutions for the three techniques. In both the relaxation solutions the EAST component is worse than the NORTH, and is probably due to the essentially north-south tracks followed by the satellites.

Three individual single day solutions were computed for three consecutive days of observations, using each of the three techniques. These results were compared in two different ways. Firstly an average solution for the whole three day period was calculated and compared to the VLBI truth, fig 3. Again, the values shown here are the root mean square differences between the VLBI and GPS solutions. Secondly, the repeatability of the results from day to day was compared and is shown in fig 4. Here the values are the root mean square differences between the various daily solutions.

Only in the case of orbit integration were the results improved by averaging. This is due to the much better repeatability of this method in comparison to the relaxation techniques. Even though the final results obtained were of similar accuracy, the repeatability of the multi-arc and of the single arc relaxation results was degraded by the EAST component, and again this may be due to the poor east-west constraint caused by the north-south tracks of the satellites.

Surprisingly, the repeatability in the height component, usually considered to be the worst, was by far the best for all three techniques, with root mean square repeatabilities of 5 cm or better.

Current developments

These tests using the US Standard Data Set have demonstrated that with the current processing techniques and software, baseline accuracies of the order of 0.1 ppm are being obtained routinely. With a number of current developments the aim is to improve this to a routine accuracy of 0.01 ppm. These include the modelling of the tropospheric zenithal delay parameter. A time varying model is being developed which will allow for the un-modelled variations in the tropospheric delay during the period of observations.

It has been shown (Blewitt, 1989 and Dong, Bock, 1988) that baseline accuracy and repeatability may be substantially improved if the integer biases can be resolved and constrained to their correct integer values.

When processing continental scale baselines of the order of several thousand kilometres the differential effects of Earth body tides on the receiver station coordinates are no longer negligible and must be modelled. The simplified Wahr body tide model is currently being installed in the Nottingham GPS processing software. The effects of deficiencies in the force model of the GPS satellites used during the orbit integration procedure are increased as the duration of the ephemeris prediction in increased. Improvements to the solar radiation pressure model are currently being investigated and will include the Rock IV model together with an improved Y-bias model.

CONCLUSIONS

1. The accuracy of static GPS relative positions may be increased by up to two orders of magnitude by using fiducial GPS control networks and releasing the satellite ephemerides.

2. Two methods of orbit determination have been developed at the University of Nottingham, namely 'orbit integration' and 'orbit relaxation'.

3. These two techniques have been compared using a standard continental scale data set.

4. As expected the orbit integration results are better than those determined using orbit relaxation. Both techniques appear to be capable of producing results at the 0.1 ppm level, for single day solutions using a continental network of fiducial stations.

5. With a number of small enhancements to the software, baseline accuracy of the order of 0.01 ppm, or better, should be obtained routinely.

Acknowledgments

Research on high precision applications of GPS at Nottingham University is sponsored by the Science and Engineering Research Council's Marine Technology Directorate and Halliburton Geophysical Services. Acknowledgements are also due to Dr Y Bock and Dr S Lichten for their help and advice.

REFERENCES

Agnew, D.C., Y. Bock, T.H. Jorden, R.W. King, T.H. Dixon, B.H. Hager, D.D. Jackson, "GPS measurements in central and southern California, submitted to *CSTG Bulletin,* 11 March 1988.

Aquino, M., "Improving GPS Positioning Accuracies by Orbit Relaxation". MPhil Thesis under preparation at the University of Nottingham, 1989.

Ashkenazi, V., Moore, T., Whalley, S., 1989., "Orbit Determination for GPS Satellites", presented at IAG, Edinburgh, 1989.

Ashkenazi, V and Moore, T., "The Navigation of Navigation Satellites". *The Journal of Navigation Vol. 39 No. 3, September 1986.*

Blewitt, G. "Carrier Phase Ambiguity Resolution for the Global Positioning System Applied to Geodetic Baselines up to 2000 km", submitted to the *Journal of Geophysical Research, 1989.*

Dong, D.N. and Y. Bock; "A demonstration of 1-2 parts in 10^8 accuracy with network mode bias-fixing using GPS", submitted to the *Journal of Geophysical Research, 1988.*

Moore, T. and Ashkenazi, V., "Precise Orbit Determination at Nottingham", *Astrodynamics, Vol 65, Advances in Astronautical Sciences,* American Astronautical Society, 1987.

Whalley, S., "Precise Orbit Determination for GPS Satellites", PhD thesis under preparation at the University of Nottingham",1989.

GPS BROADCAST ORBITS VERSUS PRECISE ORBITS: A COMPARISON STUDY

B. W. Remondi and B. Hofmann-Wellenhof (1)
National Ocean Service, NOAA
National Geodetic Survey
Rockville, MD 20852

ABSTRACT

How accurate are the Global Positioning System (GPS) broadcast orbits? This question is important to the GPS community. The survey community, for example, would like to know if field processing of GPS baselines can be considered final. There is widespread belief that for accurate surveys (e.g., 1 ppm) post-processing with the precise ephemerides is required.

Precise-versus-broadcast orbit comparisons are given in terms of along-track, cross-track, and radial plots and statistics based on hundreds of actual broadcast orbit comparisons. The proper use of the broadcast orbits is discussed, and a tutorial example where the extrapolation goes beyond the fitted interval is shown.

A strong measure of orbital accuracy is the effect of orbit errors on baseline vector computations. Numerous baseline vector solutions, using both precise orbits and broadcast orbits, are presented.

INTRODUCTION

Broadcast Orbits

The GPS Operational [master] Control Segment (OCS) has global tracking and monitor sites located at Colorado Springs, Diego Garcia, Ascension, Kwajalein, and Hawaii. These sites collect pseudorange, Doppler, and other tracking measurements for all GPS satellites. The most recent week of these data is used to compute a reference orbit for the GPS satellites. With a reference orbit, thus established, the most recent tracking data (e.g., 12 hours to 24 hours) is used in a batch sequential (or similar) algorithm to improve the orbit. Refer to Wells (1987) for further details. The definitive orbit, thus established, is subsequently used for orbit extrapolation. Although predicted for many days, we are herewith primarily interested in the first single day of prediction. Over these 24 hours a broadcast orbit is computed for each hour using an arc from the extrapolated trajectory spanning from -2 hours to +2 hours with respect to that hour. The orbital parameters (e.g., a, e, i, etc., plus perturbations) computed are uploaded to the GPS satellites by the OCS for subsequent use until the next upload -- possibly 12 hours later.

(1) On leave from the Technical University Graz, Austria

The user, for example the GPS surveyor, collects broadcast message data with his receiver at a 50 bits per second rate. These messages contain relevant information about the satellites: orbits, clocks, health, etc. It is not the purpose of this paper to discuss the contents of the broadcast message, rather, we intend to restrict this paper to the accuracy of the broadcast orbits which are extracted from these messages.

Precise Orbits

Approximately 4-8 weeks after the collection of the tracking data, the Naval Surface Warfare Center (NSWC), in cooperation with the Defense Mapping Agency (DMA), generates what is known as the precise ephemerides.

The tracking data are collected by DMA at five global tracking stations (Australia, Ecuador, England, Argentina, and Bahrain). These tracking data are combined with the data from the five Air Force tracking stations discussed in the previous section. Approximately 8 days of data from these 10 stations are smoothed and edited and then used as input to a sophisticated orbit determination program (Swift, 1985). These orbits are referred to as "fitted" orbits because the final orbit must fit the entire data set in the least squares sense and no extrapolation is required.

Initially the broadcast orbits go through a similar process but ultimately those orbits are extrapolated. Extrapolated orbits naturally have larger errors than fitted orbits simply because they are more subject to the vagaries of unmodeled perturbations.

Description of this Comparison Study

The primary purpose of this paper was to study the accuracy and consistency of the broadcast orbits with respect to the precise orbits. Our approach to this was multifaceted, but primarily we used two standard techniques.

In the first method, we simply computed satellite positions (using both the precise and the broadcast orbits) and compared them with respect to along-track, cross-track, and radial components.

In the second method, we computed baseline vectors from 1 km to 85 km in length, using broadcast orbits and again using precise orbits, to compare them in an actual application. The surveyor, after all, does not care about actual orbit error per se; he wants to know the extent that his baseline vector will be in error when using the broadcast ephemerides. In this method we also isolate the satellite having the largest differences (precise minus broadcast). We then remove this satellite and reprocess the data. One gets from this technique additional indicators with regard to the effect of broadcast orbit errors on the baseline solution, and, as a byproduct, a limited measure of the accuracy of the precise orbits. After removing the worst satellite, the broadcast baselines agree better with the precise baseline. Another aspect of this method is the "integer" ambiguities. For this study, the exact integer values were always established. One is thus able to glean additional information about the accuracy of the broadcast and precise orbits by noting how close to perfect integers are the float-double-difference ambiguities. All solutions given here were computed in both float-double-difference and fixed-double difference modes. From the float-double-difference solutions only the double difference ambiguities are presented. All other baseline vector solution information comes from the "integer-fixed" double-difference solutions.

BROADCAST VERSUS PRECISE EPHEMERIDES

The first part of this study comprised the comparisons of 451 broadcast messages gathered from August 1987 and over the period from May 1988 to March 1989. We used those that were most readily available to us. We only required that they cover reasonably well the past 2 years and, especially, the past year to reflect our concern about the increased ionospheric activity. These messages came from many different receivers (including TI-4100, Trimble 4000SX (10-channel), Mini-Mac 2816, and Ashtech XII). It should be noted that these broadcast messages came from Europe, Australia, Japan, and North America. No significant differences were observed in their accuracy when compared to the precise ephemerides.

Each of these 451 broadcast orbits were computed to determine the maximum difference with respect to the corresponding precise orbits (over the interval from TOE - 2 hours to TOE + 2 hours, where TOE is the time of ephemeris -- which is at the center of the period of validity of the broadcast orbital data). The maximum along-track, cross-track, and radial components were tabulated. Table 1 summarizes the means and standard deviations of these three maxima over the 451 messages. PRN 3 appears twice because it used a cesium clock before January 1989 and a rubidium thereafter. It is important to realize that the statistics were computed on maxima rather than means. Had the means been computed the statistics in Table 1 would have been considerably smaller. Only three broadcast messages, out of 451, were encountered where the precise and the broadcast differed greatly. These are clearly operational blunders or data anomalies. Two of the three (210 m) are consecutive PRN 8 broadcast messages and are obviously attributable to the same problem. The third (102 m) is a similar PRN 11 outlier. These anomalies are not included in the statistics given here but are included in Remondi and Hofmann-Wellenhof (1989). Such anomalies would easily be detectable on, say, a 30 km baseline because the ambiguities would differ greatly from exact integers.

Associated with Table 1 is Figure 1 which also gives maxima for all satellites. Along the abscissa is maximum difference between precise and broadcast orbits. Along the ordinate is frequency of occurrence. One reads Figure 1 as follows: with a probability of 50%, the broadcast along-track error component of PRN 3 (cesium) has a maximum less than 6 m.

Table 1 and Figure 1 present the same data in different forms. In either case it is clear that, for GPS satellites operating with cesium clocks, the broadcast and precise orbits nearly always differ by less than 10 meters. (The cesium satellites are emphasized because the future of GPS will be almost completely based on cesium clocks.) In fact, with a probability of 84%, the maximum difference between the broadcast and the precise orbits is approximately the sum of the mean and standard deviation (e.g., 8.1 m, 7.5 m, 7.9 m, and 6.6 m for the along-track components of 3, 11, 12, and 13, respectively).

It should be noted that the broadcast orbits are sometimes used outside of their prescribed time interval. In Figure 2 we show, for tutorial reasons, the precise minus broadcast differences where a single message was used from TOE - 12 hours to TOE + 12 hours instead of the prescribed TOE - 2 hours to TOE + 2 hours. It is obvious that this should be avoided for precise baseline estimation.

On the other hand, Figure 3 presents an example where the 24 broadcast messages listed in Table 2 were combined in a sequential manner so as to construct 35 hours of orbital data. This concatenation reflects the reality of actual field usage. The 35-hour arc was compared with the precise ephemerides to generate these plots. One can see, from Table 2, that there are two gaps (08:00-14:00 and 21:00-01:00). The latter was not a problem since TOE = 21:00 can be extrapolated forward to 23:00 and the TOE = 01:00 can be extrapolated backward to 23:00. The former, however, is a 6-hour gap. In Figure 3 we extrapolated TOE = 08:00 forward to 11:00 and TOE = 14:00 backwards to 11:00. In general this is not recommended and is unnecessary because users cannot collect a broadcast message when the satellite is not visible and, when the satellite is not visible they also cannot collect geodetic observations to process. These broadcast orbits were recorded by the GPS receivers of the National Geodetic Survey (NGS), and PRN 11 was not visible to the NGS receivers during this period. One can see a strong downward spike in the along-track plot of Figure 3 during this period.

As a final note on Figure 3, we computed the mean and standard deviation based on a 15-minute sampling interval. (Throughout this paper we compute means of absolute values.) In this case we were not interested in the maxima. These statistics are given in Table 3.

How do the means and the maxima compare for a given broadcast message over the intended 4-hour period? Very roughly, we note a factor of 1.5 to 2.0 for along-track and cross-track and a factor of 1.0 to 1.5 for radial.

COMPARISONS OF BASELINES REDUCED WITH PRECISE AND BROADCAST ORBITS

In the next phase of this evaluation of broadcast orbits we processed eight baselines (four of them on consecutive days) using the broadcast and, again, using the precise. Furthermore, we repeated the computations after removing the satellite with the largest precise minus broadcast differences. All occupation times are indicated in the attached tables. In general, 3 or more hours of data were collected.

On August 17, 1987, the five stations depicted in Figure 4 were occupied. On August 18, 1987, they were reoccupied. First we discuss the August 17, 1987 data. These data were collected by Dr. Al Evans of NSWC using TI-4100 GPS receivers.

First we created a broadcast orbit file in the format of the precise orbits. This file was created by concatenating broadcast orbit files as was done in Figure 3. This reflects the actual field usage of broadcast orbits. These broadcast orbit files were compared with the precise orbits and a 1-minute sampled mean and standard deviation were generated. Refer to Table 4 to see the results of this comparison. Note that the PRN 8 mean is significantly larger than the others. For this reason the data were first processed with PRN 8 and then reprocessed without PRN 8.

Table 5 lists the baseline vector differences, components and lengths, between broadcast and precise orbits in millimeters and in units of hundredths of ppm (hppm). In the first row, MBRE to NSWC, all PRN's were used; in the second row PRN 8 was omitted. Refer to Table 6 for baseline lengths in meters. They range from 1.5 km to 8.4 km. When PRN 8 is omitted in the processing, the baseline lengths between precise and broadcast agree in the mean at the 0.1 ppm level (maximum of 0.16 ppm) and the components agree in the mean at the 0.1 ppm level (maximum 0.29 ppm). Table 6 shows the ambiguities from the float-double-difference solutions. Notice that we show only the fractional part times 1000 for simplicity. Note how well the integers agree between broadcast and precise processing with or without PRN 8.

Tables 7, 8, and 9 show similar results for August 18, 1987 data. The integers are somewhat degraded but the degradation is similar when using either broadcast or precise orbits. The precise minus broadcast orbit comparisons are similar to the previous case as can be seen from Table 7. When PRN 8 is omitted in the processing, the baseline lengths, between precise and broadcast, agree in the mean at the 0.1 ppm level (maximum 0.28 ppm) and the components agree in the mean at the 0.1 ppm level (maximum 0.33 ppm). An examination of Table 9 indicates that the ambiguities were farther from exact integers than the previous day (e.g., MBRE to BOM2 and MBRE to RAD9).

A similar comparison study was made using Ashtech XII data collected during the August 1988 Federal Geodetic Control Committee (FGCC) test. The baselines from NBSO to MDPT and MDPT to ASWP were chosen on the basis of length (23 km and 85 km; Figure 5 and Table 12). The broadcast orbits were combined as explained above and the results compared to the precise. The comparisons are given in Table 10. The statistics are similar to the two cases above. Table 11 shows the baseline vector comparisons. In the case where PRN 8 is omitted the lengths agree in the mean at the 0.1 ppm level (maximum of 0.16 ppm) and the components agree in the mean at the 0.2 ppm level (maximum 0.44 ppm).

Table 12 comprises the ambiguities from the floating point double difference solution. Except for PRN 8, in the NBSO to MDPT baseline (computed with the broadcast orbits), where the ambiguity was 0.330 from an exact integer, the integers were quite good for both the broadcast and precise orbits.

A final comparison study was performed for data collected by NGS on February 17, 1989 using Trimble 4000 SX (10-channel receivers). The data were chosen on the basis of length (38 km and 75 km; Figure 6 and Table 15). In this case PRN 8 data were not collected. One can see from Table 13 that the orbit comparisons are similar to previous cases. This time PRN 6 has the largest precise minus broadcast difference. For this reason the comparison studies were repeated with PRN 6 omitted. These results are given in Tables 14 and 15. Notice that the cases where PRN 6 is omitted yield precise minus broadcast baseline (lengths and components) agreement at the 0.1 ppm level. Notice also that the ambiguities (Table 15) for broadcast and precise are in good agreement.

It should be noted that a quantity to reflect the satellite geometry has been included in Tables 5, 8, 11, and 14. The reader should notice that the geometries used in this study were intentionally similar. These geometries reflect that five or six satellites were used for a period of 2 to 4 hours. The figure of geometry is the square root of the trace times an arbitrary scale factor. This figure has been normalized in the sense of equal data epochs. Otherwise, closer data epochs would give the false indication of stronger geometry. The surveyor who is interested in 0.1 ppm survey accuracies should have no reluctance to observe for 4 hours.

Those four comparison studies yield similar results. They indicate that under the conditions of these studies the mean orbit error in meters when compared to 26 x 10^6 meters (i.e., semimajor axis) gives a good indication of the baseline vector differences between precise and broadcast orbits. We are well aware that with more cesium satellites (e.g., 12) tracked over a longer interval (e.g., 8 hours) the rule of thumb would be too pessimistic. Stated differently, the broadcast ephemerides of the cesium satellites appear to be consistent with 0.1 ppm differential survey activities when compared with precise ephemerides. In the future, the cesium GPS satellites will likely have broadcast orbits which will agree with the precise in the mean at 2-3 m or better (see Tables 4, 7, 10, and 13).

Based on this and related studies (not all of which have been discussed here), it is clear that the cesium satellites (e.g., PRN 13) approach this level of accuracy already. By tracking, say, 12 Block II cesium satellites over a period of, say, 8 hours, the broadcast ephemerides should be consistent with 0.1 ppm baseline lengths and components in comparison with the precise. This is not to suggest that 0.1 ppm surveys will be easy to achieve; it only suggests that orbital data, precise or broadcast, should not be an obstacle to the achievement of 0.1 ppm static (differential) GPS baselines.

SUMMARY

The results of a comparison of 451 broadcast ephemerides with the NSWC/DMA precise orbits are given. They have been compared in the along-track, cross-track, and radial components. We have used maxima at times and means and standard deviations other times. They tend to be related over the 4-hour broadcast period by maximum = mean + 1.5*sigma. The cesium GPS satellites precise minus broadcast differences appear to have [magnitude] means of 3 - 4 m (i.e., 0.1 ppm). We emphasize the cesium satellites since the future will be based on them. The results for the other satellites are less interesting but have been tabulated.

We have computed baseline vectors of eight baselines for four different case studies where two of them were on consecutive days and are likely to be correlated. The number of satellites used and the period of data collection for the four studies were similar. This is reflected in the geometries which were nearly always within a factor of two (with one exception in Table 5). In all cases, the baseline vector components and lengths agree to better than 1 ppm and, in general, when the satellite with the largest precise minus broadcast difference is removed, the baseline vectors agree in the mean at the 0.1 ppm to 0.2 ppm range. The integer ambiguities were nearly as good using broadcast or precise ephemerides.

CONCLUSION

First, the broadcast ephemerides of the cesium satellites agree with the precise in the mean at about 3-4 meters in the worst component (along-track, cross-track, or radial). This conclusion comes from Tables 4, 7, 10, and 13 which list means and not Table 1 or 3 or Figure 1 which give maxima. However, this conclusion is consistent with Table 1 and Figure 1 when the relationship between means and maxima is applied.

Second, the baseline vector solutions computed with broadcast orbits indicate that in a future based upon cesium satellites, the broadcast orbits should be consistent with 0.1 ppm baseline lengths and components when compared to the precise. This conclusion assumes an improvement factor of two when six simultaneous cesium satellites become available. Another improvement factor will come from using more cesium satellites (e.g., 12) for a longer data collection period (e.g., 8 hours).

Third, and less precisely, this study implies (by the closeness of the ambiguities to integers) that the precise orbits are better than 5 meters in the mean. Otherwise all the double difference ambiguities could not be within 0.061 cycles of exact integers (1.2 cm) over an 85 km baseline. This should be a conservative estimate considering that L1-only processing is being considered and numerous other error sources contribute to the ambiguities not being integers.

Fourth, the broadcast orbits are remarkably consistent as can be seen by Figure 1 and the standard deviations shown in Tables 1, 3, 4, 7, 10, and 13.

Fifth, the maximum along-track difference between broadcast and precise ephemerides (over the 4-hour period) is 5.3 m, 13.3 m, and 21 m in the mean for cesium, rubidium, and quartz satellites, respectively. The maxima for the other two components are smaller (3.3/2.0; 7.4/6.2; 20.0/10.0, respectively, -- cf. Table 1).

The mean along-track differences (based on 1-minute sampling epochs) for the four study cases used in this research are 2.9 m, 7.5 m, and 13.9 m for cesium, rubidium, and quartz satellites, respectively. The means for the other two components are smaller (1.7/1.4; 1.7/4.7; 10.2/12.0, respectively -- cf. Tables 4, 7, 10, 13). We note that means, thus computed, appear to be more consistent than maxima in estimating the proportional orbit error effect on the baseline vector.

REFERENCES

Remondi, B. W. and Hofmann-Wellenhof, B. (1989). On the accuracy of the GPS broadcast orbits for relative survey, NOAA Technical Report NOS 132, NGS 45 National Geodetic Information Branch, NOAA, Rockville, MD.

Rockwell International Corporation: Navstar GPS Space Segment Navigation User Interface. ICD-200, Sept. 26, 1984.

Swift, E. R. (1985): NSWC's GPS Orbit/Clock Determination System. Proceedings of the First International Symposium on Precise Positioning with the Global Positioning System. Rockville, MD, April 15-19, Vol. 1, pp. 51-62.

Wells, D. (Editor) (1987): Guide to GPS Positioning. Canadian GPS Associates, Fredericton, New Brunswick, Canada.

PRODUCT DISCLAIMER

Mention of a commercial company or product does not constitute an endorsement by the National Oceanic and Atmospheric Administration. Use for publicity or advertisement purposes of information from this publication concerning proprietary products or the tests of such products is not authorized.

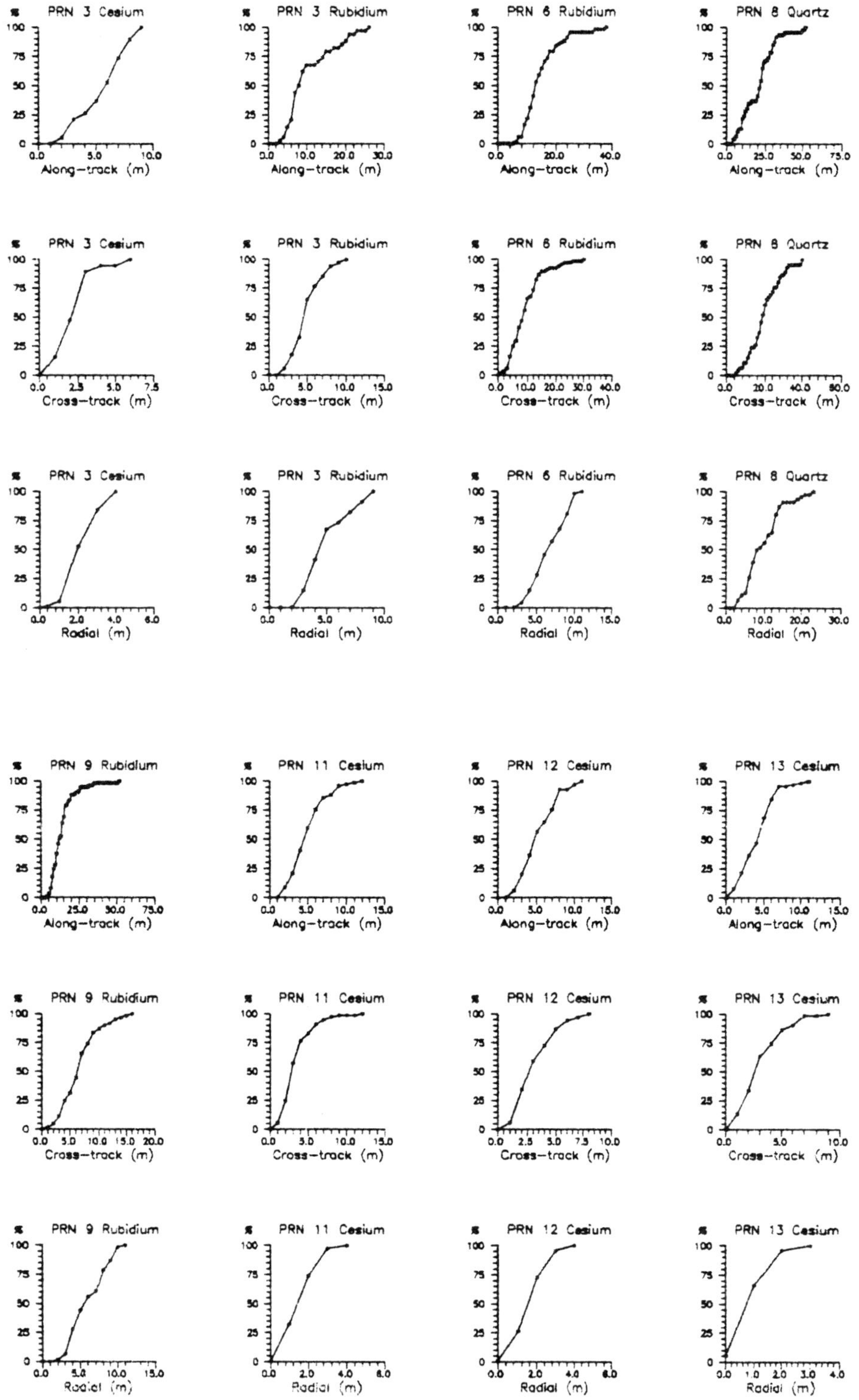

Figure 1. Maximum precise minus broadcast component error versus probability of occurence

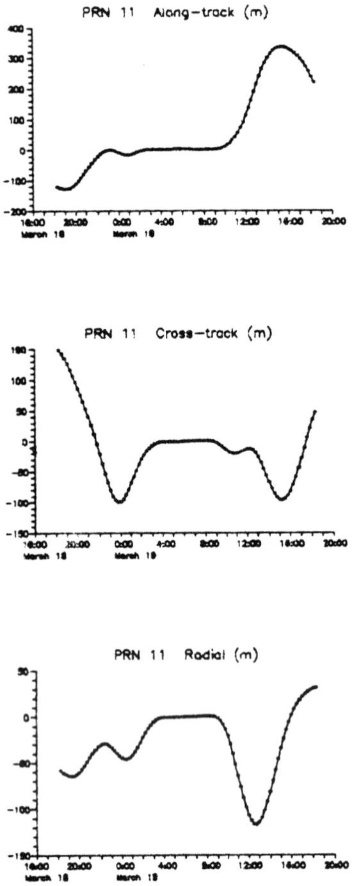

Figure 2. March 19, 1989 broadcast orbit used beyond its prescribed interval.

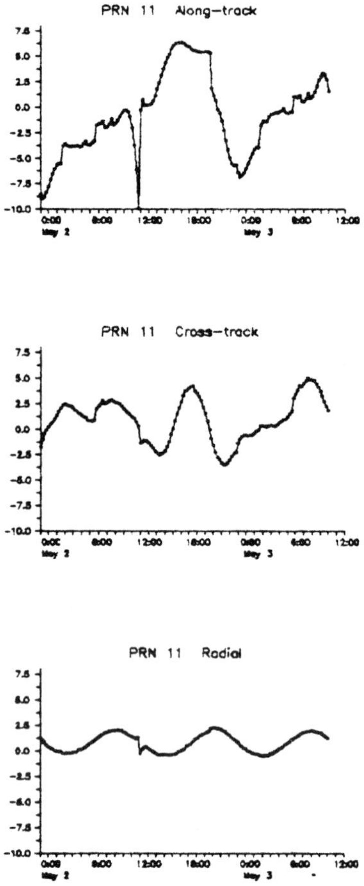

Figure 3. Concatenation of 24 broadcast messages

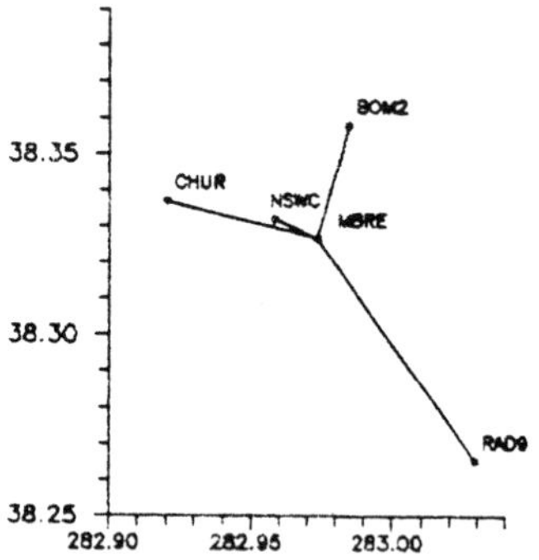

Figure 4. August 17-18, 1987 test network

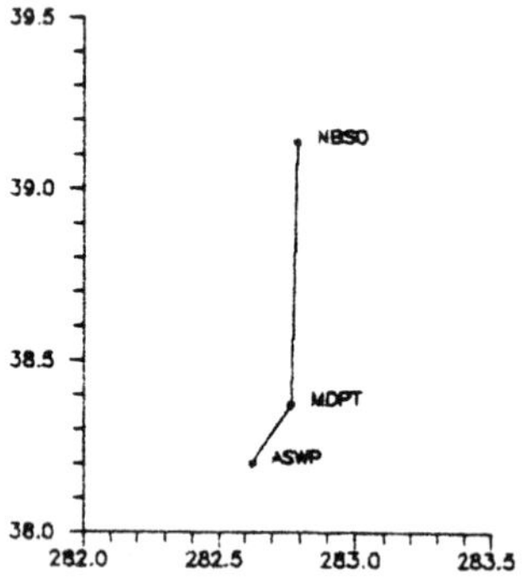

Figure 5. August 15, 1988 test network

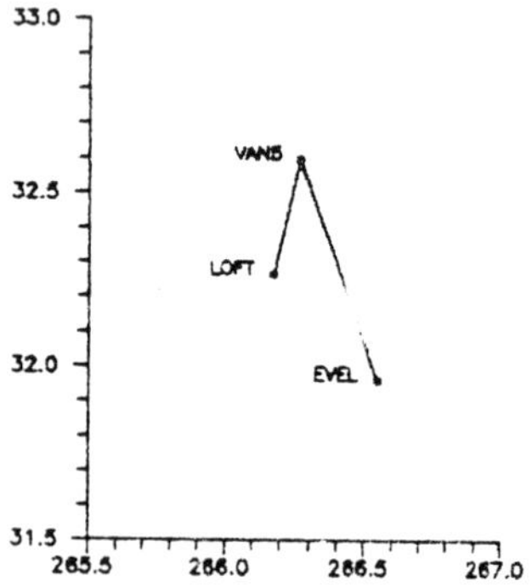

Figure 6. February 17, 1989 test network

Table 1. Means and standard deviations of precise minus broadcast component maxima

PRN	Oscillator	Along-track (meters) mean/sigma	Cross-track (meters) mean/sigma	Radial (meters) mean/sigma
3	cesium	5.9/2.2	2.6/1.2	2.6/0.8
3	rubidium	10.8/6.1	5.3/1.9	5.3/1.9
6	rubidium	15.1/6.5	9.8/5.7	7.0/2.2
8	quartz	21.0/10.9	20.0/8.3	10.0/4.8
9	rubidium	14.0/8.1	7.0/3.2	6.4/2.3
11	cesium	5.3/2.2	3.7/2.0	2.0/0.8
12	cesium	5.6/2.3	3.5/1.7	2.0/0.9
13	cesium	4.4/2.2	3.4/1.8	1.3/0.6

Table 2. 24 TOEs used to create the concatenated file referenced in Figure 3.

May 2, 01:00	May 2, 14:00	May 3, 01:00
May 2, 02:00	May 2, 15:00	May 3, 02:00
May 2, 03:00	May 2, 16:00	May 3, 03:00
May 2, 04:00	May 2, 17:00	May 3, 04:00
May 2, 05:00	May 2, 18:00	May 3, 05:00
May 2, 06:00	May 2, 19:00	May 3, 06:00
May 2, 07:00	May 2, 20:00	May 3, 07:00
May 2, 08:00	May 2, 21:00	May 3, 08:00

Table 3. 15-minute sampled mean and standard deviation of precise minus broadcast plot in Figure 3.

PRN	Along-track (meters) mean/sigma	Cross-track (meters) mean/sigma	Radial (meters) mean/sigma	Comparison Start	Interval End
11	3.2/2.4	1.9/1.3	0.9/0.7	May 1, 23:00	May 3, 10:00

Table 4. Precise minus broadcast orbit differences for August 17, 1987.

PRN	Along-track (meters) mean/sigma	Cross-track (meters) mean/sigma	Radial (meters) mean/sigma	Comparison Start	Interval End
3	6.7/1.0	2.5/0.1	2.6/0.2	Aug 17, 23:20	Aug 18, 00:05
6	3.9/2.7	0.7/0.4	2.6/0.6	Aug 17, 18:44	Aug 17, 21:20
8	13.9/5.5	8.9/4.9	11.0/2.0	Aug 17, 18:44	Aug 17, 22:34
11	1.5/1.2	2.3/1.1	1.1/0.8	Aug 17, 18:44	Aug 18, 00:05
12	2.1/0.9	0.9/0.5	2.1/0.3	Aug 17, 20:57	Aug 17, 23:25
13	3.3/0.9	0.5/0.2	1.8/0.4	Aug 17, 20:41	Aug 18, 00:05

Table 5. Precise minus broadcast baseline vector differences for August 17, 1987.

FROM	TO	DX	DY	DZ	D-LENGTH	GEOM.	NOTES
MBRE	NSWC	-0.1/7	-0.6/41	0.0/0	-0.2/14	819	
MBRE	NSWC	0.0/0	0.2/14	-0.3/20	-0.1/7	1481	w/o 8
MBRE	BOM2	-0.2/6	-0.4/11	-0.7/19	-0.8/22	819	
MBRE	BOM2	0.2/6	0.4/11	-0.7/19	-0.2/6	1502	w/o 8
MBRE	CHUR	-0.7/15	-2.3/48	1.2/25	1.1/23	1288	
MBRE	CHUR	-0.1/2	1.0/21	-0.8/17	0.2/4	5757	w/o 8
MBRE	RAD9	0.8/10	2.6/31	1.2/14	-1.2/14	856	
MBRE	RAD9	-0.5/6	-1.7/20	2.4/29	-1.3/16	1541	w/o 8

Table 6. Ambiguities (fractional part times 1000) for August 17, 1987.

FROM	TO	DATE	ORBIT	LENGTH	03	06	08	09	11	12	13
MBRE	NSWC	Aug 17, 87	PR	1478	798	080	078	---	Ref	905	922
"	"	"	BC	"	804	077	075	---	Ref	909	925
MBRE	NSWC	Aug 17, 87	PR	1478	844	072	---	---	Ref	921	937
"	"	"	BC	"	843	073	---	---	Ref	921	937
MBRE	BOM2	Aug 17, 87	PR	3604	853	023	984	---	Ref	903	952
"	"	"	BC	"	861	026	995	---	Ref	904	958
MBRE	BOM2	Aug 17, 87	PR	3604	838	030	---	---	Ref	895	951
"	"	"	BC	"	847	032	---	---	Ref	895	955
MBRE	CHUR	Aug 17, 87	PR	4771	---	919	913	---	Ref	158	889
"	"	"	BC	"	---	919	916	---	Ref	149	894
MBRE	CHUR	Aug 17, 87	PR	4771	---	926	---	---	Ref	127	907
"	"	"	BC	"	---	923	---	---	Ref	126	907
MBRE	RAD9	Aug 17, 87	PR	8372	017	026	003	---	Ref	983	014
"	"	"	BC	"	020	019	983	---	Ref	984	014
MBRE	RAD9	Aug 17, 87	PR	8372	968	042	---	---	Ref	956	995
"	"		BC	"	965	033	---	---	Ref	957	992

Table 7. Precise minus broadcast orbit differences for August 18, 1987.

PRN	Along-track (meters) mean/sigma	Cross-track (meters) mean/sigma	Radial (meters) mean/sigma	Comparison Start	Interval End
3	4.4/0.9	1.9/0.9	1.3/0.7	Aug 18, 22:46	Aug 18, 00:05
6	6.4/3.8	0.8/0.5	3.1/0.9	Aug 18, 18:44	Aug 18, 21:16
8	16.2/11.2	8.3/5.1	12.1/1.8	Aug 18, 18:44	Aug 18, 22:21
11	1.8/1.5	2.0/1.0	1.1/0.7	Aug 18, 18:44	Aug 19, 00:05
12	3.1/2.8	0.7/0.4	2.2/0.7	Aug 18, 20:53	Aug 18, 23:21
13	0.6/0.4	2.4/1.4	1.1/0.5	Aug 18, 20:37	Aug 19, 00:05

Table 8. Precise minus broadcast baseline vector differences for August 18, 1987.

FROM	TO	DX	DY	DZ	D-LENGTH	GEOM.	NOTES
MBRE	NSWC	-0.1/7	-1.0/68	0.2/14	0.2/14	806	
MBRE	NSWC	0.0/0	0.0/0	-0.1/7	-0.1/7	1399	w/o 8
MBRE	BOM2	0.0/0	-1.5/42	0.1/3	-0.9/25	748	
MBRE	BOM2	0.1/3	-1.2/33	-0.3/8	-1.0/28	1330	w/o 8
MBRE	CHUR	-0.6/13	-3.1/65	0.7/15	1.0/21	773	
MBRE	CHUR	0.1/2	0.2/4	-0.4/8	-0.1/2	1399	w/o 8
MBRE	RAD9	0.8/10	7.0/84	-0.8/10	-1.5/18	783	.
MBRE	RAD9	-0.3/4	2.3/27	1.0/12	-1.7/20	1415	w/o 8

Table 9. Ambiguities (fractional part times 1000) for August 18, 1987.

FROM	TO	DATE	ORBIT	LENGTH	03	06	08	09	11	12	13
MBRE	NSWC	Aug 18, 87	PR	1478	822	960	985	---	Ref	949	949
"	"	"	BC	"	823	960	989	---	Ref	950	943
MBRE	NSWC	Aug 18, 87	PR	1478	849	967	---	---	Ref	961	961
"	"	"	BC	"	850	966	---	---	Ref	963	962
MBRE	BOM2	Aug 18, 87	PR	3604	618	048	917	---	Ref	823	822
"	"	"	BC	"	615	054	917	---	Ref	817	821
MBRE	BOM2	Aug 18, 87	PR	3604	647	057	---	---	Ref	828	843
"	"	"	BC	"	645	063	---	---	Ref	823	843
MBRE	CHUR	Aug 18, 87	PR	4771	939	006	034	---	Ref	990	012
"	"	"	BC	"	939	007	044	---	Ref	994	013
MBRE	CHUR	Aug 18, 87	PR	4771	975	000	---	---	Ref	990	037
"	"	"	BC	"	939	007	---	---	Ref	994	013
MBRE	RAD9	Aug 18, 87	PR	8372	723	030	961	---	Ref	789	840
"	"	"	BC	"	707	038	974	---	Ref	782	835
MBRE	RAD9	Aug 18, 87	PR	8372	711	012	---	---	Ref	790	824
"	"	"	BC	"	709	024	---	---	Ref	788	828

Table 10. Precise minus broadcast orbit differences for August 15, 1988.

PRN	Along-track (meters) mean/sigma	Cross-track (meters) mean/sigma	Radial (meters) mean/sigma	Comparison Start	Interval End
6	2.1/1.5	2.4/0.8	5.2/0.8	Aug 15, 19:09	Aug 15, 20:15
8	11.7/7.1	13.5/7.6	12.9/0.9	Aug 15, 19:09	Aug 15, 21:10
9	9.6/3.4	1.9/0.8	3.9/0.4	Aug 15, 19:09	Aug 15, 21:15
11	1.0/0.5	1.0/0.4	0.9/0.5	Aug 15, 19:09	Aug 15, 22:00
12	5.5/1.1	1.0/0.8	3.0/0.4	Aug 15, 20:00	Aug 15, 22:00
13	1.9/0.5	2.4/0.1	1.0/0.4	Aug 15, 20:00	Aug 15, 22:00

Table 11. Precise minus broadcast baseline vector differences for August 15, 1988.

FROM	TO	DX	DY	DZ	D-LENGTH	GEOM.	NOTES
MDPT	ASWP	-3.4/15	11.2/49	-2.1/9	-4.2/18	715	
MDPT	ASWP	-6.9/30	6.2/27	0.0/0	-1.1/5	1670	w/o 8
NBSO	MDPT	2.5/3	73.7/87	-27.2/32	-23.4/28	1169	
NBSO	MDPT	-37.6/44	-4.1/5	14.9/18	-13.6/16	2019	w/o 8

Table 12. Ambiguities (fractional part times 1000) for August 15, 1988.

FROM	TO	DATE	ORBIT	LENGTH	03	06	08	09	11	12	13
MDPT	ASWP	Aug 15, 88	PR	22712	---	945	939	994	Ref	065	071
"	"	"	BC	"	---	953	910	053	Ref	000	022
MDPT	ASWP	Aug 15, 88	PR	22712	---	955	---	994	Ref	052	061
		"	BC	"	---	953	---	051	Ref	000	022
NBSO	MDPT	Aug 15, 88	PR	84599	---	898	046	060	Ref	881	962
"	"	"	BC	"	---	904	330	817	Ref	054	094
NBSO	MDPT	Aug 15, 88	PR	84599	---	881	---	061	Ref	909	978
"	"	"	BC	"	---	896	---	820	Ref	063	099

Table 13. Precise minus broadcast orbit differences for February 17, 1989.

PRN	Along-track (meters) mean/sigma	Cross-track (meters) mean/sigma	Radial (meters) mean/sigma	Comparison Start	Interval End
3	6.3/1.2	3.9/0.2	3.9/0.5	Feb 17, 08:50	Feb 17, 09:42
6	18.1/5.4	0.4/0.2	7.2/2.0	Feb 17, 06:40	Feb 17, 08:40
9	6.4/3.8	1.8/0.4	7.3/1.1	Feb 17, 06:40	Feb 17, 09:42
11	3.5/1.5	3.8/1.1	1.0/0.7	Feb 17, 06:40	Feb 17, 09:42
12	2.7/0.6	0.9/0.6	0.5/0.2	Feb 17, 07:15	Feb 17, 09:42
13	2.3/1.5	2.1/1.4	0.5/0.4	Feb 17, 06:40	Feb 17, 09:42

Table 14. Precise minus broadcast baseline vector differences for February 17, 1989.

FROM	TO	DX	DY	DZ	D-LENGTH	GEOM.	NOTES
VAN5	LOFT	5.2/14	-6.6/17	-4.0/11	5.3/14	812	
VAN5	LOFT	1.1/3	2.2/6	0.2/1	-1.6/4	1056	w/o 6
VAN5	EVEL	11.7/16	-15.0/20	-12.8/17	21.7/29	823	
VAN5	EVEL	3.2/4	-1.8/2	-2.1/3	3.7/5	1082	w/o 6

Table 15. Ambiguities (fractional part times 1000) for February 15, 1989.

FROM	TO	DATE	ORBIT	LENGTH	03	06	08	09	11	12	13
VAN5	LOFT	Feb 17, 89	PR	37897	126	035	---	006	Ref	013	964
"	"	"	BC	"	127	079	---	078	Ref	993	983
VAN5	LOFT	Feb 17, 89	PR	37897	188	---	---	001	Ref	054	941
"	"	"	BC	"	117	---	---	076	Ref	989	986
VAN5	EVEL	Feb 17, 89	PR	75112	107	918	---	016	Ref	961	978
"	"	"	BC	"	054	019	---	137	Ref	006	027
VAN5	EVEL	Feb 17, 89	PR	75112	114	---	---	026	Ref	960	976
"		"	BC	"	979	---	---	151	Ref	060	055

OPTIMIZATION OF GPS THEORY, TECHNIQUES AND OPERATIONAL SYSTEMS: PROGRESS AND PROSPECTS

D. Delikaraoglou and F. Lahaye
Canada Centre for Surveying, 615 Booth Street
Ottawa, Ontario
Canada K1A OE9

INTRODUCTION

The objectives of future GPS-based geodetic systems have often been summarized as being twofold: (a) to provide accurate positions (up to 100s of kilometers with sub-decimeter accuracy, and at 10s of kilometers to <1-3 cm accuracy level) for the thousands of horizontal and vertical control points needed for the support of the topographic mapping efforts and the growing need for better geodetic control for multipurpose cadastre systems; and (b) to measure vector baselines up to intercontinental distances with an accuracy better than 0.01 part per million and to apply this capability to the study of tectonic and crustal deformations.

In this paper, an overview is presented of the status of current investigations relating to the optimization of methods with respect to the measurement operations, the general mathematical models, and the data analysis techniques applicable to GPS and its combination with other measuring systems. Reference to practical results is made to illustrate how the continuous refinement of such methodologies can lead to the continuous improvement in the accuracy of baseline determinations and satellite orbits, achievable under most observing conditions.

Over the last few years the Global Positioning System (GPS) has been used with increasingly more accurate results for precise relative positioning using the pseudorange and carrier beat phase signals. Three-dimensional accuracies at the level of several millimeters for short (<10 km) baselines and a few parts per million (ppm) for baselines of several tens of kilometers are routinely achievable today, in spite of the limitations imposed by the current partial satellite constellation. With the use of dual frequency measurements and improved satellite orbits several investigators (e.g. Beutler et al., 1986; Bock et al, 1986) have demonstrated that baseline accuracies at the 0.1 ppm level or better are possible, although the observing and processing strategies to achieve these accuracies in a routine fashion still remain much of a developing technology.

Undoubtedly, as contemporary GPS receiver technology improves and the number of available satellites increases, GPS-based systems will be used for a wide variety of surveying and geodetic activities ranging from efforts to support uniform topographic mapping and the growing need for better geodetic control for multipurpose cadastre systems (Ethridge, 1988) to precise geodynamic measurements and precise low-Earth satellite tracking (Lichten and Border, 1987). Any conceivable strategy to meet such diverse needs must be essentially three-tiered: (a) the highest level being the optimization of methodologies and strategies for high precision orbit and baseline determinations for GPS-based geodetic measurements, including the optimal network configurations for the unified determination of orbits, station coordinates and other geodetic parameters from global, regional and local networks of various scales and geometries; (b) the intermediate level being the compensation for a variety of natural and intentional errors that may affect positioning accuracy and the provision of mechanisms to monitor the integrity of the GPS system performance for potentially abnormal measurement situations which might occur during routine operations; and (c) the lower level, the extension of classical network observation strategies involving simultaneous deployment of several GPS receivers and the consolidation of various GPS receiver projects or subnetworks into homogeneous networks.

The wide scope of these topics has been treated within the activities of the International Association of Geodesy Special Study Group 4.117 on the "Optimization of Modern Positioning Techniques" (chaired by D. Delikaraoglou). This paper is in many ways a reflection of the various research results of this SSG to date, as well as a synthesis of what the Group has perceived to be the concepts which have some bearing on the subject of optimization of measurement operations, the general mathematical models, and the data analysis techniques applicable to GPS and its combination with other measuring systems.

SETTING ACCURACY AND RELIABILITY OBJECTIVES

Following the classical approach to the optimization of terrestrial geodetic networks, the following three targets are usually considered in order to judge the quality of an overall GPS network design: good precision, high reliability and low cost. These general criteria have often a contradictory influence on the optimal design of a GPS network. In general, the higher the reliability and the better the precision of the network, the more expensive its realization. On the other hand, precision and reliability are not completely independent either. Hence in principle, the objectives of achieving maximum precision **P** and reliability **R** at minimum cost **C** can be sought in the solution of an *objective function*

$$\mathcal{R} = a_P\,\mathbf{P} + a_R\,\mathbf{R} + a_C\,\mathbf{C}^{-1} = max \qquad (1)$$

for suitably chosen coefficients a_P , a_R and a_C by which the influence of each requirement can be weighted adequately.

The first logical step in the design of a GPS network is to specify the desired accuracy in the three-dimensional coordinates of the points to be positioned by the GPS survey. This is normally based, in part, on the covariance matrix for the coordinates to be estimated. This matrix is obtained by the covariance law applied to the underlying model thereby propagating the random errors in the observations into the unknown parameters. In a direct extension of the classical *type I error* quality measure, one may define a (1-a)% (point or relative) confidence sphere, where (1-a) is stipulated, for example, to be in the range 0.95-0.99, depending upon the type of the survey (Wells et al., 1986). The main shortcoming of this approach is that only random errors are considered. However, in practice with GPS random errors are usually very small and have little effect on the adjustment solution due to the high measurement redundancy. By contrast, the influence of systematic errors is generally greater than the effect of random errors.

Following the lead e.g. of Grafarend (1985) and Schaffrin (1985), Jaeger (1988) has recently suggested that *spectral criterions* or *criterion matrices* based upon an "*ideal*"

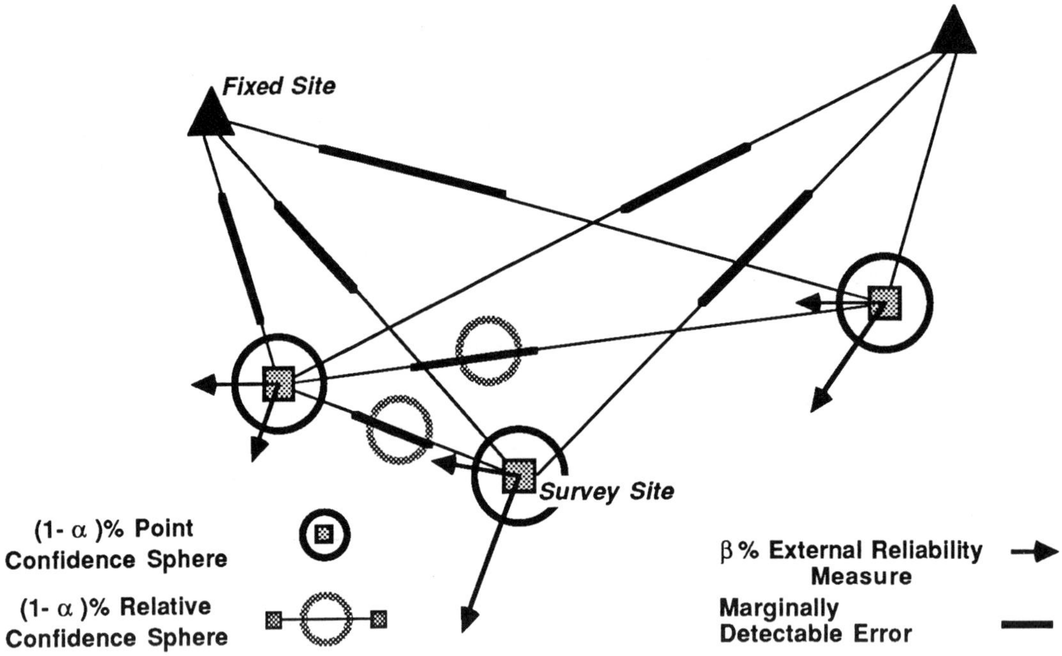

Figure 1 - Setting internal and external reliability standards

variance-covariance matrix $\mathbf{C_X}$ of the network points and/or of the estimable quantities $\mathbf{C_b}$ (e.g. baselines) be used to set objective quality measures. That is, the analysis and optimization of a given network be based on target fuctions which can then be described by eigenvalues and eigenvectors of the coordinates of the network points. Optimality can then be sought, for example, by (a) an *A-optimality*, i.e minimizing the mean variance tr($\mathbf{C_X}$), which is computed as the sum of the eigenvalues λ_i of $\mathbf{C_X}$; (b) a *D-optimality*, i.e. minimizing the generalized variance Det($\mathbf{C_X}$), which is the product of the eigenvalues proportional to the volume of a n-dimensional confidence hyper-ellipsoid; or (c) an *E-optimality* leading to a minimum for the maximum eigenvalue λ_i of $\mathbf{C_X}$, i.e. when all eigenvalues are equal; that is, the confidence hyper-ellipsoids degenerate into hyper-spheres, so that one ends up with a homogeneous and isotropic situation throughout the network (Schmitt, 1985).

Another important target specification concerns the reliability of the GPS survey, i.e. the ability of the network to self-check against blunders or systematic errors. The latter can occur in the observations or in other parameters such as satellite or station coordinates or both. *Internal reliability* can be characterized by the size of the marginally detectable error in an observation ℓ_i, which is given as

$$\nabla_o \ell_i = \sigma_o \sqrt{\omega_o} / \sigma_{\hat{\varepsilon}_i}, \qquad (2)$$

where $\sigma_{\hat{\varepsilon}_i}$ is the i-th diagonal element of the matrix $\mathbf{C}_{\hat{\varepsilon}}\, \mathbf{C}_\ell^{-1}$, $\mathbf{C}_{\hat{\varepsilon}}$ is the covariance matrix of the estimated residual vector, $\mathbf{C}_\ell$ is the covariance matrix of the observations, and ω_o is a non-centrality parameter (computed once α and β (see below) are known) corresponding to the alternative hypothesis governing the situation of having the presence of blunders (Kavouras, 1982; Teunissen, 1985a). Similarly, *external reliability* can be quantified in terms of the β% acceptable amount (type II error) by which the unknowns (the coordinates) can be affected by an undetected blunder of size equal to the internal reliability, i.e.

$$\nabla_0 \mathbf{x} = -(A^T C_\ell^{-1} A)^{-1} A^T C_\ell^{-1} \nabla_0 \ell_i \tag{3}$$

where $\nabla_0 \mathbf{x}$ is a vector of changes in the unknowns, and A is the network design matrix. The type II error is usually expressed as a point shift vector (whose acceptable magnitude would be a fraction of the size of point confidence sphere) as shown in Figure 1. By analyzing which observations produce the largest shifts, networks can be designed to reduce these values. Experiences with terrestrial networks suggest that it is uneconomical to design surveys with $\beta < 0.20$; however, in the case of GPS, this may be possible largely because of the high redundancy usually provided by GPS measurements.

There has been little effort to date to bring cost criteria into the mathematical form of **C**. Formal investigations into logistics design were carried out at the early days of GPS. For instance, Hothem et al. (1984) examined some practical aspects of two-receiver strategies, while three- and four-receiver strategies were considered by Snay (1986). Janes et al. (1986) attempted to deal with the problem of optimizing campaign logistics so as to make the most effective use of available resources. Their logistics design was formulated around the selection of the most efficient mapping of vehicles to receivers, and receivers to stations, so as to minimize the total distance travelled over the period of the campaign, while maintaining the integrity of a particular configuration design.

These efforts were prompted mainly by the need to optimize receiver deployment in light of the limitations of the early (Block I) GPS constellation. In the near future, as the number of available satellites increases, and as new GPS receivers of substantial lower cost and capable of tracking many (e.g. eight or more) satellites become available, these early GPS survey design considerations may not be as critical. Instead, GPS network design and optimization involving the simultaneous deployment of multiple GPS receivers will likely be confined to adhering: (a) to specific principles capable of providing various levels of accuracy associated with well established multi-station, multi-satellite observation strategies; and (b) to associated general mathematical models and data analysis techniques. In practice, the design of a particular survey will then, in large part, be determined by the way in which the measurement operations are to be carried out and the observations be processed, and will involve some trade-offs between rigour and economy.

Having specified the goals of GPS design and optimization, let us now discuss the steps that will help one to achieve them.

CONSIDERATIONS IN GPS NETWORK DESIGN

Evidently designing a GPS survey to an a priori specified level of accuracy and reliability, while minimizing the cost of a GPS field campaign, and at the same time adhering to these specifications is a complicated process involving various levels of decision making. Following the classical approach to the optimization of terrestrial geodetic networks, the overall GPS design process may be described in four major parts:

- *datum definition* (zero-order design)
- *network configuration* (first-order design)
- *specification of observation weights* (second-order design)

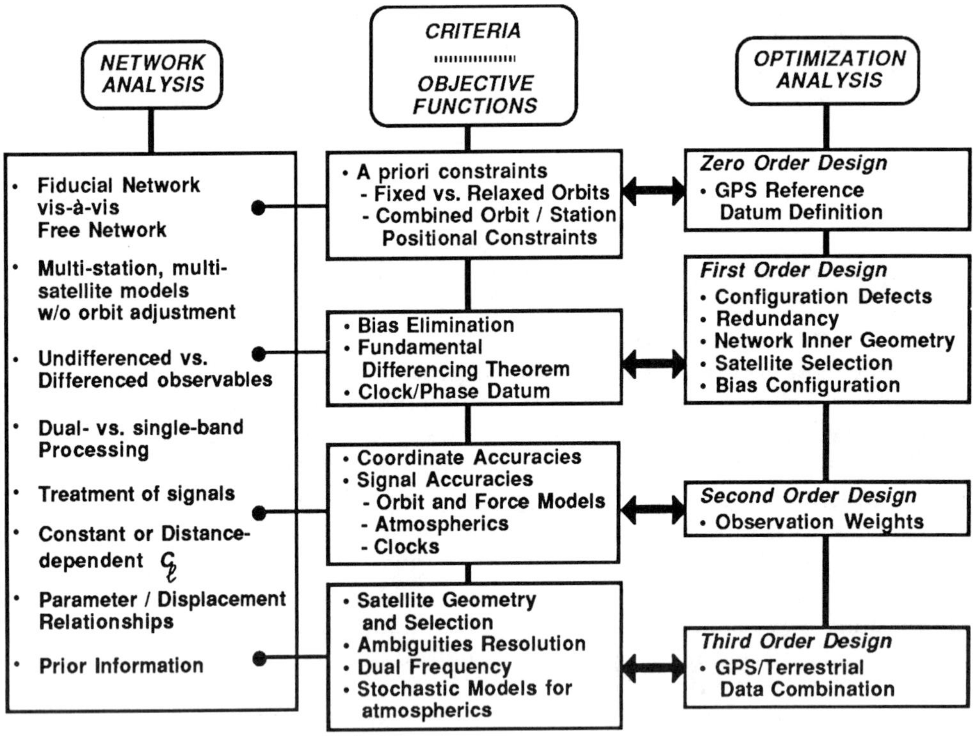

Figure 2 - Aspects of network and optimization analysis

- *integration with existing terrestrial networks* such as triangulation, trilateration, levelling, etc. (third-order design).

As Figure 2 illustrates, there is a direct correspondence between the underlying principles of network design and the optimum (functional as well as stochastic) models that relate the GPS observations to a finite set of unknown parameters determined in the data analysis stage. All the design optimization problems are related to the adjustment of the relevant observations, and are in fact formulated with the help of the same terms (i.e. design matrix, weight matrix, and covariance matrix of unknown parameters) which ultimately appear in the adjustment of the collected observations. In the case of GPS, the unknown parameters are almost exclusively the coordinates of the satellite and ground points that make up the network, and other nuisance parameters (clock errors, atmospheric delays, carrier phase ambiguities, etc.) that describe the inconsistencies between the available observations and the corresponding observables (i.e. the outcomes of the underlying functional model). This connection is much less obvious with GPS than with terrestrial observations, so that an understanding of the GPS adjustment models along with the assessment of the role and likely influence of systematic errors is an essential ingredient of any GPS survey design.

GPS Reference Datum Definition

The uniqueness in coordinate system definition plays a fundamental role when identifying, interpreting, qualifying and comparing results from GPS network adjustments. As with the conventional terrestrial networks, GPS zero order design is concerned with the problem of datum defects which arise when the collected observations contain insufficient information to uniquely define the reference coordinate frame.

It is well known that GPS observations, be they pseudoranges or carrier beat phases, are actually biased ranges rather than geometric distances. Consequently, each measurement implies its own scale, which in turn is provided only to the extent that the contributing biases (e.g. clock, ambiguities, atmospheric errors etc.) can be modelled or sufficiently accounted for. Resolution of the remaining origin and orientation datum defects are then defined by minimum contraints on some combination of the satellite and terrestrial coordinates that make up the network. Unlike the terrestrial networks where origin is defined by merely holding fixed the coordinates of one point, and orientation is defined by zenith distance and azimuth observations, the GPS datum definition is generally more complex and generally overconstrained. The nature of the constraints applied and the assumptions upon which they are based will have a direct influence on the nature of the reference system realized. Generally, these may take the form of *fixed constraints* (i.e. specified "known" positions), *conditional constraints* (i.e. through additional relationships between certain points within the network) or *weighted constraints* (i.e. weights assigned to some or all (station or satellite) coordinate values involved in the adjustment.

For most surveying applications, the "network" of the three-dimensional satellite position vectors defining the satellite constellation is generally assumed known as defined by the adopted (e.g. broadcast) ephemerides of the GPS satellites used in the data reduction. Consequently, constraints on the orbital parameters which model the orbital motion of the satellites provide the mechanism for reference frame definition. That is, the reference system is defined by the set of global tracking stations used to generate the broadcast ephemerides. The GPS datum is then transfered to the stations of the survey network via these ephemerides without the need of direct observations between the survey and the global tracking stations. In theory, fixing the orbit of one satellite would suffice to define the origin, orientation and scale of an earth-centered, quasi-inertial reference frame; and, if the associated earth rotation parameters are assumed known, to define an earth-centered, earth-fixed reference system in which the positions of the stations of the survey network may be estimated . In practice, however, the orbital parameters of more than one satellite are held fixed thus providing an over-constrained datum (Grant,1988). The drawbacks of such practice are rather obvious. If the satellite ephemerides have been generated by a complement of stations close to the survey network, these separate datum definitions will likely be consistent with one another and with the tracking network. Otherwise, if the satellite ephemerides have been generated by a distant tracking network, these separate datum definitions will likely yield greater discrepancies, in addition to network distortions likely to be introduced in the survey network in case that large errors exist between the orbits implied by the adopted fixed ephemerides and the true satellite orbits.

The above two-step process may effectively be combined in a single, multi-satellite, multi-station adjustment known as the *Fiducial Network* concept (Davidson et al., 1985) whereby both the satellite ephemerides and the station positions are defined in the framework of a few fiducial stations whose coordinates are known to a few cm by some method independent of GPS, such as *Very Long Baseline Interferometry* (VLBI) or *Satellite Laser Ranging* (SLR). Tying a GPS network to a reference system relying on VLBI or SLR ensures that it can be utilized for very precise geodetic measurements on a regional or local scale, since the coordinate system orientation and scaling should be potentially self-consistent at a level of 1 part in 10^8 or better. To the extent that orbital dynamics, clocks and atmospheric effects are adequately modelled, the accuracy of the survey stations will then reflect that of the fiducial stations. In particular, by locating the GPS fiducial sites at VLBI locations, the GPS orbits and survey stations are inherently expressed in the coordinate system of the extragalactic sources which in itself has the appeal of unifying GPS and VLBI results in one and the same system with all the practical benefits of permitting their direct comparison and combined use in geophysical interpretation. If the same fiducial network is used in a given region, an additional advantage of the fiducial network approach is that the datum can be defined in the same way between surveys. This is particularly important for deformation studies relying solely on repeated GPS surveys, since any distortions caused by incorrect fiducial station coordinates (being the same for all epochs) will tend to cancel.

The overall fiducial station geometry is the most important factor affecting the sensitivity of baselines solutions to errors in the fiducial station locations. A network with a global extent is conceivably stronger than a network of only continental size, mainly because the satellites are tracked over a longer portion of their orbit, so that the random errors in individual fiducial station locations tend to average out of the solution. The *Cooperative International GPS Network* (CIGNET) is currently providing a continuous fiducial data set (available on request from the U.S. National Geodetic Survey, Rockville, Maryland) that is used for orbit computations and research (e.g studies of compatibility between VLBI/SLR and GPS/WGS84 reference frames). In the near future, by extending the geodetic fiducial network globally (e.g. through Continuously Active Remotely Monitored Arrays [Mueller and Zerbini, 1989]) may well provide improved reliability which will undoubtedly prove beneficial to further development and utilization of GPS for the practice of geodesy and geodynamics, including significant contributions for the tracking of future low-earth satellites (e.g. TOPEX/POSEIDON) equipped with onboard GPS receivers.

Presently, the fiducial network approach has been proven useful especially in the U.S.A. where there is an extensive VLBI/SLR network of stations with accurately determined coordinates. In Canada, a three-week campaign for the *GPS Orbit Tracking EXperiment* (GOTEX-1) held in November '88 provided a unique opportunity to carry out first-epoch measurements over a 28-station Canada-wide network (including the VLBI sites at Yellowknife, N.W.T., Penticton, B.C. and Algonquin,Ont.) which will serve as the basis for the Canadian *Active Control System* (ACS) currently under development by the Canadian Centre for Surveying and the Geological Survey of Canada (Delikaraoglou, 1989a). Elsewhere in regions where perhaps only Doppler coordinates (with uncertainties of a few meters or, at best, up to a few decimeters) are available to provide a geocentric datum, the *Free Network* concept (Beutler et al, 1985) may be more appropriate. In this approach a combination of station and satellite coordinates are allowed to vary simultaneously in a Bayesian estimation scheme by way of imposing relatively strong a priori constraints on the orbits and allowing the network of GPS stations (except one) to adjust free. Defining these a priori constraints plays an important role, since normally the experience of the investigator is the most reliable tool to define the quality of the given orbits. In the near future, it is hoped that routine availability of information on orbit quality will be readily available as a result of regional continuous GPS networks, so that the investigator's subjectivity problem is only a short-term one. Unlike the fiducial network approach, the free network approach obviously leads to a datum definition which is less repeatable since the a priori orbit errors are likely to change from epoch to epoch. On the other hand, this approach is less susceptible to distortions due to a priori coordinate errors.

GPS Network Configuration Considerations

As with the conventional terrestrial networks, GPS first-order design deals with the optimization of the number, combination and configuration of the observations relating the ground and satellite points that make up the network. First-order design may be considered as a two part process involving redundancy and configuration design. The automation of a first order design process is practically limited by the more mundane considerations of station placement, accessibility and permanence which rarely provide the GPS network designer with the flexibility to follow idealized geometries in the site selection. Economic considerations generally place additional restrictions on network configurations. Hence, logistics design usually becomes part of the first order design.

GPS redundancy design. Redundancy design is concerned with the establishment of a minimum network configuration that allows a number of unknown parameters to be uniquely defined [Delikaraoglou, 1985; Lindlohr and Wells (1985)]. GPS observations, being differences between satellite-generated time/phase signals (e.g. P-, C/A-code, carrier phase and observation time tags) and local receiver oscillator signals, are sensitive to differences between satellite and receiver clock errors and various causes that affect the propagation of these signals (Wells et al., 1986). The influence of these biases may be treated in various ways. In general, the type of the model employed depends on the a priori knowledge or assumptions concerning the structure of the bias. If a bias is assumed to have a stable and well understood structure, its temporal or spatial variation can be treated as unknown and solved for (together with station coordinates or other parameters of interest) in the form of nuisance parameters (e.g. clock offsets, carrier phase ambiguities, orbital biases, etc.). Alternatively, additional observations can be employed allowing either to directly estimate the bias (e.g. dual frequency ionospheric measurements), or to scale the bias model (e.g. tropospheric delay anchored to surface meteorology and/or water vapour radiometer measurements). Finally, if the bias is assumed to be perfectly correlated across some combination of measurements, it can be removed or greatly reduced through differencing of one way observations (e.g. between receivers, between satellites, and between epochs or combinations thereof).

Undifferenced carrier beat phase observables are favoured for orbit determination applications because they offer the flexibility of modelling rigorously the clock errors while still allowing the adjustment of the observations in a manner equivalent to double differencing. One additional advantage of using undifferenced data lies in the fact that the simultaneous observations are independent of each other and therefore no correlation modelling is necessary as in differencing schemes. By contrast, the latter are favoured because each level of differencing removes or greatly reduces errors related to the satellites, the propagation medium, and station instrumental effects. Schaffrin and Grafarend (1986) and Lindohr and Wells (1985) have shown that both approaches give identical results provided that the *Fundamental Differencing Theorem* applies, i.e.

"*Linear biases can be accounted for either by reducing the number of observations so that the biases cancel, or by adding an equal number of unknowns to model the biases. Both approaches give identical results*" (ibid).

The fundamental differencing theorem provides a mechanism for analyzing the rank defect in the clock/ambiguity parameters. Assuming that differences are formed first between receivers, then between satellites, then time, the total number of independent non-differenced and differenced observations in a given observation session is

R S T	undifferenced observations,
(R-1) S T	single differenced observations,
(R-1)(S-1)T	double differenced observations,
(R-1)(S-1)(T-1) . . .	triple differenced observations,

with RT receiver clock errors, ST satellite clock errors and RS ambiguities to be determined, where R, S, and T denote the number of receivers, satellites and observation epochs respectively. In practice, as Lindlohr and Wells (1985) have demonstrated, numerous minimum configurations exist for each observation type and parameters, depending upon whether one chooses to minimize R, S or T. In the undifferenced model, a simple way of eliminating the clock datum defect is to hold the T clock errors of a reference oscillator fixed, and choosing R+S-1 ambiguities to eliminate the remaining rank defect. The remaining problem is then to provide the phase link between the reference oscillator and every other oscillator in the network. Holding the ambiguities between a *base receiver* and all satellites fixed, as well as the ambiguities between a *base satellite* and all receivers fixed, is a common approach of achieving this link. Using this strategy, resolving the ambiguities for stations closer to the base station to their correct integer values ("bias fixing") can be done more accurately than for the ambiguities of stations further away. Hence, the choice of the base receiver is an important element particularly for the design of networks spanning progressively longer baselines.

GPS configuration design. In designing a GPS survey campaign, the choice of station locations will be mostly dictated by the need to occupy (or re-occupy) as many stations as possible, while maintaining an optimal duration and timing of the data collection interval. Two approaches are usually possible:

a. to establish those geometric station/satellite configurations which result in the optimum dilution of precision over the ground and satellite points which comprise the entire network; that is, leading to the minimization of the dispersion in the diagonal elements (including ambiguities) of the parameter covariance matrix. Thereafter, the degree of achievable accuracy is mainly a matter of choice of appropriate observation weights.
b. to establish those geometric station/satellite configurations which result in the maximum number of independent baselines over the network, thereby increasing redundancy and reliability.

Even with receivers capable of tracking all satellites in view, the timing and duration of the data collection interval can be significant, especially for short data spans and short baselines, whereby the accuracy of the results depends on whether the whole cycle carrier phase ambiguities can be resolved. This capability is, in turn, determined by which satellites are selected and the timing and duration of the tracking interval.

In terms of satellite geometry and selection, the *geometric dilution of precision* (GDOP) is often used to select satellite tracking priorities and observation times. GDOP, based on the volume of the tetrahedron formed by four satellites and a single receiver, is essentially a measure of the optimum satellite geometry for an instantaneous point position using pseudoranges. As such, it does not take advantage of the ability of current receivers to track more than four satellites. Hence, it is not necessarily the best indicator of the optimum geometry involving carrier phase observations which are also biased by satellite oscillator errors and integer ambiguities. Vanicek and Craymer (1986) have investigated the development of more generalized methods for differential positioning involving multiple receivers and, with the introduction of newer receivers, of more than four satellites. Furthermore, Landau and Eissfeller (1986) have attempted to account for the impact of atmospheric delays into the satellite selection problem, whereas Grant (1988) following the lead of Merminod et al. (1988) has introduced the use of *bias dilution of precision indicators* to indicate the relative precision of baseline components with or without cycle ambiguity bias fixing. In either case, differential position of dilution indicators are derived by minimization of the trace of the covariance matrix $\mathbf{C_X}$ of the unknown coordinates and biases, i.e.

$$\mathbf{C_X} = (\mathbf{A^T P A} + \mathbf{P_X})^{-1} = \mathbf{N_X}^{-1}, \tag{4}$$

where $\mathbf{P}$ is the weight matrix of observations (e.g. undifferenced phase, single, double or triple differences), $\mathbf{P_X}$ is the a priori matrix of the parameters, and $\mathbf{A}$ is the appropriate design matrix for the type of observable used. For the differenced models, both the weight matrices and design matrices may be easily derived from that of the undifferenced model by using the concept of a differencing operator (Lindlohr and Wells, 1985; Schaffrin and Bock,1988). For example, a *Relative Dilution of Precision* indicator based on the minimization of the trace of the covariance matrix $\mathbf{C}(\mathrm{B}_{f,m})$ for the baseline components between fiducial and mobile or new stations to be surveyed, i.e.

$$\mathbf{RDOP} = \sqrt{\mathrm{tr}[\mathbf{C}(\mathrm{B}_{f,m})]}, \quad (5a)$$

can provide a measure of relative precision of baseline components between fiducial and survey stations in a bias free solution (i.e. in a solution whereby the carrier phase ambiguities are let unresolved at their real values). Similarly, a *Biased Relative Dilution of Precision* indicator

$$\mathbf{BRDOP} = \sqrt{\mathrm{tr}[\mathbf{Q}_b]}, \quad (5b)$$

where $\mathbf{Q}_b$ denotes the covariance matrix of the adjusted ambiguity parameters, provides a measure of relative precision of the ambiguities in a bias free solution, albeit in a non-unique way since the exact definition of BRDOP depends on the ambiguity mapping used to construct an independent set of biases over an independent path through the network. Evidently, determining whether the optimal satellite selections are obtained, while at the same time ensuring an optimal configuration of ground points so that ambiguities can be resolved progressively from the shorter baselines to the longer ones, it is also more complicated (cf. Schaffrin and Bock,1988; Dong and Bock, 1989). Of primary concern is the effect of significant mathematical correlations introduced particularly for double difference observations. In practice, it is not necessary to solve for cycle ambiguities or their differences. Instead a triple difference procedure can be employed; then the position coordinates can be back substituted into the original equations to solve for the cycle ambiguities and clock parameters. It can then be argued (cf. Hutch and Avery, 1988) that equations (4) and (5), when formulated for the case of a triple difference solution, are directly related to whether or not the whole cycle ambiguities can be successfully resolved. It follows that a *Bias Free Dilution of Precision* (BFDOP) indicator defined by equation (5b) for a triple difference solution provides a good relative measure of successfully resolving the cycle ambiguities-assuming a constant measurement noise. Thus, the choice of which satellites should be used, along with the choice of duration and timing of the data collection can be based on the relative magnitude of the so-obtained BFDOP values. Naturally, as one gains experience with the typical measurement noise levels involved with various interstation distances, judgements can be made on what level of BFDOP is required for cycle ambiguity resolution. The most significant complication that remains is that BFDOP is a function of, not only the duration of the observing session implied in its definition, but also of the specific satellite geometry at which data collection is to commence.

Selection of GPS Observing Accuracy (Second Order Design)

Given a predetermined network configuration, the selection of observing accuracy is one of the most crucial parameters which control the accuracy of GPS determined positions. To date, with the improved instrumental resolution of the newer receivers (e.g. low noise, better antennas etc.) the typical magnitude of GPS instrumental errors is continually declining. The positioning accuracy depends on several other items which contribute to the

GPS error budget. Namely, GPS results may be biased by effects which are primarily attributable to:

- imperfect location of the observations, both in time (e.g., imperfect synchronization of the satellite and receiver clocks) and location (e.g., satellite orbital errors)
- anomalous tracking system performance (e.g., due to multipath and antenna phase centre variations), and
- signal propagation delays (e.g., residual tropospheric and ionospheric propagation effects).

Generally the longer the GPS satellites are tracked, the more accurate is the resulting relative positioning likely to be, mainly due to the more extensive sampling of the atmosphere (and hence randomization of atmospheric delays) and more varying geometry (and resulting randomization of orbital errors). Dependence of accuracy on sampling rate is very weak since frequent sampling only reduces the random part of the errors but does nothing to reduce the biases. There are several approaches that can be taken to deal with these biases as already pointed out. Consequently, the accuracy requirements for a priori information on satellite orbits, clock behaviour and atmospheric influences will depend upon the underlying model employed. The residual errors of both propagation medium and orbital origin are approximately proportional to the interstation distance, at least for distances up to several tens of kilometers. They may also depend on orientation due, for instance, to gradients of the atmosphere. Hence from the point of view of relative weighting, the choice of which approach to use depends solely on circumstances and applications. To this end, the bias modelling aspects in geodetic applications of GPS become an integral part of the selection of observing accuracy. To illustrate the point, we shall discuss briefly these influences and exemplify their impact with some typical practical results.

Atmospheric modelling Influences. GPS technology, like all space geodetic positioning technologies, has a critical dependency on the monitoring of local meteorological and environmentally induced phenomena which have a non-predictable behaviour over measurement averaging intervals and/or short time scales that affect the transmission and propagation of the satellite signals.

Tropospheric Delay Modelling. Tropospheric delay mismodelling can be one of the significant causes of error in GPS positioning in general, and a major one for small networks. Calculation of the tropospheric contributions require a model of the dry and wet troposphere and the use of the satellite/station geometry (mainly by elevation angle). By observing over many azimuthal and elevation directions, sufficient redundancy is obtained to estimate the tropospheric contribution by either:

a. combining surface measurements of atmospheric temperature (T_S), barometric pressure (P_S), and relative humidity (RH) with simple atmospheric models;
b. direct measurements of atmospheric water vapour using water vapour radiometers (WVRs); and/or
c. stochastic estimation according to the apparent temporal behaviour of the wet troposphere.

Various models (e.g. Hopfield's, Yionoulis', Chao's, Black's and others) for the tropospheric correction on range measurements express the tropospheric delay as a linear combination of the dry and wet components. The dry component is qualitatively more important while the wet component is more difficult to estimate due to its dependence on water vapour in the lower (up to about 12 km) part of the atmosphere. Usually, the dry part of the troposphere is well modelled from surface measurements with an accuracy of about 2% (Wells et al., 1986). The evaluation of the wet term (which under certain conditions can account for as much as 10% of the total effect) depends on the nature of the data available to characterize the humidity in the various layers of the atmosphere above the station and along the signal path. Hence, even with precise surface meteorological data and well-calibrated instruments, tropospheric delay estimates can still be in error by as much as 10% since the conditions along the signal path are not necessarily correlated with surface conditions. The variability of the water vapour content also implies that, for interstation distances larger than a few tens of kilometers, the tropospheric effects cannot effectively be removed using differential observations.

To avoid the use of expensive WVR instrumentation, estimation of a tropospheric scale factor at each station and each session is usually added to the processing algorithm to deal with this observation error. It has been previously shown (Santerre, 1987) that the estimation of a single tropospheric scale factor for each station during a long observation period can introduce unexpected variations in the height and baseline length components, especially if standard meteorological data are used at networks with large (e.g. >100 m) station elevation differences or where significant atmospheric changes occur. Consequently, it has been concluded that such variations are propably due to the constraints imposed on the tropospheric scaling parameters. Empirical correction algorithms and filtering techniques have been developed lately (e.g. Treuhaft and Lanyi, 1987) for some data analyses strategies which assume that the tropospheric delay has a stochastic character with a specified power and decorrelation time. Constructing such stochastic models for the permissible rate of change of the tropospheric conditions has shown tremendous promise (e.g. Bertiger et al., 1988) in estimating the troposphere's behaviour over time and consequently reducing errors in station positions, as well as improving the repeatability of the estimated GPS orbits.

Ionospheric Delay Modelling. By contrast to the troposphere, the ionosphere has a variable index of refraction which is a function of both the frequency of the electromagnetic signal and the electron density along the signal path.

Ionospheric effects, being frequency dependent, can be evaluated accurately using dual-frequency measurements. Alternatively, for stations that are closely spaced, differencing between stations can reduce large-scale ionospheric effects. Typically, residual ionospheric errors can amount to about 0.5 ppm of the distance between stations during nighttime observations, and about 2 ppm during the day. However, there currently exist unanswered questions regarding the effects of increased ionospheric variability on GPS carrier phase signals at the equatorial regions and in the northern geomagnetic latitudes. Over the last few months the effect of ionospheric disturbances has been noted clearly (through routine daily data quality testing) in the performance of the two most-northerly located sites of the CIGNET network (i.e. Yellowknife, N.W.T., Canada and Tromso, Norway). In addition, experiences gained while processing GPS observations collected at the Canadian aurora region (Heroux and Kleusberg, 1989; Delikaraoglou, 1989b) have demonstrated that ionospheric disturbances at high-latitude regions can indeed be a critical factor for precise positioning with GPS. Even at periods of minimum solar activity, baseline errors up to several ppm can be attributed to the time delays imposed on the GPS signals upon their passage through the ionosphere. The effect is likely to be more pronounced during the periods of the upcoming maximum of the solar activity cycle. Naturally, if single frequency receivers are deployed in such a situation, such difficulties will likely be further complicated.

In an effort to enhance the modelling of the ionospheric delay signals in the GPS measurements it has been suggested (ibid) that the ionospheric-induced variations between stations may be interpreted as governed by a stochastic process. In Delikaraoglou (1989b), the L_1 and L_2 GPS carrier signals and additional parameters were utilized, and after model identification between several possible stochastic process models (i.e. autoregressive, moving average, or combinations thereof), parameter estimation and diagnostic checking, it was demonstrated that it is possible to derive relatively simple autoregressive stochastic models of the ionospheric delays of the form

$$(d\mathbf{I}_r{}^k)_i = \sum_{j=1}^{p} a_j \, (d\mathbf{I}_r{}^k)_{i-j} + w_i \qquad (6)$$

in which each element of the ionospheric delay $(d\mathbf{I}_r{}^k)_i$ is expressed as a weighted average of its previous p values plus a noise component w_i. That is, even in the presence of complex ionospheric irregularities, it is possible to predict, albeit not perfectly, the future behaviour of the ionospheric delays from a perturbed version of their past history-at least for the short time periods of typical GPS observing sessions (e.g. 2-4 hours) and over distances less than 100 km. Hence, the usual GPS two frequency-correction model can be extended to a dynamically "tuned" functional model in the realm of which the coefficients a_j of the AR process can be solved for in the same way as other parameters. However, in practice, they can be obtained from a residual analysis, and introduced as constants in a repeat adjustment of the data.

Influence of Satellite Position Uncertainty. We have already seen how the uncertainty in satellite positions becomes important in the GPS datum definion and network configuration design. In connection with the second order design, there are two major aspects one has to deal with: (a) the minimization of the effect of orbital errors, and (b) the characterization of these errors in statistical terms (e.g. via the definition of the weight matrix of the a priori satellite orbit information).

Minimization of the effect of the orbital errors through modelling is in itself a difficult problem which has attracted considerable attention, for instance, by Beutler et al. (1986),

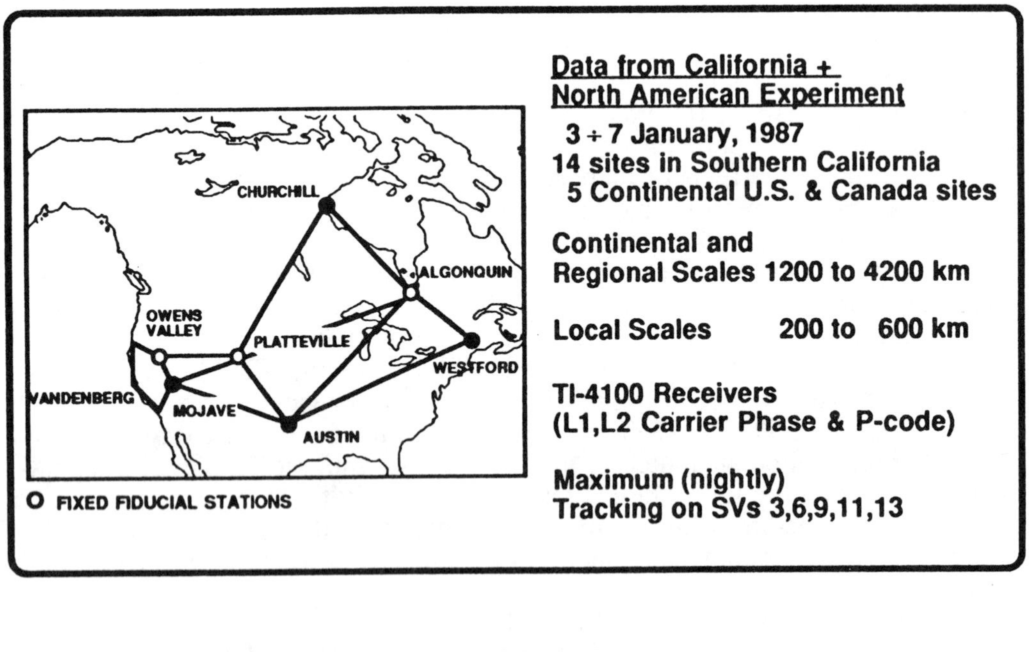

Figure 3 - Main characteristics of the 1987 Standard GPS Data Set.

Colombo (1986, 1989), Lichten and Border (1987), and Delikaraoglou (1989a) among others. Orbital errors have a diversity of origins of which the imperfect force models of gravity, solar radiation, the so-called y-bias force etc. are the most dominant. Theoretical developments based on analytical orbit perturbation theory and experimental findings (e.g. by Colombo, 1986; 1989) have suggested that orbital errors, in spite of the complexity of their causes, can be corrected-at least for satellite arcs during which no orbit manouvers have taken place-by adjusting a few parameters representing simple empirical accelerations in addition to adjusting the initial state conditions for each satellite.

Figures 4 to 6 show examples of typical levels of GPS orbit determination performance obtained from preliminary computations at the Canadian Geodetic Survey with the GPS

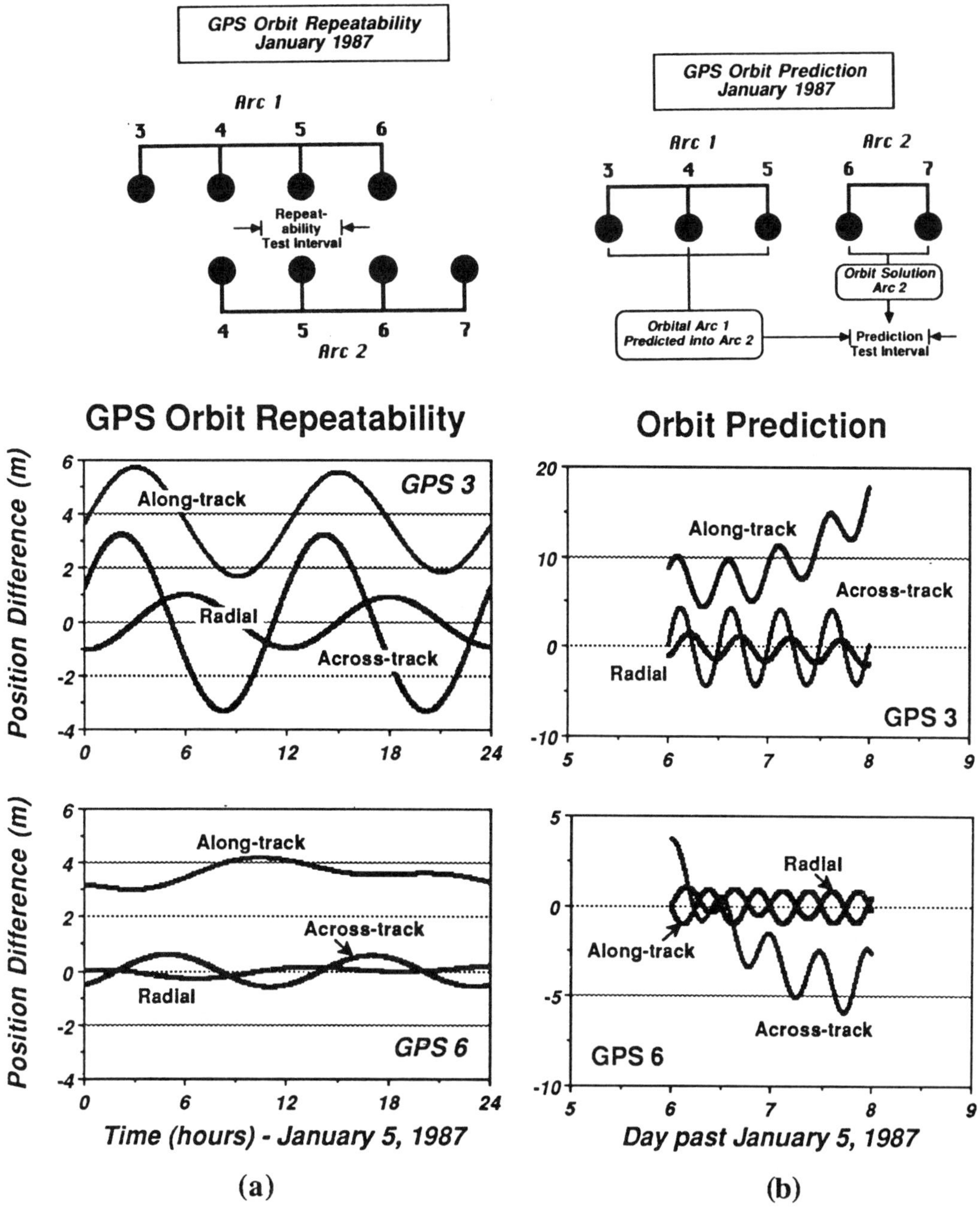

Figure 4 - GPS orbit repeatability and prediction tests

Standard Data Set (Bock, 1988) currently being studied by members of the IAG Special Study Groups 1.104 and 4.117. This data set was collected in January 1987, as part of an ambitious experiment of crustal deformation studies in California, over a network of U.S. and Canadian sites designed to span a wide-range of spatial scales (Figure 3). Four criteria are used in these examples to give an assessment of the quality of GPS orbits determined from multi-day arcs and to illustrate the circumstances necessary to solve for orbit parameters: (1) formal errors from the combined orbit/network solutions; (2) orbit repeatability; (3) orbit prediction; and (4) daily baseline repeatability.

In general, for well tracked satellites, 1σ formal errors (not shown here) as obtained in each position component through various solutions using 2-, 3- and 4-day long arcs respectively were at the level of 1 m which corresponds to about 5 parts in 10^8. These uncertainties correspond to the estimated initial conditions for each satellite at the beginning of the corresponding arc. The accuracies towards the end points of a satellite arc are not necessarily as accurate as the above values. They are a function not only of the accuracy of the estimated initial conditions and non-gravitational parameters, but also on the completeness of the physical model used to integrate the equations of motion and the variational equations. These formal error levels are indicative of the level of precision of the computed orbits.

Since, in general, formal errors tend to underestimate orbit accuracy, further comparisons between solutions of two overlapping arcs can be used to better assess the quality of multi-day orbits. *Orbit repeatability* indicates the precision, and to some extent, the accuracy of the GPS orbits. It can be quantified by comparing overlapping orbits determined independently without any common measurements and computing the rms difference over a time interval during which no data were used for either of the two solutions. This constitutes a demanding trial on the quality of the computed orbits, because the longer the arc, the more difficult to keep orbit errors small and to achieve good repeatability without readjusting the orbits. Figure 4a illustrates the repeatability test applied to the multi-day arc

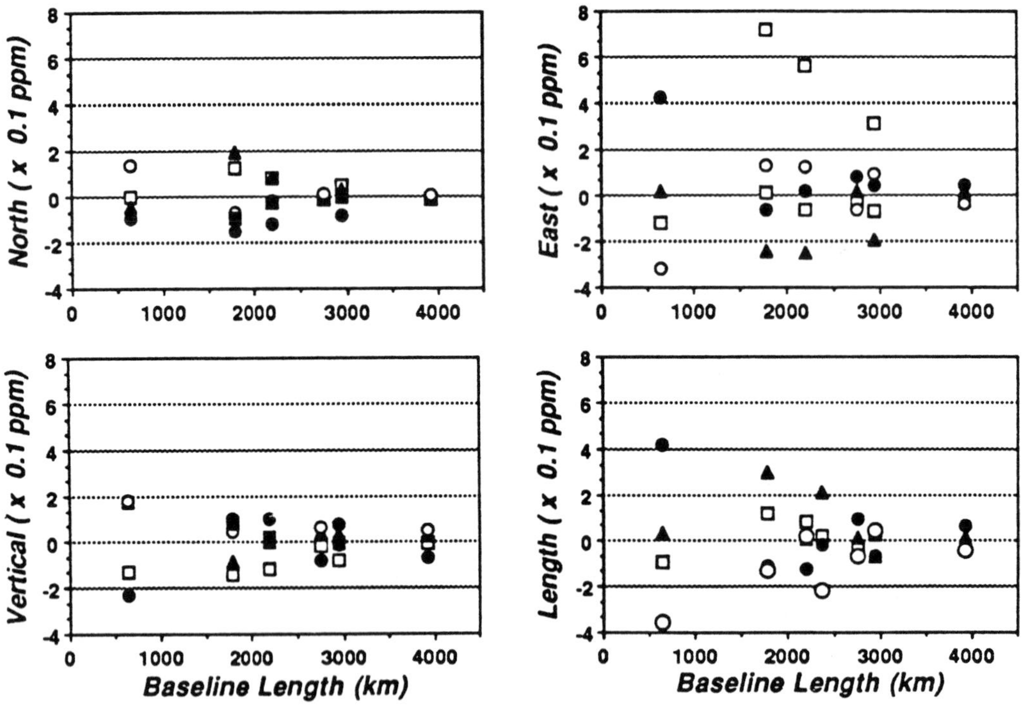

Figure 5 - Baseline repeatabilities of 4 daily solutions
[six continental baselines (N. America)]

orbits for the GPS satellites 3 and 6, the two satellites with the most data and best ground viewing geometry. Also shown is the time history of the position differences computed over a 24-hour interval in January 5, 1987 when no measurements were used in the orbit solutions.

Orbit prediction is a more stringent test of orbit accuracy, since the estimated satellite state vector is mapped outside the measurement interval to give predicted satellite position and velocity. Such tests also tend to test the robustness of the orbit models used for the orbit prediction, since orbit errors tend to be magnified in this mapping process. Figure 4b shows the prediction test applied to an interval corresponding to January 6 and 7, 1987.

Generally, radial errors remain at the level of 1-2 m, whereas across-track errors are the level of 1-5 m. Along-track errors, on the other hand, soon grow in a fashion overwelmed by a linear trend (or slightly quadratic bend) likely caused by the y-bias force. More importantly, all plots show fluctuations with a frequency close to once per revolution which support Colombo's (1986,1989) assertion that any unmodelled or improperly modelled orbital effects in GPS ephemerides have a resonant character and that they can indeed be lumped together and be dealt with simultaneously without using elaborate models for radiation pressure, gravity or other complex forces.

To further assess the GPS orbit accuracy determined with multi-day arcs, Figure 5 shows the results of four daily determinations for the baselines between the non-fiducial sites (Westford and Churchill) to each of the fiducial sites (Algonquin, Plateville and Owens Valley) used in these solutions. Baseline repeatability is quantified by the rms (scatter) about the mean of the four daily solutions and is given for the north, east, vertical components and the length of the baseline vector. Generally, in most cases the results indicate achievable repeatabilities in the north and vertical components, and in the baseline length at the rms level of 2 parts in 10^7 or better. The rms scatter for the east component seems to be slightly larger, at the level of 4 parts in 10^7 or better which is likely due to the

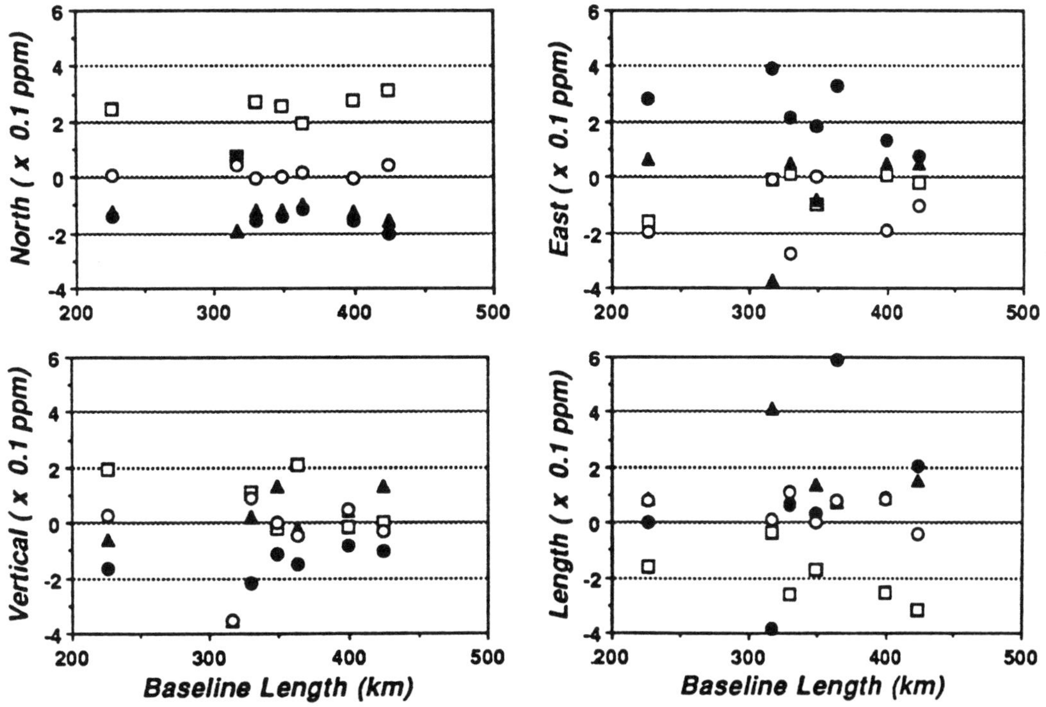

Figure 6 - Baseline repeatabilities of 4 daily solutions
[seven local baselines (Southern California)]

relative geometry of the satellite constellation and the fiducial sites. Since the nominal schedule for the observations was the same each day, a large deviation for a given day implies a loss of data and/or larger data noise (which is usually reflected also in the a posteriori variance factor from the least squares adjustment).

Figure 6 shows the results from the analysis of seven baselines in the local network of southern California estimated in a grand adjustment where the GPS observations from the central and southern California network were processed together with the data from the continental network. For baselines of this length, an important factor in the achievable accuracy is the ability to resolve the integer-cycle ambiguities in the carrier phase observations. The MICROCOSM software used for these tests does not allow in its present form for integer-bias fixing, and this is reflected in the somewhat higher scatter (at about the level of 4 parts in 10^7) of the daily determinations in all components, except perhaps the vertical where a scatter at the level of 2 parts in 10^7 or better is noted. The results obtained for baselines of this length are probably slightly inferior to those which would likely be obtained from VLBI or SLR, but nevertheless are invaluable (especially the east component of the recovered baselines) in assessing the impact of the errors in the satellite orbits. Using the same data set, Dong and Bock (1989) have shown that by applying a sequential ambiguity resolution algorithm and orbit improvement scheme they were able to improve further the repeatabilities of these California baselines by one order of magnitude, thus illustrating the important interplay of these biases and how critical it is to be able to separate their influences by proper network design and solution strategies.

From the discussion above it is evident that there is a wide range of factors defining observing accuracy. For example, a reasonable empirical formulation for the variance of double difference phase observations

$$\sigma_D^2 = k\,(\alpha^2 + \beta^2 L_{AB}^2) \tag{7}$$

where L_{AB} is the baseline length between stations A and B, and α, β and k are characteristic constants. k for instance takes the value of 4, resulting from propagating carrier phase error into double differenced phase error. In good observing conditions, α, which is primarily a function of multipath and residual tropospheric refraction error, should not exceed 5 mm. The distance dependent term β is primarily a function of the residual error in the ionospheric propagation correction (which can vary for nighttime vs. daytime observations, and with solar activity), and the quality of the satellite ephemeris information. It may vary from less than 2 ppm (for $L_{AB}<50$ km) when using single-frequency receivers and broadcast ephemerides to 0.5-0.01 ppm when using dual-frequency receivers with orbits determined from fiducial stations. Schaffrin and Bock (1988) have elaborated in a particular scheme that starting from equation (7) arrives at an elegant and computationally efficient construction of full covariance matrices for double difference phases and for weighted constraints on the ionosphere in order to improve phase ambiguity resolution. Such a scheme is clearly preferable to using identity weight matrices, and by being applicable to any spatial and temporal distribution of stations and satellites, it should be intuitively reflecting the physical situation closer.

INTEGRATION OF TERRESTRIAL AND GPS NETWORKS

The last few years have shown that GPS is rapidly developing into a versatile and relatively inexpensive tool for establishing geodetic control on local and regional scales. However, in spite of the successes of GPS techniques in establishing ultra precise surveys, there are different levels of accuracies associated with GPS which make it quite unlikely that as the use of GPS becomes more widespread it will lead to the development of completely separate GPS networks. The latter would likely continue to be justifiable for such specialized purposes as scientific studies of crustal movements, earth orientation, monitoring of global sea level, etc. This reasoning has led to date to a significant trend of combining GPS measurements with terrestrial observations and the improvement of existing networks by additional GPS measurements. Consequently, the third order design problem has assumed a renewed importance lately, especially with regard to addressing the issues of combining GPS with conventional data, particularly over networks established for the study of crustal deformations. Grant (1988) has studied extensively the optimum procedures for detecting earth deformation from a combination of terrestrial and GPS observations and noted that new technologies such as GPS, while offering the possibility of faster detection of deformation, also present special problems stemming primarily from the fact that systematic errors in GPS are different from those in conventional triangulation or precise trilateration and hence, if ignored, would likely be misconstrued as fictitious deformations of geophysical origin. This is the reason mainly why to date the combination of terrestrial and GPS surveys has been required mostly for the purpose of ensuring both an accurate assessment of systematic errors and calibration of the respective techniques with respect to each other (e.g. Kleusberg et al, 1988).

The proper combination of GPS and terrestrial data is ultimately associated with (a) a hybrid functional model with respect to a common reference system, and (b) a stochastic model (e.g. estimation of variance components) allowing the inclusion of all data in one grand least squares adjustment without distorting or degrading the superior quality of the GPS data. The first task is often treated by well-known Gauss-Helmert type *similarity* (S-) *transformations* (Teunissen, 1985b; Welsh, 1986, Beutler et al., 1989)

$$X_{Terr} - X_{GPS} + G\,p = 0 \tag{8}$$

of the GPS determined network coordinates X_{GPS} and their associated covariance matrices into the terrestrial coordinate system X_{Terr} refering to a given national geodetic datum, where G is the coefficient matrix of the transformation and the S-transformation parameters p are usually 3 rotations, 3 translations and a scale change common to all network points. Recent developments both within the realm of conventional geodesy (e.g. Chrzanowski et al., 1989; Mainville and Veronneau, 1989) and *integrated* or *operational geodesy* (Landau et al., 1988) have extended the formulation of combining GPS data with gravity field information (e.g. geoid heights) which would likely lead in the near future to a new major adjustment strategy for the significant replacement of the (expensive) traditional spirit levelling. The second task relates to assigning proper weights to the various observation groups, or at least their weight relationships and their mutual correlations. In the case of terrestrial networks, this is a problem that has been treated extensively in the past with the combination for instance of various distance, horizontal and vertical directions, Doppler data, etc. The respective situation with GPS data arises because

- here are frequently many different types of observables involved (e.g. single vs. dual-frequency observations);
- the same observables might be obtained with different receiver instrumentation;
- each type of observable may be contaminated with errors having several components corresponding to different characteristics (e.g. day vs. night observations, multipath, etc.); or
- solutions may be obtained from undifferenced, single, double or triple differenced observations.

Conventionally, this requirement is fulfilled by

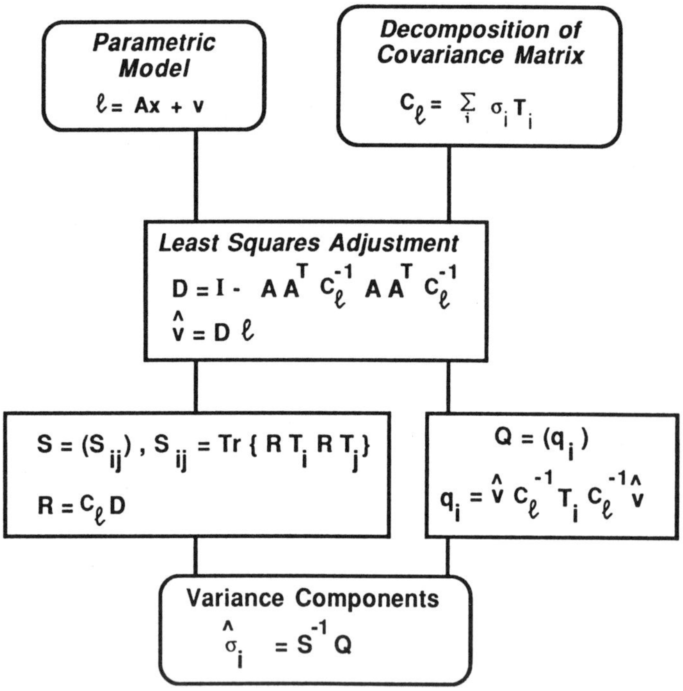

Figure 7 - MINQUE Variance Component Estimation

- some a priori knowledge of the stochastic behaviour of the observations as claimed by the GPS receiver manufacturers (cf. equation (14)),
- separate adjustments of the network using individual groups of observations (e.g. single frequency-only adjustments, dual frequency-only adjustments), and/or
- empirical definition of the suspected variances.

In this conection, *Variance component estimation* (VCE) is a general method which warrants further study with respect to its applicability to the terrestrial/GPS data combination (Welsh, 1986). Based on *Minimum Norm Quadratic Unbiased Estimation* (MINQUE) principles, VCE's major attraction is that it provides results independent of the reference datum or any systematic external influences (e.g. instrumentation and propagation medium, ephemeris errors etc.). Grant (1988) has emphasized the usefulness of VCE to testing the significance of deformation parameters derived from the combination of terrestrial and GPS data, particularly when one is attempting to evaluate the effects of systematic errors on the estimates of strain from subsets of combined observations. The possibility of correlation between campaigns can also be investigated (Secord, 1985). The computational scheme for the parametric model is shown in Figure 7. The functional model and the covariance structure $\mathbf{C}_\ell$ have to be selected prior to the VCE computations. $\mathbf{C}_\ell$ can be diagonal, a full matrix or anything in between. It is usually introduced by selecting a proper component basis σ_i, and the corresponding structure matrix $\mathbf{T}_i$. The computational procedure, as outlined e.g. by Chen (1983), requires that a least squares estimation of the coordinates $\mathbf{X}$ must be performed first using the usual parametric model $\ell = \mathbf{AX} + \mathbf{v}$. The outcome of this adjustment, specifically the residuals v_i, the a posteriori variance factor, and the possibility of detectable outliers or systematic errors would affect VCE. Hence in practice, iterative MINQUE methods are usually applied in these two steps.

SUMMARY AND CONCLUSIONS

We have presented a relatively detailed overview of the recent progress and near-future prospects relating to the optimization of the GPS theoretical methodologies, observational strategies and data processing techniques. As stated at the outset, various criteria can be adopted for the purpose of optimizing the design of a GPS network to an a priori specified level of accuracy and reliability. Drawing from analogies of the optimization and design of conventional (terrestrial) networks, similar aspects arising in the context of GPS network design have been examined. The usefulness of the fiducial network concept both as a means of dealing with the GPS datum definition problem, and as an observing strategy for simultaneous precise determination of the GPS orbits and geodetic baselines has been exemplified with practical results obtained with the 1987 Standard GPS Data Set. It was shown that both the observing strategy (from the viewpoint of which stations will be fixed or secondary, and which satellites to be observed) and the stepwise buildup of a GPS network will be closely related to the choice of the observational model to be employed. Unlike the terrestrial networks, where the mere optimal distribution of observational weights in a fixed network configuration is usually sought, second order design in GPS networks is intimately related to the quality of the satellite orbital information, clock behaviour and atmospheric influences on the propagation of the GPS signals, and subsequently the influence of any unmodelled or improperly modelled biases. The integration of existing terrestrial networks and GPS subnetworks is only recently beginning to attract considerable attention. The majority of classical networks such as the national horizontal and vertical networks of all orders have already been designed and observed. As far as optimization is concerned one can only look forward to their improvement by additional GPS densification and/or additional observations. The situation is somewhat different for geodetic networks established for the study of crustal deformation. In spite of the potential of GPS to establish ultra precise networks with relative accuracy higher than that achievable by terrestrial observations, historical terrestrial data can hardly be disregarded in favour of GPS repeated surveys. Combined terrestrial and GPS surveys will continue to be required at least in the immediate future (e.g. the next 10 years) in order to ensure both an adequate assessment of systematic errors in each observational type, and a known basis for verification and calibration of the respective techniques.

REFERENCES

Bertiger, W.I., S.M. Lichten and E.C. Katsigris (1988) - "A demonstration of sub-meter GPS orbit determination and high precision user positioning", Proc. Position Location and Navigation (PLANS) Symp., Orlando, Fl., Nov.

Beutler, G. W. Gurtner, M. Rothacher and I.B. Schildknecht (1985) - "Evaluation of the March 1985 High Precision Baseline Test: Fiducial Point Concept versus Free Network solutions", EOS Transactions, American Geophysical Union, Vol. 67, Nov.

Beutler, G., W. Gurtner, M. Rothacher, T. Schiddknect and I. Bauersima (1986) - "Determination of GPS Orbits using Double Difference Phase OBservations from Regional Networks", *Bull. Geod.*, 60, pp. 205-220.

Beutler, G., I. Bauersima, S. Botton, W. Gurtner, M. Rothacher and T. Schildknecht (1989) - "Accuracy and biases in the geodetic application of the Global Positioning System", *Manuscripta Geodetica*, Vol. 14 (1), pp. 28-35.

Bock, Y, R.I. Abbot, C.C. Councelman III and R.W. King (1986) - "A demonstration of 1-2 parts in 10^7 accuracy using GPS", *Bull. Geod.*, 60, 241-254.

Bock, Y, S.A. Gourevitch, C.C. Councelman III, R.W. King and R.I. Abbot (1986) - "Interferometric analysis of GPS phase observations", *Manuscripta Geodetica*, 11, 282-288.

Bock, Y. (1988) - "Availability of Standard GPS Data Set", unpublished.

Chen, Y. (1983) - "Analysis of deformation surveys - a generalized method", Tech. Rep. No. 94, Dep. of Surveying Eng., Univ. New Brunswick, Fredericton, N.B., April.

Colombo, O.L. (1986) - "Ephemeris errors of GPS satellites", *Bull. Geod.*, 60 (1), pp. 64-84.

Colombo, O.L. (1989) - "The dynamics of Global Positioning System orbits and the determination of precise ephemerides", *J. Geoph. Res.*, (in press).

Davidson, J.M., C.L. Thornton, C.J. Vegos, L.E. Young and T.P. Yunck (1985) - "The March 1985 demonstration of the Fiducial Network concept for GPS geodesy: a preliminary report", Proc. of the 1st Int'l Symp. on Precise Positioning with GPS, Rockville, MD, April.

Delikaraoglou, D. (1985) - "Estimability analyses of the free networks of differential range observations to GPS satellites", in Optimization and Design of Geodetic Networks, Spinger- Verlag, New York, pp. 196-220.

Delikaraoglou, D. (1989a) - "On Principles, Methods and Recent Advances in Studies Towards a GPS-Based Control System for Geodesy and Geodynamics", *NASA Tech. Mem.* 100716, Febr.

Delikaraoglou, D. (1989b) - "On the stochastic modelling of GPS ionospheric delays", *Manuscripta Geodetica,* Vol. 14, pp 100-109.

Dong, D. and Y. Bock (1989) - "GPS network analysis with phase ambiguity resolution applied to crustal deformation studies in California" , *J. of Geoph. Res.*, Vol. 94 (B4), pp. 3949-3966.

Ethridge, M.M. (1988) - "The National Geodetic Reference System of the Future", preprint, P.O.B. Publishing Co., Canton, Michigan.

Grafarend, E. (1985) - "Criterion Matrices for Deforming Networks", in Optimization and Design of Geodetic Networks, Spinger-Verlag, New York, pp. 363-428.

Grant, D.B. (1988) - "Combination of terrestrial and GPS data for earth deformation studies in New Zealand", Ph.D. Thesis, School of Surveying, Univ. New South Wales, Kensington, NSW, Australia.

Heroux, P. and Kleusberg, A. (1989) - "GPS Precise reative positioning and ionosphere in aurora regions", to appear in the Proc. of the Fifth Int'l Symp. on Geodetic Satellite Positioning, Las Cruces, New Mexico.

Hothem, L.D., C.C. Goad and B.W. Remondi (1984) - "GPS Surveying - Practical Aspects", *The Canadian Surveyor*, 38 (3), pp. 177-192.

Hutch, R.R. and E.V. Avery (1988) - " A strategic planning tool for GPS surveys", presented at the ASCE Specialty Conference GPS-88 on engineering applications of GPS satellite Surveying technology, Nashville, Tenn., May 11-14.

Jaeger, R. (1988) - "Analyse und Optimierung geodaetischer Netze nach spektralen Kriterien und Mechanische Analogien", Doctoral Thesis, University of Karlsruhe, FRG.

Janes, H., K. Doucet, B. Roy, D.E. Wells, R. Langley, P. Vanicek and M. Craymer (1986) - "GPSNET: A program for the interactive design of geodetic networks", Geodetic Survey of Canada CR 86-003, Sept.

Kavouras, M. (1982) - "On the detection of outliers and the determination of reliability in geodetic networks", Tech. Rep. No. 87, Dep. Surv. Eng., Univ. New Brunswick, Fredericton, N.B.

Kleusberg, A., Y. Georgiadou and H. Dragert (1988) - "Establishment of crustal deformation networks using GPS: a case study", CISM Journal, Vol. 42 (4), pp. 341-351.

Landau, H. and B. Eissfeller (1986) - "Optimization of satellite selection for high precision differential positioning", in GPS Research 1985 at the Inst. of Astronomical and Physical Geodesy (Ed. H. Landau and G. Hein), Schriftenreihe, Heft 19, Univ. der Bundeswehr, Munchen, pp. 65-105, March.

Landau, H., K. Hehl, B. Eissfeller, G.W. Hein and W.I. Reilly (1988) - "Operational Geodesy software packages", Schriftenreihe, Heft 34, Univ. der Bundeswehr, Munchen.

Lichten, S.M. and J.S. Border (1987) - "Strategies for high precision GPS orbit determination", *J. of Geoph. Res.*, Vol. 92, No. 10, pp.12751-12762, Nov.

Lindlohr, W. and W.E. Wells (1985) - "GPS design using undifferenced carrier beat phase observations", *Manuscripta Geodetica*, 10 (4), pp. 255-295.

Mainville, A. and M. Veronneau (1989) - "Orthometric Heights using GPS", presented at the Annual Meeting of the Canadian Institute of Surveying, Halifax, N.S., June.

Merminod, B. , D.B. Grant and C. Rizos (1988) - "A new approach to satellite selection and observation scheduling for GPS surveys", School of Surveying, Univ. New South Wales, Kensington, NSW, Australia.

Mueller, I.I. and S. Zerbini, Eds. (1989) - "The Interdisciplinary Role of Space GEodesy", Springer-Verlag Lecture Notes in Earth Sciences No. 22, Berlin.

Parrot, D. (1989) - "Ambiguity and cycle slip software development", Canadian Geodetic Survey CR 89-001.

Santerre, R. (1987) - "Tropospheric refraction effects in GPS positioning", GS SE 6910, Dept. of Surveying Eng., Univ. New Brunswick, Fredericton, N.B., Dec.

Schaffrin, B. (1985) - "Aspects of Network Design", in Optimization and Design of Geodetic Networks, Spinger-Verlag, New York, pp. 548-595.

Schaffrin, B. and E. Grafarend (1986) - "Generating classes of equivalent linear models by nuisance parameter elimination - application to GPS observations",Proc. of the Fourth Inter'l Geod. Symp. on Satellite Positioning, Austin, Tx, April 28-May 2.

Schaffrin, B. and Y. Bock (1988) - "A unified scheme for processing GPS dual-band phase observations", *Bull. Geod.*, 62, 142-160.

Schmitt, G. (1985) - "Second Order Design", in Optimization and Design of Geodetic Networks, Spinger-Verlag, New York, pp. 74-121.

Secord, J.M. (1985) - "Implementation of a generalized method for the analysis of deformation surveys", Tech, Rep. No. 117, Dept. of Surveying Eng., Univ. of New Brunswick, Fredericton, N.B.

Snay, R.A. (1986) - Network design strategies applicable to GPS surveys using three or four receivers", *Bull. Geod.*,Vol. 60 (1), pp. 37-50.

Teunnisen, P.J.G. (1985a) - "Quality control in geodetic networks", in Optimization and Design of Geodetic Networks, Spinger-Verlag, New York, pp. 526-547.

Teunnisen, P.J.G. (1985b) - "Zero order design: generalized inverses, adjustment and datum problem and S-transformations", in Optimization and Design of Geodetic Networks, Spinger- Verlag, New York, pp. 11-55.

Treuhaft, R.N. and G.E. Lanyi (1987) - "The effects of the dynamic wet troposphere on radio interferometric measurements", *Radio Science*, Vol. 22 (2), pp. 251-265.

Wells, D. , N. Beck, D. Delikaraoglou, A. Kleusberg, E.J. Krakiwsky, G. Lachapelle, R.B. Langley, M. Nakiboglu, K.P. Schwarz, J.M. Tranquilla, P. Vanicek (1986) - "Guide to GPS Positioning", Canadian GPS Associates, P.O. Box 3184, Post. St. B, Fredericton, N.B. Canada E3A 5G9.

Welsch, W.M. (1986) - "Problems in accuracies in combined terrestrial and satellite control networks", Proc. of the Fourth Int'l Geodetic Symp. on Satellite Positioning, Austin, TX, pp. 181-196.

Vanicek, P. and M. Craymer (1986) - "Satellite selection methods for differential GPS positioning", Appendix D in Janes, H., K. Doucet, B. Roy, D.E. Wells, R. Langley, P. Vanicek and M. Craymer (1986) - "GPSNET: A program for the interactive design of geodetic networks", Geodetic Survey of Canada CR 86-003, Sept.

Strain/analysis of Systematic Distortions of GPS-Networks

Alain Geiger
Institute of Geodesy and Photogrammetry
Federal Institute of Technology
CH-8093 Zürich, Switzerland

ABSTRACT

In this paper emphasis is put on the effects of systematic errors causing distortions of the GPS-solution. The method of continous strainanalysis can be applied to GPS-networks. By this method rotational as well as scaling effects are clearly worked out. As an example the effect of the difference between standard correction and a more realistic tropospheric model has numerically been calculated. The physical effect of refraction can clearly be isolated from the effect of the observed satellite constellation. With the presented analysis one could try to detect and interpret or to predict systematic network distortions in a physically and geometrically (constellation) meaningful way.

INTRODUCTION

In many cases the distortions of GPS-Networks show clearly a systematic behaviour. In these cases one would of course try to reduce the systematic effects to more stocastic remainders. Systematic errors will distort the network in very distinct way. The basic idea consists in treating this systematic distortions by continous formulation. This results in a method which is directly related to the strain analysis of continuum mechanics. As an example the effect of tropospheric ranging error will be treated by this continuous approach of error analysis.

BASIC EQUATIONS

We consider a GPS network where the station **y** is fixed and serves as reference point. If we introduce baselines from the fixed station **y** to the points **x**, we can write for the error in position of station **x**

$$\delta \mathbf{x} = \mathbf{N}^{-1} \mathbf{A}^T (\delta \mathbf{r}_\mathbf{x} - \delta \mathbf{r}_\mathbf{y})$$

$\delta\mathbf{x}$ is a 4-vector, since the error in time is included too.

This formula has been derived in (Geiger, 1988a, 1988b) and represents a continuous formulation of the error $\delta\mathbf{x}$ at the point $\mathbf{x}$.

$\delta\mathbf{r}_\mathbf{x}$, $\delta\mathbf{r}_\mathbf{y}$ are the errors in range at the sites $\mathbf{x}$ and at the fixed site $\mathbf{y}$, respectively. $\mathbf{N}^{-1}$ denotes the inverse of the normalmatrix of the GPS-net,

$\mathbf{A}^T$ is the design-matrix.

We define the gradient matrix in the following way

$$\mathrm{U} = \frac{\partial\, \delta\mathbf{x}}{\partial \mathbf{x}}$$

This formula points directly towards the strain-analysis of the error $\delta\mathbf{x}$.

If we restrict ourselves to small scale networks, the design-matrix $\mathbf{A}$ and the normalmatrix $\mathbf{N}$ will not depend on the site location $\mathbf{x}$. In this case the derivative reduces to:

$$\mathbf{U} = \mathbf{N}^{-1} \mathbf{A}^{T} \frac{\partial\, \delta\mathbf{r}(\mathbf{x})}{\partial \mathbf{x}} \qquad 1)$$

In many cases the range error can be split into two parts. One part is discribing the geometric or the constellation effect. The other one gives the physical contribution to the ranging error.

In this case we may write:

$$\delta\mathbf{r} = \delta \mathrm{s}(\mathbf{x})\, \mathbf{c} \qquad 2)$$

where the elements of $\mathbf{c}$ represent the constellation for the corresponding epoch of observation and δs represents a scalar-function depending on the coordinates of the points under consideration. The tropospheric effect can very well be described by this type of function. We will discuss this case later on.

Introducing this equation into 1) we will get

$$\mathbf{U} = \mathbf{N}^{-1} \mathbf{A}^{T} \mathbf{c} \left(\frac{\partial \delta \mathrm{s}}{\partial \mathbf{x}} \right)^{T} \qquad 3)$$

Note that **c** doesn't depent on station coordinates **x** since the dimension of the GPS-network is assumed to be small.

Denoting the derivative by **d** which depends on the coordinates **x** only (not on the constellation), and substituting

$$\mathbf{N}^{-1}\mathbf{A}^{T}\mathbf{c} = \mathbf{u} \qquad 4)$$

which depends on the satellite constellation only we can write:

$$\mathbf{U} = \mathbf{u}\,\mathbf{d}^{T} \qquad 5)$$

It is to say that the Matrix **U** has the dimension 4 x 3. This follows from the dimensions of the vector **u** (4 x 1) and $\mathbf{d}^T$ (1 x 3). **v** denotes the constellation dependent part of the distortion. It includes as forth componend the error in time. In case we are interested in distortions of the solution of station coordinates, we will consider the three space components of **u** only. So we may rewrite 5) by new matrices **G** (3 x 3) and **v** (3 x 1) as

$$\mathbf{G} = \mathbf{v}\,\mathbf{d}^{T} \qquad 6)$$

This formula gives the distortion matrix which will be analysed in the next chapter.

EIGENVALUE-PROBLEM AND STRAINANALYSIS

As usual we form two new matrices by symmetric-antisymmetric decomposition

$$\Omega_{\pm} = \frac{1}{2}\left(\mathbf{G} \pm \mathbf{G}^{T}\right) \qquad 7)$$

The Eigenvalue-problem:

$$\Omega_{\pm}\,\mathbf{x} = \lambda\,\mathbf{x} \qquad 8)$$

can now be solved explicitly by taking into account that 6) and consequently

$$\Omega_{\pm} = \frac{1}{2}\left(\mathbf{v}\,\mathbf{d}^{T} + \mathbf{d}\,\mathbf{v}^{T}\right) \qquad 9)$$

holds.

For the symmetric part Ω_+ we find the following not normalized three eigenvectors:

$$\mathbf{v}^0 + \mathbf{d}^0, \quad \mathbf{v}^0 - \mathbf{d}^0, \quad \mathbf{d} \wedge \mathbf{v} \tag{10)}$$

with the corresponding eigenvalues

$$\mathbf{v}^T \mathbf{d} + v\,d, \qquad \mathbf{v}^T \mathbf{d} - v\,d, \quad 0 \tag{11)}$$

The antisymmetric part Ω_- has, as it is well known only one real eigenvector and one real eigenvalue. The real eigenvector can be interpreted as the axis of an infinitesimal rotation. The roational angle can easily be found. Finally we get by the isomorphism between vector and antisymmetrix matrix representation the rotation vector

$$\omega = \frac{1}{2}\,(\,\mathbf{v} \wedge \mathbf{d}\,) \tag{12)}$$

To summarize: We found a straintensor of diagonal form:

$$\gamma = \frac{1}{2}\begin{bmatrix} \mathbf{v}^T\mathbf{d} + vd & 0 & 0 \\ 0 & \mathbf{v}^T\mathbf{d} - vd & 0 \\ 0 & 0 & 0 \end{bmatrix} \tag{13)}$$

and a rotation about an axis perpendicular to **v** and **d** (see 12)). From 13) it is seen that no strain will appear in the third eigenvector direction, which corresponds following 10) to $\mathbf{d} \wedge \mathbf{v}$.

As an example assume that the error effect of the satellite-constellation and the physical error effect are parallel and e.g are pointing to the zenith, then we immedately see that no rotation will occur ($\omega = \mathbf{0}$) and that there exists only one scale factor in the first eigenvector direction, in our case the zenith or height -direction. This example corresponds exactly to the effect of errors in a horizontally layered troposphere during a homogenous satellite constellation.

APPLICATION TO TROPOSPHERIC EFFECTS

We now would like to apply the above formulas to the tropospheric effect. For this purpose we introduce the simple range correction model

$$\delta r = \frac{\delta r_0}{\cos \theta} \qquad 14)$$

δr_0 is the zenith pathdelay.

θ is the zenith angle of the path.

This formula holds for horizontally layered models. If we now consider a variing pathdelay $\delta r_0(\mathbf{x})$ depending on the coordinates of the station $\mathbf{x}$ under consideration and if all observations at $\mathbf{x}$ are taken into account we can write the following equation:

$$\boldsymbol{\delta r} = \delta r_0(\mathbf{x})\, \mathbf{c} \qquad 15)$$

where $\mathbf{c}^T = (\ldots, 1/\cos \theta, \ldots)$ depends only on the constellation.

Equation 15) takes exactly the form of equation 2). We can directly identify $\delta r_0(\mathbf{x})$ by the physical function $\delta s(\mathbf{x})$ in 2). $\mathbf{c}$ takes the form above. Following the formula 3) the gradient of $\delta r_0(\mathbf{x})$ has to be performed.

$$\mathbf{d} = \frac{\partial\, \delta r_0(\mathbf{x})}{\partial \mathbf{x}} \qquad 16)$$

For the determination of the gradient we need a function describing $\delta r_0(\mathbf{x})$. Such formula can be found e.g. in Saastamoinen (1973). Simplifying the well known Saastamoinen correction formula results in:

$$\delta r_0 = 0.002277 \left[p + \left(\frac{1255}{T} + 0.05 \right) e \right] \qquad 17)$$

Forming the gradient of 17) leads to:

$$\mathbf{d} = \frac{\partial\, \delta r_0}{\partial p} \nabla p + \frac{\partial\, \delta r_0}{\partial T} \nabla T + \frac{\partial\, \delta r_0}{\partial e} \nabla e \qquad 18)$$

The Nabla-operator denotes the gradient of the corresponding pressure, temperature and humidity field, respectively. From 17) it follows for the coefficients of the gradients (standard values assumed):

$$\delta r_p \approx 2 \text{ mm/mbar}$$
$$\delta r_T \approx -14 \text{ mm/°C}$$
$$\delta r_e \approx 4 \text{ mm/\%} \qquad 19)$$

The components of gradients of pressure and temperature are in horizontal direction:

$$p_h \leq 10 \text{ mbar/100km}$$
$$T_h \leq 10 \text{ °C/100km} \qquad 20)$$

and in vertical direction:

$$p_v \approx -1 \text{ mbar/10m}$$
$$T_v \approx -0.6 \text{ °C/100m} \qquad 21)$$

The humidity reveals much larger variations. Thinking of a baseline crossing a front a systematic horizontal change of 20% to 30% humidity could be realistic. The distribution of humidity is not as predictible as the pressure and temperature field.

Again thinking of a front we assume that the major atmospheric horizontal changes will occur in the same distinct direction. This allows us to put e.g. our local y-axis into the same direction.Consequently the gradient in such a coordinate system will only have the components y and z. Normally a first order approximation of the z-component of the gradient is taken into account when meteocorrections with standard models are applied. Not so for the horizontal component. We consider the difference between the actual gradient and the standard gradient, which includes a z-component only, and we note that the standard gradient approximates the actual z-component well enough. With this in mind we see immedately that in the difference only the y-component will remain.

$$\delta \mathbf{d} = \mathbf{d} - \mathbf{d}_0 = \begin{bmatrix} 0 \\ d_y \\ 0 \end{bmatrix} \qquad 22)$$

Introducing equations 18) and the values given in 19) we find:

$$d_y = 2\,p_h - 14\,T_h + 4\,e_h \tag{23}$$

With the values given above the gradient will be:

$$\delta \mathbf{d} = 2.4\ \text{ppm} \begin{bmatrix} 0 \\ 1 \\ 0 \end{bmatrix} \tag{24}$$

For a homogenous constellation we find the following constellation effect (Geiger, 1988):

$$\mathbf{u} = \mathbf{N}^{-1}\,\mathbf{A}^{T}\,\mathbf{c} = \begin{bmatrix} 0 \\ 0 \\ 2.7 \\ 3.5 \end{bmatrix} \quad \text{or} \quad \mathbf{v} = \begin{bmatrix} 0 \\ 0 \\ 2.7 \end{bmatrix} \tag{25}$$

According to formulas 12) and 13) we find a rotation of the net by -3 μrad about the x-axis. The straintensor (in the eigenvector system) looks like

$$\gamma = 3\ \text{ppm} \begin{bmatrix} 1 & 0 & 0 \\ 0 & -1 & 0 \\ 0 & 0 & 0 \end{bmatrix} \tag{26}$$

If we combine both, the rotation and the strain, applying the rotation R given by the strain eigenvectors, then

$$\delta \mathbf{x} = \omega \wedge \mathbf{x} + \mathbf{R}\,\gamma\,\mathbf{R}^{T}\,\mathbf{x} \tag{27}$$

denotes the error of the baseline **x**. With the given values and the height difference $\Delta h = h_x - h_{ref}$ of the baseline we only get an error in the y-component ($\delta x = \delta z = 0$).

$$\delta y \approx 6\ \text{ppm}\ \Delta h \tag{28}$$

We see that rotation, compression and dilatation will destort the net. Errors in the horizontal coordinate in the order of 6 ppm of the height difference can occur even if standard corrections have been applied and even if a completely symmetric constellation has been observed. Note also that the distortion of the baseline will depend upon its orientation. E.g. lines at the same altitude as the reference site will not be affected. It also should be noted that the above figures are valid only for medium scale baselines of the length of about 10 km to 100 km.

CONCLUSIONS

The method of continous strainanalysis can be applied to GPS-networks. Explicit formulas are derived for special error functions which consist of a multiplication of a constellation term and physically determined error term. Refracting phenomenon can be described by such function types. The effect of the difference between standard correction and a more realistic tropospheric model has numerically been calculated. Rotation and strain (up to 6ppm of height difference) are completely described by 12) and 13). The results are valid for medium scale nets. For long baselines we have to take into account the dependency of the normal and design matrices on the station coordinates. In this case the meteoparameter cannot linearly be extrapolated from the medium length values. Also for small scale nets the meteoparameter must be choosen appropriently. With the presented analysis one could try to detect and interpret or to predict systematic network distortions in a physically and geometrically (constellation) meaningful way.

REFERENCES

Geiger, A. (1988a): Simulating Disturbances in GPS by Continuous Satellite Distribution. *Journal of Surveying Engineering, Vol 114, No.4, 1988.*

Geiger, A. (1988b): Modeling of Phase Center Variation and its Influence on GPS-Positioning. *In: (eds) Groten, Strauss: GPS-Techniques Applied to Geodesy and Surveying, Lecture Notes in Earth Sciences, Vol. 19, Springer-Verlag, 1988.*

Saastamoinen, I.I. (1973): Contribution to the theory of atmospheric refraction. *Bull.Geod. (107),1973,pp.13-34.*

OPTIMISING THE MOVEMENT OF GPS RECEIVERS

Peter Dare and Brian Whiting
Department of Land Surveying
Polytechnic of East London, Dagenham, Essex, England

INTRODUCTION

The purpose of this paper is to present the progress made in applying methods present in the field of Operations Research to the optimisation of GPS logistics. So far, effort has concentrated on the route the receivers should follow through the network to minimise some predifined variables.

GPS network design is also being investigated by others. Snay (1986) has produced algorithms based on the connectivity structure of the network. They enable observing schedules to be obtained for up to four receivers and stations to be reoccupied. Because the algorithms are based on the connectivity structure of the network, the optimal observing schedule is unlikely to be obtained. However, the networks are efficient in that no two stations are jointly occupied for more than one observing session. Morgan (1987) further developed Snay's ideas based on experiences in South Australia. Concentric Circle and Concentric Arc models are developed which reduce the redundancy and eliminate long "crossing" lines.

The second order design of GPS networks has been studied. Borutta (1985) used dynamic programming to minimise the trace of the covariance matrix of the adjusted parameters. Grafarend et. al. (1985) concentrated on the examination of errors due to orbits, clocks and the atmosphere.

The optimal location of GPS stations is not covered in this paper. However, Parkinson and Fitzgibbon (1986) have studied the optimal location of pseudolites based on the worst possible satellite outage in 24 hours and GDOP analysis. Both single and dual pseudolite systems are considered.

For a publication covering many aspects of GPS design (including redundancy, configuration, logistics and second order design) see Wells et. al. (1987). Some practical aspects of GPS surveying are well documented in Hothem et. al. (1984). Vanicek et. al. (1985) present algorithms for solving for the required number of receivers while differentiating between those receivers making observations and those moving between stations.

Although GPS logistics has not been investigated to a great extent, progress has been made in the optimization of costs in trilateration networks using graph theory (Kortesis and Dermanis, 1987). The "approximate" solution is based upon transforming the problem into a more complex version of the travelling salesperson problem (the travelling salesperson problem and its possible solutions are described in more detail in the next section).

THEORY

This section deals with the theory of optimal movement both generally and within the GPS context.

We assume that we would like to be able to move the receivers around the network in the most optimal manner. Here, optimal may mean minimising the time of travel between stations or maybe minimising the distance travelled around the network. As an example, consider Fig. 1 which shows a four station network. The numbers between the stations refer to the cost of moving the receivers between those two points; they could be, for example, time or distance. It does not matter which of those two it represents as the method of solution is the same. If we let the cost represent distance, then we would like to find the route around the network which will minimise the total distance travelled. In other words, we wish to determine the shortest route through the network. We assume that each station is to be visited only once (although visiting a station three times would facilitate the detection of centering errors (Snay, 1986)). This problem then becomes the travelling salesperson problem: an example of a problem which is easy to state but difficult to solve.

The travelling salesperson problem is stated as follows: A salesperson, starting in one city, wishes to visit each of $n-1$ other cities once and only once and return to the start. In what order should he/she visit the cities to minimize the total distance travelled? Often the word 'nodes' is used rather than 'cities' as the solution of the travelling salesperson problem has widespread applications in sequencing problems. We prefer to use the word 'stations' when considering just one receiver.

If we assemble all the costs into a cost matrix, we can distinguish two types of travelling salesperson problems. If the cost of travelling from station i to station j is the same as travelling from station j to station i then we have a symmetrical cost matrix, the alternative is an asymmetrical cost matrix. With n cities and an asymmetrical cost matrix there are $(n-1)!$ possible routes through the network one or more of which must give the minimum cost. In principle, the optimum route could be found by enumerating all possible tours but, even for a comparatively small number of cities, it becomes a computationally impossible task.

Solutions to the problem fall into two classes: "exact" (in the sense that the optimal solution is obtained) and "approximate" (in the sense that the solution obtained may not be the optimal). Eilon et. al. (1971) provide a good review of many different types of algorithms which fall into both classes. An exact algorithm which has been extensively referred to in the Operational Research literature, due to its flexibility, is a branch and bound technique first published by Little et. al. (1963). Due to its existing popularity, the method was selected as a possible start to determining the optimum movement of GPS receivers.

METHOD

The method is based upon constructing a cost matrix. For one receiver the cost matrix for the network in Fig. 1 would be as follows.

$$
\begin{array}{c} \\ A \\ B \\ C \\ D \end{array}
\begin{array}{c} \begin{array}{cccc} A & B & C & D \end{array} \\
\begin{bmatrix} -1 & 5 & 6 & 3 \\ 5 & -1 & 4 & 1 \\ 6 & 4 & -1 & 3 \\ 3 & 1 & 3 & -1 \end{bmatrix} \end{array}
$$

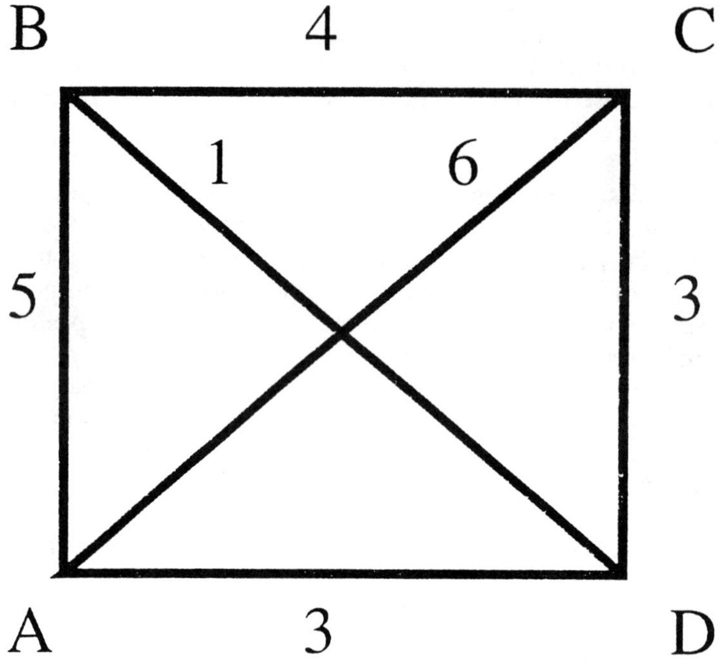

Fig. 1. Example network with movement costs.

The matrix elements coded as '-1' represent routes that should not be investigated by the algorithm as we do not require routes that go from, for example, station C to station C.

Of more practical use is to find the optimal route for 2 receivers through the network. We are not requiring one receiver to follow the other receiver through the network, but rather to simultaneously be on certain different stations to enable the required differential measurements to be made.

In this context, therefore, the nodes of the travelling salesperson problem are not the stations but the pairs of stations between which differential measurements are required. We therfore end up with a much larger cost matrix in which the individual elements are the cost of moving the receiver(s) from one pair of stations to another pair; the cost matrix of the example in Fig. 1 is shown in Fig. 2. The required station pairs appear twice (e.g. AC and CA) to allow for all possible receiver movements. For example, the optimal route may require receiver 1 to move from station B to D and receiver 2 to move from station C to B, i.e. the movement BC to DB must be permitted. However, the method implemented ensures that only one of the duplicated pairs appears in the optimal solution.

RESULTS

If we have just one receiver and we wish to start and finish at station A (which may in effect be the project office) the two optimal routes obtained (using the previous cost matrix) are

A - D - B - C - A

and

A - C - B - D - A.

The second route is simply the first route in reverse as a symmetrical cost matrix was used; each route costs 14 units. The units could be kilometres if the costs represent distance or minutes if the costs represent time.

For 2 receivers, using the cost matrix in Fig. 2, the optimal routes for the receivers are:

Receiver 1	A	A	A	D	B	D	A
Receiver 2	B	D	C	C	C	B	B.

The cost is only 16 units (2 more than one receiver). Three other optimal routes are obtained by reversing the observational order (as the cost matrix is symmetrical) and reversing all the station pairs (i.e receiver 1 follows the route of receiver 2 and receiver 2 follows the route of receiver 1).

DISCUSSION

The ability to optimise the movement of more than one receiver will enable GPS surveys to be carried out more efficiently. The results show that by considering the problem as a travelling salesperson problem optimum routes (with limitations) can be obtained.

As can be seen, one current limitation is that stations A and B are jointly occupied twice unlike the technique developed by Snay (1986). If certain stations are not to be jointly occupied (e.g. stations a long way apart as suggested by Morgan (1987) since errors increase on long lines but kept in by Snay (1986) on reliability grounds) then the relevant rows and columns of the cost matrix can simply be deleted. Further research into the efficient movement of GPS receivers is continuing.

	AB	BA	AC	CA	AD	DA	BC	CB	BD	DB	CD	DC
AB	-1	-1	4	11	1	8	9	6	6	3	7	7
BA	-1	-1	11	4	8	1	6	9	3	6	7	7
AC	4	11	-1	-1	3	9	5	10	8	7	9	3
CA	11	4	-1	-1	9	3	10	5	7	8	3	9
AD	1	8	3	9	-1	-1	8	7	5	4	6	6
DA	8	1	9	3	-1	-1	7	8	4	5	6	6
BC	9	6	5	10	8	7	-1	-1	3	5	7	1
CB	6	9	10	5	7	8	-1	-1	5	3	1	7
BD	6	3	8	7	5	4	3	5	-1	-1	4	4
DB	3	6	7	8	4	5	5	3	-1	-1	4	4
CD	7	7	9	3	6	6	7	1	4	4	-1	-1
DC	7	7	3	9	6	6	1	7	4	4	-1	-1

Fig. 2. Cost matrix for network shown in Fig. 1 with 2 receivers.

CONCLUSIONS

The optimal route for a GPS receiver to follow through a series of stations can be determined using an exact solution to the travelling salesperson problem.

For the optimal route of 2 receivers, a solution to the travelling salesperson problem can be used if the nodes represent the pairs of stations to be jointly occupied. The solution has limitations which further research should eliminate.

Acknowledgement. Dr. Michael Peel of the Department of Mathematical & Physical Sciences at the Polytechnic of East London made some valuable suggestions.

REFERENCES

Borutta, H. (1985). Discrete Dynamic Optimization of Relative GPS Positioning. Proceedings of the 7th International Symposium on Geodetic Computations, Cracow, Poland, June 18-21, pp. 159-174.

Eilon, S., C.D.T. Watson-Gandy and N. Christofides. (1971). *Distribution management: Mathematical modelling and practical analysis.* Charles Griffin & Company Ltd., London.

Grafarend, E.W., W. Lindlohr and A. Stomma. (1985). Improved Second Order Design of the Global Positioning System - Ephemerides, Clocks and Atmospheric Influences. Proceedings of the First International Symposium on Precise Positioning with the Global Positioning System, Rockville, Maryland, USA, April 15-19, pp. 273-284.

Hothem, L.D., C.C. Goad and B. Remondi. (1984). GPS Satellite Surveying - Practical Aspects. *Canadian Surveyor,* Vol. 38, No. 3, Autumn.

Kortesis, S. and A. Dermanis. (1987). An application of graph theory to the optimization of costs in trilateration networks. *Manuscripta Geodaetica,* Vol. 12, No. 4, pp. 296-308.

Little, J.D.C.., K.G. Murty, D.W. Sweeney and C. Karel. (1963). An Algorithm for the Travelling Salesman Problem. *Operations Research,* Vol. 11, pp. 972-989

Morgan, P. (1987). Models for GPS Network Design. *Australian Journal of Geodesy, Photogrammetry and Surveying,* Nos. 46 & 47, pp. 41-55.

Parkinson, B.W. and K.T. Fitzgibbon. (1986). Optimal Locations of Pseudolites for Differential GPS. *Navigation,* Vol. 33, No. 4, pp. 259-283.

Snay, R.A. (1986). Network Design Strategies Applicable to GPS Surveys using Three or Four Receivers. *Bulletin Geodesique,* Vol. 60, No. 1, pp. 37-50.

Vanicek, P., G. Beutler, A. Kleusberg, R.B. Langley, R. Santerre and D.E. Wells. (1985). DIPOP: Differential Positioning Program Package for the Global Positioning System. Department of Surveying Engineering Technical Report No. 115, University of New Brunswick, Fredericton, New Brunswick, Canada.

Wells, D.E., W. Lindlohr, B. Schaffrin and E. Grafarend. (1987). GPS Design: Undifferenced Carrier Beat Phase Observations and the Fundamental Differencing Theorem. Department of Surveying Engineering Technical Report No. 116, University of New Brunswick, Fredericton, New Brunswick, Canada.

OPTIMUM POSITIONS FOR GPS–POINTS AND SUPPORTING FIX–POINTS IN GEODETIC NETWORKS

Reiner R. Jäger
Geodetic Institute, University of Karlsruhe
Englerstraße 7, D-7500 Karlsruhe, West–Germany

INTRODUCTION

GPS is actually becoming a standard tool for determining discrete point-configurations within geodetic networks, and the combination of GPS with usual terrestrial measurements is presently coming to be a typical kind of geodetic network. For such GPS–terrestrial as well as for free engineering networks which are to be linked to some fix points e.g. of a superior network, there arises the question of selecting the best point–positions for their fixing with respect to a given discrete number of supporting-points in the sense of an optimality criterion. In this context JÄGER and DRIXLER, 1989 show that the optimum positions for fix–points are not receivable and identical, respectively to those points which get the number of maximum weights in the weight–hierarchy of an 'all–point Third Order Design aproach'. The following recently developped optimization method (JÄGER and DRIXLER, 1989) for optimum supporting point–positions is founded in the concept of spectral network–analysis (JÄGER, 1988; WIESER, 1988) and spectral optimization of geodetic networks (JÄGER, 1988), both referring to spectral criterions and target functions respectively. The decisive spectral target function of a maximum contraction of the eigenvalue–spectrum of the covariance–matrix of network–coordinates provides for the best possible homogenization of the error situation and the best possible stability or strength respectively of the regarding network. The latter criterion is similar to an optimum stiffness of the network in the sense of valuable mechanical analogies (JÄGER, 1988). The target function of a maximum spectrum–contraction guarantees to restrain as well as possible the so called 'weak forms' (JÄGER, 1988) which are likely to lead to dangerous network deflections (as pointed out in BILL et al., 1984), if there is given no support or any non optimum support to the network. In the same sense the connection of a geodetic control network, e.g. by GPS-points, to the very optimum points of maximum spectrum contraction is appropriate with a maximum increasing of the networks sensitivity – or for avoiding misinterpretations, respectively – due to global long waved deformations in the network area, as described, for example, in NIEMEIER, 1982 or in SCHMITT et al., 1989.

THEORY OF OPTIMIZATION

Spectral analysis of geodetic networks – mechanical analogies as useful tool for the interpretation of the analysis results

The spectral analysis of geodetic networks by means of the eigenvalue problem of the normal equation matrix N=A'PA

λ_j/x_j=j–th eigenvalue and –vector; j=(1,u=dimension N,I,dx)
A=design–matrix; P=observation weight–matrix
$(N-\lambda\cdot I)\cdot x=0$ $N=\Sigma N_i=\Sigma A_i{}'P_iA_i$, i=i–th observation, i=(1,n=total number)   (1)
x=coordinate–increments (unknows) referring to initial values x_0 and x to be represented in the spectral components λ_j/x_j.

is closely linked to the kinetics, especially to the eigen–vibrations of elastical mechanical structures with node–points x_0 (fig. 1,(c1),(c2)). Proceeding from the elastic continuum the complete analogy foundation between elastomechanics and geodetic network–compensation by least squares can be carried out in the step of its finite element discretisation (JÄGER, 1988). With respect to the discrete analogies, the normal equation matrix N and the contributions N_i of the observations are corresponding to the total stiffness–matrix K of and the elementwise stiffnesses K_i respectively of the discrete mechanical system or the finite element discretization and the elements respectively of an elastic continuum.

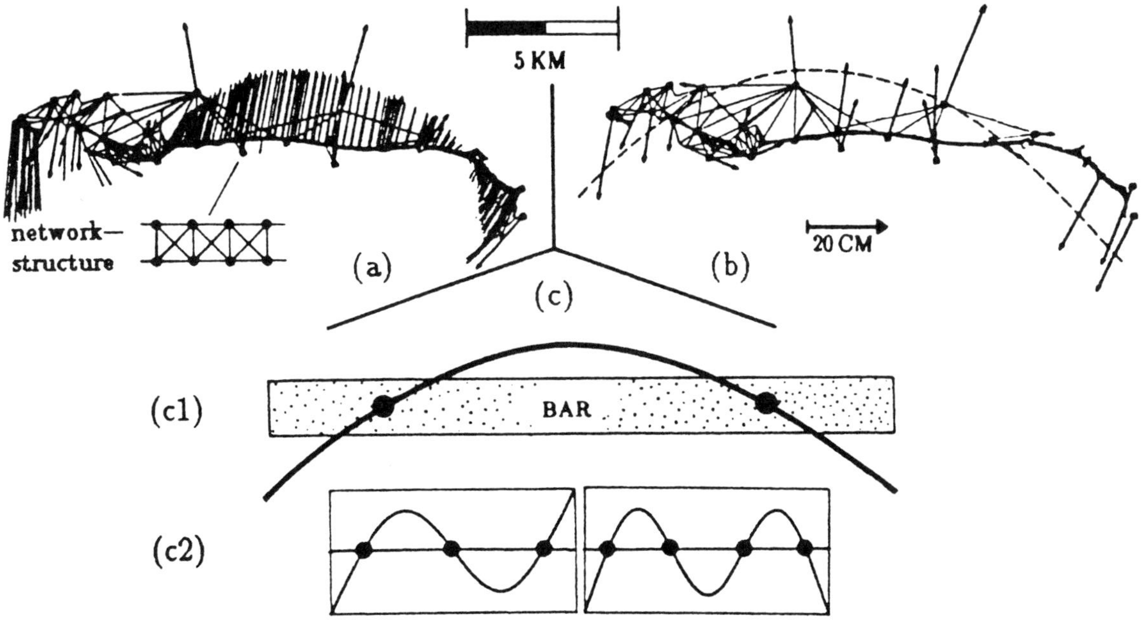

Fig.1: Spectral analysis of a geodetic network and mechanical analogies (Network example: Railway–net–chain of the Deutsche Bundesbahn)

(a) The shape of the most likely deflection/analysis–model (1): the pointwise eigenvector–components of the lowest non–zero eigenvalue–vector x_m (3).

(b) The network realization: True residuals with respect to known control points.

(c) 'Two in one': The networks mechanical analogy – a vibrating bar in the vibration mode of the basic frequency (c1) – and the eigenfunction x_m corresponding to spectral analysis model (2). (c2) shows the vibration of the 1st and the 2nd overtones.

With regard to mechanical analogies, the eigenvalues λ_j and eigenvectors x_j are identificable with the vibration frequenices and the vibration modes respectively of the mechanical structure associated to the geodetic network (fig. 1(c)). Starting now from the finite element discretization, as a common base within the analogy–foundation, and going the way back vice versa to a geodetic (network)–continuum the matrix equations (1) change to the network's differential equations:

in (1): λ_j=eigenvalues, x_j=–functions $j=(1,\infty)$

$x=x(\Delta)$ and $\Delta \to 0$ $\Rightarrow$ $(L-\lambda\cdot I)\cdot x=0$ L=d–dim. differential operator (2)

$N=N(\Delta)$ d=dimension of the network

The explicit calculation of the differential equation (2) which is a continuous substitute for (1) – L especially being the continuous substitute of N – is performable by the analytical transition–procedure of turning the finite structural mesh–size Δ of the considering network–structure towards zero (theory and examples, see JÄGER 1988). The calculation of L due to the given network structure is a recommendable and practicable tool for the identification of a geodetic network structure with its'mechanical partner'. This is advantageous because an identification in the discrete domain is not always directly possible by regarding the elements and observations respectively, as the finite element discretization of a given continuum is nonunique and can be carried out by different elements and –designs. In case of an identical L, what means identity in the structural continuum properties, a geodetic network and a mechanical structure show of course identical eigenvalue–spectra and eigenvectors/–functions, if additionally shape and boundary conditions coincide (JÄGER, 1988).

From the description of the analogies in the eigenvalues and – vectors/functions between geodetic networks and mechanical structures we now arrive at the explanation of the essential meaning and importance of the eigenvalue–problems (1) and (2) as the models of a spectral geodetic network analysis. It is easy to prove (see e.g. JÄGER, 1988; NIEMEIER 1982) that starting from

$$\lambda_1 < \lambda_2 < \ldots < \lambda_{i-1} < \lambda_i < \lambda_{i+1} < \ldots, \text{ with} \qquad (3)$$
$$\lambda_m = \text{lowest non–zero eigenvalue of (1) or (2) and}$$
$$x_m = \text{eigenvector/–function of } \lambda_m \text{ (m= maximum–size weak form)}$$

the inverse $\mu_m=1/\lambda_m$ and the following inverse eigenvalues $\mu_i=1/\lambda_i$ give each – in descending order – the likelihood and size, which means the variance of expectable deflections between the form of the adjusted network and its, of course desired, correct true form (JÄGER, 1988). The deflection–shapes ("weak forms") are shown out by the pointwise components of the x_i or the eigenfunctions x_i. An occurring discrepancy between the desired 'true shape' of the network and its 'truly realized shape' is most likely to appear in the shape of x_m (3) which shows maximum variance $1/\lambda_m$, see fig. 1(a),(b).

Optimum positions and the target function of a maximum spectrum contraction

Taking the ordered eigenvalues of a geodetic net–continuum (2)

$$\lambda_1 < \lambda_2 < \ldots < \lambda_{i-1} < \lambda_i < \lambda_{i+1} < \ldots.. \qquad \text{, see (3)}$$

one can ask for that new spectrum, which may result due to M (further) supports of the geodetic network at M fix–coordinates. Regarding the general case of the i–th eigenvalue λ_i, before and after this fixing, this question is generally answered by the inequality relation (BATHE and WILSON, 1976; GANTMACHER and KREIN, 1960):

$$\lambda_i^{(-)} < \lambda_i^{(M)} \leq \lambda_{i+M}^{(-)}, \text{ with } \begin{array}{l}(-)\text{=unfixed or a precedent fixed state}\\ (M)\text{=state after M new supports}\end{array} \qquad (4a)$$

The case of equality on the right side of (4a)

$$\lambda_i^{(M)} = \lambda_{i+M}^{(-)} \qquad (4b)$$

is achieved onlay by the very fixing of the network at those point–positions which are simply indicated and characterized by the eigenforms' x_i nodal–points of number

M/d, d=network–dimension (GANTMACHER/KREIN, 1960).

An M–coordinate support for a given geodetic network, in the nodal points is therefore called 'optimum support' as it provides for the maximum of any possible contraction of the eigenvalues $\mu_i = 1/\lambda_i$ of the resulting covariance–matrix of network coordinates in the following way:

Spectrum of covariance matrix: $$\mu_1 > \mu_2 > ... > \mu_{i-1} > \mu_i = 1/\lambda_i > \mu_{i+1} > ... \qquad (4b)$$

Non–optimum of an arbitrary support: $$\mu_i^{(M)} > \mu_{i+M}^{(-)} \qquad (4c)$$

Optimum 'nodal–point–support': $$\mu_i^{(M)} = \mu_{i+M}^{(-)} \qquad (4d)$$

According to (4d) the new weak forms (3) given by the new eigenforms $x_i^{(M)}$ after an optimum supporting show the property of having each a minimum expectable size/ - variance and likelihood of appearance respectively, fulfilling the optimality criterions of the spectral analysis models (1) and (2).

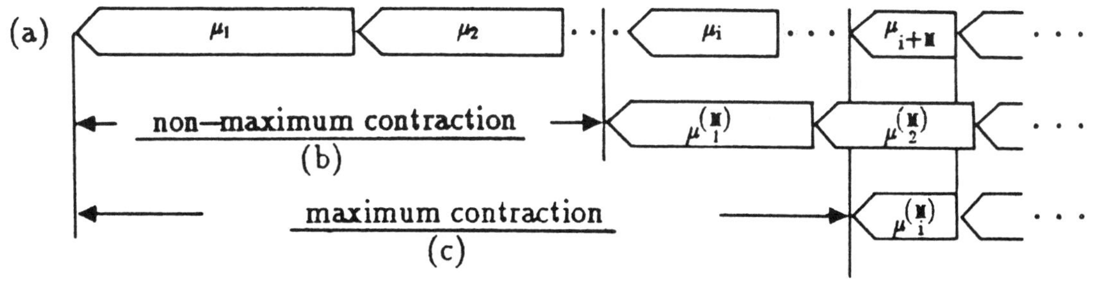

Fig.2: Behaviour of the eigenvalue–spetrum of the covariance matrix coordinates (a)precedent spectrum (4b); (b)arbitrary support (4c); (c)opt. support (4d)

The property (4d) is to be characterized as an effect of global stabilisation of the network–geometry. Apart form this (4d) provides also for an optimum homogenization of the networks error situation as the eigenvalues (4b) of the high indices in (4b) are closer together than the first ones in (4b) with lower indices.

Further on the precedent first M weak forms which were occurring before the optimum fixing are completely extinguished. The resulting new weak forms (3) of an optimum nodal–point support are given by the relation (GANTMACHER/KREIN, 1960)

$$x_i^{(M)} = x_{i+M}^{(-)} \qquad (5)$$

The mimimum size property (4d) acts in this context favourably together with the fact that higher order shapes x_{i+M} (5) achievable by the optimum support show generally a shorter wavelength than x_i, the lower ancient ones (BORRE, 1987; JÄGER, 1988). The maximum short–waved eigenfunctions (5) which are acting the short range of the relative observations contribute to an optimum local stabilisation of geometry of the resulting network.

OPTIMUM POSITIONS– REAL AND SYNTHETIC NETWORK–EXAMPLES

Optimum Choice of a Network Datum

For a free network with datum–defect r (=rigid body movements) the above treated theory allows the solution of the problem of 'how to carry the datum–parameters (e.g. orientation, scale etc.) to a network' by optimum choosen fiducial datum–points (problem of O. Order Design). Starting with the eigenvalue spectrum according to (1) or (2) of a free network in the dimension d

$$\underbrace{0,\cdots,0}_{r} \le \lambda_{r+1} \le \lambda_{r+2} \le \ldots \le \lambda_{r+i} \le \ldots \tag{6}$$

and applying the rules above the optimum datum in the sense of (4d) and (5) is transferred to the network by supporting it at the r/d nodal–points of the corresponding eigenfunction(s) x_i. Because of the interaction of (4d) and (5) with the wellknown relation

$$N^{+} = \sum_{r+1}^{u} 1/\lambda_i \cdot x_i' \cdot x_i \quad ,\lambda_i \text{ (6) according to (1) and (2)} \tag{7}$$

(u being the numer of coordinates in the discrete case (1) or '∞' for a continuum (2) respectively) the optimum datum covariance matrix – resulting from few d/r fiducal datum–points – will be identical with the wellknown minimum–trace solution, the so called pseudoinverse $N+$. That means that the very datum of a highest precision–

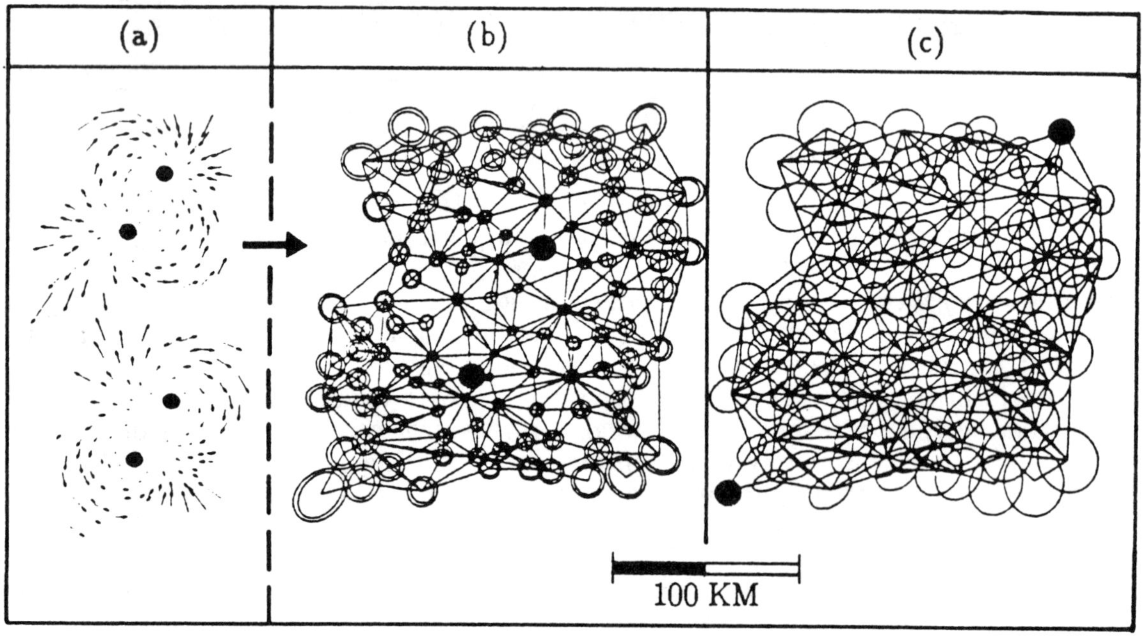

Fig. 3 Optimum and arbitrary datum – comparing the resulting error situations (the example shows the 'Baden–Württemberg Hauptdreiecksnetz' as an example for a regional land surveying network based on direction–measurements, r=4)
(a) nodal–points of an optimum datum, pointed out by x_5 and x_6
(b) Error situation of an optimum datum (outer ellipses) compared to the error situation of the inner solution N+
(c) The – much worse – error situation of another arbitrary datum

quality, which is also wellknown as the datum of the so called 'inner–solution' exists twice: first the datum–transfer implying the covariance–matrix N+ can be carried out by a Helmert–transformation over all coordinates belonging to the desired datum and second this high–quality datum is obtainable only by r/d (namely the optimum) datum points.

The small discrepancies in the size of the error ellipses belonging to the optimum datum–points with respect to those of the pseudoinverse N+, fig. 3, (b) are easily explainable: they are due simply and only to the fact that the above mentioned identities (4d) and (5) refer principally to the network–continuum model (2) requiring exact zero–components (=nodes) in the so called nodal–points of the eigenfunctions. This 'zero–requirement' does not hold for the discrete eigenvectors. The optimization–theory however suffers with respect to this fact in general only from a 'discretization–error', without being concerned in principle or being influenced in the optimization results respectively.

Optimum positions for a number of interesting fiducial points

The following optimization examples take a look to different types and shapes of terrestrial networks with respect to the determination of optimum supporting–point configurations within a given networks. With respect to the determination of fiducial points, e.g. by the actually progressing observation type 'GPS', this problem is to be classified as 1st Order Design – the task of finding the optimum locations within a network for performing a special observation type.

As a continuation to fig. 3, fig 4. shows the results concerning the optimum positions for supporting the Baden–Württemberg Hauptdreiecksnetz by 4 and 6 points respectively. The nodal–points of optimum support are to be found as the "calm centers of local eddies" within the eigenvector–fields.

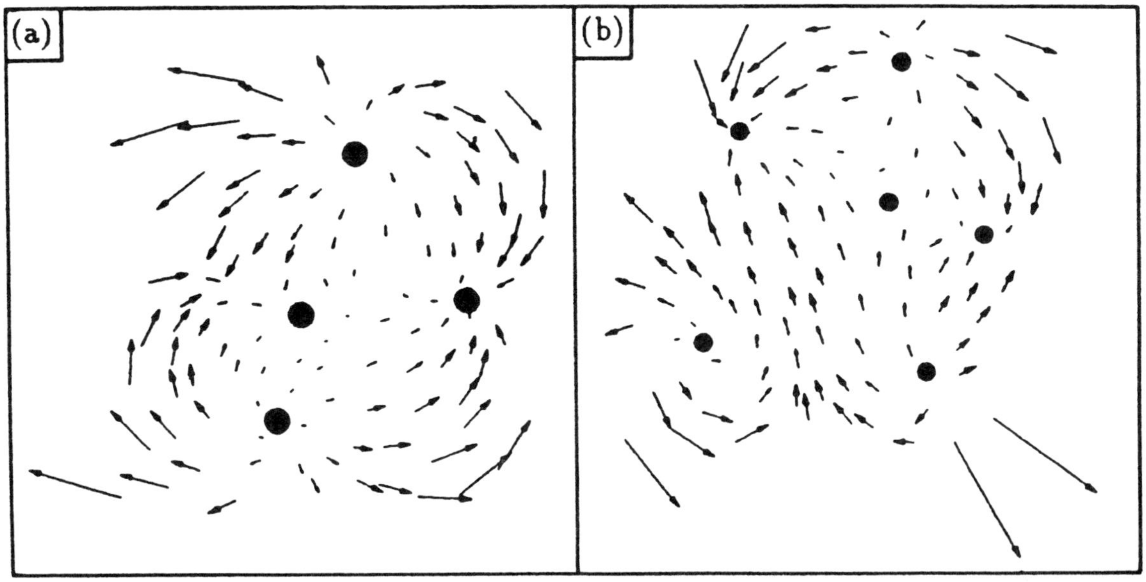

Fig. 4: Baden–Württemberg–net: (a)Optimum positions for 4, (b) for 6 supports

As a second example fig. 5 shows the places for an optimum supporting of a free and a left–side prefixed height– or gravity–traverse. Corresponding to the fact that this type of network implies the discretization of a vibrating membrane (BORRE, 1987) with respect to (2), the results show nodal–lines instead of –points. For an optimum support those lines are again in need of being discretized into points.

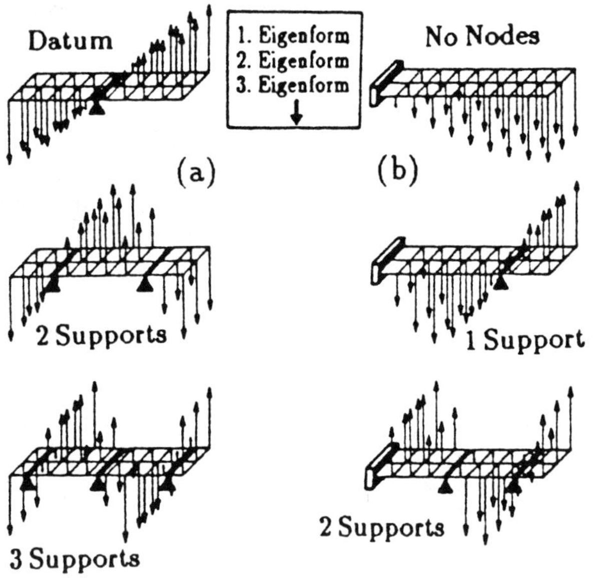

Fig. 5:
Principal look to the optimum support of a height– or gravity–netchain: (a) a free networks'–, (b) a fixed networks' additional supports

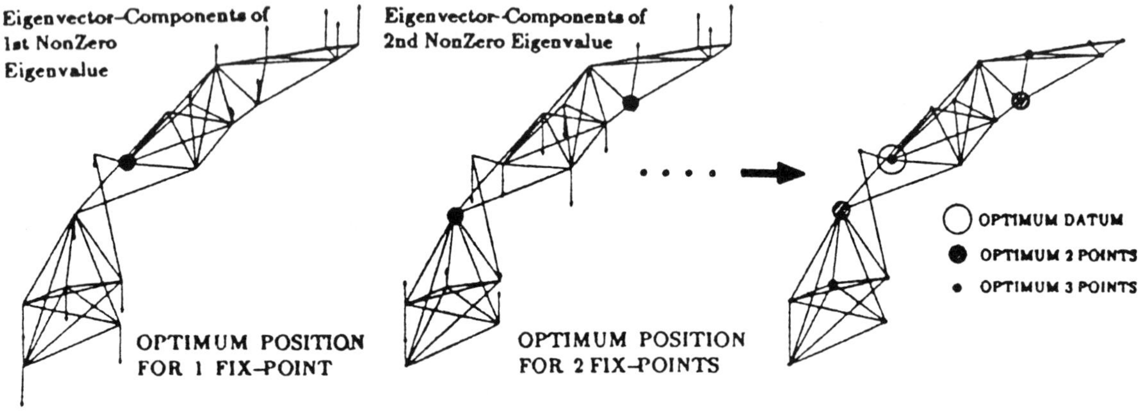

Fig. 6: Optimization for height–support: very discrete free deformation–network

Fig. 6 shows the optimization results of an existing deformation network (SCHMITT et al., 1989) with respect to a height–support. Although the network has a very discrete character the continuum based optimization principle still works.

The last example, fig. 7 takes a principal look to a network–chain consisting of direction measurements. For such a chain, e.g. a railway traverse (BILL et al., 1984)

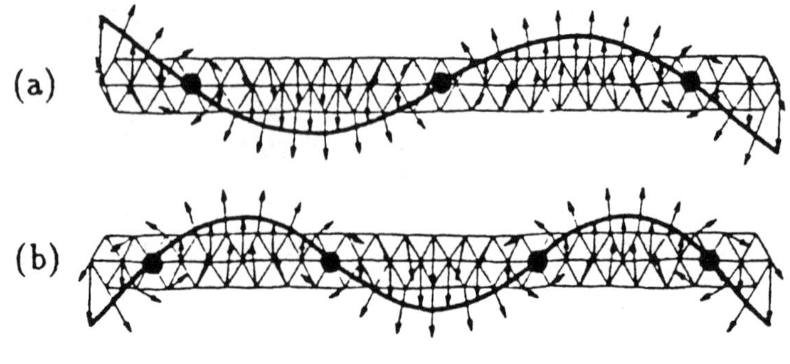

Fig. 7:
Principal look to the optimum support of a network–chain by 3 and 4 points (referring to fig. 1,(c2))

the optimum datum is situated in the so called 'Bessel–points' (JÄGER, 1988), see fig. 1, (c1). While for an ellipsoidal shape of edge, the ellipses focus–points turn out to be the optimum datum–points. Because of the existing mechanical analogies the cases of an optimum 3– or 4– point–support of the discrete chain above are referring to the nodal–points of the vibration–shapes of a continuous bar, as shown in fig. 1,(c2).

CONCLUSIONS

The optimization theory treats a new concept for solving the 0th Order Design problem of choosing the best situated points for a network datum transfer; secondly the theory holds for the 1./0. Order Design problem of selecting the optimum positions for the observation type of uncorrelated pointobservation or fix fiducial points – both applyable e.g. for optimizing GPS–point–configurations to be established within existing geodetic networks. For rather dicrete networks, experience shows that the optimum positions situated at the eigenfunctions nodal–points of a continuum are subjected to a 'discretization–error' and can't be resolved within the pointwise eigenvector–components anymore. It is helpful in this case to carry out an artifical densification of the network to be optimized, which is preserving the network–shape and boundary conditions as well the characteristics of the regarded network structure. Therfore the author proposes to investigate systematically elementary network–types, shapes, dimension and datum defects by high density models to get a deeper insight in the problem of optimum supporting. This will enable to speak out general recommendations for optimum support configurations, as well as the continuum results may serve as starting configurations for discrete search–methods to be developped.

Acknowledgements This research work has been carried out within the project '*Konfigurationsoptimierung*'. The author wishes to thank the *Deutsche Forschungsgemeinschaft (DFG)* for the opportunity to work and for its support to this project.

REFERENCES

BATHE, K., WILSON, E.L. (1976) : Numerical Methods in Finite Element Analysis. Prentice-Hall, Inc., Engelwood Cliffs, New Jersey.

BILL, R., JÄGER, R., SCHMITT, G. (1984) : Effekte in langgestreckten Netzen und ihre statischen Analogien. Zeitschrift fuer Vermessungswesen 109, S.526-540.

BORRE, K. (1987) : Untersuchung geodaetischer Netze mittels partieller Differentialoperatoren. Mitteilungen der geodaetischen Institute der TU Graz, Folge 56.

GANTMACHER, F.R. und M.G. KREIN (1960): Oszillationsmatrizen, Oszillationskerne und kleine Schwingungen mechanischer Systeme. Akademie Verlag Berlin.

JÄGER, R. (1988) : Analyse und Optimierung geodaetischer Netze nach spektralen Kriterien und mechanische Analogien. Deutsche Geodaetische Kommission, Reihe C, Nr. 342, Muenchen.

JÄGER, R. and E. DRIXLER (1989): Netzoptimierung im Design 0. und 1. Ordnung: Optimale Positionen fuer eine vorzugebende Zahl einzuschaltender Stuetzpunkte, AVN (96), No. 7.

NIEMEIER, W, (1982) : Principal Component Analysis and Geodetic Networks.– Some Baisic Considerations. In : BORRE/WELSCH (Hrsg.) : 'Deformationsanalysen' 83. Schriftenreihe

SCHMITT, G., MOLDOVEANU, T., NICA, V. and JAGER, R. (1989) : Deformation analysis of a local terrestrial network in Romania with respect to the Vrancea earthquacke of August 30.1986. Proceedings of the IAG General Meeting/Edinburgh 1989, included in this book.

WIESER, M. (1988) : Theoretische Untersuchungen spektraler Methoden zur Analyse regelmaeßiger Strukturen am Beispiel der Fouriertransformation geodaetischer Netze. Mitteilungen der Geodaetischen Institute der technischen Universitaet Graz, Folge 60, Graz.

LOGISTIC OPTIMIZATION FOR GPS SATELLITE NETWORKS BY GENERALIZED S.O.D. ALGORITHMS

P. Baldi*, F. Crosilla*, T. Russo**
*Istituto di Scienze della Terra, Universitá di Udine, Italy.
**Centro di Calcolo, Universitá di Trieste, Italy.

ABSTRACT

The paper deals with a new approach and the consequent technique to perform the analysis of a GPS network where undifferenced beat phase observation equations are considered. From the results of the analysis an original method to construct an ideal dispersion matrix is proposed. Such a matrix is constructed only for those estimable functions of the receivers' coordinates for which omogeneous and isotropic dispersion values contained within a centimetric range can realistically be requested. The solution equation of a Second Order Design problem that fulfills the ideal dispersion matrix, just constructed, is finally given.

ANALYTICAL MODEL

Let us start from the well known observation equation for a single (undifferenced beat) phase observation $\phi_r^s(t)$ as reported in Lindhlor and Wells (1985)

$$\phi_r^s(t) = \frac{f}{c}\|\underline{U}^s(t) - \underline{X}_r(t)\| + \alpha_r(t) + \beta^s(t) + \gamma_r^s \tag{1}$$

where the quantum numbers

$$r = 1.....NR; \quad s = 1.....NS; \quad t = 1.....NT$$

denote ground station receivers, satellites and measurement epochs, respectively. The observation equation is non linear in the coordinate terms $\underline{U}^s(t)$ and $\underline{X}_r(t)$. In particular $\underline{U}^s(t)$ and $\underline{X}_r(t)$ represent the three dimensional satellites and ground station (receiver) position vectors. In case of a static network $\underline{X}_r$ becomes independent of the argument (t).

The observation equation (1) is linear in the three bias terms $\alpha_r(t)$, $\beta^s(t)$, γ_r^s, where $\alpha_r(t)$ is the receiver-specific bias in ϕ (due essentially to clock errors), $\beta^s(t)$ is the satellite-specific bias in ϕ (due essentially to clock errors) and γ_r^s is the satellite-specific pair-specific bias in ϕ (essentially initial ambiguity + tropospheric correction independent in time); f and c represent the GPS L1 carrier frequency and the vacuum speed of light, respectively.

In this paper the contribution given by $\alpha_r(t)$ and $\beta^s(t)$ to equation (1) is neglected. This is done only in order to simplify the interpretation of the results of the analysis. Furthermore the components of vector $\underline{U}^s(t)$, coming from broadcast ephemeris, are considered as stochastic prior information while γ_r^s values are considered as unknown parameters to be estimated together with the elements of vector $\underline{x}_r$ which are the unknown increments to the approximated ground station coordinates contained in $\underline{X}_r^{app}$

$$\underline{X}_r = \underline{X}_r^{app} + \underline{x}_r \tag{2}$$

Differencing the observation equations with respect to the unknown non linear terms, the following linearized system is obtained

$$\phi_{obs_r^s}(t) + e = A \begin{pmatrix} \underline{x}_r \\ \underline{\gamma}_r^s \\ \underline{u}^s(t) \end{pmatrix} \tag{3}$$

where $\underline{u}^s(t)$ is the stochastic vector of increments to satellite coordinate with respect to the ephemeris values, to be predicted. $\phi_{obs_r^s}(t) := \phi_r^s(t) - \omega(t)$; $\omega(t)$ is a vector whose terms have the form $\frac{f}{c}\rho_r^s$, where ρ_r^s is an apriori distance computed from the appoximated $\underline{X}_r^{app}$ and $\underline{U}^s(t)$. The design matrix A can be partitioned in three parts $A = [A_x|A_b|A_s]$ according to the meaning of the three subvectors $\underline{x}_r$, $\underline{\gamma}_r^s$,$\underline{u}^s(t)$.

Originally matrix A_b contains only 0/1 values. In order to express the unknown values γ_r^s in meters, it is enough to multiply each coefficent by f/c. Therefore γ_r^s values assume the significance of corrections to an approximate pseudo distance satellite/receiver.

Let Q_{ss} be the dispersion matrix of the (stochastic) satellite coordinates given a priori by broadcast ephemeris. In the following examples we have assumed Q_{ss} as a diagonal matrix, with all the diagonal terms equal to σ^2 where we have fixed σ ranging from $1 \div 10$ m. (Matrix Q_{ss} is in general treated in the literature as full matrix, or at least, full for each satellite, neglecting in this way correlations existing among all the different satellite orbits). If a mixed linear model is applied (Schaffrin, 1985) it is possible to proceed to the estimation of subvectors $|\ \underline{x}_r\ \underline{\gamma}_r^s\ |^T$ and to the prediction of the subvector $\underline{u}^s(t)$.

Considering only the estimation case the dispersion matrix of the estimated parameters $|\ \underline{x}_r\ \underline{\gamma}_r^s\ |^T$ can be computed from

$$Q_{dd} = \left[A_d^T(Q_{ll} + A_s Q_{ss} A_s^T)^{-1} A_d\right]^+ \tag{4}$$

where Q_{ll} is the assumed diagonal dispersion matrix of the observations, and A_d is the matrix obtained joining together submatrices A_x and A_b.

Symbol $()^+$ stands for a pseudoinverse; this means that we suppose to perform, in general, a purely free net + bias adjustment, where the sum of the squares of the increments $\underline{x}_r$ and of the bias $\underline{\gamma}_r^s$ is minimized.

We could also consider the case of applying an S-transformation, for example fixing the datum on some receiver's coordinates only. Nevertheless, according to some theorems reported in Crosilla and Russo (1989) the final results, reported in the following, are equivalent and not dependent on the datum choice.

ANALYSIS OF A GPS NETWORK

Formula (4) is usually applied to (linear) Gauss-Morkov models where previous estimates of part of the unknowns are available. Nevertheless, in the case of the GPS, some considerations about the structure of the dispersion matrices can be done, taking into account the geometry of the observations and their precision obtainable with today's technologies.

First of all, we observe that Q_{ss} is a regular matrix with eigenvalues of the order of some squared meters. Consequently, $A_s Q_{ss} A_s^t$ is a matrix with the same rank of A_s, whose eigenvalues are of the order of several squared wave lenghts ($\lambda = 19\ cm$). Matrix Q_{ll} is regular, usually diagonal, and often scalar, and its eigenvalues are of the order of some hundreths of squared wave lenght (approximately 1 cm^2). Thus the spectral decomposition of $(Q_{ll} + A_s Q_{ss} A_s^T)$ puts clearly in evidence two sub spaces: one, of dimension equal to the rank of A_s with eigenvalues of the order of 10 m^2, *(metric space)*, the other one with eigenvalues of the order of 10^{-4} m^2 *(centimetric space)*.

In the case in which distances between receivers are small with respect to the distances between receivers and satellites, according to the particular structure of A_s, the differences between the measured phases relating to the same satellites and to the same epoch fall within the *centimetric* space of $A_s Q_{ss} A_s^T$. This is the reason for which many authors preferred to rewrite equation (1) in one of its possible differenced forms (see e.g. Lindhlor and Wells, 1985); this approach has the disadvantage that a certain number of (independent) baselines between the receivers must be chosen a priori. Using instead the undifferenced model, the global information, furnished by the observations, is preserved. So the separation between what is estimable with a precision comprised within a centimetric range and what is estimable with a worse precision can be done, as we shall see, a posteriori.

In fact, in this case, given the particular structure of A_d, the spectral structure of $(Q_{ll} + A_s Q_{ss} A_s^T)$ is transferred into Q_{dd} (although with different dimensions).

The spectral decomposition of Q_{dd} puts clearly in evidence three spaces:

- A null space (possibly empty) of the not estimable linear combinations of the non-stochastic unknowns.
- A space with eigenvalues of the order of some m^2 *(metric space)* containing those linear combinations of the unknowns which essentially depend on *sums* of the observed phases, and which are estimable with a σ value of the order of some meters.
- A space with eigenvalues of the order of $10^{-4} m^2$ (*centimetric space*) containing those non-stochastic linear combinations of the unknowns which essentially depend on the observed phase differences.

It is clear that if a GPS network is constructed for monitoring crustal deformations, the knowledge of the quantities which fall in the *metric space* is of no interest. Only those linear functions of the receivers' coordinates falling in the *centimetric space* are estimable with enough precision; the meaning of this space can be determined inspecting the spectral decomposition of Q_{dd}.

This method has the advantage of being appliable also in the case where distances between receivers are not negligible with respect to distances between receivers and satellites (regional networks). In this case (see the following example) the separation between *metric* and *centimetric* spaces becomes less evident, and the eigenvalue spectrum of Q_{dd} must be examined in order to determine which eigenvalues define the *space of interest*, where deformations can be measured with enough precision. This is the space which should be involved by design techniques of a GPS network.

Numerical example 1

Let us consider as example a North-South baseline and a satellite passing over it from West to East at 10000 km (Fig.1). There are 4 measured phases, 8 unknowns (two of which biases); Q_{dd} is an 8x8 matrix of rank 4. We assume that $Q_{ll} = 0.0025\ m^2 * I$ and $Q_{ss} = 1\ m^2 * I$.

The non-null eigenvalues have associated eingenvectors with the following meaning:

1)	$0.3\ 10^{-4}\ m^2$	$z(R_1) - z(R_2) + (\gamma_1^1 - \gamma_1^2)$	*centimetric*
2)	$0.9\ 10^{-4}\ m^2$	$x(R_1) - x(R_2)$	*space*
3)	$0.7\ m^2$	$z(R_1) + z(R_2) - (\gamma_1^1 + \gamma_1^2)$	*metric*
4)	$2.\ m^2$	$x(R_1) + x(R_2)$	*space*

Thus, the only functions of receivers' coordinates estimable by this model are the mean longitude (within 1.4 m) and the difference between longitudes (within 0.95 cm).

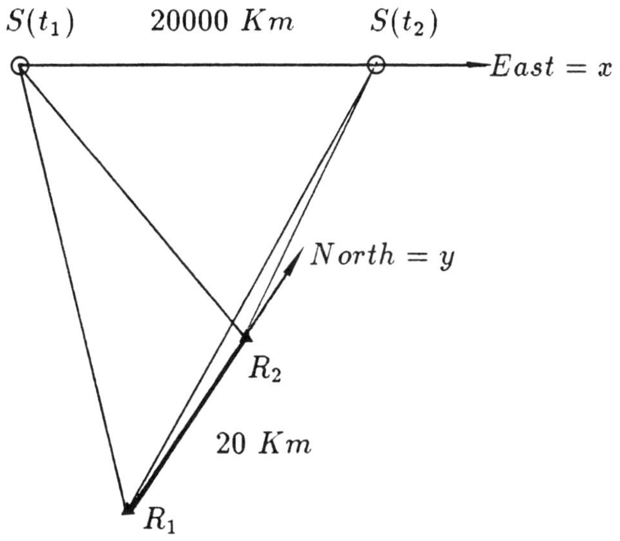

Fig.1: Example 1; scheme of the observations.

Increasing the distance between R_1 and R_2, the four non-null eigenvalues vary as follows:

$\|\|R_1 - R_2\|\|$	$20\ Km$	$200\ Km$	$2000\ Km$	$6000Km$	$20000Km$
1)	$.3\ 10^{-4}$	$.6\ 10^{-4}$	$.3\ 10^{-2}$	$.3\ 10^{-1}$	.2
2)	$.91\ 10^{-4}$	$.19\ 10^{-3}$	$.1\ 10^{-1}$	$.9\ 10^{-1}$	.4
3)	.67	.67	.67	.63	1.
4)	2.	2.	2.	2.	2.

i.e. the *centimetric* space merges gradually with the *metric* one. If, instead, the distance between $S_1(t_1)$ and $S_1(t_2)$ is reduced, the eigenvalues become:

$\|\|S(t_1) - S(t_2)\|\|$	$20000\ Km$	$2000\ Km$	$200\ Km$	$60\ Km$	$20\ Km$
1)	$.3\ 10^{-4}$	$.23\ 10^{-4}$	$.23\ 10^{-4}$	$.23\ 10^{-4}$	$.23\ 10^{-4}$
2)	$.91\ 10^{-4}$	$.46\ 10^{-2}$	.46	.5	.5
3)	.67	.5	.5	.5	.46
4)	2.	101.	$1.\ 10^4$	$1.\ 10^5$	$1.\ 10^6$

As can be seen, the eigenvalue 2) becomes *metric*, while the eigenvalue 4) tends to degenerate into the null space.

Numerical example 2

For a more realistic example, 4 true GPS satellites have been traced for 2 hours, visible from a GPS network of 5 receivers sited at an average distance of $\sim$ 40 km. Observations have been assumed every 30 minutes (see fig.2 for elevation and azimuth of each satellite at each epoch (Baldi and Crosilla, 1989)). Thus, we obtained 100 observations for 35 unknowns, of which 20 biases. Assuming the same Q_{ll} and Q_{ss} of the first example, Q_{dd} turns out regular (as it usually should be), showing so no null space, while the *metric* and *centimetric* ones are sharply distinct.

Since Q_{dd} is regular, the dispersion matrix Q_{xx} of the receivers' coordinates only can be extracted as its upper left 15x15 submatrix (rather than projecting it onto the x subspace by an R-Transformation (Schaffrin and Grafarend, 1986), as would be needed in the previous example for a singular Q_{dd}).

$Q'_{xx}s$ eigenvalues are:

1) 31.7	2) 2.05	3) 1.26	
4) $.63\ 10^{-3}$	5) $.62\ 10^{-3}$	6) $.61\ 10^{-3}$	7) $.61\ 10^{-3}$
8) $.40\ 10^{-4}$	9) $.39\ 10^{-4}$	10) $.38\ 10^{-4}$	11) $.38\ 10^{-4}$
12) $.24\ 10^{-4}$	13) $.23\ 10^{-4}$	14) $.23\ 10^{-4}$	15) $.23\ 10^{-4}$

A point worth of note is that both the *metric* and *centimetric* spaces are further subdivided into two subspaces of functions estimable with different precision:

this has been experienced also with different networks and different satellite configurations. Inspecting the related eigenvectors, it is apparent that 1-3 refer (as expected) to rigid translations of the whole set of receivers while 4-15 refer to net deformations. In particular, the eigenvalue 1) individuates roughly a *vertical* displacement, while eigenvalues 4-7 refer to *vertical deformations*, (where *vertical* does not denote strictly the direction of the earth's radius, but rather the direction from the net to some baricenter of the satellites' configuration). This accounts, for purely geometrical reasons, to the fact (usually reported in literature) that GPS networks give high accuracy in horizontal measurements while are poorer for altimetry.

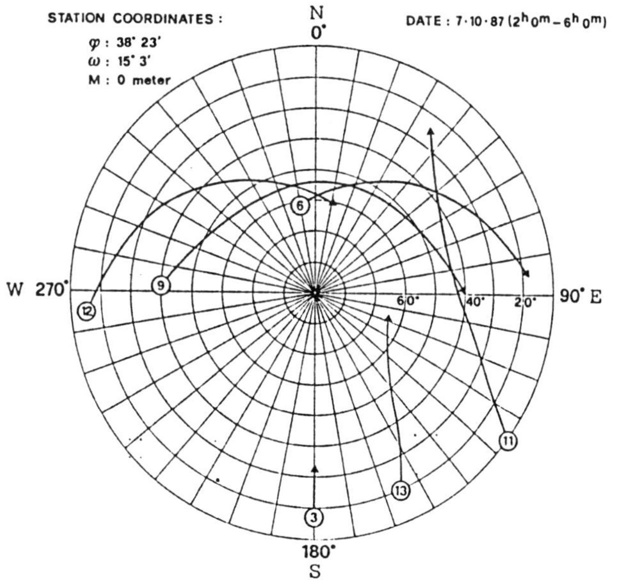

Fig.2: Elevation and azimuth of satellites.

CONSTRUCTION OF A CRITERION MATRIX

From the previous analysis it appears clearly that a GPS network (ignoring refraction problems) is able to monitor crustal deformations with horizontal dispersion values less than one centimeter and vertical dispersion values of some centimeters, while the absolute position of the receivers can be determined with dispersion values of some meters. According to these results we propose to ignore completely those information falling in the *metric space*, assimilating this space to the *null space*, which might be obtained by a geodetic free net adjustment.
In this case, Schaffrin (1985) proposes to construct a regular criterion matrix (of Taylor-Karman structure , for instance) and to project it onto the non-null space of the estimable functions of the coordinates. The criterion matrix can be then approximated by a proper choice of the observations' weights (Solution of the Second Order Design problem). In the case of a GPS network, we have preferred

to project the chosen criterion matrix onto a *space of interest* which can coincide with the *centimetric* space or with a subspace of its (for instance the subspace of the horizontal displacements, when monitoring a fault slip, or the subspace of the vertical displacements in the case of an altimetric GPS network).

Obviously the unknown biases do not fall into the *space of interest*: therefore the criterion matric $\tilde{Q}_{xx}$, so built, can be bordered with zeroes to give the criterion matrix $\tilde{Q}_{dd}$. Matrix $\tilde{Q}_{dd}$ can be then approximated, or, better, fulfilled in the *space of interest* only by a proper algorithm of S.O.D.

AN ALGORITHM FOR THE SECOND ORDER DESIGN

Fixed a criterion matrix $\tilde{Q}_{xx}$, projecting it onto the *space of interest* and bordering it with zeroes, a (singular) criterion matrix $\tilde{Q}_{dd}$ is obtained.

The proposed algorithm determine a diagonal matrix $\overline{Q}_{ll}^{-1}$ such that:

$$\overline{Q}_{dd}^{+} = A_d^T(\overline{Q}_{ll} + A_s Q_{ss} A_s^T)^{-1} A_d \geq \tilde{Q}_{dd}^{+} \tag{5}$$

in the Lowner partial ordering (Marshall and Olkin, 1979), that is, such to fulfil $\tilde{Q}_{dd}^{+}$ (Crosilla and Russo, 1988).

How to proceed: determine a matrix $\tilde{M}$ characterized by a minimum norm, such that

$$\tilde{Q}_{dd}^{+} = A_d^T \tilde{M} A_d \tag{6}$$

by means of

$$\tilde{M} = A_d^{T+} \tilde{Q}_{dd}^{+} A_d^{+} \tag{7}$$

Execute the spectral decomposition of $\tilde{M}$

$$\tilde{M} = U_M \Lambda_M U_M^T \tag{8}$$

and compute $\tilde{Q}_{ll}$ by

$$\tilde{Q}_{ll} = \tilde{M}^{+} - U_M U_M^T A_s Q_{ss} A_s^T U_M U_M^T \tag{9}$$

Matrix $\tilde{Q}_{ll}$ fulfills formula (5) but it is neither diagonal nor regular. Determine the diagonal matrix $\tilde{Q}_{ll}^{d+}$ which is the least squares approximation of $\tilde{Q}_{ll}^{+}$:

$$\tilde{Q}_{ll}^{d+} = \tilde{Q}_{ll}^{+} * I \tag{10}$$

Matrix $\tilde{Q}_{ll}^{d}$ is diagonal and regular but does not fulfil formula (5). In order to determine a matrix $\overline{Q}_{ll}$ which fulfills formula (5) it is enough to compute the maximum general eigenvalue, λ_{max}, of $\tilde{Q}_{ll}^{d}$ with respect to $(M^{+} - U_M U_M^T A_s Q_{ss} A_s^T U_M U_M^T)$ and to calculate (van Mierlo, 1982)

$$\overline{Q}_{ll} = 1/\lambda_{max}\tilde{Q}^d_{ll} \tag{11}$$

Matrix $\overline{Q}_{ll}$ is diagonal and fulfills formula (5).

RESULTS OF THE S.O.D. PROBLEM

Table 1 report the weights obtained by the algorithm just described imposing three different criterion matrices to the GPS network considered in the second numerical example. The weights relating to the receiver n.1 only are reported since the same behaviour holds for the other receivers, given the small dimensions of the network ($\sim 40\ km$). In the first case a T.K. criterion matrix, with *chaotic* structure ($1cm^2 * I$) has been projected onto the *subspace* of *vertical* deformations (eigenvalues 4-7).

Table 1. Wheights obtained imposing three different criterion matrices to the subspaces of vertical deformations (a), horizontal deformations (b) and to the entire deformations' space (c).

SATELLITE	EPOCH	WEIGHT		
		(a)	(b)	(c)
12	1	0.16	4.26	0.77
	2	0.53	1.37	0.54
	3	13.82	0.33	8.62
	4	0.97	1.54	0.85
	5	25.28	2.99	16.10
9	1	8.85	3.59	6.03
	2	0.26	1.65	0.43
	3	27.68	1.32	17.37
	4	3.07	1.35	2.11
	5	20.73	4.80	13.69
6	1	29.13	3.58	18.44
	2	0.58	1.10	0.54
	3	6.40	0.12	3.98
	4	2.19	1.05	1.50
	5	0.39	3.89	0.83
13	1	17.12	3.08	11.00
	2	1.58	0.58	1.07
	3	7.40	0.01	4.57
	4	4.28	0.59	2.72
	5	3.65	2.75	2.71

Comparing table 1(a) with figure 2, it is evident that the algorithm furnish higher weight values for:

- those observations close to the zenith;
- those observations which subtend the greatest difference in the orbit elevation;
- those observations which subtend the greatest elevation among different satellites, at the same epoch.

In the second case a criterion matrix, with chaotic structure ($0.15cm^2 * I$) has been projected onto the subspace of the horizontal deformations (requiring a traces' halving of matrix Q_{xx}). As can be seen in table 1(b), the algorithm furnishes higher weight values for the extreme positions of the satellite orbits. This because those positions subtend the greatest horizontal elongations .

In the third case a criterion matrix with chaotic structure ($1cm^2 * I$) has been projected onto the entire deformations' space, improving the precision of the vertical deformations and worsening that of the horizontal ones. An intermediate result has been obtained, with respect to those mentioned previously, characterized by a typical *saw teeth.*

CONCLUSIONS

The reported analysis shows clearly what kind of deformations can be detected by a GPS network. The proposed S.O.D. algorythm is able to determine the most important observations needed for a given purpose from a geometrical point of view. The obtained results (expecially those in table 1(b) and 1(c)) are somehow unrealistic, in that, greater importance is given precisely to those observations for which the tropospheric corrections (and hence their uncertainty) are expected to be larger. The next step will thus be the introduction of a model of tropospheric corrections to measures, in order to obtain results closer to the reality.

REFERENCES

Baldi, P. and Crosilla, F. (1989). A contribution to the optimal GPS satellite configuration for differential positioning, (in press).

Crosilla, F. and Russo, T. (1988). Fulfilment of a criterion matrix for a photogrammetric block triangolation by a Second Order Design of the control network, Manuscripta Geodaetica, 13 (4), 323-336.

Crosilla, F., Russo, T., (1989). About the irrelevance of the datum choice when searching for physically meaningful deformations (to appear).

Lindhlor, W. and Wells, D. (1985). GPS design using undifferenced carrier beat phase observations. Manuscripta Geodaetica, 10 (4), 255-295.

Marshall, W. and Olkin, I. (1979). Inequalities: Theory of majorization and its applications, Accademic Press, New York.

Mierlo(van), J. (1982). Difficulties in defining the quality of geodetic networks; in K. Borre and W.M. Welsch (eds), Survey control network, Schristenreihe, HSBw, heft 7, Munchen, 259-274.

Schaffrin, B. (1985). Network Design; in: E. Grafarend and F. Sansó (eds), Optimization and Design of geodetic Networks, Springer, Verlag: Berlin Heidelberg.

Schaffrin, B. and Grafarend E.W. (1986). Generating classes of equivalent linear models by nuisance parameter elimination; application to GPS observations, Manuscripta Geodaetica, 11 (4), 262-271.

GPS VERTICAL SURVEYING: A DISCUSSION OF SOME SPECIAL CONSIDERATIONS

Chris Rizos
School of Surveying, University of N.S.W.,
Sydney, N.S.W. 2033, AUSTRALIA

Don B. Grant
Dept. of Survey & Land Information,
Wellington, NEW ZEALAND

Robert D. Holloway
Dept. of Land Administration,
Perth, W.A. 6000, AUSTRALIA

INTRODUCTION

GPS Vertical Surveying - A Statement

The Global Positioning System (GPS) is a revolutionary satellite positioning technology that provides surveyors and geodesists with high accuracy 3-dimensional positional information. Although still in its pre-operational phase, GPS has been used from the early 1980's to derive (relative) coordinates with accuracies ranging from several parts in 10^6 (so-called "surveying" accuracies) to several parts in 10^8 ("geodetic" accuracies), as expressed in terms of the GPS receiver separation distance. Although accuracies have generally been expressed as a linear function of baseline length, the accuracies of the coordinate components are different for horizontal and vertical (or height) components by virtue of:

1. they are influenced by different error sources,
2. the error sources propagate into the various coordinate components in different ways, and
3. the satellite and ground receiver geometry has a different influence on the horizontal components, as opposed to the vertical component.

For the most part ellipsoid height information is a by-product of GPS surveys (albeit an important one in some cases), and the planning strategies, field operations and data reduction procedures have been generally developed to support horizontal GPS surveys. However, the use of GPS for vertical surveys is growing. In Australia, GPS is now seen as a viable high speed medium accuracy levelling technique. The Public Works Department of N.S.W. has used GPS, in conjunction with gravimetrically-derived geoid slope information, to check third order spirit levelling results (MacLeod et al., 1989), and further campaigns are underway to strengthen the Australian Levelling Survey with GPS ties. Concern over the longterm sea level fluctuations associated with the "Greenhouse Effect" has also highlighted the potential of GPS and VLBI for providing a high precision vertical datum. The Australian Government recently announced a special programme to measure sea level at 8 "super" tide gauges, which will themselves be monitored using a combination of GPS and VLBI. Rizos and Coleman (1989) describe a GPS survey whose principal aim is to derive high accuracy (relative) heights of a network of tide gauges in the Bass Strait, between the Australian mainland and Tasmania.

Errors in GPS Heights

The GPS literature is peppered with claims of differing precision/accuracy for GPS heights, for example:

"2-3 ppm for lines 10-70km length" Delikaraoglou et al. (1985)

"cm, or even sub-centimetre accuracy over distances of the order of 100km" Hein (1985)

"2 ppm horizontal, 4 ppm vertical" Stolz (priv. comm.)

"ellipsoidal height differences with cm order accuracy for distances up to 100km" Denker et al. (1986)

"1 ppm over baselines 50-100km and in the near future perhaps 0.1ppm" Hein (1986)

There is no doubt that with careful data processing (using algorithms not normally found in instrument manufacturers' software) that GPS heights well below the 1 ppm level can now be routinely obtained. In a number of studies of error propagation in GPS surveying, three effects have been identified as having an effect on GPS height results:

1. The vertical component is particularly sensitive to unmodelled errors in the atmospheric refraction, see, for example, Beutler et al. (1989); Grant (1989).
2. Whether the cycle ambiguities are "resolved" (to their integer values) during data processing. There is evidence that the resolution of cycle ambiguities has little or no effect on GPS heights, see, for example, Holloway (1988); Dong and Bock (1989).
3. If not all visible satellites can be observed, there may be a different strategy for selecting satellites to track, depending upon whether it is a horizontal or vertical GPS survey. This is encapsuled in a rule-of-thumb: low satellites are preferable for vertical surveys, high satellites are more appropriate for horizontal GPS surveys.

As GPS data reduction is based on the application of least squares analysis techniques, least squares simulations therefore have a useful part to play in investigating the effects of errors (or biases) <u>and</u> satellite-receiver geometry on GPS results. Furthermore, as there is less "established wisdom" for a new technique such as GPS than, for example, geodetic triangulation, simulations may be the only means by which an estimate of the final <u>accuracy</u> (as opposed to the <u>precision</u>) can be obtained for a specific observation and adjustment scheme.

There are essentially two types of simulations:

1. Simple simulations, based on the analysis of the VCV matrix of the adjusted parameters, which give the **precision** of the parameter estimates.
2. Simulations which incorporate the effect of systematic errors (in effect, <u>unadjusted</u> parameters) on the adjusted parameter estimates, and hence yield the **accuracy** of the GPS results.

Both types of simulations have a role, and examples of both will be presented in order to derive some "feel" for the perturbing effects on GPS vertical surveying.

COVARIANCE ANALYSIS AND SYSTEMATIC ERRORS

The study of systematic errors in simulated least squares adjustments is known variously as **covariance analysis**, **sensitivity analysis** or **consider analysis** (Gelb, 1974; Bierman, 1977). These names arise because the **cross covariance** matrix between systematic biases and the estimated parameters is used to determine the **sensitivity** of the solution to unadjusted parameters that are **considered** to be biased. Such analyses require assumptions to be made regarding the magnitude and behaviour of the systematic errors, but generally these are more plausible than the assumption of no systematic errors, which is the basis of standard precision analyses such as used, for example, for "optimising" terrestrial geodetic networks.

Systematic errors are considered to be errors in the *a priori* values of parameters that are not adjusted, but which are an integral part of the functional model. These may be physical constants that have been derived experimentally and which are not known exactly. Or they may be observed, calculated or assumed differences between the modelled observables and the actual measurements. They will often not appear explicitly in the functional model, but appear implicitly in assumptions on which the model is based. For example, a triangulation adjustment model may be based on the assumption that light waves travel in a straight line and that the theodolite and observed target are vertically over the station marks. In this case the unadjusted parameters (or systematic biases) may be horizontal refraction and station eccentricity.

In the case of GPS phase adjustment, systematic biases have origins in the satellite orbits (if unadjusted), atmospheric refraction (troposphere and ionosphere) and the reference system (fixed station coordinates, reference frame orientation parameters, etc.). Such techniques of covariance analysis have been used by a number of authors to investigate observation and reduction strategies as to their strengths and weaknesses, for example, Freymueller et al. (1986) using the OASIS software developed at the Jet Propulsion Laboratory, Pasadena, California, and Grant (1989) using the DASH software developed at the University of N.S.W., Sydney, Australia. In order to perform a covariance analysis of this type, the following need to be specified:

1. The observation scenario; time and length of tracking, receiver geometry, measurement precision (GPS satellite geometry is defined by the available satellite constellation).
2. The adjusted parameters and their *a priori* uncertainties.
3. The unadjusted (or consider) parameters; their magnitude and behaviour (for example, whether constant or time-varying).

We present two examples of such analyses for baseline lengths of 50km or less.

Example 1 - Fixed Orbits, Single-Frequency Instrumentation

Simulations were carried out for two baselines in the vicinity of Sydney, one of 5km length and the second 50km, using an 18 satellite constellation. The observation period was for two hours with all 6 visible satellites above 15° being tracked. The receiver measurement error (all receiver and antenna errors) was assumed 0.005m for 2 minute data. The adjusted parameters were the coordinates of one end of the baselines and the cycle ambiguities. The unadjusted, or consider, parameters were:

1. errors in the fixed station coordinates with respect to WGS84 of 10m, as it was assumed that *a priori* coordinates are usually obtained from a pseudo-range solution.
2. errors in the satellite orbits in the HCL system of 6m (radial), 6m (cross) and 18m

(alongtrack), giving a total position error of about 20m.

3. residual tropospheric delay error after the observations have been corrected *a priori* using surface met data, a tropospheric refraction model, or both. The residual error was assumed to have a zenith magnitude of 0.048m (approximately 2% of full effect), and a time-varying behaviour that was modelled in a Kalman filter.

4. error caused by the ionospheric delay on single-frequency observation of 1m at the zenith, and a time-varying behaviour that was modelled in a Kalman filter.

Two cases were investigated. The first being where the ambiguities were not resolved on the 5km or 50km baselines to their integer values. The second where the ambiguities were resolved for the 5km line. Fig.1 shows that when the ambiguities are not resolved the dominant height error is the tropospheric error and is proportionally much the same for both baselines. The receiver noise error is constant and therefore has a much greater influence on the shorter line. The total error is the RSS ("root sum squared") of the individual errors.

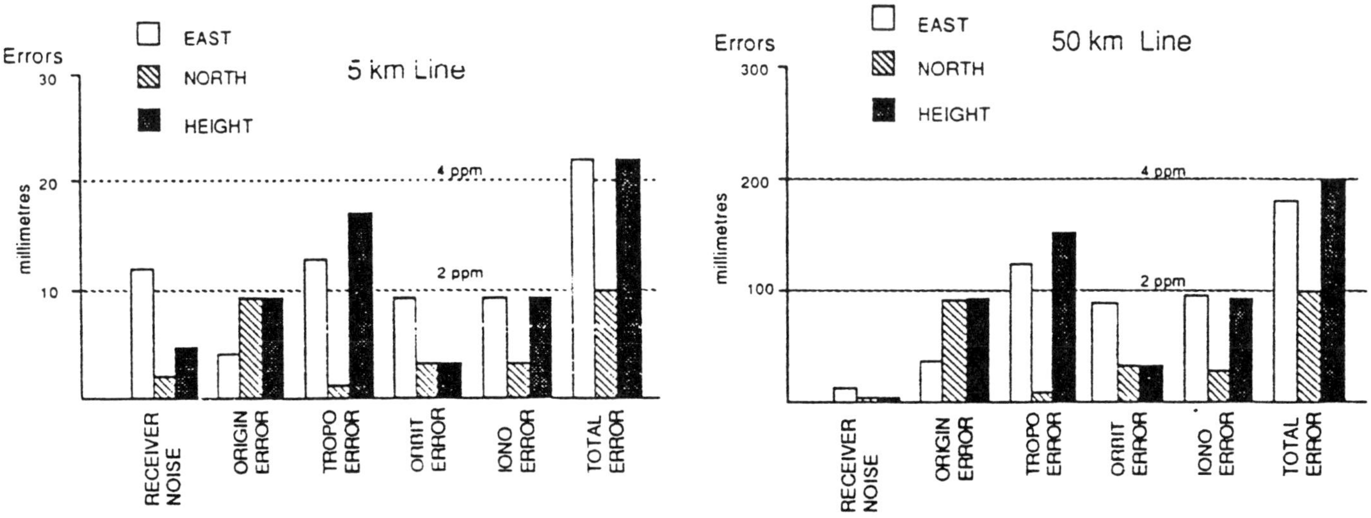

Fig.1. Effect of simulated errors in 5km and 50km baselines, assuming that ambiguities were not resolved.

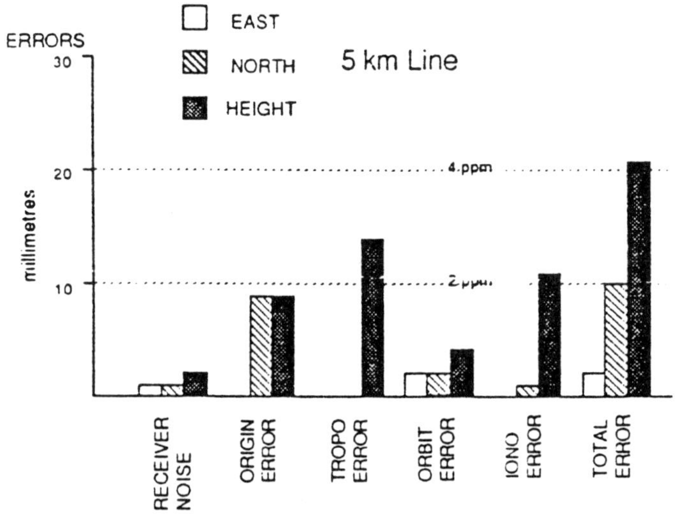

Fig.2. Effect of simulated errors in 5km baseline, assuming that ambiguities have been correctly resolved.

Fig. 2 shows that by resolving the ambiguities correctly the error in the east coordinate component is improved dramatically. The contribution of the receiver noise is also diminished. The dominant error for the height component is still the residual tropospheric error with errors in the ionosphere and the fixed station coordinates also being significant. What is most striking however is that there is no improvement in the north or height components by resolving ambiguities.

Comment. There appears to be little benefit in resolving cycle ambiguities for GPS vertical surveys. The orbit error assumed in these simulations (20m) is on the optimisitic side if we consider the possible effect of "Selective Availability" for the Block II satellite constellation. Use of dual-frequency receivers would eliminate the systematic biases arising from the ionosphere. The magnitude of each individual error source in Fig. 1 and 2 can be scaled up or down and a new RSS total error obtained, but the general conclusions are still valid.

Example 2 - Progressive Ambiguity Resolution

That ambiguity resolution does not bring any significant gains in height accuracy for baselines up to 500km was also noted by Dong and Bock (1989). This intriguing conclusion led us to look more closely at the ambiguity resolution process. A progressive ambiguity resolution scheme for GPS phase adjustment was simulated. The first adjustment has all ambiguities free (as solve-for parameters). Following each adjustment, the most accurately determined ambiguity parameter is resolved to the nearest integer (if possible) and the adjustment is run again with this ambiguity held fixed. This procedure is repeated until either all ambiguities have been resolved or no further ambiguities can be resolved. Simulated observations on a 28.4km baseline in the vicinity of Wellington, New Zealand, were used for a 2hr observing session involving satellites from the 18 satellite constellation. 6 satellites were visible. Fig. 3 illustrates how, as an increasing number of ambiguities are held fixed, the remaining ambiguities become more accurately determined. There were 6 adjustments and the ambiguity to satellite 7 has an error of nearly one cycle in the first (ambiguity-free) adjustment but, by the time we come to resolve it after adjustment 5, its error is only 0.14 cycles. The ambiguity that is most difficult to resolve is the one to satellite 9. In adjustment 2, this ambiguity has an error of 0.29 cycles. If the ambiguity resolution is to proceed it must be fixed to the nearest integer in adjustment 3. The ambiguities are therefore resolved in order: satellite 2, 9, 14, 12, 7.

We now look at the accuracy of the baseline determination at each step of the ambiguity resolution process. The total error, including the systematic effects of the origin coordinates, orbit error and troposphere, is illustrated in Fig. 4 for the 3 baseline components. As with Example 1 above, the east coordinate component improves significantly as each ambiguity is resolved. The east coordinate component error in the final ambiguity-fixed adjustment is just over one quarter of the error in the ambiguity-free adjustment. The north coordinate component error is virtually unaltered by the ambiguity resolution process and the height component in this case shows only a slight improvement.

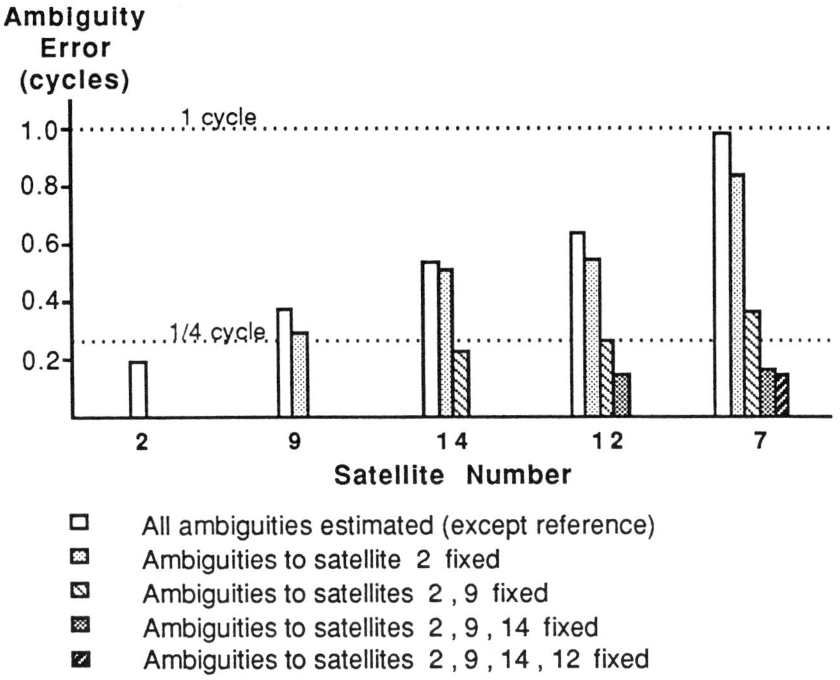

Fig. 3. Progressive resolution of cycle ambiguities and accuracy of remaining estimated ambiguities on a 28.4km line for a 2hr session 0730-0930hrs.

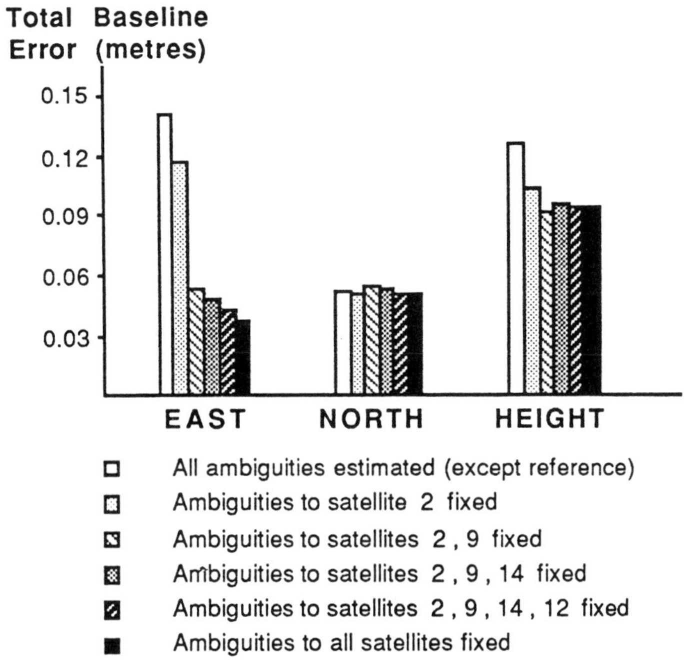

Fig. 4. Accuracy of baseline as cycle ambiguities are progressively resolved on a 28.4km line for a 2hr session 0730-0930hrs.

Resolving Ambiguities for GPS Vertical Surveys. Example 1 indicates no improvement in the accuracy of the height component after ambiguity resolution, yet Example 2 (based on the same assumptions reagrding systematic errors as Example 1) indicates some improvement (albeit small). We therefore investigated more closely the role of satellite geometry in GPS horizontal and vertical surveys.

PRECISION INDICATORS AND SATELLITE GEOMETRY

The analysis of satellite geometry is performed using only the VCV information for the adjusted parameters (coordinates and ambiguities).

Indicators of Precision: Surveying *vis a vis* Navigation

For GPS navigation, an indicator of precision which is only related to the relative satellite-receiver geometry at the instant of a navigation "fix" is the "Position Dilution of Precision" (PDOP). A low PDOP or GDOP (Geometric DOP, an indicator similar to PDOP but including the variance component of the receiver clock error) characterises "good" geometry for pseudo-range navigation. However the use of these DOP factors in GPS surveying is inappropriate because (Merminod et al., 1989):

1. The navigation solution is based on the **instantaneous** satellite-receiver geometry, whereas the GPS phase adjustment is governed by the continually changing geometry of an observing session which may last for several hours.
2. PDOP gives the geometric strength of a **point position**, whereas the GPS carrier phase adjustment provides the **relative position** of 2 or more receivers.
3. The pseudo-ranges are considered to be biased only by the errors of receiver oscillator (which changes from epoch to epoch). In a GPS phase adjustment the double-differenced observation (or equivalent) is biased only by the integer cycle ambiguities. The geometric strength of double-differenced **ambiguous range** measurements (integrated carrier phase) is quite different to that of **pseudo-range**.

Merminod (1989) proposes a set of three alternative indicators of satellite geometry that are appropriate for GPS phase adjustment. These "Bias Dilution Of Precision" (BDOP) indicators provide us with a measure of the relative precision of:

1. coordinates in an ambiguity-free solution (BDOP1),
2. ambiguity parameters in an ambiguity-free solution (BDOP2), and
3. coordinates in an ambiguity-fixed solution (BDOP3).

The BDOPs are computed by accumulating the normal matrix with the contributions of all the measurements made to all or a subset of visible satellites, for the duration of a specified observing session (Ibid, 1989). The matrix contains partial derivatives associated with station coordinates and cycle ambiguities and has dimension (3+#sat-1 x 3+#sat-1), where #sat is the number of satellites tracked. The cofactor matrix $\mathbf{Q_{xx}}$ is obtained by inversion of the resultant normal matrix, and is partitioned into four sub-matrices:

$$\mathbf{Q_{xx}} = \begin{bmatrix} \mathbf{Q_{cc}} & \mathbf{Q_{cb}} \\ \mathbf{Q_{bc}} & \mathbf{Q_{bb}} \end{bmatrix} \tag{1}$$

BDOP1 is **defined** as the square root of the trace of the coordinate submatrix, dimension (3x3), and indicates the precision of the coordinates before cycle ambiguity resolution. That is:

$$\mathbf{BDOP1} = \sqrt{\text{trace}\left[Q_{cc}\right]} \tag{2}$$

BDOP2 is **defined** as the square root of the trace of the cycle ambiguity submatrix, dimension (#sat-1 x #sat-1), and indicates their precision. That is:

$$\mathbf{BDOP2} = \sqrt{\text{trace}\left[Q_{bb}\right]} \tag{3}$$

The sub-matrix of the coordinates of the accumulated normal matrix (3x3) is isolated and inverted separately. This is equivalent to excluding the cycle ambiguities from the set of solve-for parameters. BDOP3 is then **defined** as the square root of the trace of the cofactor matrix and indicates the precision of the coordinates after all cycle ambiguities have been resolved. That is:

$$\mathbf{BDOP3} = \sqrt{\text{trace}\left[N_{cc}^{-1}\right]} \tag{4}$$

A low BDOP1 therefore indicates good geometry for GPS phase adjustment. For a discussion of the significance and examples of the use of BDOPs we refer the reader to Merminod et al. (1989).

Example 3 - Covariance Analysis with Good Geometry

The covariance analysis described in Example 2 was based on a poor BDOP1 value. The same analysis was repeated for a 2hr session starting just one hour later than in Example 2. This period corresponded to a BDOP1 value that was about half that of the previous session, indicating better satellite geometry, even though only 5 satellites were now visible, rather than the 6 as in the previous case. Fig. 5 illustrates the accuracy of the 28.4km baseline for a session with low BDOP1.

Note that in Fig. 5 there is only a slight improvement in the accuracy of the east component, and almost no change in the north and height components as the ambiguities are progressively resolved, compared with Fig. 4.

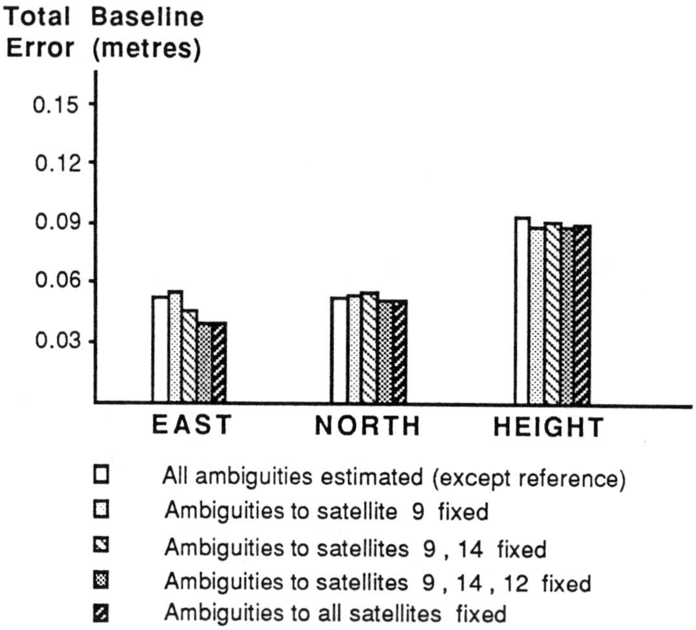

Fig. 5. Accuracy of baseline as cycle ambiguities are progressively resolved on a 28.4km line for a 2hr session 0830-1030hrs.

CONCLUSIONS

"Good" geometry for GPS surveying can be quantified by the BDOP1 factor, computed for a simulated session, from the VCV matrix containing the station coordinates and cycle ambiguities as solve-for parameters. The BDOP1 characteristics bear little or no relation to PDOP/GDOP (that is, low BDOP1 is not correlated with low PDOP/GDOP). Good satellite geometry for GPS surveying (short to moderate session lengths, and no orbit adjustment) implies that ambiguity resolution for relatively short baselines (<50km) brings very little benefit to horizontal coordinate or vertical coordinate components. Weak satellite geometry implies no, or very little, improvement in the accuracy of the vertical and north coordinate components, but a dramatic improvement in the east component. Until more simulation studies are performed, we cannot say whether this is a universal result for any length/orientation of baseline, or any adjustment scheme (e.g. orbit adjustment), or for particular subsets of GPS satellites.

Acknowledgement. The second author was supported at the School of Surveying, University of N.S.W., by a study award from the Dept. of Survey and Land Information, New Zealand.

REFERENCES

Beutler, G., Bauersima, I., Gurtner, W., Rothacher, M., Schildknecht, T. and Geiger, A. (1989). Atmospheric refraction and other important biases in GPS carrier phase observations, *Atmospheric effects on geodetic space measurements*, monograph **12**, 15-

44, School of Surveying, University of New South Wales, Australia.

Bierman, G.J. (1977). *Factorization methods for discrete sequential estimation*, Academic Press, Orlando.

Delikaraoglou, D., Beck, N., McArthur, D. and Lochhead, L. (1985). On the establishment of 3-D geodetic control by interferometry with the TI-4100 GPS receiver, in *First Int. Symp. on Precise Positioning with the Global Positioning System*, C.C. Goad (ed.), Rockville, Maryland, 15-19 April.

Denker, H., Torge, W., Wenzel, H.G., Lelgemann, D. and Weber, G. (1986). Strategies and requirements for a new European geoid determination, in *IAG Symp. on the Definition of the Geoid*, Florence, 26-30 May.

Dong, D. and Bock, Y. (1989). GPS network analysis with phase ambiguity resolution applied to crustal deformation studies in California, *J. Geophys. Res.*, **94(B4)**,3949-3966.

Freymueller, J.T., Kellogg, J.N. and Davidson, J.M. (1986). GPS geodesy in the North Andes: results of covariance analysis, *EOS Trans. AGU*, **67**, 910.

Gelb, A. (1974). *Applied optimal estimation*, MIT Press, Cambridge, Mass.

Grant, D.B. (1989). Combination of terrestrial and GPS data for earth deformation studies in New Zealand, UNISURV **S-32**, School of Surveying, University of New South Wales, Australia, in press.

Hein, G.W. (1985). Orthometric height determination using GPS observations and the integrated geodesy adjustment model, NOAA tech. rept. NOS 110 NGS **32**, National Oceanic and Atmospheric Adminstration, Rockville, Maryland.

Hein, G.W. (1986). The role of GPS data in gravity field approximation or the role of the gravity field in GPS surveys, in *IAG Symp. on the Definition of the Geoid*, Florence, 26-30 May.

Holloway, R.D. (1988). The integration of GPS heights into the Australian Height Datum, UNISURV **S-33**, School of Surveying, University of New South Wales, Australia.

MacLeod, R.T., Rizos, C. and Kearsley, A.H.W. (1989). The resolution of mean sea level anomalies along the New South Wales coastline by GPS, IAG Symp. Edinburgh, 3-12 August.

Merminod, B. (1989). The resolution of the cycle ambiguities, UNISURV S report, School of Surveying, University of New South Wales, Australia, in prep.

Merminod, B., Grant, D.B. and Rizos, C. (1989). Planning GPS surveys - using appropriate precision indicators, *CISM Journal ACSGC*, submitted.

Rizos, C. and Coleman, R. (1989). Vertical datum definition for oceanographic studies using GPS, IAG Symp. Edinburgh, 3-12 August.

STATISTICAL TESTING AND QUALITY ANALYSIS OF 3-D NETWORKS

Part I: Theory[1]

A.J.M. Kösters
H. van der Marel

Geodetic Computing Centre (LGR)
Faculty of Geodesy
Delft University of Technology
The Netherlands

Abstract

Many geodetic problems are solved by least-squares estimation of a number of unknown parameters from an overdetermined system of observation equations. From least-squares theory itself, methods have been developed in the past, to evaluate precision of observations and parameters, to check the correctness of the functional model, including the observations themselves, and to describe reliability of observations and parameters.

These general methods can be applied to any least-squares estimation problem and occur at various stages of the computational process from observations to coordinates in a higher order reference frame.

In part I of this paper, the above mentioned methods are described briefly, as an introduction to part II, [van der Marel, Kösters, 1989], where this theory is applied to GPS.

1. Introduction

In a linear Gauß-Markov model (1), a (nx1)-parameter vector x can be estimated through the least-squares algorithm. The (mx1)-vector of observables $\underline{y}$ (m>n) are stochastic variables, assumed to be normally distributed with mathematical expectation $E\{\underline{y}\}$ and covariance matrix $\sigma_0^2 Q_y$ with known variance factor σ_0^2. The (m×n)-design matrix A is the linearized description of the relation between observations and parameters. Hence, the overdetermined system of observation equations is:

$$E\{\underline{y}\} = Ax \; , \qquad E\{(\underline{y}-E\{\underline{y}\})(\underline{y}-E\{\underline{y}\})^T\} = \sigma_0^2 Q_y \tag{1}$$

- - - - -

1 *This paper is a summary of the theoretic parts in [van der Marel, 1989] and [Kösters, Kok, 1989], both handed out during poster sessions at the IAG symposium S102: "The Global Positioning System and other radio tracking systems" in Edinburgh, August 1989.*

The solution is found through orthogonally projecting $\underline{y}$ onto the range space of A: R(A), which is spanned by the columns of A. Therefore, the corrections are in the orthogonal complement of R(A), $R(Q_yB)$ (see fig. 1):

$$\hat{\underline{x}} = (A^TQ_y^{-1}A)^{-1} A^TQ_y^{-1}\underline{y} \; ; \qquad Q_x = \sigma_0^2 (A^TQ_y^{-1}A)^{-1} \tag{2a}$$

$$\hat{\underline{e}} = \underline{y} - \hat{\underline{y}} = \underline{y} - A\hat{\underline{x}} = (I - \mathbb{P}_A)\underline{y} = \mathbb{P}_{QB}\underline{y}; \qquad Q_e = \sigma_0^2 (Q_y - AQ_xA^T) \tag{2b}$$

However, the above formulae are only valid in the absence of errors in both the functional and the probabilistic model. Concerning the functional model, we can formulate tests that are capable of detecting errors or blunders in the observations (data snooping) or tests that can detect incompletenesses in the design matrix. For both tests we may like to know the maximum size of undetected model errors and their influence on estimated parameters. These quantities were defined by [Baarda,1967,1968] as internal and external reliability respectively. After adjustment, we may like to detect possible insignificant parameters or test if the deviation of their estimated value from a known value is significant. These tests can be derived, analogous to the tests mentioned above. Here too, internal reliability is defined. The assumptions we made concerning the probabilistic model of the observations can be tested too but these tests are out of scope of this paper.

2.1 Hypothesis testing and reliability

In brief, the filosofy of hypothesis testing is as follows. The null-hypothesis H_0 is that the linear Gauß-Markov model (1) is valid. In order to proof this hypothesis we formulate alternative hypotheses H_a by specifying (mxc)-matrices C that describe any possible model error ∇ we can think of. If ∇ turns out to be statistically significant for any H_a, H_0 will be rejected in favour of that particular H_a. If no alternative hypotheses H_a are preferred above H_0, H_0 will be accepted.

$$\begin{cases} H_0\colon E\{\underline{y}\} = Ax & ; \; \sigma_0 Q_y \\ H_a\colon E\{\underline{y}\} = Ax + C\nabla & ; \; \sigma_0^2 Q_y \end{cases} \tag{3}$$

with C a (mxc)-matrix containing the functional relation between the observables $\underline{y}$ and the (cx1)-vector for the model error ∇. The columns of C should be independent of the columns of A.

A geometric interpretation of hypothesis testing in least-squares adjustment is given in [v.Mierlo,1981] and [Teunissen,1986] (fig. 1).

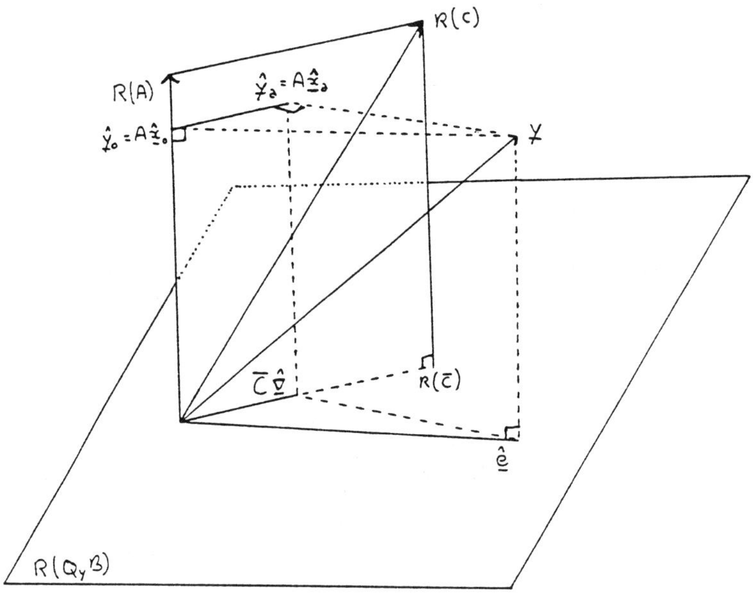

- fig. 1 -

The least-squares estimator $\hat{\underline{x}}_0$ of x under H_0 is related to the projector of $\underline{y}$ on the subspace R(A), by $\hat{\underline{y}}_0 = A\hat{\underline{x}}_0$. Under H_a, the least-squares estimator $\hat{\underline{x}}_a$ of x is related to the projector of $\underline{y}$ on the subspace R(A:C), by $\hat{\underline{y}}_a=(A:C)(\hat{\underline{x}}_a^T\ \hat{\underline{\nabla}}^T)$. $\hat{\underline{\nabla}}$ is an estimator for ∇, the model error described by H_a.

$$\hat{\underline{\nabla}} = \left(C^TQ_y^{-1}Q_eQ_y^{-1}C \right)^{-1} C^TQ_y^{-1}\,\hat{\underline{e}} \qquad\qquad Q_{\hat{\nabla}} = \left(C^TQ_y^{-1}Q_eQ_y^{-1}C \right)^{-1} \tag{4}$$

The squared norm of $\hat{\underline{\nabla}}$ is a quadratic form of $\hat{\underline{e}}$. It can be expressed in units σ_0^2, yielding a χ_c^2-distributed test-statistic for proving the validity of H_a.

$$\underline{T} = \frac{\hat{\underline{e}}^TQ_y^{-1}C\left(C^TQ_y^{-1}Q_eQ_y^{-1}C \right)^{-1}C^TQ_y^{-1}\hat{\underline{e}}}{\sigma_0^2} \sim \chi_c^2 \tag{5}$$

This statistic can be computed from adjustment under H_0 only and does not require adjustment under H_a. If T exceeds a critical value k, computed from the central χ_c^2-distribution with a chosen level of significance α, the null-hypothesis will be rejected in favoûr of the alternative hypothesis.

For an arbitrary hypothesis C_i, the non-centrality parameter λ_0 of the χ_c^2-distribution under H_a is:

$$\lambda_0 = \nabla_i^T\ C_i^TQ_y^{-1}Q_eQ_y^{-1}C_i\ \nabla_i = \nabla_i^T\ Q_{\hat{\nabla}}^{-1}\ \nabla_i = \|\ \nabla_i\|^2 = \lambda_i \qquad \forall\ C_i \tag{6}$$

The essence of the B-method of testing is that λ_0 is fixed for all alternative hypotheses. It is also a function of the level of significance α, power β and the number of degrees of freedom, so

$$\lambda_0 = \lambda(\alpha_1,\beta,1,\infty) = \lambda(\alpha_c,\beta,c,\infty) = \lambda(\alpha_{m-n},\beta,m-n,\infty) \tag{7}$$

Using (7), we can relate reliability for statistical tests of various dimensions. The effect of the model error ∇ on the observations y is called the internal reliability vector $\nabla_i y$. Given λ_0 and C_i, (cx1)-vectors $\nabla_i y$ are obtained from $C_i\nabla_i$, cf. (8). They represent the size of model errors to be found with probability β.

$$C_i\nabla_i \overset{def}{=} \nabla_i y \; ; \quad \| \nabla_i y \|^2 = \nabla_i^T C_i^T Q_y^{-1} C_i \nabla_i \overset{def}{=} \lambda_{y,i} \tag{8}$$

Note that (8) is a quadratic form of ∇_i and that for c>1, all possible $\nabla_i y$ describe a (hyper)-ellipsoid. For every vector $\nabla_i y$ its influence on the parameter estimates (external reliability) can be computed by

$$\nabla_i x = (A^T Q_y^{-1} A)^{-1} A^T Q_y^{-1} (\nabla_i y) \tag{9}$$

The scalar $\bar{\lambda}_{i,x}$ is defined as the squared norm of $\nabla_i x$, so

$$\lambda_{x,i} = \| \nabla_i x \|^2 = (\nabla_i x)^T (A^T Q_y^{-1} A) (\nabla_i x) \tag{10}$$

and relates reliability back again to precision because an upper bound for the influence $\phi^T A(\nabla_i x)$ of an undetected error in the observable vector y, with size $\nabla_i y$ on any function ϕ of the parameters, is given through:

$$\phi^T A(\nabla_i x) < \sigma_\phi \sqrt{\lambda_{x,i}} \tag{11}$$

with σ_ϕ the standard deviation of the estimator of that function of parameters.

2.2 Examples of hypotheses

The choices made for C, specify the type of errors to be detected. We will briefly mention some possible and useful options.

- For the overall model test, in which one tries out the fit of the functional model in a most general way, one chooses C so that R(A)⊕R(C) spans the complete observation space. The test-statistic is:

$$\underline{T}_{m-n} = \frac{\hat{\underline{e}}^T Q_y^{-1} \hat{\underline{e}}}{(m-n)\cdot\sigma_0^2} \sim F_{m-n,\infty} \tag{12}$$

- In order to detect errors in observations (data snooping) the C matrix is specified as:

$$\delta_i = (\underset{1}{0},\underset{2}{0},\underset{.}{0},.,\underset{.}{0},\underset{i}{1},\underset{.}{0},\dots,\underset{m}{0})^T \tag{13}$$

This results in the test-statistic

$$\underline{w}_l = \frac{\delta_l^T Q_y^{-1} \hat{\underline{e}}}{\sigma_0 (\delta_l^T Q_y^{-1} Q_e Q_y^{-1} \delta_l)^{1/2}} = \frac{\delta_l^T Q_y^{-1} \hat{\underline{e}}}{\sigma_0 \cdot \sqrt{N_{ll}}} \sim N(0,1) \tag{14}$$

while marginal detectable errors (here scalars) follow from:

$$\| \nabla_l y \| = \sigma_0 \sqrt{\frac{\lambda_0}{(\delta_l^T Q_y^{-1} Q_e Q_y^{-1} \delta_l)}} = \sigma_0 \sqrt{\frac{\lambda_0}{N_{ll}}} \tag{15}$$

- Multi-dimensional generalizations of this data snooping procedure can be considered too. If we define

$$\delta_{l,l+1,l+2} = \begin{pmatrix} 0 & 0 & 0 & \dots & 1 & 0 & 0 & \dots & 0 & 0 & 0 \\ 0 & 0 & 0 & \dots & 0 & 1 & 0 & \dots & 0 & 0 & 0 \\ 0 & 0 & 0 & \dots & 0 & 0 & 1 & \dots & 0 & 0 & 0 \end{pmatrix}^T \tag{16}$$
$$\quad 1 \quad 2 \quad 3 \quad . \quad l \quad l+1 \quad l+2 \qquad m-2 \quad m-1 \quad m$$

we have an alternative hypothesis for testing a group of 3 observations, when certain linear combinations of these observations are suspected to be erroneous. The test statistic will be similar to (6), while internal reliability can be derived from the eigenvalues and eigenvectors of:

$$(\delta_{l,l+1,l+2}^T Q_y^{-1} Q_e Q_y^{-1} \delta_{l,l+1,l+2})^{-1} \lambda_0 \sigma_0^2 \tag{17}$$

- With the same formulae, other hypotheses, describing an incompleteness in the design matrix can be tested. Depending on the problem at hand, these may be unjustly constrained transformation parameters, typical GPS errors (cf. part II of this paper), refraction hypotheses, deformation patterns, etc. The C-matrix contains the linearized function of such a hypothesis, expressed in the approximate values for the unknown parameters. All theory concerning reliability can be applied.
- Iterated data snooping is an automated iterative procedure to detect errors in a set of observations when more than one error is present. Specifying I_c in the alternative hypothesis as:

$$H_a : E\{\underline{y}\} = Ax + I_c \nabla; \quad \sigma_0^2 Q_y \qquad (I_c = I_{c-1} \vdots \delta_j) \tag{18}$$

we succesively add δ_j, belonging to the largest test statistic $\|\underline{w}_j\| = \|\underline{w}_l\|_{max}$ to I_{c-1} and make an update of the correction vector $\hat{\underline{e}}$ and variance matrix Q_e. This is repeated until the largest (updated) test-statistic $\|\underline{w}_l\|_{max}$ no longer exceeds the critical value of the one-dimensional data snooping tests, cf. [Kok, 1984], [Kösters, Kok, 1989].

2.3 Significance of estimated parameters

If an acceptable null-hypothesis has been found, i.e. no better alternative hypotheses are available, we can investigate significance of estimated parameters. A new null-hypothesis is introduced, which is an assumption about one or more parameters. The representation with observation equations of (3) becomes a mixed model representation, in the form of observation equations with conditions on the parameter vector x. From adjustment theory follows, that we can solve this system of equations in a first step without conditions and in a second step including conditions on the parameters. Considering a p-dimensional null- hypothesis, characterized by D^T, this model reads:

$$\begin{cases} H_0 : E\{\underline{y}\} = Ax \;;\; \sigma_0^2 Q_y \;;\; D^T x = d_0 \\ H_a : E\{\underline{y}\} = Ax \;;\; \sigma_0^2 Q_y \;;\; D^T x + I\,\Theta = d_0 \end{cases} \qquad (19)$$

Again, the alternative hypothesis is an extension of the null-hypothesis. Using the norm of the estimated model error $\hat{\underline{\Theta}}$, we can derive a test statistic for proving the validity of the alternative hypothesis:

$$\underline{T} = \frac{(d_0 - D^T\hat{\underline{x}})^T (D^T Q_x D)^{-1} (d_0 - D^T\hat{\underline{x}})}{\sigma_0^2} \sim \chi_p^2 \qquad (20)$$

If we choose D^T as:

$$\delta_i^T = (\underset{1}{0},\underset{2}{0},\underset{.}{0},.,\underset{.}{0},\underset{i}{1},\underset{.}{0},\ldots,\underset{n}{0}) \qquad (21)$$

we obtain:

$$\underline{v} = \frac{d_0 - \delta_i^T \hat{\underline{x}}}{\sigma_0 \sqrt{\delta_i^T Q_x \delta_i}} = \frac{d_{0,i} - \hat{\underline{x}}_i}{\sigma_{x_{ii}}} \sim N(0,1) \qquad (22)$$

This formula is easily recognized as an 'ordinary' test for significance. Note however, that critical values are computed using a pre-defined value for the power of the test β' so again, is the non-centrality parameter λ'_0 is:

$$\lambda'_0 = \lambda(\alpha'_1,\beta',1,\infty) = \lambda(\alpha'_p,\beta',p,\infty) \qquad (23)$$

Internal reliability is defined as:

$$\lambda'_0 = \Theta_i^T (D^T Q_x D)^{-1} \Theta_i \qquad (24)$$

For δ_i, this yields:

$$\| \Theta_1 \| = \sigma_{x_{11}} \sqrt{\lambda'_0} \tag{25}$$

We can also consider more-dimensional hypotheses D^T or functions of parameters. For testing the validity of any hypothesis, the statistic (20) can be used. With respect to internal reliability, a multi-dimensional hypotheses yields a quadratic form (24), which is an equation for a (hyper-)ellipsoid. Length and orientation of the axes of the defining matrix (26) of this quadratic form can be determined from eigenvalues and associated eigenvectors of this matrix.

$$(D^T Q_x D) \; \lambda'_0 \tag{26}$$

3. Discussion and conclusions

Mathematical techniques, used for statistical testing and quality analysis of observations and parameters in least-squares adjustment problems are by no means specifically geodetic. However, practical experience in geodetic works has resulted in meaningfull concepts, i.e. reliability and precision, to be used for evaluating the quality of e.g. a geodetic network. Furthermore it gave guidelines for the computational approach, e.g. computations were split up in sufficiently redundant phases in order to distinguish between possible error sources. Depending on the work at hand and the quality required, demands for reliability and precision for every phase can be formulated.

Precision is defined as the dispersion of the results in the absence of model errors. For the evaluation of precision, the variance matrix of the estimated parameters is required. Care should be taken using this matrix as all of its elements are datum-dependent. In this paper, no attention was given to the evaluation of precision.

The evaluation of reliability focusses mainly on the magnitude of specified model errors which can be detected by hypothesis tests with a certain probability. These model errors may be errors or blunders in the observations or an inadequate description of the relations between observations and parameters in the design matrix. If these errors are translated to an alternative hypothesis, by formulating a C-matrix as in (3), and if we choose acceptable probabilities for making false decisions, we are able to test the significance of a possible model error and we can compute:

- the minimum size of errors to be detected with certain probability (internal reliability)
- the influence of undetected errors of that size on the estimated parameters (external reliability)
- a multiplication factor that relates reliability and precision for every alternative hypothesis

It should be stressed that every reliability description depends on the choice of the covariance matrix of the observations and that this choice should be as realistic as possible.

References:

Baarda, W. (1967): Statistical concepts in geodesy. NGC-publications on geodesy, New series, Vol. 2, No. 4, Delft.

Baarda, W. (1968): A testing procedure for use in geodetic networks. NGC-publications on geodesy, Vol. 2, No. 5, Delft.

Kok, J.J. (1984): On data snooping and multiple outlier testing. NOAA Techical Report NOS NGS 30, Rockville, Maryland.

Kösters, A.J.M., J.J. Kok (1989): Statistical testing and quality analysis of observations and transformation parameters in combining 3-dimensional networks. Paper, handed out during poster sessions of symposium S102 of the IAG conference in Edinburg, August 1989.

Marel, H. van der (1989): Optimization and design of network computations with GPS. Paper, handed out during poster sessions of symposium S102 of the IAG conference in Edinburg, August 1989.

Marel, H. van der, A.J.M. Kösters: Statistical testing and quality analysis of 3-D networks, part II: Application to GPS. This volume.

Mierlo, J. van (1981): A review of model checks and reliability. In: Proc. of the International Symposium on Geodetic Networks and Computations of the IAG, Munich 1981. Vol. V, DGK Reihe B, Heft 258, München 1982.

Staff of the Geodetic Computing Center (LGR) (1982): The Delft approach for the design and computation of geodetic networks. In: Anniversary Volume on the occasion of Prof. Baarda's 65th. birthday, Vol. 1, p. 202-274, Department of Geodesy, Delft University of Technology.

Teunissen, P.J.G. (1984): Quality control in geodetic networks. In: Optimization of geodetic networks, p. 526-547, Eds. Grafarend and Sansò, Springer Verlag.

Teunissen, P.J.G. (1986): Adjusting and testing with the models of the affine and similarity transformation. In: Manuscripta Geodetica 11, p.214-225.

Statistical Testing and Quality Analysis in 3–D Networks (part II) Application to GPS

dr.ir. H. van der Marel, ir. A.J.M. Kösters

Geodetic Computing Centre (LGR)
Faculty of Geodesy, Delft University of Technology
Thijsseweg 11, 2629 JA Delft, The Netherlands

Abstract

In this paper reliability measures for geodetic networks with GPS measurements are discussed. Particular attention is given to the type of alternative hypotheses to be selected in order to get a realistic and meaningful concept of internal– and external reliability. For classical surveying networks it is often sufficient to consider only the so–called conventional alternative hypotheses, however, this is not true for GPS.

In part I of this paper a brief introduction to the theory was given [*Kösters, van der Marel, 1989*]. In this part the relevant alternative hypotheses for GPS, with some examples, are given. The computations are split –for conceptual reasons– into three steps: 1) multi baseline/single session adjustment, 2) network adjustment and 3) the connection adjustment. Each step has its own kind of hypotheses to be considered.

Introduction

In the design and optimization of geodetic networks two criteria are important. The ***precision*** describes the dispersion of the results in the case that model errors are not present. The ***reliability***, on the other hand, gives the magnitude of –specified– model errors which can be detected by hypotheses tests with a certain probability. In this paper we concentrate on the reliability description. Precision is not considered here explicitly, although it is assumed that the stochastic model for the GPS measurements is known "well enough" in order to make an accurate description of the reliability possible.

Consider the linear(ized) Gauß–Markov model

$$E\{\underline{\mathbf{y}}\} = \mathbf{A}\,\mathbf{x} \;\;,\;\; E\{(\underline{\mathbf{y}} - E\{\underline{\mathbf{y}}\})(\underline{\mathbf{y}} - E\{\underline{\mathbf{y}}\})^{\top}\} = \mathbf{Q}_y \tag{1}$$

with $\mathbf{A}$ the $m \times n$ matrix of partial derivatives, with $Rank(\mathbf{A}) = r$, $\underline{\mathbf{y}}$ the $m \times 1$ vector of observables with its $m \times m$ co-variance matrix $\mathbf{Q}_y$ and $\mathbf{x}$ the $n \times 1$ $(n < m)$ vector of –unknown– parameters. The hypothesis that both the functional and stochastic models are correct is called the null–hypothesis

$$H_0 : \;\; \underline{\mathbf{y}} \sim N(\mathbf{A}\,\mathbf{x},\; \mathbf{Q}_y) \tag{2}$$

This implies also that there are no gross–errors in the observations. Alternative hypotheses assume additional model errors,

$$H_{a_i}: \quad \underline{\mathbf{y}} \sim N(\mathbf{A}\,\mathbf{x} + \mathbf{C}_i\,\nabla_i,\ \mathbf{Q}_y) \tag{3}$$

with ∇_i the $b \times 1$ vector of model errors and $\mathbf{C}_i$ the $m \times b$ matrix which specifies the model error, with $Rank(\mathbf{C}_i) = b$, $1 \le b \le m - Rank(\mathbf{A})$ and $Rank(\mathbf{A} : \mathbf{C}_i) = Rank(\mathbf{A}) + b$. In fact an estimator of ∇_i can be computed in an ordinary adjustment (under H_0) from the least squares residuals $\hat{\underline{\mathbf{e}}} = \underline{\mathbf{y}} - \mathbf{A}\hat{\underline{\mathbf{x}}}$ without performing a new adjustment, i.e.

$$\hat{\underline{\nabla}}_i = \left(\mathbf{C}_i^\top[\mathbf{Q}_y^{-1} - \mathbf{Q}_y^{-1}\mathbf{A}(\mathbf{A}^\top\mathbf{Q}_y^{-1}\mathbf{A})^{-1}\mathbf{A}^\top\mathbf{Q}_y^{-1}]\mathbf{C}_i\right)^{-1}\mathbf{C}_i^\top\mathbf{Q}_y^{-1}\hat{\underline{\mathbf{e}}} \tag{4}$$

with $\mathbf{Q}_{\hat{\nabla}_i} = (\mathbf{C}_i^\top[\mathbf{Q}_y^{-1} - \mathbf{Q}_y^{-1}\mathbf{A}(\mathbf{A}^\top\mathbf{Q}_y^{-1}\mathbf{A})^{-1}\mathbf{A}^\top\mathbf{Q}_y^{-1}]\mathbf{C}_i)^{-1}$ the regular $b \times b$ co-variance matrix of $\hat{\underline{\nabla}}_i$ (see part I).

The alternative hypotheses H_{a_i} are used in two distinctive ways:

- **Testing**: estimate $\hat{\nabla}_i$ from the least–squares residuals $\hat{\underline{\mathbf{e}}} = \underline{\mathbf{y}} - \mathbf{A}\hat{\underline{\mathbf{x}}}$, which gives the test statistic $\underline{T}_b = \|\hat{\underline{\nabla}}_i\|^2$.
- **Reliability**: fix all $\|\nabla_i\|^2 = \lambda_i = \lambda_0$, then $\|\nabla_i\|^2 = \nabla_i{}^\top\mathbf{Q}_{\hat{\nabla}_i}^{-1}\nabla_i = \lambda_0$ describes a b-dimensional (hyper–)ellipsoid as the geometric position of ∇_i.

With the B–method of testing the non–centrality parameter λ_0 of the shifted distribution and the power of the test, β_0, are fixed for all alternative hypotheses (usually we choose β_0 and $\alpha_{b=1}$, and then compute λ_0, see also part I or [*Baarda, 1968, Teunissen, 1984*]).

The internal reliability gives the effect of ∇_i on the observation vector $\mathbf{y}$:

$$\mathbf{C}_i\nabla_i \doteq \nabla_i\mathbf{y} \qquad \|\nabla_i\mathbf{y}\|^2 = \nabla_i^\top\mathbf{C}_i^\top\mathbf{Q}_y^{-1}\mathbf{C}_i\nabla_i \doteq \lambda_{y,i} \tag{5}$$

For $b = 1$, ∇_i, $\nabla_i\mathbf{y}$ and $\lambda_{y,i}$ are unique. For $b > 1$, ∇_i, $\nabla_i\mathbf{y}$ and $\lambda_{y,i}$ are not unique: ∇_i and $\nabla_i\mathbf{y}$ describe a hyper–ellipsoid. In order to get a workable criterion we must parameterize ∇_i as $\nabla_i = \|\nabla\|\mathbf{d}$, with $\mathbf{d}$ a $b \times 1$ unit vector. By letting $\mathbf{d}$ be one of the eigenvectors of the matrix $\mathbf{Q}_{\hat{\nabla}_i}^{-1}$, with eigenvalue λ_k, we find $\|\nabla\| = \sqrt{\lambda_0/\lambda_k}$ to be the length of the corresponding principal axis. Similarly, the maximum value for $\lambda_{y,i}$ may be computed from the b–dimensional general eigenvalue problem $|\ \mathbf{C}_i^\top\mathbf{Q}_y^{-1}\mathbf{C}_i - \lambda\mathbf{Q}_{\hat{\nabla}_i}^{-1}\ | = 0$ with $\lambda_{y,i} \le \lambda_{max}\lambda_0$.

The external reliability gives the effect of model errors on the final results, i.e.

$$\nabla_i\mathbf{x} = (\mathbf{A}^\top\mathbf{Q}_y^{-1}\mathbf{A})^{-1}\mathbf{A}^\top\mathbf{Q}_y^{-1}\nabla_i\mathbf{y} \qquad \|\nabla_i\mathbf{x}\|^2 = \nabla_i\mathbf{x}^\top\mathbf{Q}_{\hat{x}}^{-1}\nabla_i\mathbf{x} \doteq \lambda_{x,i} \tag{6}$$

or, even better, part of the unknowns (e.g. the coordinates) by using $\overline{\mathbf{A}}_1 = (\mathbf{I} - \mathbf{P}_{A_2})\mathbf{A}_1$ instead of $\mathbf{A} = (\mathbf{A}_1, \mathbf{A}_2)$, with $\mathbf{P}_{A_2} = \mathbf{A}_2(\mathbf{A}_2^\top\mathbf{Q}_y^{-1}\mathbf{A}_2)^{-1}\mathbf{A}_2^\top\mathbf{Q}_y^{-1}$. The relation between $\lambda_{y,i}$ and $\lambda_{x,i}$ is

$$\lambda_{x,i} = \lambda_{y,i} - \lambda_0 - \|\mathbf{P}_{A_2}\nabla_i\mathbf{y}\|^2 \tag{7}$$

This formula gives a more practical way to compute $\lambda_{x,i}$, without forming the $\nabla_i\mathbf{x}$'s explicitly. Note that the $\nabla_i\mathbf{x}$ depend on the chosen S–basis, but the length $\lambda_{x,i}$ not. For the more dimensional tests $\nabla_i\mathbf{x}$ and $\lambda_{x,i}$ are not unique (the $\nabla_i\mathbf{x}$ describe an hyper–ellipsoid). Again, in a similar way as for the internal reliability, the maximum value of $\lambda_{x,i}$ can be computed from a b–dimensional general eigenvalue problem. The internal reliability can be very bad for some hypotheses, but this does not always mean that the final results are influenced with the same magnitude. The external reliability is a more relevant measure.

A good selection of H_{a_i}'s is crucial. Alternative hypotheses should reflect the type of model errors to be expected, and during the design of the network we should account for the frequency of occurrence. For classical surveying networks it is often sufficient to consider only the so–called conventional alternative hypotheses (i.e. there is an error in one of the observations). For GPS networks this is however not sufficient. In order to get a more meaningful concept of reliability, other hypotheses, which describe typical GPS errors, must be considered.

Computing Steps

The carrier–beat phase equation for the receiver–satellite pair (i, k) is:

$$\phi_i^k = -\frac{f}{c}\|\mathbf{r}^k - \mathbf{r}_i\| - f(\delta t^k - \delta t_i) - \frac{f}{c}(-d_{ion} + d_{trop}) + N_i^k \tag{8}$$

with ϕ the measured phase, $\mathbf{r}$ position vectors, δt clock errors (including an initial phase term), d ionospheric and tropospheric disturbances, N the (integer) cycle ambiguity, f the frequency and c the speed of light. The upper indices k, l refer to satellites, the subscripts i, j refer to receivers (see e.g. [*King et al., 1985*]).

In the so–called double–difference solution the clock terms and first order effect of refraction are eliminated by differencing the raw carrier–beat phase observations. This gives new observables

$$\phi_{ij}^{kl} = \phi_j^k - \phi_i^k - \phi_j^l + \phi_i^l \tag{9}$$

and new (unknown) parameters

$$\Delta\mathbf{r}_{ij} \doteq (\mathbf{r}^k - \mathbf{r}_i) \quad , \quad N_{ij}^{kl} \doteq N_j^k - N_i^k - N_j^l + N_i^l \tag{10}$$

while the clock error and refraction terms can be omitted safely from the model. The unknown cycle ambiguities N_{ij}^{kl} in the adjustment are reals. In many applications an additional solution is computed in which the N_{ij}^{kl} are constrained to be integers.

The computations are (both for conceptual and practical reasons) split in three steps:

1. **Multi base–line/Single session Adjustment**: Processing of GPS data observed during a single session; the receiver configuration does not change and there are no long interruptions (typically 1–4 hours of observation).

2. **Network Adjustment**: Combination of step I GPS data, possibly with other observations (surveying, photogrammetry) and connection to a (local) terrestrial datum. Here we may have a free network, as well as an over–constrained network, approach. Input are the results of step I: $\Delta\mathbf{r}_{ij}$ (and possibly N_{ij}^{kl}) and weight matrix $\mathbf{Q}_{\hat{\mathbf{x}}^I}^{-1}$ (the normal matrix of the preceding phase).

3. **Connection Adjustment**: Combination of coordinate sets from different sources; for example the coordinates from step II are connected to a national control network, or to global VLBI and SLR networks.

The three steps can actually be combined in a single adjustment, yielding in principle the same results (provided that the variances are propagated properly from one step to the other). For scientific work steps I and II are often combined, including orbit improvement and connection to VLBI/SLR sites, to give the most precise results. However, for surveying applications, where it is not necessary to improve the orbits, this may not be the best procedure. There may be many different stations, occupied at different times during relatively small periods of time, by different teams; so it makes sense to check first small batches of data as soon as they become available. In addition there is also an hierarchy in the alternative hypotheses to be considered, viz. some hypotheses are restricted to one phase only. Therefore, for surveying applications we advocate the step wise approach.

Moreover, hypotheses are not only used for testing. They are also a means of assessing the reliability of the network. It is important to realize that the reliability can be evaluated beforehand, before the actual measurements are carried out. The design of the network should be such that errors in the measurements can be found with a reasonable probability. What is reasonable, and what not, is purely an economic matter.

In the following sections we will propose a number of alternative hypotheses which we think are useful during each step of GPS processing.

Alternative Hypotheses for the "Multi Base–Line/Single Session" Adjustment (step I)

The most simple and often most useful hypotheses are the so–called conventional alternative hypotheses:

One of the observations is wrong (data–snooping): All observations are checked in turn, and one by one (data–snooping). There are as many conventional alternative hypotheses as there are observations, i.e.

$$\mathbf{c}_i = (\delta_i) = (0\ldots0\;1\;0\ldots0)^{\mathsf{T}} \;,\; b = 1 \;,\; i = 1,\ldots,m \tag{11}$$

In this particular case the vector $\nabla_i \mathbf{y}$ for the internal reliability has only one number, $(\nabla_i \mathbf{y})_i$, different from zero. So the complete internal reliability can be described by one m–vector. The type of solution (carrier–beat, double–difference) is not very relevant here. Outliers in single observations can be detected very well by this test, because there are so many observations (see the example).

Intermezzo (example of data–snooping): *Consider a 4 satellite 4 receiver combination. The interstation distances vary from $20-100$ km. The double difference solution, with N-biases fixed, is used. The parameters chosen for the B-method of testing are: significance level $\alpha_0 = 0.001$ and power $\beta_0 = 0.80$.*

The internal reliability (marginal detectable errors in the observations)

$$\nabla_i y \simeq 4\sigma_y \quad (when\ \sigma_y = 1cm \rightarrow \nabla_i y = \frac{1}{4}cycle) \tag{12}$$

depends only slightly on the measurement interval and number of satellites (except when less than 3 satellites are used), but it does not depend on the measurement period, interstation distances (except when σ_y increases) and is not affected too much by neglecting the mathematical correlation in the double-difference solution. The conclusion must be that neighbouring observations check each other very well.

The square root of the norm of the external reliability vector

$$\sqrt{\lambda_{x,i}} \simeq 1.3 - 2.1 \;\; (30min.) \;\; 0.6 - 1.7 \;\; (90min.) \tag{13}$$

improves with the number of observations (number of satellites and number of epochs), but it does not depend on the interstation distances. The conclusion is that marginally detectable errors in the observations hardly influence the coordinates [Joosten, 1988].

But, data–snooping only, is not sufficient to get a good description of the reliability. The reason for this is simple: other potential errors have a bigger impact. A more realistic description of the reliability is obtained using the following hypotheses.

Bias in an receiver–satellite combination: This hypothesis is useful when the cycle ambiguities N_{ij}^{kl} are constrained to integers. The corresponding $\mathbf{c}_i$ vector is

$$\mathbf{c}_i = (\delta_{lj})^\top \tag{14}$$

with $\delta_{lj} = 1$ when the observation refers to satellite (l) and receiver (j) of the selected combination.

Error in the fixed station position (station j): Both in the differenced and undifferenced solutions the coordinates of one of the stations is held fixed. The alternative hypothesis assuming an error in the fixed station coordinates is – for the double–difference solution– described by

$$\mathbf{C}_i = (\delta_j \mathbf{B})^\top \;\; , \;\; \mathbf{B} = \frac{1}{R} \begin{pmatrix} x_j - x^k & 0 & 0 \\ 0 & y_j - y^k & 0 \\ 0 & 0 & z_j - z^k \end{pmatrix} \tag{15}$$

with $\delta_j = 1$ when the double-difference observations refer to station j, else $\delta_j = 0$. The internal reliability (the marginally detectable errors) will be very bad, otherwise it would not have been necessary to fix this station in the first place. But on the other hand, it is hoped that the external reliability (which gives the effect on the coordinate differences $\Delta\mathbf{r}_{ij}$) is quite good. The external reliability in fact gives an impression of the errors which may exist in the final solution due to an error in the fixed station position, but which were not noticed by statistical tests.

Error in the satellite ephemeris: Alternative hypotheses can be specified which assume errors in one or more of the Keplerian elements, or just an along track, cross track and/or radial bias. For standard surveying applications the internal reliability will be very bad. The external reliability, which gives us the effect on the coordinate differences, will be more useful.

Cycle–slip errors: Cycle slips are a problem in processing GPS data. Though cycle slip errors are usually removed beforehand in a separate processing step (e.g. based on a triple difference solution) it is useful to check the final results from a carrier–beat phase or double difference solution also, and to give the corresponding description of the reliability. The c–vector describing a cycle–slip is

$$\mathbf{c}_i = (0 \ldots 0 \; \delta_{lj})^\top \tag{16}$$

with $\delta_{lj} = 1$ when the observation refers to satellite (l) and receiver (j) of the selected combination, from a certain epoch on; otherwise zero. In the triple difference solution conventional alternative hypotheses can be used for detecting cycle–slips.

Atmospheric models: For more precise applications parameters for atmospheric and ionospheric refraction are included in the model. Alternatively the columns in the design matrix corresponding to these parameters may be omitted from the model under H_0 and included as alternative hypothesis instead.

Alternative Hypotheses for the 3–D Network Adjustment (step II)

The objective of this step is to combine GPS–sessions into a single network, possibly with other observations (surveying, photogrammetry), and to connect it to a (local) terrestrial datum. We may have a free network, as well as an over–constrained network, approach. Input are the results of step I: viz. $\Delta\mathbf{r}_{ij}$ (and possibly N_{ij}^{kl}) and weight matrix $\mathbf{Q}_{\hat{\mathbf{x}}^I}^{-1}$ (the normal matrix of step I).

When the GPS data is combined with other data it may be necessary to introduce scale and rotation parameters in the model, depending on the nature of the auxiliary data. In this section only alternative hypotheses which can be associated to GPS data are considered. Conventional alternative hypotheses are again the most straightforward:

Error in a GPS coordinate difference (data–snooping): An error in one GPS coordinate difference is assumed. This type of test is good at finding registration errors (e.g. typing errors).

Identification errors: In this case one GPS vector is assumed to be wrong (e.g. due to an identification error in one of the stations), i.e.

$$\mathbf{C}_i^{\mathsf{T}} = \begin{pmatrix} 0\ldots0 & 1 & 0 & 0 & 0\ldots0 \\ 0\ldots0 & 0 & 1 & 0 & 0\ldots0 \\ 0\ldots0 & 0 & 0 & 1 & 0\ldots0 \end{pmatrix} \tag{17}$$

with dimension 3. Another possibility is that an error in the measured antenna height or eccentricity is made, i.e.

$$\mathbf{C}_i^{\mathsf{T}} = \begin{pmatrix} 0\ldots0 & t_{hx} & t_{hy} & t_{hz} & 0\ldots0 \end{pmatrix} \tag{18}$$

$$\mathbf{C}_i^{\mathsf{T}} = \begin{pmatrix} 0\ldots0 & t_{lx} & t_{ly} & t_{lz} & 0\ldots0 \\ 0\ldots0 & t_{bx} & t_{by} & t_{bz} & 0\ldots0 \end{pmatrix} \tag{19}$$

of respectively dimension one and two, with $t_{\alpha\gamma}$ $(\alpha = l, b, h\ ,\ \gamma = x, y, z)$ elements from linearized transformation matrix from cartesian coordinates x, y, z to ellipsoidal coordinates l, b, h.

Re–evaluation of hypotheses from step I: Another possibility is to re-evaluate earlier formed hypotheses of step I. The redundancy is now much higher, which results in an improved reliability. The $\mathbf{C}_i$–matrix is simply

$$\mathbf{C}_i = (0\ldots0\ \nabla_i\mathbf{x}_I^{\mathsf{T}}\ 0\ldots0)^{\mathsf{T}} \tag{20}$$

with $\nabla_i\mathbf{x}_I$ the external reliability vector from the previous adjustment step corresponding to the coordinate differences of a single phase I adjustment only. Useful hypotheses to re–evaluate are: bias in receiver/satellite combination, errors in the fixed station positions, errors in the satellite ephemeris and atmospheric models. Re–evaluating the data–snooping and cycle–slip tests from step I is not as useful.

Alternative Hypotheses for the Connection Adjustment (step III)

The objective of this step is to connect two or more coordinate sets by means of a connection adjustment [*Kok, 1985, Teunissen, 1986, Brouwer et al., 1988, Kösters, 1988*]. The purpose may be two–fold: merging of coordinate sets in order to get improved results, or the comparison of coordinate sets. GPS results may successfully be merged with VLBI or SLR stations, or to a local geodetic datum, provided that sufficient stations are in common. Already in the previous step of our adjustment process the coordinates of some of the stations had to be fixed. Usually we take here the minimum possible number of constraints, most of the time well known stations.

Consider a general mathematical model for a connection adjustment, with two 3–D input coordinate sets, resulting from separate adjustments, which are mutually completely overlapping. The second set is transformed into the first by a seven parameter similarity transformation. These parameters, and an improved set of coordinates are estimated. The probabilistic model of both sets is assumed to be completely known; no correlation between these sets exists. This mathematical model is given as:

$$H_0: \quad E\left\{ \begin{array}{c} \Delta \underline{\mathbf{y}}_1 \\ \Delta \underline{\mathbf{y}}_2 \end{array} \right\} = \begin{pmatrix} \mathbf{I} & \mathbf{0} \\ \mathbf{I} & \mathbf{T} \end{pmatrix} \begin{pmatrix} \Delta \mathbf{x} \\ \Delta \mathbf{t} \end{pmatrix} \quad ; \quad \mathbf{Q}_y = \sigma_0^2 \begin{pmatrix} \mathbf{Q}_1 & \mathbf{0} \\ \mathbf{0} & \mathbf{Q}_2 \end{pmatrix} \tag{21}$$

with $\Delta\underline{\mathbf{y}}_i$ the (observed minus computed) coordinates of the ith set, $\Delta\mathbf{x}$ the unknown corrections to the approximate coordinates, $\Delta\mathbf{t} = (t_x, t_y, t_z, r_x, r_y, r_z, \lambda)^{\mathsf{T}}$ the unknown corrections to the approximate values of the transformation parameters, consisting of 3 translations along and 3 rotations around the chosen coordinate axes and one scale factor. $\mathbf{T}$ is the linearized similarity transformation for 3–dimensional cartesian coordinate sets, e.g. derived in [*Kösters,1988*].

Data–snooping techniques can be applied in order to find possible observational errors and to quantify reliability as in step I and II.

Intermezzo (example of multi–dimensional data–snooping): *In [Brouwer, Kok and Kösters,1988] the outcomes of statistical tests and reliability analysis on the inclusion of the space derived data in ED87 terrestrial data were given. Thanks to multi–dimensional data snooping some errors in the space data, terrestrial data or the local ties between the reference points were detected and repaired where possible. Reliability ellipsoids showed that, taking into account the assumed probabilistic properties of the space and terrestrial data, space data was incapable of detecting distortions in the terrestrial network up to 6 meters in some regions. In other regions however, internal reliability was well below the 50 cm level.*

In this paper, we would like to pay attention to other types of hypotheses. Suppose we refrain from estimating some of the 7 transformation parameters. In other words, these parameters are constrained to their a priori known value. In order to check the correctness of this restriction we choose for the $\mathbf{C}_i$–matrix the columns of the design matrix corresponding to these parameters in the extended model, compute the test statistic $\underline{\mathrm{T}}_b$ and compare it with its critical value. The internal reliability gives then an indication of how much these parameters may differ from their a priori known values before the above mentioned test will be rejected with a certain probability. The external reliability indicates the maximum influence that constraining these parameters has on other parameters.

Alternatively, we could have estimated all transformation parameters and test their significance afterwards. The necessary test-statistics and reliability indicators were derived in part I of this paper. It should be noted that these parameters need only be constrained when there are sufficient theoretical reasons to do so, as datum definitions themselves are arbitrary.

Conclusions

The reliability of a network is described by the magnitude of -specified- model errors which can be detected by hypotheses tests with a certain probability. Therefore, the set of model errors, or alternative hypotheses, selected for the description is crucial. In classical surveying networks we know, from experience, that conventional alternative hypotheses (data-snooping) are almost always sufficient. However, this is not true anymore for GPS networks.

It turns out that neighbouring carrier-beat observations check each other very well, and that the marginally detectable errors in these observations hardly influence the coordinates. Other alternative hypotheses, which describe typical GPS errors, such as a bias in an receiver-satellite combination, station eccentricity, antenna height and other identification errors, etc., give a more realistic impression of what the reliability is really like.

References

[1] Baarda, W. (1968): *A testing procedure for use in geodetic networks.* NGC-publications on geodesy, Vol. 2, No. 5, Delft.

[2] Brouwer, F.J.J., J.J. Kok and A.J.M. Kösters (1988): *Quality control of ED87.* Paper presented at the symposium of the IAG-subcommission for RETrig, Lisbon, May 1988. Proceedings in print.

[3] Joosten, P. (1988). *On the Precision and Reliability of GPS Networks using the Double Difference Observable.* Graduate thesis, Fac. of Geodesy, TU Delft.

[4] King, R.W., E.G. Masters, C. Rizos, A. Stolz (1985). *Surveying with GPS.* Lecture notes prepared for workshop "Surveying with GPS", Univ. of New Soth Wales, Sydney, 13–14 May 1985.

[5] Kok, J.J. (1985): *A proposal for statistical tests and reliability computations in junction points.* In: Report on the symposium of the IAG-subcommission for RETrig, The Hague, november 1985, publication no. 16, edited by K. Poder, Mtnchen 1986.

[6] Kösters, A.J.M. (1988): *On the connection of geodetic point fields in RETrig and related tests for model errors.* Graduate thesis. Reports of the Faculty of Geodesy, Mathematical and Physical Geodesy 88.1, Delft University of Technology.

[7] Kösters, A.J.M., H. van der Marel (1989): *Statistical Testing and Quality Analysis in 3-D Networks, Part I: Theory.* This volume.

[8] Teunissen, P.J.G. (1984): *Quality control in geodetic networks.* In: Optimization of geodetic networks, p. 526–547, Eds. Grafarend and Sansò, Springer Verlag.

[9] Teunissen, P.J.G. (1986): Adjusting and testing with the models of the affine and similarity transformation. In: Manuscripta Geodetica 11, p. 214–225.

KINEMATIC APPLICATIONS OF GPS AND GPS/INS ALGORITHMS, PROCEDURES, AND EQUIPMENT TRENDS

G. Lachapelle and K.P. Schwarz
Department of Surveying Engineering
The University of Calgary
2500 University Dr., N.W.
Calgary, Alberta, T2N 1N4
Canada

INTRODUCTION

The current capabilities of unaided GPS and integrated GPS-INS for land, marine and airborne kinematic positioning are reviewed. Advances and current trends related to real-time and post-mission algorithms, procedures and equipment are discussed. Aspects such as the use of differential GPS, the utilization of code and carrier phase measurements, the effect of multipath and cycle slips, the use of several monitor stations to counteract the effect of Selective Availability and the use of multi-antenna receivers to determine attitude and azimuth are reviewed. Major GPS limitations are discussed. Current software and hardware developments related to integration of GPS with INS to overcome unaided GPS limitations are presented. Resulting accuracy and reliability performance enhancements are analysed. Operational aspects related to the use of GPS and GPS-INS techniques are addressed.

Today, the bulk of geodetic measurements is made in stationary mode, stationary with respect to the earth, that is. The only notable exceptions are marine are airborne applications, such as hydrographic surveying and airborne and shipborne gravimetry. With the advances in satellite and inertial techniques, a considerable portion of the geodetic work can potentially be performed in kinematic mode, i.e. from a platform which moves with respect to the earth. The advantages of such an approach are quite obvious. The survey can be performed much faster and is therefore more economical. The systems used are all-weather systems, work day and night, and do not require intervisibility between stations. Since in principle any type of carrier can be used, airborne applications can often overcome problems encountered in surveys on the surface of the earth. The wide range of existing and potential applications will be indicated by enumerating some typical examples.

Semi-kinematic methods have been applied for control point densification using INS as well as GPS; see for instance Huddle (1977), Schwarz (1983), Rueger (1986) for INS applications, and Remondi (1985), Goad (1985), Landau (1989), and Cannon (1989) for GPS applications. Photogrammetry without ground control using fully kinematic methods was proposed in Schwarz et al. (1984). Results using GPS have been published by Mader (1986) and Keel et al. (1988). Results combining GPS and INS measurements can be found in Baustert et al. (1989). An obvious extension of this application is the use of an integrated GPS/INS for airborne line imaging. In this case, the inertial attitude data are essential to the solution. Laser profiling from helicopter or fixed wing aircraft requires precise kinematic modelling of the carrier trajectory. Inertial platforms have been used for this task, combining trajectory determination with the pointing of the laser, see Moreau & Jeudy (1986). However, differential GPS can be used equally well for the first task. Trajectory determination is also required in many marine applications and the capabilities of GPS for hydrographic surveying have been thoroughly investigated, see for instance Wells (1983), Lachapelle et al. (1984, 1988), Seeber et al. (1986). The estimation of the gravity field vector or of the relative geoid in kinematic mode is a typical application which requires both GPS and INS measurements, see for instance Schwarz (1987) and Knickmeyer and Schwarz (1989). A number of related applications, such as highway and railway monitoring systems, automatic vehicle location systems, pipeline monitoring by INS and precise navigation, all require trajectory determination as an essential step. Although they may not fit into geodesy proper, they fit very well into the framework of kinematic modelling discussed in the following and should therefore be considered as belonging to the potential range of applications.

The paper will survey the state of the art in what may be called kinematic geodesy. Since many of the applications can be done by either using inertial sensors or GPS, the underlying methodology will be treated first. Specific questions related to algorithms, procedures and equipment trends in either GPS or GPS/INS will then be discussed in separate chapters. This will give some insight into the suitability of each sensor configuration for a specific application.

KINEMATIC MODELLING

Modelling the movement of a vehicle in three-dimensional space requires either the knowledge of the force causing the motion or the measurement of vehicle motion in a given three-dimensional coordinate system. The first type of modelling is called dynamic, the second kinematic. A typical case of dynamic modelling is satellite orbit determination in a known gravitational field. A typical case of kinematic modelling is the determination of a vehicle path from range measurements to satellites, assuming the satellite position to be known.

Motion can be represented by different kinematic models. The most obvious one is by way of a kinematic position vector

$$\mathbf{r}(t) = \{x(t), y(t), z(t)\}^T \quad . \tag{1a}$$

However, a kinematic velocity vector

$$\mathbf{v}(t) = \frac{d\mathbf{r}(t)}{dt} = \left\{ \frac{dx}{dt} , \frac{dy}{dt} , \frac{dz}{dt} \right\}^T \tag{1b}$$

or a kinematic acceleration vector

$$\mathbf{a}(t) = \frac{d^2\mathbf{r}(t)}{dt^2} = \left\{ \frac{d^2x}{dt^2} , \frac{d^2y}{dt^2} , \frac{d^2z}{dt^2} \right\}^T \tag{1c}$$

may be used as well. In theory, the three representations are equivalent, provided the appropriate initial values for (1b) and (1c) are known and the time history is continuous. In practice, this is often not the case and the representation which most closely models the measurement is chosen.

The choice of the kinematic model is rather important because measurements are used to derive the parameters of the model. Since all measurements have errors, it is critical for the final results to minimize the effect of these errors on the model parameters. Going from model (1a) to model (1c) involves differentiation which is an unstable process when done numerically. Thus, noise will be amplified and, without proper precautions, results may become meaningless. This is important when using GPS as a kinematic system. In this case, pseudo-range and phase measurement provide positioning information. Thus, the appropriate model is (1a). If these results are used to derive vehicle accelerations, as for instance required for airborne gravimetry, strong amplification of the measurement noise should be expected, see Kleusberg et al. (1989) for some numbers.

On the other hand, going from model (1c) to (1a) involves integration which is a stable process. White noise in the data will be reduced. However, integration is very sensitive to errors in the initial values. This is important when using INS as a kinematic system. In this case, gravity corrected specific force measurements provide information on vehicle acceleration. The appropriate model is therefore (1c). Integrating accelerations twice with respect to time to obtain positions will result in large time-dependent errors. Fortunately, these errors change slowly and are very systematic in nature. They can therefore be controlled if update measurements are available.

If trajectory modelling is required, as is the case in many geodetic applications, discrete, noisy data are used to derive position as a continuous function of time. Thus the problem of interpolation between discrete position fixes arises. The general formula is

$$\mathbf{r}_{i+1} = \mathbf{r}_i + \dot{\mathbf{r}}_i\Delta t + \ddot{\mathbf{r}}_i\Delta t^2/2 + \dddot{\mathbf{r}}\Delta t^3/6 + \ldots\ldots \quad (2a)$$

where the number of dots indicates the order of time differentiation and

$$\Delta t = t_{i+1} - t_i \quad .$$

Equation (2a) indicates that all time derivations of the position vector are needed for position interpolation. In practice, a truncated form of (2a) is used resulting in the constant velocity model

$$\mathbf{r}_{i+1} = \mathbf{r}_i + \dot{\mathbf{r}}_i\Delta t \quad (2b)$$

or the constant acceleration model

$$\mathbf{r}_{i+1} = \mathbf{r}_i + \dot{\mathbf{r}}_i\Delta t + \ddot{\mathbf{r}}_i\Delta t^2/2 \quad . \quad (2c)$$

In these approximations, the vectors $\dot{\mathbf{r}}_i$ and $\ddot{\mathbf{r}}_i$ are considered constant for the time interval Δt. The accuracy of the representation obviously depends on the dynamics of the vehicle and the length of Δt. For strapdown inertial measurements where the data rate is between 16 and 64 Hz, there is essentially no interpolation problem. For GPS where the data rate is typically 1 Hz or more, the difference between model (2b) and (2c) may be important. For a more detailed discussion which also presents some numerical results, see Schwarz et al. (1989). The choice of an adequate model is obviously dependent on the application and the measurements available.

GPS AS A KINEMATIC SYSTEM

The GPS constellation, with its 21 satellites distributed over six planes inclined at 55° as shown in Figure 1, is designed to provide near-uniform coverage on a worldwide basis. Its two RF carrier frequencies in the UHF part of the spectrum, namely 1575.43 (L1) and 1227.6 (L2) MHz, are practically unaffected by weather conditions and available on a 24 hour basis. In this frequency range, the ionospheric group delay is usually smaller than 20 m, with extremes up to 100 m in rare cases when solar activity, characterized by the sunspot number, is high as at present. Residual tropospheric errors rarely exceed a few meters, except in extreme cases when satellites below an elevation angle of 5° are observed. GPS observables consist of code (pseudo-range) and carrier phase measurements. Orbital parameters are referenced to WGS84, a geocentric reference system whose relationship to many local datum is well determined (White et al. 1989). Instantaneous single point positioning, based on pseudo-ranges, results in absolute accuracies of 20 - 100 m, depending whether P or C/A code is used and if Selective Availability is implemented. The additional use of phase measurements

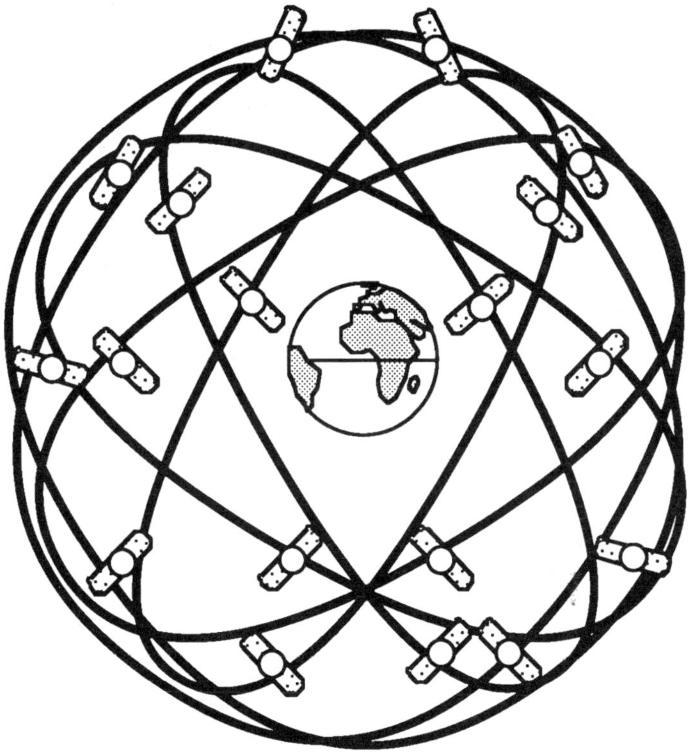

Fig. 1. Configuration of Full GPS Constellation

will improve the relative accuracy of the mobile trajectory but not the absolute accuracy quoted above, due to the presence of orbital and other errors.Orbital, satellite clock and ionospheric errors can be reduced significantly using differential GPS (DGPS). Tropospheric errors can also be somewhat reduced using DGPS. The use of P or C/A code in DGPS mode will generally result in accuracies of the order of 3 to 10 m (e.g., Lachapelle et al.1984) while the use of pure carrier phase or carrier phase combined with pseudo-range measurements will result in a higher level of accuracy, namely of the order of 10 cm to 2 m, depending on a variety of operational procedures and constraints.

Pseudo random noise (PRN) code modulation of the carrier, together with the precise synchronization of satellite-borne atomic time standards to a common time scale, makes pseudo-ranging on GPS possible with an accuracy of several meters in favorable circumstances. L1 is modulated by both the C/A and P code while L2 is modulated by the P code only, for U.S. national security reasons. The chipping rate of C/A code, at 1.023 MHz, is 0.1 of that of the P code. making the nominal measuring accuracy on the C/A code approximately 10 times worse than the P code. With averaging techniques, the measuring accuracy on the C/A code can usually be maintained at a level of a few meters, provided the measuring environment is free from multipath. The P code will, from time to time, be encrypted into the Y code, rendering it useless for unauthorized civilian users. The C/A code on L1 will however remain available in degraded mode through the implementation of Selective Availability (S.A.). This will limit the single point absolute accuracy of GPS to 100 m (2DRMS) or 76 m (SEP). S.A. is believed to be implemented on the Block II satellites launched since early 1989 (Kremer et al. 1989). The effect of S.A. on the range of DGPS is not clear at present.

Code multipath may be the most important error source which will not be reduced using DGPS. The maximum effect of multipath on a single range can theoretically reach 29.3 m on P code and 293 m on C/A code. More typically, in the static case, the effect may reach some tens of meters on C/A and 10 to 20 m on P code with periods ranging from a few to several minutes. In the kinematic case, multipath is not generally a problem as its period is so short that it can be readily treated as random noise. The major multipath effect will therefore be at the monitor station. Several multipath countermeasures exist, such as careful monitor site selection and the use of an absorbent ground plane. Figure 2, extracted from (Lachapelle et al. 1989), shows the effect of multipath on P code with and without an absorbent ground plane. The absorbent ground plane consisted of a piece of carbon impregnated foam cut in the shape of an inverted cone. Multipath was induced by the proximity of convex reflective surfaces within a radius of 20 to 30 m from the antenna. The peak-to-peak amplitude of the multipath effect reaches 8 m with periods of 5 to 8 minutes when no ground plane is used. The use of a ground plane reduces the amplitude to less than 3 m.

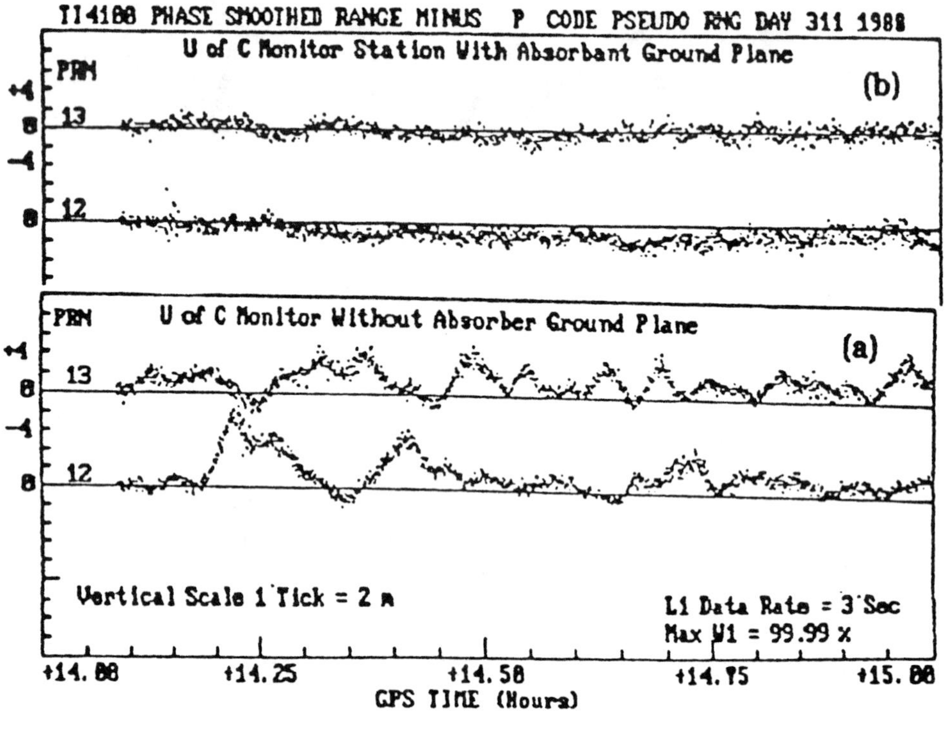

Fig. 2. P code Multipath Signature (a) Without and (b) With an Absorbent Ground Plane

Carrier phase measurements are at least two orders of magnitude less noisy than code measurements. Their accuracy is a function of several parameters, the most important one being the dynamics of the mobile receiver. In most kinematic applications, an accuracy of 10° of phase or better can be maintained, which is equivalent to a range difference accuracy of 5-6 mm. The effect of multipath on carrier phase measurements is much smaller than on code measurements but can occasionally reach several cm. Once the initial position of the mobile with respect to the monitor is established, phase measurements can, in principle, provide the position changes of the mobile with an accuracy of 10 cm or better in DGPS mode. The determination of the initial position of the mobile with respect to the monitor, referred to as the *initial baseline problem* in the sequel, can be achieved in many ways and will be discussed in section 3.2. If the mobile cannot be held fixed for a certain period of time or cannot be collocated directly with the monitor or with another station precisely known with respect to the monitor, code measurements must be used to determine the initial baseline. If these are used solely to determine the initial baseline and pure carrier phase measurements are subsequently used, the initial baseline error, which could be of the order of several meters, will propagate on all the subsequent trajectory as shown in Figure 3.

An example of pure kinematic DGPS carrier phase positioning in a multipath environment is shown in Figure 4 (Lachapelle et al. 1989). The distance between the monitor and remote was of the order of 25 m. An oscillating mast was used to generate motion in an East-West direction. The initial baseline was known *a priori* and used to fix the carrier phase ambiguities. The relative accuracy in any of the three coordinates is better than 10 cm. The peak-to-peak variation of approximately 15 cm in the height coordinate is likely the result of carrier phase multipath at the monitor. The monitor site was the same site which produced the code multipath shown in Figure 2.

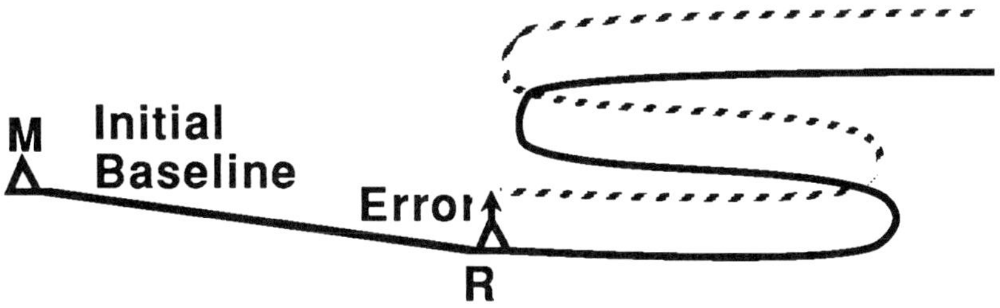

Fig. 3. Effect of Error in Initial Baseline on DGPS Operations

Carrier phase measurements, which contain an inherent ambiguity, are preferred for estimating the position changes of the mobile. Provided that the satellite geometry is satisfactory, the relative positioning accuracy which can be obtained using carrier phase measurements with DGPS is, in principle, better than 10 cm. In practice, atmospheric effects and cycle slips will make this accuracy difficult to maintain operationally. The combination of code and carrier phase measurements is used to bound the errors resulting from cycle slips and to estimate the initial baseline.

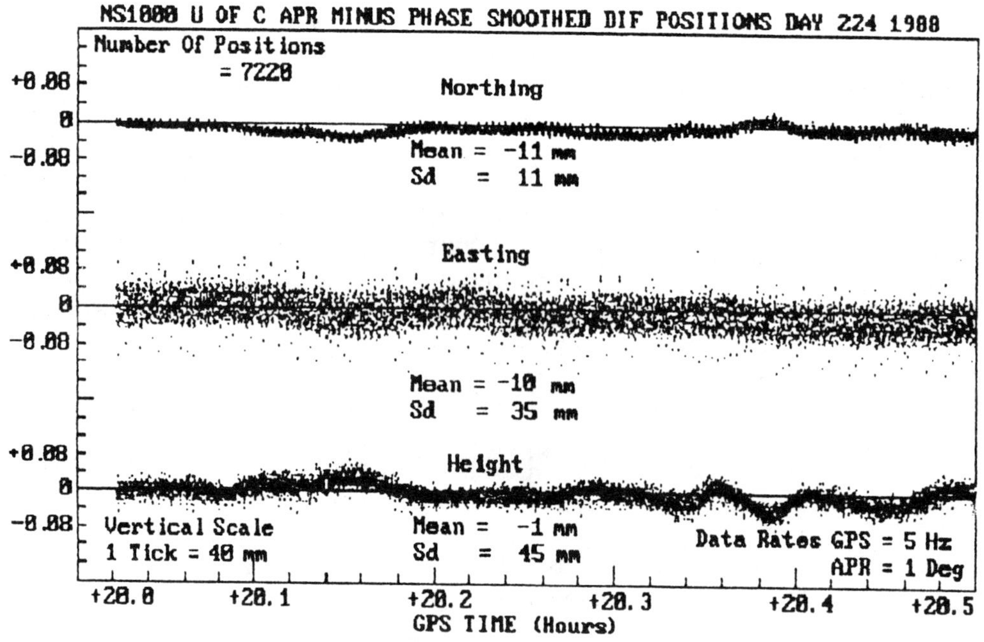

Fig. 4. DGPS Carrier Phase Positioning in a Multipath Environment

Algorithms

A panoply of algorithms has been developed and tested for absolute and differential GPS during the past ten years. In the early years, many algorithms were based solely on the use of code measurements (e.g., Lachapelle et al. 1984). Today, high performance algorithms are almost invariably based on pure carrier phase measurements or on the combination of code and carrier phase measurements. Several constraints impose restrictions on the choice of an algorithm. While many of the hardware constraints are disappearing with the emergence of a new generation of powerful receivers, many of the operational constraints remain a major factor in the selection of a suitable algorithm.

Major hardware and operational limitations affecting algorithms are:

- GPS Hardware limitations: number of satellites simultaneously observable, availability of code and carrier phase measurement, code/carrier tracking bandwidth, and raw data output frequency.

- Operational limitations: dynamics of the mobile, real-time requirements, capacity of data link available in the case of DGPS operations, physical access to the monitor station, and effect of Selective Availability.

Many of the above limitations will be discussed in the subsequent sections.

Early airborne and marine GPS experiments conducted using code measurements resulted in accuracies of 25 m in absolute mode and 5 to 10 m in differential mode (e.g., Lachapelle et al. 1984). In DGPS, the use of either differential range or differential position corrections produced similar results. With the launching of additional satellites and the publication of recommendations for the broadcast of differential corrections by RTCM SC104 (Kalafus et al. 1986), the differential range correction method, more flexible and more precise, was quickly adopted for real-time operations. The 50 bits per second format recommended by RTCM SC 104 was developed to counteract the effect of Selective Availability with sufficient accuracy. The selection of an over-the-horizon data link in an available RF band is still a major concern. Recent tests conducted for the U.S. Coast Guard indicate that the use of existing radiobeacons, which operate in the medium frequency band, may provide a cost effective answer to marine DGPS users (Ruane et al. 1989).

The use of pure carrier phase measurements in DGPS mode was successfully used by Mader (1986) in post-mission mode and resulted in accuracies in the dm range. A sequential least squares adjustment method implemented by Cannon (1987) to process the phase measurements resulted in a similar level of accuracy; in this case, code measurements were used to determine the initial baseline and to bound carrier phase errors arising from cycle slips. An 8-state Kalman filter suggested by Baustert et al. (1989) to process carrier phase measurements has the potential to yield a relative accuracy of better than 10 cm; also, see Landau (1988). Several Kalman filter formulations suggested by Schwarz et al. (1989) using various dynamic models have resulted in accuracies at the dm level. A method to combine code and carrier in real-time was proposed by Lachapelle et al. (1988) for the general case when the initial baseline is unknown. Accuracies of 2 m (rms) for the shipborne case and 40 to 80 cm for the airborne case (Keel et al. 1988) were confirmed. When code and carrier measurements are combined however, multipath on code could result in biased positions, especially if multipath is present during the initial period when code measurements are more heavily weighted. The use of parallel filters are being designed as a countermeasure to this problem (e.g., Lachapelle et al. 1989, Ashjaee et al. 1989)

A detailed review of the research related to the development and testing of algorithms to derive velocities and/or accelerations from unaided GPS is beyond the scope of this paper. Results obtained up to now suggest that the velocity components of the antenna can be recovered with accuracies ranging from a few cm to a few mm per second (e.g., Evans 1986; Kleusberg et al. 1985, 1989; Schwarz et al. 1989; Hasanuddin & Wells 1989)

Much progress on the detection of cycle slips has been accomplished during the past few years. Real-time and post-mission methods will result in different levels of effectiveness. The use of dual frequency P code and carrier phase receivers such as the TI4100 has resulted in the most effective method (Goad 1985; Baustert et al. 1989). The reliability of cycle slip detection using single frequency equipment with no redundant satellite depends on the dynamics of the mobile and is not better than several cycles at present (e.g., Lachapelle et al. 1988). However, the use of one redundant satellite improves the situation considerably (Baustert et al. 1989).

Procedures

In kinematic DGPS mode, one of the most important criteria is whether accessibility to the monitor station is possible or not. This will have an impact on procedures, algorithm selection and accuracy achievable. Accessibility to the monitor can be divided into two classes, namely direct accessibility to the location of the monitor and pseudo-accessibility through another station whose coordinate differences with respect to the monitor station are precisely known. The initial baseline can be determined with a precise static differential GPS survey. The carrier phase ambiguities are then implicitly known and a pure carrier phase algorithm can be used provided cycle slips do not occur or can be detected and recovered. This method is used successfully in semi-kinematic mode (e.g., Cannon & Schwarz 1989)

The use of DPGS implies the availability of a data link if real-time positioning is required. The selection of an appropriate data link is a challenge in many cases as discussed earlier in this section. The situation is further complicated by the implementation of S.A. on Block II satellites. It is not clear how S.A. affects the range over which DGPS can be conducted accurately and effectively. A possible solution to this limitation may be the use of a network of monitor stations deployed over a large area to estimate optimal differential corrections as a function of location. A complex infrastructure would be required to support such a network but the benefits to users may be enormous. This concept has been recently investigated by Brown (1989) and Steeves et al. (1989).

Equipment Trends

For carrier phase measurements to significantly contribute to kinematic positioning, phase lock must be maintained on several satellites simultaneously. The accuracy and reliability of the solution increases with the number of satellites observed, provided the geometry is satisfactory. This calls for a receiver which can track a minimum of four satellites simultaneously. The first commercially available receiver with this capability was the TI4100. With slow switching sequential receivers, positioning was performed using code measurements only (e.g., Wells et al. 1983). On all survey grade GPS receivers currently available, both code and carrier phase measurements can be made simultaneously on up to 12 satellites. The number of satellites which will be simultaneously available when the constellation is complete will vary between 5 and 9. The use of redundant satellites will have a large impact on the algorithms currently being used. Real-time cycle slip detection will become more effective and real-time correction will become possible. The real-time implementation of statistical reliability techniques to detect blunders and biases will likely be implemented and will result in a further positioning accuracy and reliability improvement.

Bandwidth variability, which is important in adapting to a changing dynamic environment while maximizing phase measuring accuracy and minimizing the number of cycle slip occurrences, is now available on most receivers. The raw data output frequency will affect the ability of the algorithm to model the dynamics of the mobile and improve cycle slip detection. Most receivers now have a raw data output rate of 1 Hz. However, the use of yet a higher data rate would improve further dynamics modelling and cycle slip detection. This is especially the case in the shipborne environment in which case the estimation of roll, pitch and heave motion is desired for precise hydrographic measurement corrections (e.g., Lachapelle et al. 1988, Hasanuddin & Wells 1989).

The use of unaided GPS to determine attitude components with a multi-antenna receiver is becoming realistic, thanks to the small size and low cost of parallel channel hardware. Tests conducted up to now indicate that accuracies better than 1 degree for the attitude components and the azimuth are well within the capability of current hardware and software technologies (e.g., Kruczynski et al. 1989; Purcell et al. 1989).

As the P code will be encrypted, this leaves civilian users with the less accurate C/A code. Fortunately, this does not affect the accuracy of phase measurements on the L1 carrier, although carrier phase measurements are evidently affected by Selective Availability. The development of squaring techniques to measure phase on the L2 carrier will account for the effect of the ionosphere on carrier phase measurements, at least in static mode. It is not clear at this time as to the level of performance which can be expected from L2 carrier tracking using the squaring technique in kinematic mode. L2 squaring will at least be useful at the monitor station to measure the level of ionospheric activity at that station.

GPS Limitations

Line-of-sight between receiver and satellite will remain the major limitation of GPS. Antennae will always have to be installed at or near the highest point of the vehicle to avoid masking effects. This is often logistically difficult if not impossible to achieve, e.g., on an aircraft fuselage. As a consequence, a sub-optimal position which results in additional losses of phase lock is often selected. For instance, in airborne applications, the most critical part of a mission is during sharp manoeuvres when losses of phase lock may occur on several satellites simultaneously due to shading effects and vehicle dynamics.

In a hydrographic survey launch environment, the roll motion can easily reach 30° under normal operational conditions (e.g., Lachapelle et al. 1988). The dynamics of an antenna mounted on a mast several meters above water can reach a few g under such conditions and maintaining phase lock for long periods of time becomes a challenge. Solving for the accurate position of the sensor in the ship's hull requires attitude aiding to relate the position of the antenna to that of the sensor. However, the use of multi-antenna receivers is likely to alleviate this problem to a certain extent.

The velocities and/or accelerations determined at the antenna are not the same as those required at the sensors located in a different part of the mobile. For missions where velocity and/or acceleration components are required, e.g., shipborne and airborne gravimetry, this can become a quasi-insurmountable problem without external aiding.
Finally, the accurate and reliable detection of cycle slips in high dynamics remains a challenge. The use of all--in-view satellites will certainly resolve the problem to a certain extent. The correction of cycle slips in real-time, as required for a highly accurate and cost effective use of GPS in kinematic mode, remains a problem to be solved.

GPS/INS AS A KINEMATIC SYSTEM

Integrating GPS with an inertial system extends the range of potential applications and overcomes some of the GPS limitations mentioned at the end of the last chapter. The advantages of an integration become quite obvious when one looks at the characteristics of each system separately.

GPS provides high positional accuracy when used in differential mode. The accuracy is affected by satellite geometry, receiver dynamics, and atmospheric conditions but will in principle be uniform worldwide. Cycle slips in the carrier phase data are a major problem in kinematic applications when single frequency receivers are used. Similarly, interpolation in high dynamic mode is a problem for receivers with output rates of one second or more. The system does not provide accurate attitude and is not sensitive to gravity field variations.

An inertial strapdown system provides position, velocity, and attitude with high accuracy over short time spans. The accuracy degrades quickly, however, as a function of time and results can only be used for geodetic work if the INS is regularly updated. The output rate is high, typically 32 to 64 Hz, and interpolation is not a problem, even in high dynamics. The system is not affected by atmospheric conditions but is sensitive to gravity field variations.

An integrated system adds the following advantages to the 'GPS only' solution. It solves the cycle slip and interpolation problem and has the potential for GPS noise reduction by feeding INS velocity back to the carrier tracking loop. It also provides accurate attitude which is needed in many applications. A low to medium accuracy inertial strapdown unit will be sufficient for these tasks if updated regularly by GPS positions. If gravity vector and geoid estimation are also required, a high accuracy INS is needed.

The following discussion will essentially look at differential GPS and INS as stand-alone systems which generate a time-tagged data stream each. The two streams are brought together in a central computer where all processing is done, either in real-time or post mission. A portable computer with the capacity of the Compaq 386 is fully sufficient for this task. The post-mission mode has advantages for most geodetic applications where accurate real-time results are not needed but a detailed data analysis is essential. With inexpensive mass storage devices available, post-mission processing has numerous advantages for this group of applications. It should be noted, however, that for most navigation applications, the fully integrated box is the preferable solution. It contains the inertial sensors, a GPS receiver card, processor cards and other electronics, and outputs optimal real-time position, velocity and attitude. For a discussion of this solution, see for instance Hartmann (1988) or Tazartes and Mark (1988).

Algorithms

To perform GPS/INS integration four major problem areas have to be addressed. First, the most suitable coordinate frame for integration has to be defined. Second, the level of integration and thus the question of the integration model has to be addressed. Third, a cycle slip detection and correction algorithm has to be designed. Finally, the control of the carrier phase loop by INS-feedback has to be addressed. The first three problems can be either solved in real time or post mission. The last one is obviously a real-time problem. In high dynamics, a data link to the stationary receiver may be required to get sufficiently accurate updates for the INS. In low dynamics, this may not be necessary. Data link requirements have been reviewed in earlier and will not be discussed here.

The choice of the coordinate frame for GPS/INS integration is not a trivial problem because it strongly affects processing time requirements. It is thus especially important when real-time solutions are considered. Typically, GPS position computations are done in the Conventional Terrestrial Frame, while INS strapdown computations are done in the local-level frame. The outputs, three-dimensional Cartesian coordinates in one case and, ellipsoidal geodetic coordinates in the other, are related by unique transformations and integration can therefore take place at the coordinate level. However, strapdown inertial computations can in principle be done in an arbitrary coordinate frame. It makes therefore sense to perform them in the same frame as the GPS computations. Recent investigations by Wei and Schwarz (1989) have shown that strapdown computations in the Conventional Terrestrial Frame are considerably faster and lead to a reduction in computer time of about 40%. Besides the obvious advantages for real-time computation, the algorithm is also more suitable for GPS/INS integration.

Since all parameters of interest in GPS/INS integration are functions of time and are therefore adequately described by a system of first-order differential equations, a state space model appears to be the most appropriate one for the integration of the two data streams. The solution is linearized about the reference trajectory which is determined by integrating the inertial measurements from initial values provided by GPS updates. The state vector for the system errors has position and velocity states common to both GPS and INS. The two data streams can therefore be combined at the measurement level, i.e. the state vector can be directly updated by GPS measurements. This approach has been applied with good success by Wong et al. (1988). Cycle slip recovery was done in this case by an adaptive filtering scheme. It might be advantageous, however, to perform cycle slip detection and correction before updating the state vector.

Another approach which has been advocated by Carlson (1988) is the use of a decentralized filter. In this case, separate filters are designed for the two data streams. As shown in Figure 5, the INS filter is considered as the master filter, while the GPS filter is one of the local filters. Other local filters could contain models for pressure barometer measurements or Doppler velocities. The local filters interact only occasionally with the master filter, providing filter updates when needed, and accepting covariance updates in return. In this case, the integration takes place at the level of the state variables.

Mathematical details of this filtering approach are given in Speyer (1979) and Hashemipour et al. (1988). Advantages and drawbacks of the two approaches for this specific application are currently under study.

The algorithm for cycle slip detection and correction is closely linked with the filtering algorithm. Figure 6 shows the principle. INS measurements from the last GPS update are used to predict the change in cycle count which is then compared to the actual cycle count. If they are within half a cycle, the actual cycle count is used to update the state vector. If they are not, the cycle count is corrected by an integer value. The method presupposes that the INS prediction is always correct to the nearest half cycle. This will usually be the case if the time span to the most recent update is short, typically up to half a minute for a medium accuracy system and about one minute for a high accuracy system. These time intervals are correct for real-time applications when filtering is applied. They can increase to several minutes for post-mission applications where smoothing is possible. Implementation of such an algorithm is best done by using statistical tests for the difference in cycle counts. A procedure that could be easily adapted to this case has recently been published by Teunissen and Salzmann (1988).

To avoid loss of lock, the tracking loop bandwidth of GPS receivers is opened up when operating in a dynamic environment. This leads to an accuracy degradation of an order of magnitude or more in a high dynamics environment, see for instance (Lachapelle et al. 1986). Aiding the tracking loops with high frequency velocity and differential position from an INS will allow a tighter control of the bandwidth and will thus contribute to noise reduction in the GPS receiver. This technique has been thoroughly investigated for GPS/INS integration with a single receiver, see for instance Cox (1978) for an overview and Buechler and Foss (1987) for some implementation details. The same basic technique can be applied to differential GPS positioning as long as bandwidth control is the only real-time requirement and precise position and attitude can be computed post mission. If all results are required in real time, the data link transfer rate will obviously have a major effect on the design of the filter and the tracking loop control.

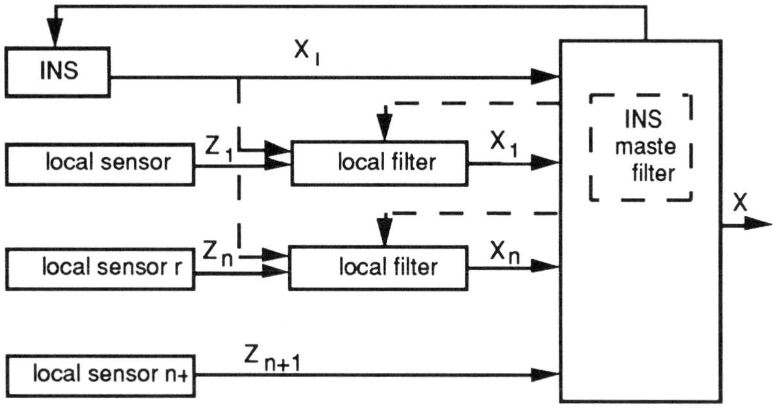

Fig. 5. Decentralized Filter Concept for GPS-INS Integration

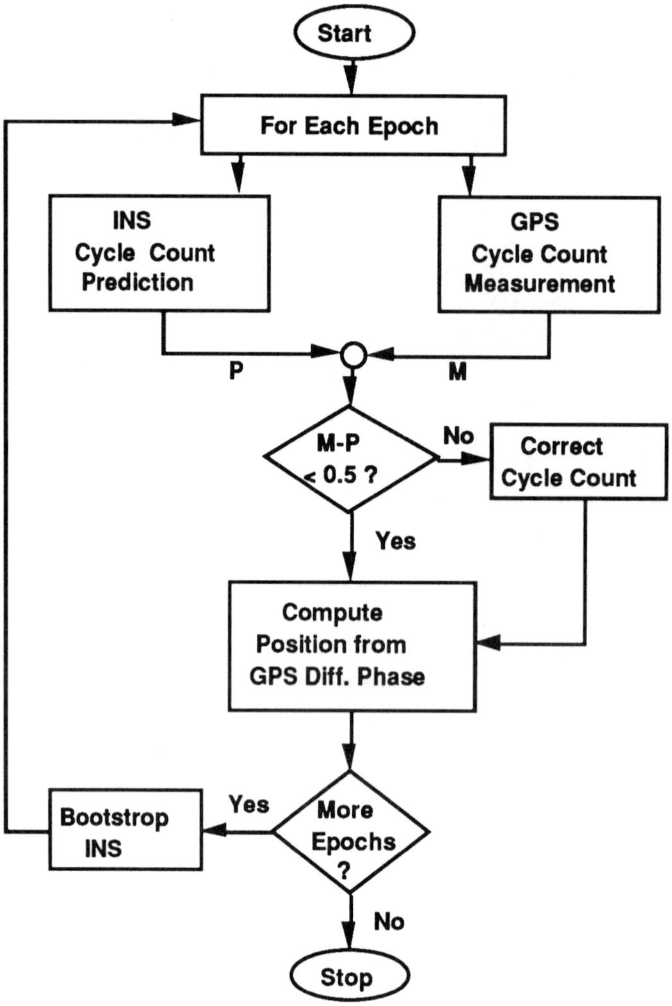

Fig. 6. Cycle Slip Detection Using Integrated GPS-INS

Procedures

Procedures to optimize differential GPS positioning have been discussed earlier and will not be repeated here. Attention will instead be focussed on procedures specific to the INS component of the integrated system and to the integration itself. Synchronization, offset-measurements, alignment and in-flight calibration will be the topics discussed here.

Synchronization of the two data streams is an obvious necessity with the approach taken here. To keep the synchronization error at velocities of 500 km/h to the cm-level, time-tagging must be accurate to 0.1 milliseconds. This is not as simple as it may sound. GPS receivers operate either in GPS-time or in UTC obtained from the receiver oscillator. INS measurements are recorded with time stamps obtained from the internal computer clock. There will be an offset and a drift between the two time systems which has to be calibrated. Since the relative drift of the two clocks is most likely not linear, calibration has to be repeated from time to time. Simultaneously reading of both clocks is usually not possible. Thus, time delays for instrument interrogation and response have to be considered and added as corrections. It appears that time tagging is possible to the millisecond level. Whether it can be improved by one order of magnitude is currently under study.

Offset measurements between the GPS-antenna and the center of the inertial measuring unit must be made in the body frame of the INS. In this way transformation of the offset measurements to another coordinate frame are easy to make. Usually, other sensors, such as a photogrammetric camera, a laser altimeter or a pressure barometer have also to be related to the INS. Although straightforward in principle, these measurements are often a headache, especially in small and crowded aircraft. If they are not done with sufficient accuracy, they result in errors which are tedious to eliminate in the post-mission processing.

The alignment of the inertial system at the beginning of the survey, i.e. its orientation with respect to the local-astronomic system, can be done in about ten minutes with a high accuracy strapdown system. This time is also needed to make the initial GPS baseline measurements. The major INS system errors remaining after the alignment are the initial azimuth misalignment, gyro drifts, accelerometer scale factors and biases. Most of these errors can be estimated if precise position updates are available and specific trajectory conditions are observed. This means that for an integrated GPS/INS, alignment and in-flight calibration should be seen as two sides of the same coin. The advantage of continual in-flight calibration is obviously the higher stability of the inertial system during times of GPS-satellite outages. Although it will not always be possible to follow an optimal flight pattern for INS-calibration, awareness of some advantageous configurations during the planning stage, will often provide better observability conditions and therefore better error resolution during the survey.

Equipment Trends

Equipment trends for GPS receivers and antennas have already been discussed earlier. This section will therefore concentrate on inertial equipment trends. Among them, the replacement of stable platform systems by strapdown units for a wide variety of applications, the increasing availability of individual inertial sensors and sensor blocks, and the potential development of solid state inertial systems are the most important trends for the applications discussed here.

Stable platform systems have for a long time been the mainstay in inertial navigation. When about twenty years ago the prototype development of new inertial gyros dovetailed with the rapid increase in processing speed, the development of inertial strapdown systems became a realistic possibility. Towards the end of the seventies, the first production systems arrived on the market and today strapdown systems are competing in many areas with conventional platform systems. Currently dry tuned rotor gyros and ring-laser gyros are both well represented on the strapdown market. It is very likely, however, that ring-laser gyros will dominate strapdown technology for the next ten years. Towards the end of that period, fibre optics gyros will become a competition, especially at the low to medium accuracy end of the market. Advantages of the strapdown configuration for the applications discussed here are access to the unfiltered sensor outputs, greater flexibility provided by a fully digital system, replacement of a largely mechanical system concept by a fully electronic concept, and reduction in weight, size, and power consumption. For a more detailed discussion, see for instance AGARD (1978).

A development that goes beyond replacing the stable platform INS by a strapdown INS is the trend towards sensor blocks, consisting of accelerometers and gyroscopes, the minimum supporting electronics and the power supply. They can be adapted to specific applications by using dedicated software. This will increase production numbers for the basic unit without limiting the number of distinct products offered. These units are either called Inertial Measuring Units (IMU), Inertial Reference Units (IRU), or Multisensor

Blocks in the literature. The major advantage for small user groups, as for instance those in geodesy and remote sensing, is the potential to accommodate new applications by software changes rather than by a redesign of the system. Another advantage is the ability to make use of the newest technology by simply replacing the basic sensor unit.

In some applications, there is no need to use a full inertial sensor block but single inertial sensors or combinations of them would be fully adequate. Similarly, there are other applications which would profit from a redundancy of sensors. In the past, it has been difficult to acquire such sensors and to operate them. The current trend towards more flexibility on the component market, although still somewhat tentative, is an encouraging sign. It appears that solid state sensors are a distinct possibility, especially with the current developments in silicon accelerometers, using processing methods originally developed for integrated circuits, and the potential of fibre optics gyros and perhaps ring-laser gyros for solid state design. The availability of relatively inexpensive sensors and electronics would certainly stimulate novel approaches to a variety of problems.

CONCLUSIONS

The current state of making geodetic measurements from a moving carrier has been explored and potential applications have been outlined. Tests conducted in a variety of environments have shown that many of them are currently feasible. An effort is underway to develop a variety of cost effective operational systems. Most of these systems will either use differential GPS or an integration of INS and GPS to solve the kinematic problem.

GPS has already proven to be a very powerful three-dimensional positioning tool. The availability of the full constellation from 1992 onwards and the emergence of powerful receivers will further enhance the level of performance currently achieved. A full constellation will result in 2 to 3 redundant satellites at nearly all times in most parts of the world. Algorithms to take advantage of redundant measurements will be developed. Cycle slip detection and correction in real-time will be possible in many cases. Receiver hardware advances are resulting in the packaging of more channels in a portable unit than the number of satellites visible with the full constellation. High data output rates will result in a more accurate modelling of the vehicle dynamics. The development of multi-antenna receivers will yield attitude components and azimuth with a level of accuracy sufficient for many applications. Powerful internal microprocessors and storage devices will be capable of supporting the increasingly complex real-time functions. The major limitation will be the line-of-sight requirement which will limit the use of unaided GPS when considering highly sophisticated kinematic tasks where accurate position, velocity and acceleration components are required.

Adding an inertial system or sensor block to a set of GPS receivers has a number of advantages. It solves the cycle slip and interpolation problem, adds precise attitude information to the output, and has the potential for GPS noise reduction by feeding INS velocity back to the carrier tracking loop. By differencing GPS and INS results it provides the capability of resolving the medium wavelength part of the gravity field spectrum and thus a means of determining local and regional geoid features from the air. Current development in algorithms and procedures maximizes the greater flexibility of the strapdown output for specific applications and uses the position accuracy of the GPS output for inflight calibration of the inertial unit. Current equipment trends clearly show that the strapdown concept will dominate the future market. Thus, access to unfiltered measurements in digital form will be characteristic for future systems. There is also a trend towards inertial sensor blocks as basic units which should result in major price reductions, especially if current efforts of developing solid state sensors fulfil their promise.

Current kinematic GPS positioning accuracies achievable in a production environment are of the order of 0.5 to 1.0 m (1s) if the initial baseline has been properly resolved. If not, accuracies of 2 to 3 m (1s) are more likely. In well-controlled test situations where control data are known beforehand and are used to eliminate data shifts, RMS values of 0.1 to 0.2 m (1s) have been achieved when using code and carrier, and values of better than 0.1 m when using cycle slip free carrier data. It appears that the data consistency under ideal conditions, i.e. no cycle slips, stable ionosphere, almost constant velocity, is about 0.03 m or better over short time intervals. It should be noted, however, that data consistency is not equal to accuracy, especially not in a moving platform. Semi-kinematic results over short traverses are of the order of 5 cm (1s) if no cycle slips occur. Improvements can be expected once more satellites are available.

For an integrated GPS/INS positioning accuracies of better than 0.1 m are achievable if cycle slips do not occur for long periods of time. Relative to an initial point about 50 km away, accuracies of 10" (1s) for attitude, 5 cm (1s) for the geoid, and 1" (1s) for deflections of the vertical seem to be feasible.

REFERENCES

Agard (1978). Strapdown Inertial Systems, AGARD Lecture Series No. 95, Neuilly Sur Seine.

Ashjaee, J., R. Lorenz, R. Sutherland, J. Dutilloy, J.-B. Minazio, R. Abtahi J.-M. Eichner, J. Kosmalska, and R. Helky (1989). New GPS Developments and Ashtech M-XII. Proceedings of the Satellite Division's International Technical Meeting (GPS89), The Institute of Navigation, Washington, D.C.

Baustert, G., G.W. Hein and H. Landau (1989). On the Use of GPS in Airborne Photogrammetry, Hydrographic Applications, and Kinematic Surveying. Proceedings of 5th International Geodetic Symposium on Satellite Positioning, Physical Science Laboratory, New Mexico State University, pp. 1029-1040.

Baustert, G., M.E. Cannon, E. Dorrer, G. Hein H. Krauss K.P. Schwarz (1989). German-Canadian Experiment in Airborne INS-GPS Integration for Photogrammetric Applications. General Meeting of the IAG, Edinburgh, August 3-12, 1989.

Brown, A. (1989). Extended Differential GPS. Proceedings of the National Technical Meeting, The Institute of Navigation, Washington, D.C., pp. 125-136.

Buechler, D. and M. Foss (1987) Integration of GPS and Strapdown Inertial Subsystems into a Single Unit. *Navigation,* Vol. **34**, 2, pp. 140-159.

Cannon, M.E. (1987). Kinematic Positioning Using GPS Pseudorange and Carrier Phase Observations. Publ. No. 20019, Dept. Surv. Eng., The University of Calgary.

Cannon M.E. (1989). High Accuracy GPS Semi-Knematic Positioning: Modelling and Results. Presented at the GPS '89 Satellite Division's International Technical Meeting of The Institute of Navigation, Colorado Springs.

Cannon, M.E., and K.P. Schwarz (1989). GPS Semikinematic Positioning along a Well-Controlled Traverse. Proceedings of 5th International Geodetic Symposium on Satellite Positioning, Physical Science Laboratory, New Mexico State University, pp. 621-630.

Carlson, N. A. (1988). Federated Filter for Fault-Tolerant Integrated Navigation Systems. IEEE Plans '88, pp. 110-119.

Cox D.B. (1978). Integration of GPS with Inertial Navigation Systems. *Navigation,* Vol. **25**, 2, 236-245.

Evans, A.G. (1986). Use of GPS Phase Measurements for Dynamic Relative Positions and Velocity Estimation: Laboratory Test Results. Proceedings of 4th International Geodetic Symposium on Satellite Positioning, Applied Research Laboratories, The University of Austin at Texas, pp. 1361-1375.

Goad, C. C. (1985). Precise Positioning with the Global Positioning System. Proceedings of the Third International Symposium On Inertial Technology for Surveying and Geodesy, Publ. No. 60005, Dept Surv. Eng., The University of Calgary, pp. 745-756.

Hartmann, R.G. (1988). An Integrated GPS/IRS Design Approach. *Navigation,* Vol. **35**, 1, pp. 121-134.

Hasandunn, Z. A., and D.E. Wells (1989) Real-Time Marine Instantaneous Velocity Determination Using Differential GPS. Proceedings of 5th International Geodetic Symposium on Satellite Positioning, Physical Science Laboratory, New Mexico State University, pp. 1009 - 1018.

Hashemipour, H.R., S. Roy, A.J. Laub (1988). Decentralized Structures for Parallel Kalman Filtering. *IEEE Trans. Automat. Control,* Vol. **AC-33,** pp. 88-94.

Huddle J.R. (1977). The Theoretical Principles for Design of the Inertial Surveyor for Position and Gravity Determinations. Proceedings of the First International Symposium on Inertial Technology for Surveying and Geodesy, Ottawa, Canada, October 12-14, 1977.

Kalafus, R. M., A.J. Van Dierendonck, and N.A. Pealer (1986). Special Committee 104 - Recommendations for Differential GPS Service. *Navigation*, Vol. **33**, No. 1.

Keel, G., H. Jones, and G. Lachapelle (1988). A Test of Airborne Kinematic GPS Positioning For Aerial Photography. Proceedings of the Satellite Division's International Technical Meeting (GPS88), The Institute of Navigation, Washington, D.C., pp. 147-151.

Kleusberg, A., A. Goodacre, and R. J. Beach (1989). On the Use of GPS for Airborne Gravimetry. Proceedings of 5th International Geodetic Symposium on Satellite Positioning, Physical Science Laboratory, New Mexico State University, pp. 977-986.

Kleusberg, A., S. H. Quek, D.E. Wells, and J. Hagglund (1985). Comparison of INS and GPS Ship Velocity Determination. Proceedings of the Third International Symposium On Inertial Technology for Surveying and Geodesy, Publ. No. 60005, Dept Surv. Eng., The University of Calgary, pp. 791-805

Knickmeyer, E.T., and K.P. Schwarz (1989). Geoid Determination Using GPS-Aided Inertial Systems. General Meeting of the IAG, Edinburgh, August 3-12, 1989.

Kremer, G.T., R.M. Kalafus, P.V.W. Loomis, and J.C. Reynolds (1989). The Effect of Selective Availability on Differential GPS Corrections. Procedings of the Satellite Division's International Technical Meeting (GPS89), The Institute of Navigation, Washington, D.C.

Kruczynski L.R., P. C. Li, and A. G. Evans (1989). Using GPS to Determine Vehicle Attitude: USS Yorktown Test Results. Procedings of the Satellite Division's International Technical Meeting (GPS89), The Institute of Navigation, Washington, D.C.

Lachapelle G., J. Hagglund, H. Jones, and M. Eaton (1984). Differential GPS Marine Navigation. Proceedings of PLANS 84, I.E.E.E., New York, pp. 245-255.

Lachapelle, G., J. Hagglund, W. Falkenberg, P. Bellemare, M. Eaton, and M. Casey (1986). GPS Land Kinematic Positioning Experiments. *The Hydrographic Journal*, No. **42** , pp. 45-55.

Lachapelle, G., W. Falkenberg,J. Hagglund D. Kinlyside, M. Casey, P. Kielland and H. Boudreau (1988). Shipborne GPS Kinematic Positioning for Hydrographic Applications. *Navigation*, Vol. **35**, No. 1, The Institute of Navigation, Washington, D.C.

Lachapelle, G., W. Falkenberg, D. Neufeldt and P. Kielland (1989). Marine DGPS Using Code and Carrier in a Multipath Environment. Proceedings of the Satellite Division's International Technical Meeting (GPS89), The Institute of Navigation, Washington, D.C.

Landau, H. (1988). Zur Nutzung des Global Positioning Systems in Geodäsie und Geodynamik: Modellbildung, Software-Entwicklung und Analyse. Studiengang Vermessun- swesen, Universität der Bundeswehr München, Heft 36.

Landau, H. (1989). Precise Kinematic GPS Positioning. *Bulletin Géodésique*, **63**, 1, 85-96.

Mader, G. (1986). Dynamic Positioning Using GPS Carrier Phase Measurements. *Manuscripta Geodaetica*, Vol. **11**, No. 4, pp. 272-277.

Moreau, R., et L.M.A. Jeudy (1986). Profilomètre Laser-Inertiel (PLI): un essai qualitatif en terrain boisé. *Le Géomètre canadien/The Canadian Surveyor,* Vol. **40**, No. 1, pp. 23-28.

Purcell Jr., G.H., J. M. Srinivasan, L.E. Young, S.J. Dinardo, E.L. Hushbeck Jr., T.K. Meehan, T.N. Munson, and T.P. Yunck (1989). Measurement of Aircraft Position, Velocity, and Attitude Using Rogue GPS Receivers. Proceedings of 5th International Geodetic Symposium on Satellite Positioning, Physical Science Laboratory, New Mexico State University, pp. 1019-1028.

Remondi B. (1985). Performing Centimeter Accuracy Relative Surveys in Seconds Using GPS Carrier Phase. Proceedings of the First International Symposium on Precise Positioning with the Global Positioning System, Rockville, Md., April 15-19.

Ruane, M. F., P. K. Enge, K. Olson, and D. Pietraszewski (1989). Sea Trials, Message Delay, and Network Design for DGPS Radiobeacons. Procedings of the Satellite Division's International Technical Meeting (GPS89), The Institute of Navigation, Washington, D.C.

Rueger, J.M. (1986). High Accuracy in Short ISS Missions. *Bull Géod.*, **60**, 2, 155-167, 1986.

Schwarz,, K.P. (1983). Inertial Surveying and Geodesy, *Reviews of Geophysics and Space Physics*, **21**, 4, 878-890.

Schwarz, K.P. (1987). Geoid Profiles from an Integration of GPS Satellite and Inertial Data. *Boll. di Geod. e Scienze Affini,* **46**, 117-131, 1987.

Schwarz, K.P., C.S. Fraser, P.C. Gustafson (1984). Aerotriangulation without Ground Control. Int. Arch. of Photogrammetry and Remote Sensing, **25**, Part A1, Rio de Janeiro, June 16-29, 1984.

Schwarz, K.P., M.E. CANNON, and R.V.C. WONG (1989). A Comparison of GPS Kinematic Models for the Determination of Position and Velocity along a Trajectory. *Manuscripta Geodaetica*, Vol. **14,** No. 5, pp. 345 - 353.

Seeber, G., A. Schuchardt and G. Wübbena (1986). Precise GPS-Aided Marine Positioning Developments and Results. Proceedings of International Symposium on Marine Positioning, D. Reidel Publ. Co., pp. 187-196.

Speyer, J.L. (1979). Computation and Transmission Requirements for a Decentralized Linear-Quadratic-Gaussian Control Problem. IEEE Trans. Automat. Control, Vol. AC-24, pp. 266-269.

Steeves, R. R., C. A. Chamberlain, and D.J. McArthur (1989). Cost Effectiveness of a Canadian Active Control System. Proceedings of 5th International Geodetic Symposium on Satellite Positioning, Physical Science Laboratory, New Mexico State University, pp. 825-830

Tazartes, D.A. and J.G. Mark (1988). Integration of GPS Receivers into Existing Inertial Navigation Systems. *Navigation*, Vol. **35**, 1, pp. 105-119.

Teunissen, P.J.G. and M.A. Salzmann (1988). Performance Analysis of Kalman Filters, Proceedings Hydro 88, November 15-17, 1988.

Wei, H. and K.P. Schwarz (1989) A Discussion of Models for GPS/INS Integration. General Meeting of the IAG, Edinburgh, Scotland, August 3-12, 1989.

A DISCUSSION OF MODELS FOR GPS/INS INTEGRATION

M. Wei and K.P. Schwarz
Department of Surveying Engineering
The University of Calgary

ABSTRACT

Models and algorithms for GPS/INS integration are discussed for applications which require highly precise trajectory determination in near real time. The paper formulates the inertial integration model in the same earth-fixed Cartesian frame in which GPS positioning computations are normally performed. This formulation is compared to the standard approach which performs the integration in geographical coordinates, using the local-level frame. A state space model is then used to integrate the two data streams via a Kalman filter. Implementation of the two algorithms and first test results indicate that the model in earth-fixed Cartesian coordinates has decisive advantages in simplicity of the mathematical formulation and efficiency of the numerical computation.

1. INTRODUCTION

The integration of GPS and INS has been proposed for a variety of applications, see for instance Cox (1980), Schwarz et al (1984), Buechler and Foss (1987), Tazartes and Mark (1988), Kleusberg (1989), Lachapelle and Schwarz (1989). During the last few years the rapid development of inertial strapdown systems for a wide range of accuracies has made it possible to trailor the inertial unit to specific applications and thus make the integration more economical. Even a low cost strapdown system will add a number of advantages to the 'GPS only' solution. It will solve the cycle slip problem, give improved interpolation accuracy in high dynamics, reduce GPS noise by feedback control of the carrier tracking loop and upgrade system integrity. A medium accuracy system will add precise attitude information, and a high accuracy system can in addition be used for gravity field determination, see Schwarz (1987) and Knickmeyer and Schwarz (1989).

For commercial aircraft navigation, the aiding of navigation grade INS by GPS pseudo-range data from a single receiver is the most important application. For kinematic applications in geodesy the integration of inertial data with differential GPS carrier phase data is of far more interest. Only this integration will be considered here. It will be assumed that single frequency receivers are used.

The paper looks at algorithms to achieve the integration. While for platform systems, the mechanization prescribes the coordinate frame for the integration, this is not so for strapdown systems. The choice can therefore be made based on the speed of computation, the ease with which the two data streams can be combined, and the effectiveness of filtering methods applied. To find an algorithm which optimally combines these requirements is the goal of this study.

2. MODELLING EQUATIONS

2.1. COORDINATE FRAMES

The coordinate frame in which the integration is performed is in principle arbitrary. Coordinate frames, which are used for strapdown computations, are the inertial frame, the earth-fixed frame and the local-level frame. Brief definitions of these frames are given in the following, for more details see (Britting,1971).

i) Operational inertial frame (x^i, y^i, z^i) ··· i-frame

The inertial frame is a Cartesian frame, in which Newton's laws of motion are valid. A frame which satisfies this condition within bounds given by the measurement accuracy is called an operational inertial frame. Its origin is at the centre of mass of the earth. The x-axis passes through the vernal equinox and the y-axis completes a right-hand orthogonal frame. The z-axis coincides with the mean axis of earth rotation, and points towards the Conventional International Origin (CIO).

ii) Earth-fixed frame (x^e, y^e, z^e) ··· e-frame

The earth-fixed frame is a Cartesian frame with its origin at the mass centre of the earth. Its z-axis coincides with the mean axis of earth rotation. The x-axis points toward the mean Greenwich meridian and the y-axis completes a right-handed orthogonal frame. In geodesy, this frame is commonly called the Conventional Terrestrial System. Positions derived from GPS data are usually referenced to this frame.

iii) Local-level frame (E,N,U) ··· n-frame

The local-level frame is a local Cartesian frame with its origin at the centre of the inertial measuring unit. Its axes are aligned with the east, north and up directions. The up direction is defined by the normal to the reference ellipsoid pointing outward. The local-level frame is thus related to the geographic coordinate frame. The local-level frame is commonly used for integrating inertial measurements.

The relationship between different coordinate frames and their transformations are discussed for instance in Britting (1971) or Wong (1988).

2.2. STATE SPACE MODELS AND MECHANIZATION EQUATIONS

A strapdown inertial system consists of two sensor triads which are fixed to the body of the system. The sensors measure three components of the specific force vector f^b and three components of the body rotation rate ω^b_{ib}. The latter are measured as angular velocities with respect to the inertial frame coordinated in the body frame.

The coordinate frame, in which the integration is performed, is arbitrary in principle, and is usually chosen such that it agrees with the coordinate frame for the output. The measurements are first transformed into the computation frame and then integrated. Results are position, velocity and attitude of the body with respect to the computation frame. Models for integrating measurements from a strapdown inertial system in various coordinate frames by using the state space model approach are given by Schwarz and Wei (1988) and Teunissen (1988). In this section, the state space models for a strapdown inertial system in the earth-fixed frame and the local-level frame are given and compared.

i) Model in the earth-fixed frame

The basic principle for inertial measurements is Newton's second law of motion in the gravitational field of the earth

$$\ddot{\mathbf{r}}^i = \mathbf{f}^i + \mathbf{G}^i \tag{1}$$

where $\mathbf{r}^i$ is the position vector, $\mathbf{f}^i$ the specific force vector and $\mathbf{G}^i$ the gravitational vector, all measured in the inertial frame.

For a rotating reference frame, equation (1) has to be transformed to take the effect of rotation into account. Transforming to the earth-fixed frame leads to

$$\ddot{\mathbf{r}}^i = \mathbf{R}^i_e \left(\ddot{\mathbf{r}}^e + 2\, \Omega^e_{ie}\, \dot{\mathbf{r}}^e + \Omega^e_{ie}\, \Omega^e_{ie}\, \mathbf{r}^e \right) \tag{2}$$

where $\mathbf{r}^e$ is the position vector from the centre of mass in the earth-fixed frame, $\mathbf{R}^i_e$ is the transformation matrix from the e-frame to the i-frame and Ω^e_{ie} is the skew-symmetric matrix of the angular velocities of earth rotation; for more details, see Britting (1971).

The set of second-order differential equations (2) can be reformulated to a set of first-order differential equations. Together with a set of first-order differential equations for the transformation matrix, the nonlinear mechanization equations in the earth-fixed frame are of the form

$$\begin{pmatrix} \dot{\mathbf{r}}^e \\ \dot{\mathbf{v}}^e \\ \dot{\mathbf{R}}^e_b \end{pmatrix} = \begin{pmatrix} \mathbf{v}^e \\ \mathbf{R}^e_b\, \mathbf{f}^b - 2\, \Omega^e_{ie}\, \mathbf{v}^e + \mathbf{g}^e \\ \mathbf{R}^e_b \Omega^b_{eb} \end{pmatrix} \tag{3}$$

where $\mathbf{r}^e$ is the position vector, $\mathbf{v}^e$ is the velocity vector, all of them in the e-frame, $\mathbf{f}^b$ is the specific force vector in the b-frame, $\mathbf{g}^e$ is the gravity vector in the e-frame, $\mathbf{R}_b^e$ is the transformation matrix from the b-frame to the e-frame and Ω_{eb}^b is the skew-symmetric matrix of the angular velocity vector ω_{eb}^b of the b-frame with respect to the e-frame, coordinated in the b-frame.

Figure 1 shows the algorithmic flowchart of the mechanization equation in the earth-fixed frame.

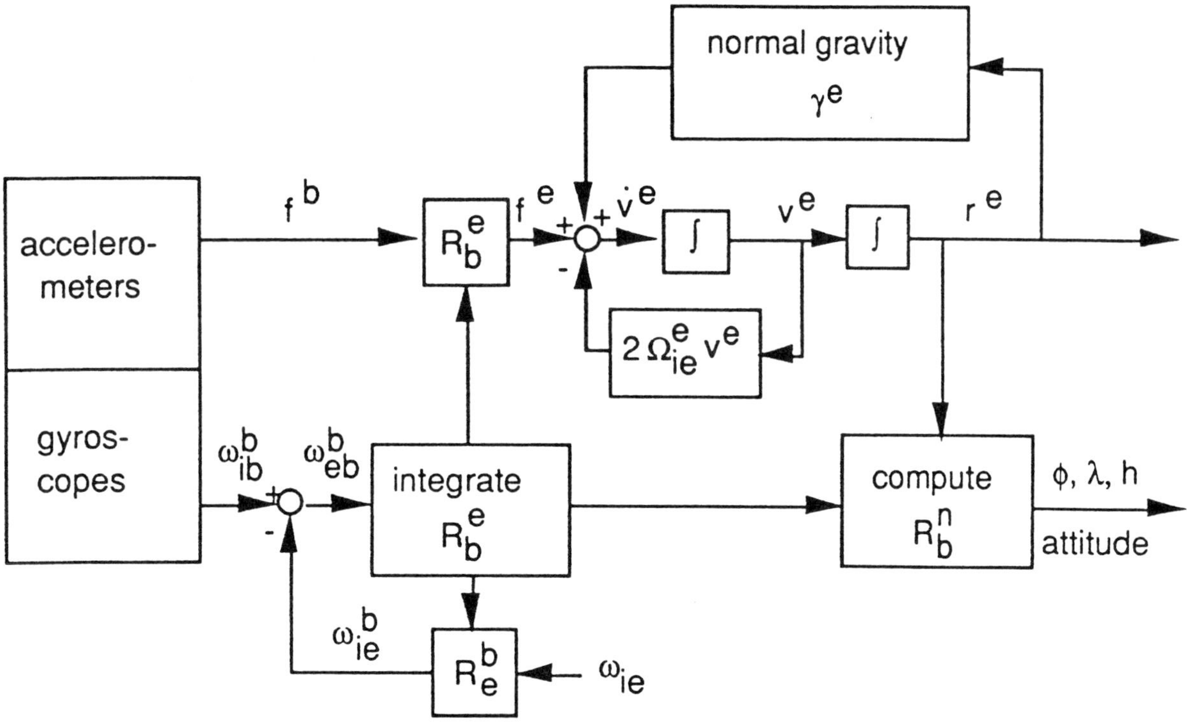

Figure 1 Mechanization equation in the earth-fixed frame

The advantages of the algorithm in the e-frame are:

- The geocentric Cartesian coordinates are directly obtained. It is thus convenient for integration with GPS.
- The mechanization algorithm is relatively simple.

Disadvantages are:

- The gravity vector has to be transformed into the earth-fixed frame;
- For navigation purposes, the geographic coordinates (ϕ,λ) and attitude with respect to the local-level frame are required.

ii) Model for the local-level frame

Transforming the mechanization equation (3) into the local-level frame leads to the representation in geographic coordinates

$$\begin{pmatrix} \dot{\mathbf{r}}^n \\ \dot{\mathbf{v}}^n \\ \dot{\mathbf{R}}_b^n \end{pmatrix} = \begin{pmatrix} \mathbf{D}\,\mathbf{v}^n \\ \mathbf{R}_b^n\,\mathbf{f}^b - (2\,\Omega_{ie}^n + \Omega_{en}^n)\,\mathbf{v}^n + \mathbf{g}^n \\ \mathbf{R}_b^n\,\Omega_{nb}^b \end{pmatrix} \tag{4}$$

where $\mathbf{r}^n = (\phi, \lambda, h)$ are geographic coordinates, $\mathbf{v}^n = (R_N \cos\dot{\lambda}, R_M \dot{\phi}, \dot{h})$ are the velocities in the local-level frame, $\mathbf{R}_b^n$ is the transformation matrix from the b-frame to the n-frame, Ω_{en}^n is the skew-symmetric matrix of the angular velocity vector ω_{en}^n of the local-level frame, Ω_{nb}^b is the skew-symmetric matrix of the body rotation rate with respect to the local-level frame.

In the mechanization equation (4), the formulas are refered to geographic coordinates (ϕ, λ, h). Figure 2 illustrates the local-level mechanization equation.

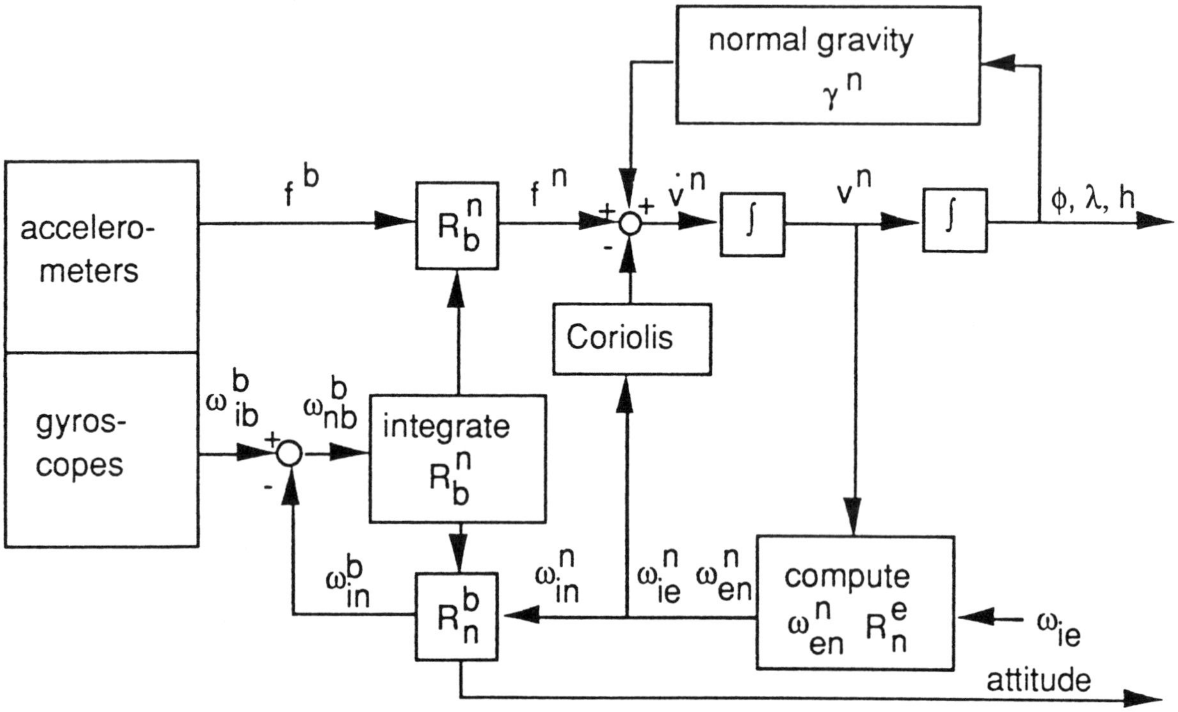

Figure 2 Local-level mechanization

The main advantages of the local-level mechanization are:

- The geographic coordinates (ϕ, λ, h) are obtained directly from the mechanization.
- The attitude of the body with respect to the local-level frame (roll, pitch, azimuth) is given directly.

Disadvantages are:

- The mechanization equations are more complicated;

- For the integration with GPS, the GPS results must be transformed to the geographic frame.

2.3. FILTER EQUATIONS AND INTEGRATION WITH GPS

The error behaviour of kinematic systems is usually modelled by systems of differential equations and can be described by a state space model of the form

$$\dot{\mathbf{x}} = \mathbf{F}\,\mathbf{x} + \mathbf{w} \qquad (5)$$

where $\mathbf{x}$ is the state vector; $\mathbf{F}$ is the dynamics matrix; $\mathbf{w}$ is the driving noise.

The Kalman filter technique is then used to estimate the errors. This model can be used for INS as well as for GPS errors. The advantage of this approach is that it allows the use of a variety of update measurements to perform the error estimation. This is important for the integration with sensors such as GPS.

A 15-state Kalman filter was developed for the estimation of the error states of a strapdown inertial system for curvilinear coordinates (Wong, 1988). The corresponding algorithm in the earth-fixed frame has a state vector of the form

$$\mathbf{x}_{INS} = (\varepsilon, \delta\mathbf{r}, \delta\mathbf{v}, \mathbf{d}, \mathbf{b})^T \qquad (6)$$

where

$\varepsilon = (\varepsilon_x, \varepsilon_y, \varepsilon_z)^e$ are attitude errors in the e-frame;

$\delta\mathbf{r} = (\delta x, \delta y, \delta z)^e$ are position errors in the e-frame;

$\delta\mathbf{v} = (\delta v_x, \delta v_y, \delta v_z)^e$ are velocity errors in the e-frame;

$\mathbf{d} = (d_x, d_y, d_z)^b$ are gyro drifts about gyro axes in the body frame;

$\mathbf{b} = (b_x, b_y, b_z)^b$ are accelerometer biases in the body frame;

The corresponding dynamics matrix is shown in the appendix. It is directly comparable to the dynamics matrix given in Wong (1988) for the local-level frame.

If GPS measurements or results are used as updates, two strategies can be followed. The first one is a common Kalman filter for both, the GPS and INS errors. In the second approach, one filter is designed for each sensor. In general, these filters work independently and they interact only occasionally. This is the so-called decentralized filter. It has advantages in terms of data integrity and speed, but it is more complex in terms of formulation.

The decentralized filter consists of a master filter and local filters (Carlson, 1988, Hashemipour, et al., 1988). The INS filter as master filter estimates position, velocity and attitude along the trajectory at a high rate. The GPS filter is one of the local Kalman filters and processes GPS data only. Results from the local GPS filter are used at a reduced rate to update the INS filter at distinct epoches to obtain an optimal estimation. Figure 3 shows the principle of the decentralized filter for GPS/INS integration in a block diagram.

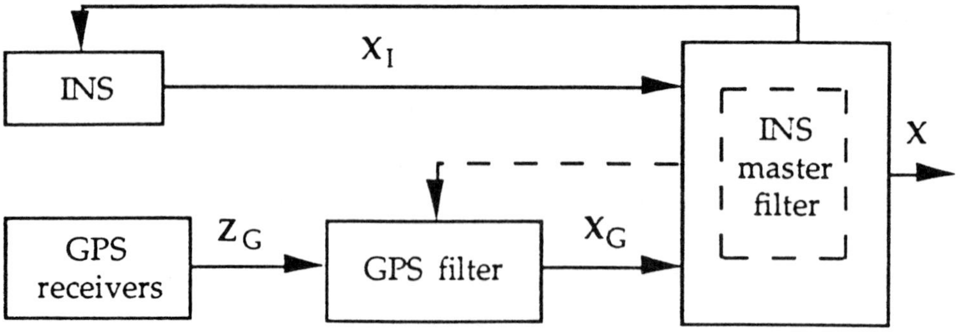

Figure 3 Decentralized filter

3. IMPLEMENTATION AND RESULTS

A strapdown inertial reference unit gives body rotation rates and specific force measurements at a high rate (32 to 64 Hz). Processing this data requires two major steps, namely integration and filtering.

3.1. IMPLEMENTATION OF MECHANIZATION EQUATIONS

In the following, implementation of the integration algorithms in both, the local-level and the earth-fixed frame, will briefly be discussed and compared.

i) Local-level frame mechanization

The integration of the raw data is done in two successive steps. First, integrating the body rotation rate leads to the Euler angles (roll, pitch, azimuth) and the transformation matrix from the b-frame to the n-frame. In the second step, the body accelerations are transformed into the local-level frame and then velocities and coordinates of the body are obtained by integrating the accelerations.

For computation of the Euler angles and transformation matrix, the quarternion approach is used. The rotation rate of the n-frame due to earth rotation and body velocities is fed back to the integration loop for the Euler angles and removed from the measured rates.

In the integration loop for accelerations, the Coriolis acceleration due to the rotation of the earth, the acceleration due to the body rotation with respect to the earth, and the gravity acceleration are removed from the measured specific force vector. The integration loops for the rotation rates and the accelerations are performed at a high rate (64 Hz).

For the gravity calculation, Somigliana's formula is used. Coordinates ϕ and h are fed back to compute normal gravity. The computation is done at a low rate (1 Hz).

Output of the algorithm are geographic coordinates (ϕ,λ,h), earth referenced velocities (v_E,v_N,v_U) and attitude of the body with respect to the local-level frame (roll,pitch,azimuth) at a 64 Hz rate.

ii) Earth-fixed frame mechanization

For the integration of the rotation rates, only the angular velocity of earth rotation is removed from the measured rates. The quarternion approach is again used to compute the transformation matrix. The measurements, rotation rates and accelerations, are integrated at a 64 Hz rate.

In order to compute the normal gravity vector in the earth-fixed frame, an approximate formula for normal gravity in the earth-fixed Cartesian frame is used (Schwarz and Wei, 1989). The geocentric coordinates (x^e, y^e, z^e) are fed back to compute the normal gravity vector. The computation is done at a low rate (1 Hz).

Output of the algorithm are geocentric coordinates (x^e, y^e, z^e), earth-referenced velocities (v_x, v_y, v_z) and the transformation matrix $\mathbf{R}_b^e$ at a 64 Hz rate. The transformation to geographic coordinates (ϕ,λ,h), velocities (v_E, v_N, v_U) and the Euler angles (roll,pitch,azimuth) is performed at a 1 Hz rate.

3.2. IMPLEMENTATION OF FILTER EQUATIONS

The Kalman filter is used to estimate system errors during the fine alignment and when the system is working in dynamic mode. The update measurements are zero velocity measurements for land surveying. For GPS/INS integration, the position and velocities estimated from GPS data are used as updates.

A 15-state Kalman filter has been modified for error estimation in the earth-fixed frame. The Kalman filter algorithm is the same for both models. The difference between the local-level frame and the earth-fixed frame is contained in the dynamics matrix. A first comparison shows that the dynamics matrix **F** for the earth-fixed frame contains more zero elements and can therefore be expected to be more efficient numerically.

3.3. COMPARISON OF MODELS

When comparing the mechanization equation in the local-level frame to the integration module in the earth-fixed frame, some advantages of the latter become apparent. The formulas and the implementation are simpler because the rotation of the local-level frame with respect to the earth is eliminated. Since the integration computations are performed at a high rate (64 Hz), the computation in the earth-fixed frame is more efficient.

The local-level mechanization has advantages for normal gravity computation because a closed formula can be used. The normal gravity vector in the earth-fixed frame is computed by using a series. An approximation accuracy of 0.05 mGal or 0.001 mGal can be achieved by using a three or four coefficient formula, respectively (Schwarz and Wei, 1989).

The local-level mechanization is prefered in navigation and land surveying because it provides directly the geographic coordinates (ϕ,λ,h) and the Euler angles (roll, pitch, azimuth). For GPS/INS integration, the mechanization in the earth-fixed frame has advantages because positions and velocities computed from the INS are directly comparable to those from differential GPS.

Computation times (CPU time) for both algorithms are shown in Table 1. Computation of the mechanization equation in the earth-fixed frame is 40% faster than in the local-level frame, similarly, Kalman filter computations for the first model are 10% faster than for the latter. The total computation for the earth-fixed frame is 32% faster than for the local-level frame (computed on VAX 750). In light of these results, integration and filtering in the e-frame seem to be preferable to that in the n-frame.

Modules	n-frame (s)	e-frame (s)	%
Mechanization equation (64 Hz)	0.60	0.36	60%
Normal gravity computation (1 Hz)	0.00036	0.00046	128%
Conversion to (ϕ, λ, h) (1 Hz)	/	0.0033	/
Kalman filter	0.21	0.19	90%
Total computation time	0.81036	0.55376	68%

Table 1 Computation time

3.4. RESULTS

To test the algorithms and to compare the accuracy of the two methods a common data set has been used. At this stage of the research only the inertial algorithms were compared. Measurements were collected for 20 minutes along a well controlled traverse of 10.5 km length. Five control points were available to analyse the positioning results. The coordinates at the first and fourth points are used for coordinate updates. Figure 7 shows the positioning errors for both methods after filtering. They agree within the system noise and show similar differences to the control points. In terms of positioning accuracy the results can therefore be considered as equivalent.

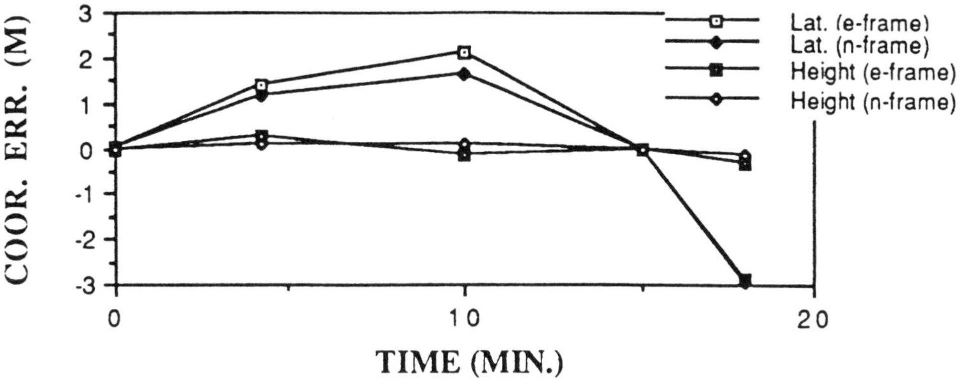

Figure 4 Errors after filtering

4. CONCLUSIONS

To formulate efficient models for GPS/INS integration, full use has been made of the fact that the integration of strapdown inertial data can be performed in a coordinate frame most appropriate to the problem at hand. In this paper, the integration in an earth-fixed Cartesian frame, the Conventional Terrestrial Frame of geodesy, is compared to the standard method of integration in a local-level frame. The new approach is simpler in terms of the mathematical formulation and about 32% faster in terms of computation speed. It has therefore major advantages for real-time strapdown applications.

The fact that the Conventional Terrestrial Frame is also used to compute GPS positions and velocities internally, offers some additional advantages for integration with GPS. Cycle slip detection and correction can be done in the simplest way possible and inertial feedback to the carrier phase tracking loop is greatly facilitated.

A brief discussion of Kalman filtering alternatives indicates that decentralized filtering is a promising method for GPS/INS integration. Studies to compare it to the standard Kalman filtering approach are currently under way.

ACKNOWLEDGMENT

Mr. H. Martell provided the program for integration and filtering in the local-level frame. His assistance is gratefully acknowledged.

REFERENCES

Britting, K.R., Inertial Navigation System Analysis. Wiley-Interscience, New York, 1971.

Buechler, D. and M. Foss, Integration of GPS and strapdown inertial subsystems into a single unit. Navigation: Journal of The Institute of Navigation, Vol. 34, No. 2, pp.140-159, 1987.

Cannon, M.E. and K.P. Schwarz, GPS semi-kinematic positioning along a well-controlled traverse. Proc. Fifth International Geodetic

Symposium on Satellite Positioning, Las Cruces, New Mexico, March 13-17, 1989.
Carlson, N.A., Distributed Filtering I. Tutorial notes, Integrity System Inc, 1988.
Cox, D.B., Integration of GPS with inertial navigation systems. Global Positioning System I, The Institute of Navigation,Washington, D.C., pp.144-153, 1980.
Hashemipour, H.R., et. al., Decentralized structures for parallel Kalman filtering. IEEE Trans. Automat. Contr., Vol. AC-33,pp. 88-94, 1988.
Kleusberg, A., Integration of INS and GPS measurements. High Precision Navigation, Springer-Verlag, Berlin, 1989.
Knickmeyer, E.T. and K.P. Schwarz, Geoid determination using GPS-aided inertial system. Presented at the IAG General Meeting '89, Edinburgh, Aug. 3-12,1989.
Lachapelle, G. and K.P. Schwarz, Kinematic applications of GPS and GPS/INS — Algorithm, processing, equipment trends. Presented at the IAG General Meeting '89, Edinburgh, Aug. 3-12, 1989.
Schwarz, K.P., Geoid profiles from an integration of GPS satellite and inertial data. Bolletino di geodesia a scienze affini, No. 2, 1987.
Schwarz, K.P., G.S. Fraser, and P.C. Gustafson, Aerotriangulation without ground control. International Archives of Photogrammetry and Remote Sensing, 25, Part A1, Rio de Janeiro, June 16-29, 1984.
Schwarz, K.P. and M. Wei, A framework for modelling kinematic measurements in gravity field applications. Proc. AGU Chapman Conference on Progress in the Determination of the Earth's Gravity Field, Ft. Landerdale, Sep. 12-16, 1988.
Schwarz, K.P. and M. Wei, Cartesian representation of the normal gravity model for inertial navigation. Submitted to Navigation: Journal of The Institute of Navigation, 1989.
Tazartes, D.A. and J.G. Mark, Integration of GPS receivers into existing inertial navigation system. Navigation: Journal of The Institute of Navigation, Vol. 35, No. 1, pp.105-119,1988.
Teunissen, P.G., Strapdown INS. Internal Report, Department of Surveying Engineering, The University of Calgary, Calgary, 1988.
Wong, R.V.C., Development of a RLG strapdown survey system. UCSE Reports No. 20027, Department of Surveying Engineering, The University of Calgary, Calgary, Canada, 1988.
Wong, R.V.C., K.P. Schwarz, and M.E. Cannon, High-accuracy kinematic positioning by GPS-INS. Navigation: Journal of The Institute of Navigation, Vol. 35, No. 2, pp.275-287, 1988.

APPENDIX

The dynamics matrix of the 15-state Kalman filter in the earth-fixed frame corresponding to equation (6) is of the form

$$
\begin{pmatrix}
0 & \omega_e & 0 & 0 & 0 & 0 & 0 & 0 & 0 & R_{11} & R_{12} & R_{13} & 0 & 0 & 0 \\
-\omega_e & 0 & 0 & 0 & 0 & 0 & 0 & 0 & 0 & R_{21} & R_{22} & R_{23} & 0 & 0 & 0 \\
0 & 0 & 0 & 0 & 0 & 0 & 0 & 0 & 0 & R_{31} & R_{32} & R_{33} & 0 & 0 & 0 \\
0 & 0 & 0 & 0 & 0 & 0 & 1 & 0 & 0 & 0 & 0 & 0 & 0 & 0 & 0 \\
0 & 0 & 0 & 0 & 0 & 0 & 0 & 1 & 0 & 0 & 0 & 0 & 0 & 0 & 0 \\
0 & 0 & 0 & 0 & 0 & 0 & 0 & 0 & 1 & 0 & 0 & 0 & 0 & 0 & 0 \\
0 & f_z & -f_y & N_{11} & N_{12} & N_{13} & 0 & 2\omega_e & 0 & 0 & 0 & 0 & R_{11} & R_{12} & R_{13} \\
-f_z & 0 & f_x & N_{21} & N_{22} & N_{23} & -2\omega_e & 0 & 0 & 0 & 0 & 0 & R_{21} & R_{22} & R_{23} \\
f_y & -f_x & 0 & N_{31} & N_{32} & N_{33} & 0 & 0 & 0 & 0 & 0 & 0 & R_{31} & R_{32} & R_{33} \\
0 & 0 & 0 & 0 & 0 & 0 & 0 & 0 & 0 & -\zeta & 0 & 0 & 0 & 0 & 0 \\
0 & 0 & 0 & 0 & 0 & 0 & 0 & 0 & 0 & 0 & -\zeta & 0 & 0 & 0 & 0 \\
0 & 0 & 0 & 0 & 0 & 0 & 0 & 0 & 0 & 0 & 0 & -\zeta & 0 & 0 & 0 \\
0 & 0 & 0 & 0 & 0 & 0 & 0 & 0 & 0 & 0 & 0 & 0 & -\beta & 0 & 0 \\
0 & 0 & 0 & 0 & 0 & 0 & 0 & 0 & 0 & 0 & 0 & 0 & 0 & -\beta & 0 \\
0 & 0 & 0 & 0 & 0 & 0 & 0 & 0 & 0 & 0 & 0 & 0 & 0 & 0 & -\beta
\end{pmatrix}
$$

where

ω_e is the earth rotation rate,

f_x, f_y, f_z are components of the specific force in the earth-fixed frame,

R_{ij} are elements of the transformation matrix $\mathbf{R}_b^e$,

N_{ij} are elements of the coefficient matrix $\mathbf{N}_G$ for the normal gravity error.

ζ is 1/(correlation time) of the first-order Gauss-Markov process describing the gyro drifts,

β is 1/(correlation time) of the first-order Gauss-Markov process describing the accelerometer biases.

GERMAN-CANADIAN EXPERIMENT IN AIRBORNE INS-GPS INTEGRATION FOR PHOTOGRAMMETRIC APPLICATIONS

G. Baustert[1]
E. Cannon[2]
E. Dorrer[3]
G. Hein[1]
H. Krauss[4]
H. Landau[1]
K.P. Schwarz[2]
Ch. Schwiertz[3]

1. Institute of Astronomical and Physical Geodesy (IAPG), University FAF Munich
2. Department of Surveying Engineering, The University of Calgary
3. Institute of Photogrammetry and Cartography, University FAF Munich
4. Rheinische Braunkohlenwerke AG, Cologne, Germany

INTRODUCTION

This paper briefly reports preliminary results from a joint project between the University FAF Munich, The University of Calgary and the (photogrammetric department of the) company Rheinbraun, Cologne, Germany, on photogrammetric aerial camera positioning using the Global Positioning System (GPS) in combination with an inertial navigation system (INS).

In detail, the objectives are:

- Verify that 'photogrammetry without ground control' can be used for monitoring deformations in an open pit mining area
- Compare camera positions from kinematic GPS to positions computed from photogrammetry
- Evaluate the contribution of a ring laser gyro INS to interpolation between GPS updates
- Determine the capability of INS for precise relative attitude estimation

TESTS

The company Rheinbraun operates an open pit coal mining area near Cologne (Hambach) which has to be controlled regularly, say once per month, with respect to horizontal and vertical deformations. For this work centimetre accuracy is needed. This requires strong efforts in establishing the necessary ground control and appropriate analysis of the photogrammetric aerotriangulation. The determination of the camera position in flight during the time of the exposure of the photos would substantially save time and money in minimizing the ground control.

In August/September 1988 several flights were carried out using the CESSNA aircraft of Rheinbraun (Fig. 1) equipped with Texas Instruments TI 4100 GPS receivers, TRIMBLE 4000SX GPS receivers, the photogrammetric camera ZEISS RMK 15a/23 with electronic shutter impulse, and the strapdown ring laser gyro inertial navigation system LITTON LTN 90-100. The instruments were time-synchronized via the GPS-impulse, an external oscillator and a PC. The flying altitude was approx. 800 m above ground and photo scale of the order 1:5400. Fig. 2 shows one example of flight tracks.

MODELS

GPS. For the kinematic analysis of the GPS (raw undifferenced carrier phases) the software TOPAS was used developed at IAPG by LANDAU (1988), see also HEIN et al (1989), HEIN (1989). It uses sequential filtering techniques and automatic cycle slip analysis by receiver clock modelling (BAUSTERT et al, 1989) and use of ionospheric residuals (so far dual-frequency GPS data were available). Independent solutions for L1 and L2 were determined. The camera positions are computed by linear interpolation between the GPS updates (1 Hz for TI 4100, 0.25 Hz for TRIMBLE 4000SX).

INS/GPS. The hybridization is realized in a Kalman filter containing 15 states where the computed GPS positions and velocities are used as updates to the INS filter. Future integration plans include cycle slip detection and correction of the GPS phase data as outlined in Fig. 3. For more information on INS modelling see WONG, SCHWARZ (1988), and for a discussion of the integration techniques see WONG et al (1988) and WEI, SCHWARZ (1988).

Fig. 1: Photogrammetric Aircraft CESSNA

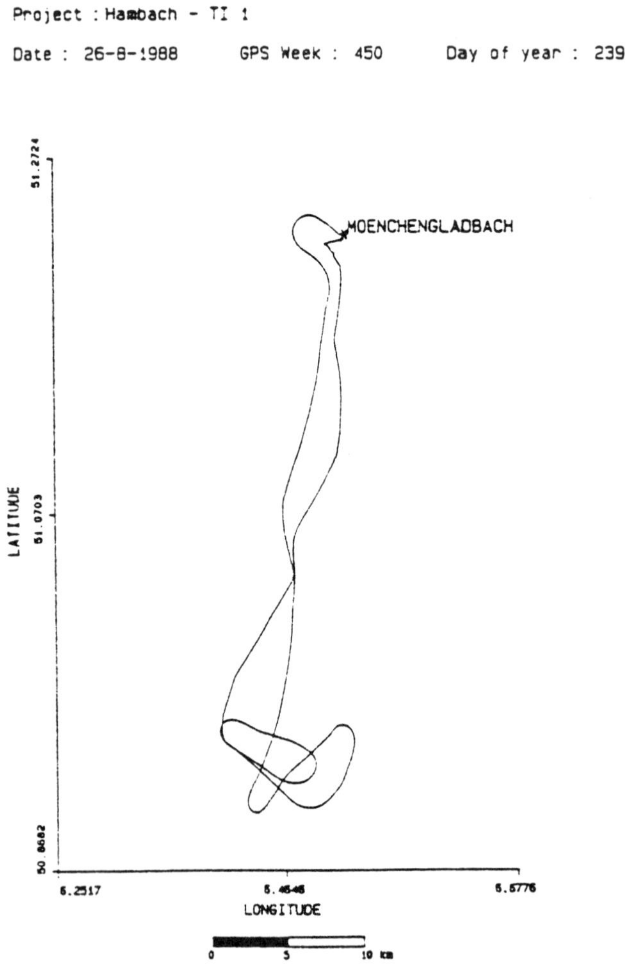

Fig. 2: Path of Photogrammetric Aircraft

PHOTOGRAMMETRY. For comparison the photogrammetric determination of the 3d-camera positions (using sufficient ground control) was carried out by an extended bundle block adjustment modified for variable camera-GPS antenna offset and the consideration of different reference systems defined by control and reference points.

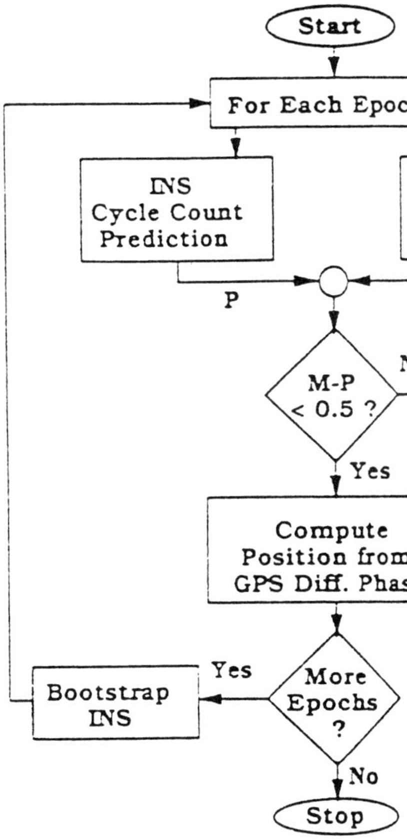

Fig. 3: GPS–INS Integration

STATISTICAL SUMMARY OF RESULTS

The analysis has shown the following results:

- Mean RMS differences between GPS-derived and photogrammetric camera positions (after a Helmert transformation was applied):

x	=	3.6 cm
y	=	6.0 cm
z	=	3.8 cm

- Mean RMS errors derived from bundle block adjustment:

x	=	3.8 cm
y	=	3.6 cm
z	=	7.9 cm

- Mean RMS errors derived from kinematic GPS positioning algorithm:

L1 (mean rms)	x	=	3.5 cm	L2 (mean rms)	x	=	4.5 cm
	y	=	6.1 cm		y	=	7.8 cm
	z	=	3.6 cm		z	=	4.7 cm

L1-L2 (mean difference)	x	=	3.9 cm
	y	=	6.2 cm
	z	=	4.7 cm

CONCLUSION

Discussing the results above, leads to the following conclusions:

- Differences between GPS-derived (TI 4100) and photogrammetric positions are at the level of 5 cm RMS per coordinate after a Helmert transformation is applied

- This corresponds to the accuracy of the photogrammetric determination

- Linear interpolation of camera position between GPS updates is sufficient as long as the observation interval is $\leq$ 1 second and the velocity is nearly constant

- INS improves cycle slip detection and interpolation for one frequency receiver (Trimble 4000SX)

Future plans for a further analysis of the data will be concerned with

- Attitude determination from INS and comparison with corresponding values from the photogrammetric bundle

- Combination of bundle block adjustment with kinematic GPS positions.

A more detailed discussion will be found in a future issue of manuscripta geodaetica.

REFERENCES

Baustert, G., G.W. Hein, H. Landau (1989) *On the Use of GPS in Airborne Photogrammetry*, Hydrographic Applications. and Kinematic Surveying. Proc. 5th International Geodetic Symposium on Satellite Positioning, Las Cruces, March 13-17, 1989, 1029-1040

Hein, G.W. (1989) *Precise Kinematic GPS/INS Positioning: A Discussion on* the Applications in Aerophotogrammetry. Proc. 42nd Photogrammetric Week, Sept. 11-16, 1989, 261-282, Stuttgart

Hein, G.W., G. Baustert, B. Eissfeller, H. Landau (1989) *High-Precision Kinematic GPS Differential Positioning and Integration of GPS with a Ring Laser Strapdown Inertial System*. J. Navigation, in print

Landau, H. (1988) *Zur Nutzung des Global Positioning Systems in Geodäsie und Geodynamik: Modellbildung, Software-Entwicklung und Analyse.* Schriftenreihe Studiengang Vermessungswesen, Universität der Bundeswehr München, Heft 36

Wei, M., K.P. Schwarz (1989): *A Discussion of Models for GPS/INS Integration.* Presented at the General Meeting of the International Association of Geodesy, Edinburgh, Scotland, 3–12 Aug., 1989

Wong, R.V.C., K.P. Schwarz (1988) *Development and Field Testing of a RLG Strapdown Inertial Survey System*. Proc. IEEE PLANS 88, 480-487

Wong, R.V.C., K.P. Schwarz, M.E. Cannon (1988) *High Accuracy Kinematic Positioning by GPS–INS*. J. Navigation, Vol. 35, No. 2, 275–297

THE USE OF PHASE FOR KINEMATIC POSITIONING BY GPS

J M Westrop, M E Napier and V Ashkenazi
Institute of Engineering Surveying and Space Geodesy,
University of Nottingham

INTRODUCTION

Positioning on the move is a task that is often required, eg for offshore positioning, airborne photogrammetry and land based surveys. GPS was designed primarily for military mobile positioning needs, and is consequently ideally suited for the types of application listed above.

Positioning whilst moving can be achieved in a variety of ways using GPS. These methods use pseudo range and/or carrier phase observations. Techniques for positioning using solely pseudo ranges have been sufficiently well-developed for commercial application. These include single point pseudo range positioning with an accuracy of 15-30 m and differential pseudo range positioning, accuracy 5-10 m.

The use of phase data for positioning can undoubtedly improve these accuracies. However, techniques employing the phase observations are still being investigated. Why is this? The major drawback for utilising phase in a moving environment is the occurrence and frequency of cycle slips which corrupt the measurements. Another limiting factor is that access to on-line differential phase data is not yet available.

This paper firstly examines the techniques for kinematic positioning where phase measurements are included. This is followed by an examination of cycle slips.

PHASE SMOOTHED PSEUDO RANGES

Phase smoothed pseudo ranges are observables that have been derived from both code and carrier measurements. They have a lower standard deviation than their observed counterparts, and offer enhanced positioning accuracies when used for single point positioning or pseudo range differencing (Lachapelle et al 1986).

A phase smoothed pseudo range PR_{smooth} at the current epoch (i) is computed from the pseudo range at some arbitrary initial epoch (o) and the change in range, implied by the phase measurements, between the two epochs:-

$$PR_{smooth}(i) = PR(o) + \lambda(\emptyset(i) - \emptyset(o)) \qquad (1)$$

where

PR is the pseudo range

$\emptyset$ is the phase measurement

λ is the GPS signal wavelength

The initial pseudo range PR (o) is computed by averaging a number of observed pseudo ranges. These are either directly observed at the initial point or implied by equation (2), from pseudo ranges taken on the move. Equation (2) is a rearrangement of equation (1) where the initial pseudo range is now the subject.

$$PR\,(o) = PR_{observed}\,(i) - \lambda\,(\emptyset\,(i) - \emptyset\,(o)) \tag{2}$$

where

$PR_{observed}\,(i)$ is the observed pseudo range at the current epoch.

The exact number of measurements used is left to the discretion of the user. However, systematic errors caused by mismodelling the changing atmospheric effects can be introduced when using a large number of observations. The initial pseudo range can also be computed from the known coordinates of an initial ground station.

The change in range, between initial and current epochs, monitored by the phase measurements, is not correct if a cycle slip occurs. If the cycle slip cannot be repaired, then a new initial pseudo range is computed, and the process is re-started at the current epoch. There is a downgrading of the positioning accuracy when uncorrected cycle slips occur. According to Lachapelle et al (1986), the phase smoothed positioning accuracy using differential pseudo range GPS is:-

< 50 cm (no slips)

$\sim$ 2 m (with slips)

for a static application and 5 hours of observations.

COMBINATION OF PHASE & PSEUDO RANGE IN A KALMAN FILTER

Independent least squares observation equations which relate the unknown receiver coordinates to GPS pseudo range or carrier phase measurements can be solved simultaneously in one least squares system. For kinematic applications where the unknown coordinates change with time, an efficient solution is to use an optimising filter, such as a Kalman filter.

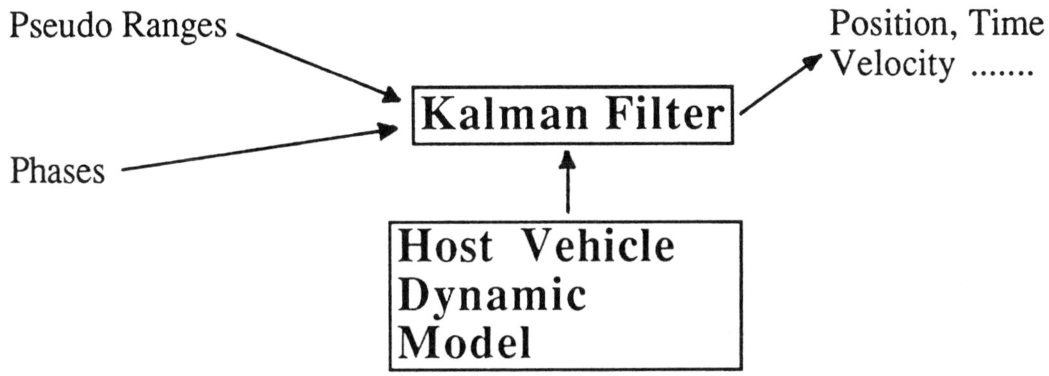

Fig 1
Input and Output for Kalman Filter

There are two models in a Kalman filter:-

1 Primary or Measurement Model

This relates the observations to the unknowns eg a pseudo range observation to the unknown receiver coordinates

$$PR = R + c(dt_R - dt_S) + ion + trop \tag{3}$$

where $$R = [(X_R - X_S)^2 + (Y_R - Y_S)^2 + (Z_R - Z_S)^2]^{1/2} \tag{4}$$

and
- PR is the pseudo range measurement
- R is the geometrical satellite - receiver range
- X_R, Y_R, Z_R are the unknown receiver coordinates
- X_S, Y_S, Z_S are the satellite coordinates
- dt_R, dt_S are the receiver and satellite clock offsets
- ion,trop are the atmospheric effects on the observations

2 Secondary or Dynamic Model

This describes how the unknowns vary with time, eg linear velocity of the host vehicle

$$X_R\,(i + 1) = X_R\,(i) + \dot{X}_R\,(i)\ .\ dt \tag{5}$$

where
- $X_R\,(i + 1)$ is the predicted X coordinate at epoch i + 1
- $X_R\,(i)$ is the known X coordinate at the current epoch
- $\dot{X}_R\,(i)$ is the known X velocity
- dt is the time interval between current and predicted epochs

The nature of the Kalman filter equations means that unknowns can be estimated even though there are no direct measurements related to them. For example, consider a system where there are only pseudo range measurements. Estimates for velocity and acceleration can be output as well as for position.

The combination of GPS measurements in a Kalman filter offers the most efficient use of the observations for kinematic positioning because there is no absolute reliance on either type of measurement. Input from other sensors eg INS and radio location systems can also be incorporated.

GPS observations can be 'differenced' in a number of ways: by station; by satellite; by time. Each of these differenced observables can be incorporated into the Kalman filter measurement model. Various research groups have tried different combinations and, generally, results indicate that kinematic positioning to sub-metre accuracies is possible. However, these results are very dependent upon the relative weights of the measurement and dynamic models within the Kalman filter. This should be dictated by the frequency of measurement data and the nature of the host vehicle motion, ie can the dynamic model adequately describe the movement, or are the measurements better?

There are cases when it may be desirable to alter the relative Kalman filter model weights during the processing, in order to more closely model the actual movement. This may be required when there are distinct changes in the nature of the movement of the mobile receiver. Such a case has occurred at Nottingham, where data collected for semi-kinematic trials is being processed by kinematic positioning techniques. During these trials, one TRIMBLE SLD receiver was the monitor station whilst the other was moved manually to various other stations in the immediate vicinity. The mobile receiver remained at each station for about 5 minutes. Consequently, the trajectory of this mobile receiver combines long stationary periods and short low-dynamic movements.

Kinematic positioning tests with this data at Nottingham use pseudo range and phase time differenced observables. The Kalman filter dynamics model is a simple constant velocity one, as shown in equation (5). The relative model weights are altered during the processing to control the influence of the dynamics model. The decision to change the weights is based upon the failure of statistical tests performed at each epoch within the Kalman filter. Teunissen & Salzmann (1988) give a detailed explanation of the statistical test procedures. Initial results indicate that the Kalman filter output, ie position and velocity, varies significantly for different relative model weights, and that the dynamics model is not well suited for this application. It is worth remembering that the internal statistical tests can fail for a number of reasons, eg a gross observational error, an unsuitable dynamics model or the incorrect relative model weights. The operator must decide which is the most likely cause or causes, and implement action to remedy it.

PHASE ONLY SOLUTIONS

The determination of position using carrier phase observations can be achieved by an independent least squares solution at every epoch. Raw, single difference, double difference or triple difference phase observables can be used. The observation equations are the same as those used in static applications, eg the double difference observable, $\varnothing_{DD}$:-

$$\varnothing_{DD} = \frac{1}{\lambda} R_{AB}^{ij} - N_{AB}^{ij} \qquad (6)$$

where $R_{AB}^{ij} = R_A^i - R_B^i - R_A^j + R_B^j$

and $N_{AB}^{ij} = N_A^i - N_B^i - N_A^j + N_B^j$

where R is the geometrical satellite to receiver range
N is the unknown number of cycles at lock-on
subscripts represent the stations
superscripts represent the satellites

Any satellite terms, eg satellite coordinates, may be computed from the broadcast ephemeris. Each observable, except the triple difference, requires the initial integer ambiguities (N's) to be computed prior to the determination of kinematic positions. This may involve starting the survey from a known point, revisiting the point many times or performing an 'antenna swap'. More detail about these techniques may be found in Hofmann-Wellenhof and Remondi (1988). There are four remaining unknowns, the receiver coordinates and its clock offset from GPS time. Consequently, at each epoch there must be at least four phase measurements.

If cycle slips do not occur, then phase only kinematic positioning can give sub-decimeter accuracies. Although the presence of cycle slips in carefully controlled low-dynamic experiments is rare, the likelihood of cycle slips in high dynamic situations and commercial applications is very great. In these instances, it is necessary to detect and correct for cycle slips. The subject of cycle slips will be examined in the following section.

CYCLE SLIPS

The phase observable is a measurement of the beat frequency phase, ie the difference between the satellite oscillator phase and the receiver oscillator phase. This can be considered as two parts. The first is the fractional part and the second is the integer component. This integer component is the number of complete beat frequency cycles that have been observed since the first measurement. Cycle slips occur when the receiver tracking loops fail to continuously monitor the number of cycles. During these periods, any number of cycles pass through the system undetected.

There are three potential causes of cycle slips:-

a) obstruction of satellite signal

b) irregular receiver oscillator output

c) irregular satellite oscillator output

Cycle slips usually occur for the first reason. The nature of the obstruction may be physical (eg buildings and trees), atmospheric (ie severe signal refraction caused by high ionospheric activity) or electronic (ie the signal tracking bandwidth is too narrow). The tracking bandwidth covers the range of most likely frequencies needed for the acquisition of the satellite signal. This range is based on the expected motion of the satellite and GPS antenna. Sudden antenna movements cannot be predicted and may result in a loss of lock on the satellite signal. Even receivers whose bandwidth can automatically widen experience cycle slips as a result of sudden movements.

Reliable cycle slip detection is very important because the high accuracy of the phase measurements depends upon them being cycle slip free. Detection when moving is difficult to achieve because of the problem of distinguishing between the effects of a cycle slip and real movement of the antenna. Once detected, the user can decide whether to correct for the cycle slips or to ignore the affected data. Several approaches for cycle slip detection and correction are described in Westrop et al (1989).

CONCLUSIONS

It is likely that the problem of cycle slips during Kinematic GPS surveys will be resolved by improved receiver design with the capability for cycle slip detection and correction, and the implementation of Block II satellites which will provide a better constellation geometry. In which case, the use of GPS phase techniques for kinematic applications will become widespread. The current status of these techniques is given in Table 1.

Table 1 Comparison of Kinematic Techniques

Technique	Accuracy	Advantages	Disadvantage
Phase smoothed pseudo ranges	1-2 m	• easy to implement in existing systems	• highly dependent on accuracy of initial pseudo range
Combination by Kalman filter	~1 m	• efficient use of all data • allows other information to be incorporated • considers trajectory positions	• sensitive to Kalman filter model weights
Phase only solutions	< 0.1 m	• highest achievable accuracy	• may require prior knowledge of parameters eg initial integer ambiguity • need a minimum of 4 measurements per epoch

It is most probable that the second method, Kalman filtering, will become favoured for high dynamic situations. It has flexibility with regard to the type and amount of input data. It is still able to function during periods when no measurement data is available, and also allows input from sensors other than GPS, eg INS and radio location systems. This aspect will become increasingly important as kinematic positioning systems tend towards 'amalgamated systems' where all available data is input to one filter.

Acknowledgements

Research on Kinematic GPS positioning at Nottingham is sponsored by the UK Science and Engineering Research Council, British Petroleum and Polytechnic Electronics plc. Acknowledgements are also due to Dr A Pratt of Polytechnic Electronics plc for his help and advice.

REFERENCES

Hofmann-Wellenhof, B. and Remondi, B. (1988)., *'The antenna Exchange: One Aspect of High-Precision GPS Kinematic Surveying'*. Paper presented at the Int GPS Workshop, Darmstadt.

Lachapelle, G., Hagglund, J., Falkenberg, W., Bellemare, P., Casey, M. and Eaton, M. (1986). *'GPS Land Kinematic Positioning Experiments'. 4th Int Geodetic Symp on Satellite Positioning, Austin, pp 1327-1344.*

Teunissen, P. and Salzmann, M. (1988). *'Performance Analysis of Kalman Filters' Delft University of Technology, Faculty of Geodesy Report No 88.2.*

Westrop, J., Napier, M. and Ashkenazi, V. (1989). *'Cycle slips on the Move: Detection and Elimination' ION GPS 89, 2nd Technical Meeting, Colorado Springs.*

HIGH PRECISION KINEMATIC POSITIONING USING GPS AT THE IGN: RECENT RESULTS

P. Willis, C. Boucher
Institut Géographique National

INTRODUCTION

Since 1985, the Institut Géographique National has acquired GPS receivers and has also developped its own GPS software, called GDVS, for geodetic static positioning. This software has been upgraded and is now capable of processing phase observables in a kinematic mode.

After some basic simulations of precise GPS phase kinematic positioning, two field tests were performed: one in Saint-Mandé (baselines of 100 m) and one near Châlons (baselines of 5 km), showing results of 1 to 3 cm precision. In the second test, a Doppler Integrated solution was also obtained using SERCEL software, showing results of 30 cm.

SIMULATIONS OF PSEUDO-KINEMATIC POSITIONING

In order to test the precision of kinematic GPS, simulations were performed at IGN. Actual data, recorded during a test of performance of static GPS conducted by SERCEL in 1988, were used. A period of 30 minutes, while five GPS satellites were simultaneously tracked, was selected for a baseline of about 2.7 km. We took into account only the phase measurements of the first epoch and those of a different epoch separated from the first epoch by a gap of 0 to 30 minutes.

A first calculation was performed by estimating simultaneously the positions and the ambiguities. Results displayed in table 1 show that the period ellapsed between the first and the last epoch is a key factor of the precision of the positioning results.

In a second run, we fixed the ambiguities to their integer values (knowned from the GPS static phase processing). We can see in table 2 that the results are less affected by the duration of the gap between the first and the last epoch. We tried also to determine if it was possible to fix the ambiguities to their integer values using only the phase data of these two epochs. Results showed that at least 15 minutes are requiered between these two epochs.

This means that the precision of the kinematic positioning will be greatly dependant with the number of reobservations of the points to be located rather than with the duration of the stay on each point (one epoch is enough!). On the other hand, at least 15 minutes are needed between the points reobservations in order to be able to fix the ambiguities in a pseudo-kinematic mode.

TEST OF SAINT MANDE

Relying on the conclusions of these simulations, a field test was conducted on the site of IGN at Saint Mandé (Fig 1). Five points, distant of about 100 m, were located during this two-days test, in may 1988, using two SERCEL NR52 receivers. The purpose of this test was to confirm the results of the simulation studies by accumulating a large number of independant determinations of each point in order to obtain a reliable statistical data set (75 baselines were observed).

First of all, a combined adjustment of all the kinematic data (taking into account only one epoch by point) was performed using GDVS software in a classical double difference mode. Results displayed in table 3 show that the agreement with the coordinates obtained from a combined terrestrial adjustment is coherent at better than 1 cm.

Comparaisons between the epoch by epoch solutions and the terrestrial solution (table 4) show that the differences are at the centimeter level. To synthetize these results a global histogramm was obtained. This histogramm, presented in figure 2, shows that there is 82% chance that the difference between the kinematic GPS coordinate and the combined terrestrial coordinate is smaller than 2 cm.

TEST OF CHALONS

In order to complete these conclusions for longer baselines (of the order of 5 km, see figure 3), a second field test was conducted near Châlons, in december, 1988. For this set of data, two different solutions were computed.

Precise kinematic solution using GDVS software

The global adjustment of all the epoch solutions obtained, in a double difference mode, using GDVS software, show aggreement at few centimeters level (table 5).

The epoch by epoch solutions (table 6) present systematic differences of few centimeters with the terrestrial solution (standard NTF coordinates transformed in WGS84). A further investigation is being conducted in order to determine what part of this difference comes from the uncertainties of the legal NTF coordinates.

Doppler Integrated solution using SERCEL software

On the other hand, an independant calculation was realized, in a Doppler Integrated mode, using a dedicated SERCEL software. A combined adjustment was computed and is presented in table 7. The differences between these coordinates and the terrestrial coordinates is at fe centimeters level.

The epoch by epoch solutions (table 8) present differences at the order of 30 cm (in average for all the points).

REFERENCES

Brossier R., Million C., Reynes A. (1987), Photogrammetric applications of the SERCEL GPS TR5S-B receiver at the Institut Géographique National, ISPRS.

Mader G. (1986), Dynamic positioning using GPS carrier phase measurements, Manuscripta Geodetica, vol 11/4, pp 272-277.

Remondi B. W. (1986), Performing centimeter accuracy in seconds with GPS carrier phase: initial results, Proceedinds of the 4th international geodetic symposium on satellite positioning, Austin (USA).

Remondi B. W. (1988), Kinematic and pseudo-kinematic GPS, Institute of navigation, GPS 88 technical meeting, Colorado Springs (USA).

Willis P. (1988), 2ème compte rendu d'expérimentation du GPS cinématique: tests GPS entre la tour et le toit de la logistique à l'IGN, IGN/SGN, CR/G n° 10.

Willis P. (1989), Utilisation du GPS en mode cinématique de précision, IGN/SGN, CC/G n° 539, published in "revue Géomètre", 132, n° 3.

Willis P., Carnino M., Barasz O. (1989), Compte rendu de mission GPS à Châlons, IGN/SGN, to be published.

length of the "gap"	RMS n_{obs}	DX σ_X	DY σ_Y	DZ σ_Z
30 mn	0.002 8	-0.042 0.023	0.016 0.025	-0.032 0.007
25 mn	0.005 8	-0.177 0.059	0.183 0.063	0.014 0.020
20 mn	0.001 8	-0.113 0.006	0.133 0.007	-0.027 0.002
15 mn	0.004 8	0.066 0.076	-0.055 0.080	-0.062 0.027
10 mn	0.004 8	-0.004 0.109	-0.055 0.113	-0.077 0.040
5 mn	0.004 8	-0.414 0.263	0.359 0.268	-0.073 0.101
1 mn	0.004 8	0.067 1.406	0.791 1.412	0.359 0.553
0 mn	0.002 8	0.436 2.802	2.475 2.810	0.120 1.106

Table 1
Test SERCEL (January-February, 1988)
Baseline: SERCEL - MATRA (2.7 km)
Simulation of the existence of lacking data
estimated ambiguities

all units are in meters

length of the "gap"	RMS n_{obs}	DX σ_X	DY σ_Y	DZ σ_Z	n_{iter}
30 mn	0.008 8	0.001 0.009	0.007 0.007	-0.008 0.009	1
25 mn	0.010 8	0.003 0.012	-0.002 0.009	-0.002 0.011	-
20 mn	0.009 8	0.015 0.012	0.009 0.009	0.005 0.011	3
15 mn	0.005 8	0.011 0.007	0.009 0.005	-0.005 0.006	3
10 mn	0.005 8	0.004 0.007	0.006 0.005	-0.008 0.006	-
5 mn	0.007 8	0.010 0.011	0.005 0.007	0.000 0.008	-
1 mn	0.004 8	0.012 0.007	0.002 0.004	-0.006 0.005	-
0 mn	0.005 8	0.017 0.007	0.004 0.004	-0.002 0.005	-

Table 2
Test SERCEL (January-Febraury, 1988)
Baseline: SERCEL - MATRA (2.7 km)
Simulation of the existence of lacking data
Ambiguities fixed to their integer value

all units are in meters

Point	DX(m)	DY(m)	DZ(m)
2	-0.005	-0.009	-0.003
3	-0.003	-0.009	-0.007
4	-0.004	-0.006	0.000
6	0.006	-0.009	-0.011
11	0.000	0.000	0.000

Table 3
Test IGN: Saint Mandé (May, 1988)
Comparaison between the combined kinematic GPS coordinates and the terrestrial coordinates

Day	Epoch	satellite	DX(m)	DY(m)	DZ(m)
24/5/88	14:06	6,8,9,11,12	-0.035	-0.004	0.046
24/5/88	14:18	6,8,9,11,12	0.036	0.026	0.009
24/5/88	14:29	6,8,9,11,12	-0.007	0.004	0.033
25/5/88	13:50	3,9,11,12,13	0.009	0.007	0.002
25/5/88	14:02	3,9,11,12,13	0.005	0.008	-0.005
25/5/88	14:12	3,9,11,12,13	0.029	0.007	-0.007
25/5/88	14:21	3,9,11,12,13	-0.008	-0.001	-0.027
25/5/88	15:57	3,9,11,12,13	0.013	0.001	0.013
25/5/88	16:08	3,9,11,12,13	0.000	-0.007	-0.002
25/5/88	16:18	3,9,11,12,13	-0.011	-0.023	-0.017
25/5/88	16:28	3,9,11,12,13	0.002	-0.011	0.014
25/5/88	16:38	3,9,11,12,13	-0.002	-0.023	0.006

Table 4
Test IGN: Saint Mandé (May, 1988)
Baseline: 2 - 11 (0.1 km)
Comparaison between the kinematic GPS coordinates obtained by epoch and the terrestrial coordinates

Point	DX(m)	DY(m)	DZ(m)
JUVIGNY II	0.000	0.000	0.000
VRAUX II	0.119	0.044	0.049
VRAUX III	0.053	0.020	0.033
ISSE I	0.020	-0.029	-0.028

Table 5
Test IGN: Châlons (December, 1988)
Comparaison between the combined kinematic GPS coordinates
obtained by epoch and the terrestrial coordinates

Day	Epoch	satellite	GDOP	DX(m)	DY(m)	DZ(m)
20/12/88	23:35	3,9,11,12,13	5.4	0.051	0.007	0.031
21/12/88	00:00	3,9,11,12,13	5.0	0.065	0.021	0.043
21/12/88	00:21	3,9,11,12,13	4.7	0.064	0.026	0.039
21/12/88	00:45	3,9,11,12,13	4.7	0.036	0.019	0.018

Table 6
Test IGN: Châlons (December, 1988)
Baseline: JUVIGNY II - VRAUX III (3.8 km)
Comparaison between the kinematic coordinates
obtained by epoch and the terrestrial coordinates

Point	DX(m)	DY(m)	DZ(m)
JUVIGNY II	0.000	0.000	0.000
VRAUX II	0.007	-0.059	-0.148
VRAUX III	0.002	0.022	-0.095
ISSE I	0.051	-0.108	0.207

Table 7
Test IGN: Châlons (December, 1988)
Comparaison between the combined Doppler Integrated GPS coordinates and the terrestrial coordinates

Day	Epoch	satellite	GDOP	DX(m)	DY(m)	DZ(m)
20/12/88	21:22-21:33	6,9,11,12	5.5	-0.138	0.016	0.156
20/12/88	22:01-22:11	6,9,11,12,13	4.5	-0.143	0.016	0.091
20/12/88	22:37-22:47	6,9,11,12,13	4.3	-0.103	0.007	-0.030
20/12/88	23:35-23:40	3,9,11,12,13	5.4	0.254	0.119	0.050
21/12/88	00:00-00:05	3,9,11,12,13	5.0	0.159	0.002	0.044
21/12/88	00:21-00:26	3,9,11,12,13	4.7	0.119	-0.307	-0.001
21/12/88	00:45-00:50	3,9,11,12,13	4.7	0.142	-0.342	0.051

Table 8
Test IGN: Châlons (December, 1988)
Baseline: JUVIGNY II - VRAUX III (3.8 km)
Comparaison between the Doppler Integrated GPS coordinates obtained by session and the terrestrial coordinates

Figure 1

TEST OF KINEMATIC GPS

SAINT MANDE , MAY , 1988

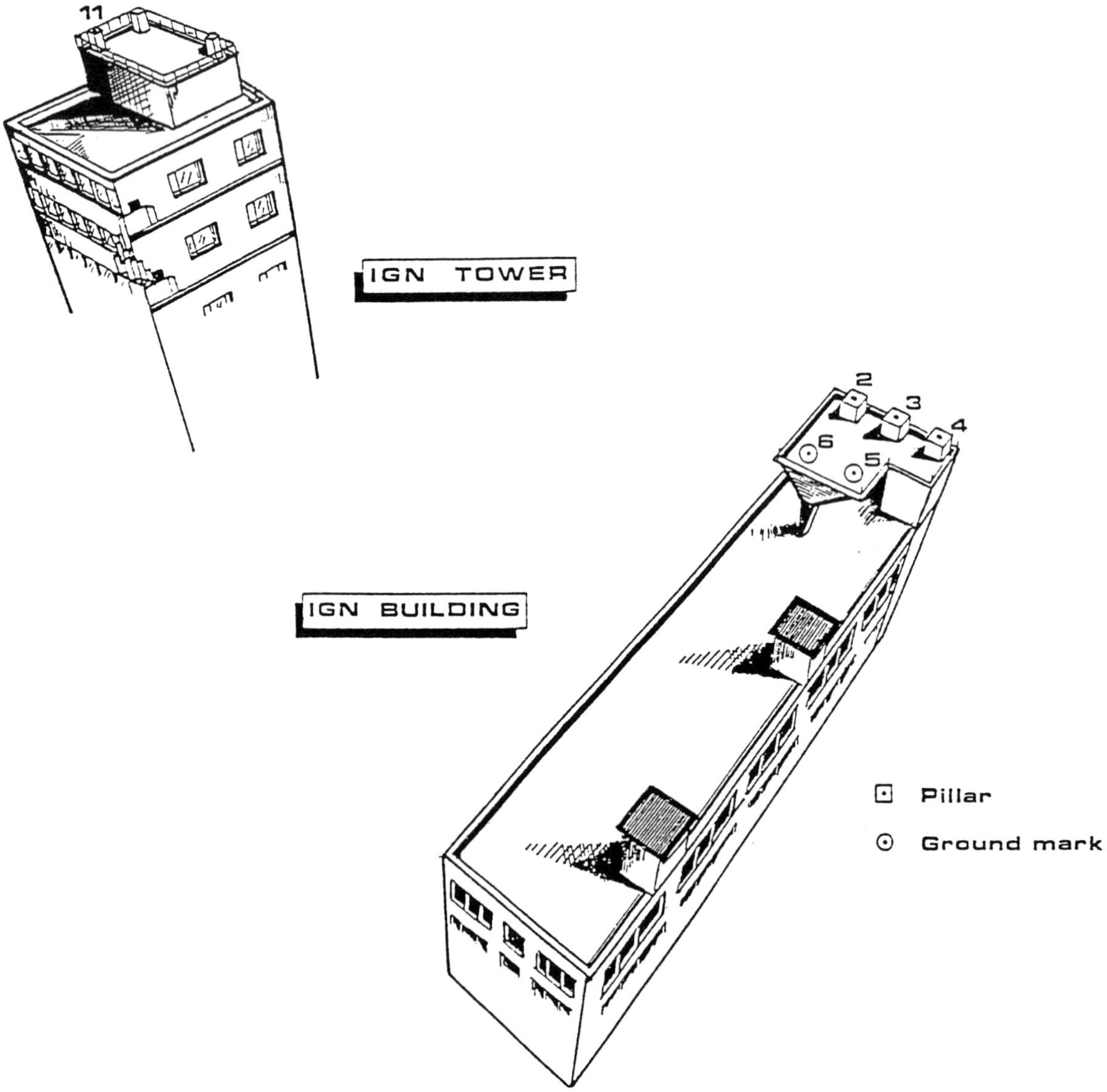

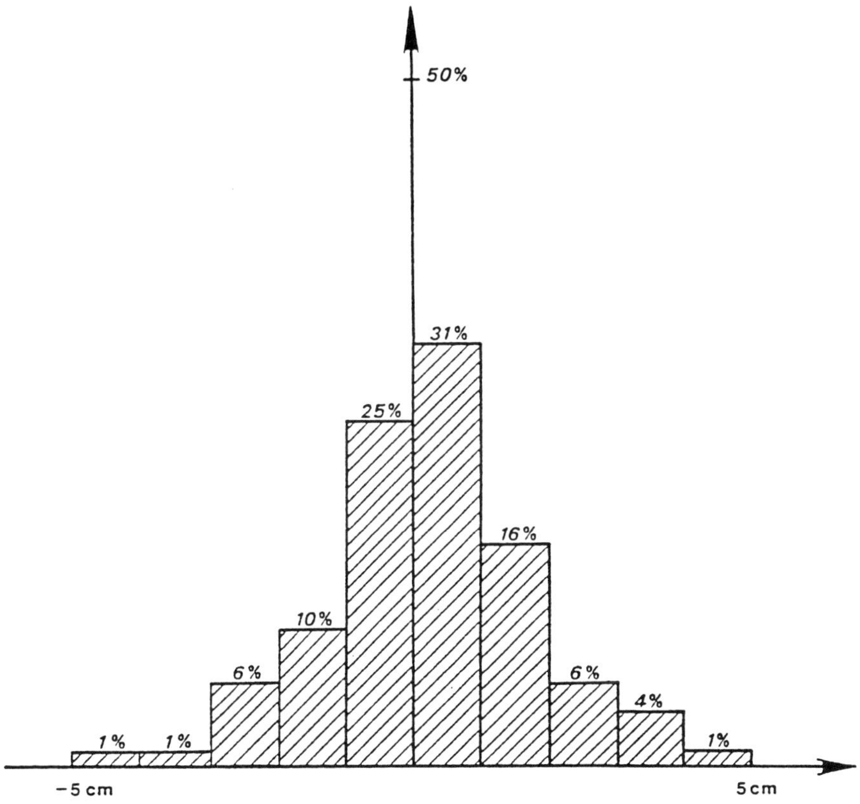

Figure 2

HISTOGRAM OF THE DIFFERENCES IN X,Y,Z BETWEEN THE POSITION BY EPOCH ESTIMATED FROM GPS AND THE TERRESTRIAL COORDINATES SAINT MANDE , 1988 , MAY

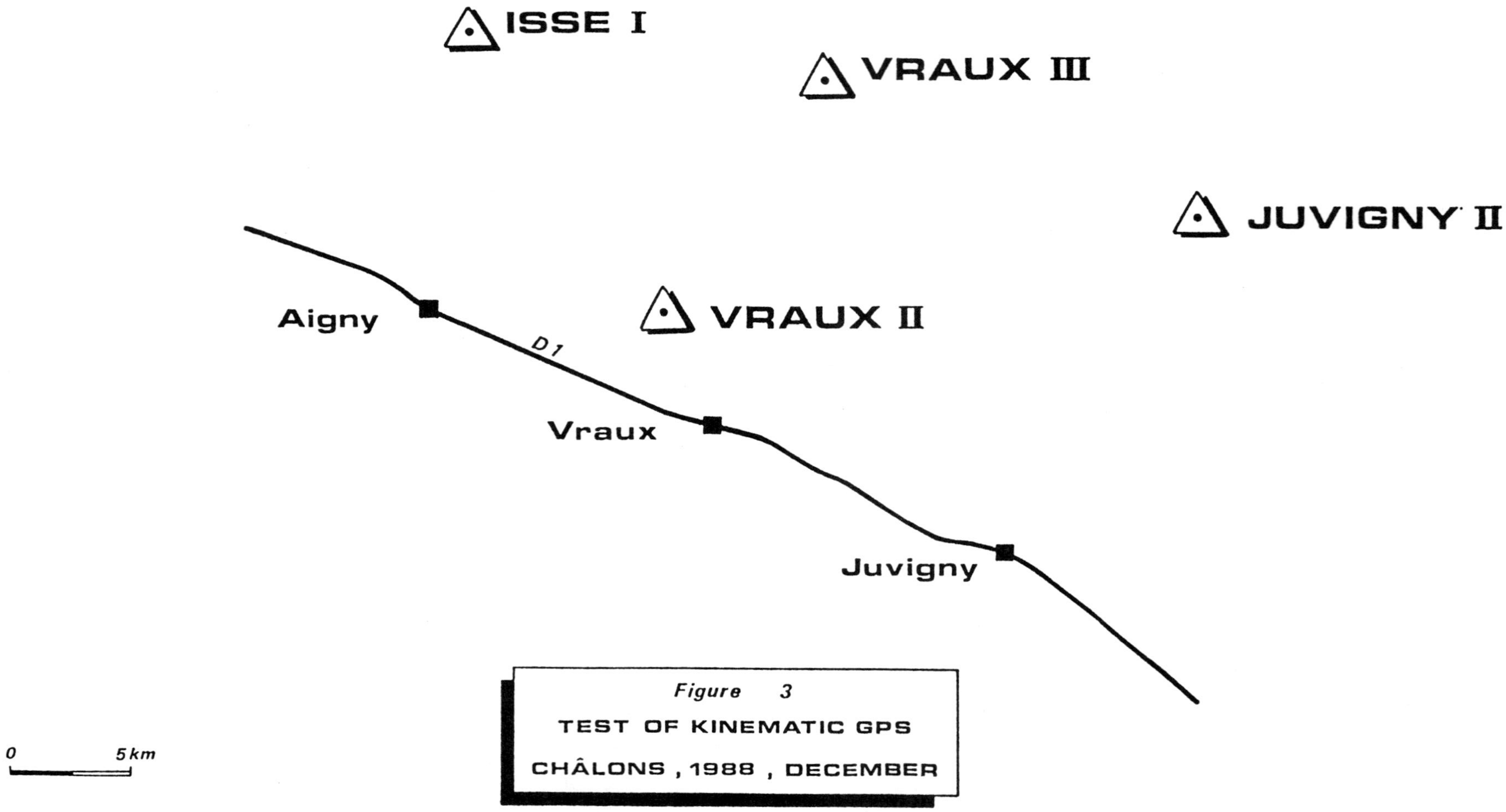

Figure 3
TEST OF KINEMATIC GPS
CHÂLONS, 1988, DECEMBER

GEOSCIENCE FROM GPS TRACKING BY EARTH SATELLITES

Thomas P. Yunck and William G. Melbourne
Jet Propulsion Laboratory, California Institute of Technology
4800 Oak Grove Drive, Pasadena, CA 91109

INTRODUCTION

As the Global Positioning System enters its operational phase, efforts to exploit GPS for earth science are coming into sharper focus. Some of these activities have taken new directions, away from the positioning and navigation tasks familiar to GPS users into areas of direct earth observation. Perhaps the most visible GPS earth science applications today are precise orbit determination of remote sensing satellites—primarily ocean altimetric satellites—and direct measurement of earth baselines for studies in solid earth dynamics. The mid-1980s saw the development of specialized GPS-based techniques to determine the orbits of any GPS-equipped low earth satellite with sub-decimeter accuracy [Yunck and Wu 1986; Yunck et al 1990]. These will have their first full-scale demonstration on TOPEX/POSEIDON, the joint US/French oceanographic satellite set for launch in 1992. Other future space missions that plan to carry GPS receivers include:

- ARISTOTELES — Planned for 1995; this joint ESA/NASA mission will use GPS for positioning and to complement the satellite's gravity gradiometer by enhanced mapping of the earth's gravitational field at longer wavelengths.
- GRAVITY PROBE B — Proposed for the late 1990's; this drag-compensated NASA mission will use GPS for precise positioning to support the its general relativity experiment, and to further improve mapping of the gravitational field.
- EOS-A & EOS-B — Planned for 1997 & 1999; these NASA platforms of the Earth Observing System will use GPS for positioning, high temporal resolution geodesy, atmospheric temperature profiling, and studies in ionospheric physics.
- EUROPEAN POLAR PLATFORM — Planned for the late 1990's; this European EOS platform will use GPS primarily for precise positioning, though the data will also be useful in global geodesy.
- SPACE STATION *FREEDOM* — Planned for the late 1990's; this international (predominantly NASA) platform will use GPS for precise positioning and for the same science objectives as EOS-A and EOS-B.
- COLUMBUS FREE FLYING LABORATORY — Planned for late 1990's; this European science laboratory will fly separately from but in conjunction with NASA's Space Station and carry its own GPS positioning capability.
- SPACE SHUTTLE — This NASA workhorse is expected to adopt GPS sometime in the next decade for navigation and to support precise Earth observation experiments.

- OTHER — Other proposed GPS-equipped platforms include NASA's Orbit Maneuvering Vehicle and Orbit Transfer Vehicle, the Navy's SPINSAT, and, from Europe, the Hermes Space Shuttle, and the Space Courier.

On the ground we have recently seen techniques for GPS-based measurement of earth baselines refined to deliver accuracies of a part in 10^8 over distances up to 3000 km [Dong and Bock 1989; Blewitt 1989; Lichten and Bertiger 1989]. That translates into baseline accuracies of 1-3 cm for the longer baselines and better than 1 cm for regional baselines of under 1000 km. This allows us to observe crustal plate motion and deformation on a time scale of a year or less and is comparable to the accuracies now achieved by the more cumbersome and costly techniques of very long baseline interferometry (VLBI) and satellite laser ranging (SLR).

Even this performance, however, will not satisfy the coming needs of geodynamics and ocean science. At a 1988 workshop on The Interdisciplinary Role of Space Geodesy, held in Erice, Italy, panels of scientists and engineers set down long term objectives for a number of areas of earth science and related technology [Mueller and Zerbini 1989]. Two of the Erice recommendations will be important drivers of precise positioning and navigation technology for years to come. The panel on ocean physics established an accuracy goal of 1 cm for continuous geocentric altitude knowledge of orbiting ocean altimeters, such as the one that will fly on EOS; the panel on solid earth physics set a goal of 1 mm accuracy for regional land-based geodesy. At the 1989 NASA-sponsored Coolfont conference on Solid Earth Sciences [Bosworth 1990], the 1 mm accuracy goal for regional geodesy was reaffirmed, and a goal of 1 cm accuracy for global geodesy, averaging down to 1 mm at 3-month intervals, was also proposed. These goals pose a formidable challenge across a range of technologies. We believe, however, that they can be achieved with GPS in the next decade.

In 1987 an activity was begun at the Jet Propulsion Laboratory (JPL) that can help to achieve these goals by exploiting GPS receivers to be placed on several orbiting science platforms, including the two NASA EOS vehicles and the Space Station, in the late 1990s. The objectives of this effort have now been expanded to include science investigations in geodynamics, climatic change, and ionospheric science, and the team has been joined by other groups at JPL, the Massachussets Institute of Technology, and the University of Illinois. Opportunities for atmospheric and ionospheric science will be created by tracking GPS satellites from orbiting platforms as the signals are occulted by the intervening media.

The full investigation will encompass precise orbit determination, with a goal of 1-3 cm accuracy for each platform; studies in solid earth dynamics, exploiting the enhanced GPS geodesy made possible with orbiting receivers, with a focus on the redistribution of strain following major earthquakes; examination of long term trends in global climate, charting the changing thermal structure of the atmosphere and searching for definitive signs of a greenhouse warming; studies of the origin and propagation of acoustic gravity waves in the atmosphere and ionosphere and their role in the flow of energy throughout the earth system; and 3-dimensional mapping of localized ionospheric structures, such as equatorial bubbles and the mid-latitude trough, by computerized tomography.

GENERAL BACKGROUND

Baseline Measurement with GPS

Precise observation of GPS signal delay and/or delay rate between receiving points on earth enables precise measurement of the baseline vectors between those points. Accurate knowledge of the GPS satellite orbits is essential for an accurate baseline solution. In the early 1980's what has come to be known as the "fiducial concept" was developed to address GPS satellite orbit error in geodetic measurements [Thornton, Fanselow and Renzetti 1986]. A regional network of reference receivers (fiducial sites) observes the GPS satellites along with

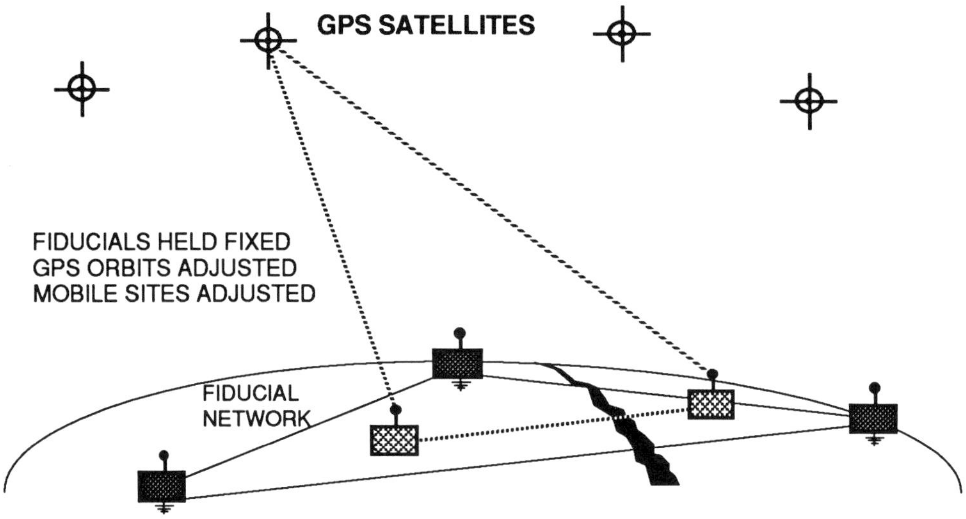

Fig. 1. The fiducial approach to GPS-based precise geodesy.

receivers at other points of geodetic interest (Fig. 1). The fiducial site locations are established in advance with extremely high accuracy by an independent technique, usually VLBI. With the fiducial sites held fixed, all GPS satellite orbits and all other receiver positions are simultaneously adjusted using an extended arc of data. Orbits and baselines are thereby obtained in the reference frame defined by the fiducial network. For highest accuracy, a "dynamic" solution is used requiring accurate modeling of receiver and GPS satellite motion over the length of the solution interval, which may be hours or days. In the solution process, fiducial accuracy is transferred to the adjusted ground sites; the corresponding GPS orbit errors scale roughly in proportion to their distance from the fiducial network. U.S. fiducial networks, established by years of VLBI observation, are now accurate to 1-2 cm, enabling comparable accuracies in continental baseline measurements and sub-meter GPS orbits.

TOPEX/POSEIDON Precise Orbit Determination (POD)

In 1981 a group at JPL was asked to look for new ways to track TOPEX/POSEIDON, a mission dedicated to radar mapping of ocean topography, with an expected precision of about 2 cm. To separate orbit variations from ocean topography, continuous independent determination of orbiter geocentric altitude with an accuracy approaching that of the altimeter is desired (Fig 2). For the TOPEX/POSEIDON study, a goal of better than 10 cm geocentric altitude accuracy was set, representing a five-fold improvement over the best tracking performance achieved on Seasat, the TOPEX/POSEIDON predecessor. The technique that emerged was a straightforward analogue of the GPS fiducial concept. In this variation, a GPS receiver is flown on the orbiter, which takes on the role of one of the adjusted ground sites, and the ground network is extended globally. This concept, first put forward in 1982, was accepted as a flight experiment on TOPEX/POSEIDON in 1984, and has been under development since then.

For the TOPEX/POSEIDON experiment, NASA's Deep Space Network (DSN) tracking sites in California, Spain, and Australia will serve as fiducials, establishing a reference frame known geocentrically to better than 5 cm; three geographically complementary sites will fill out the reference network. Using several hours of data the experiment team will solve for all GPS orbits and the "mobile" receivers, which now include both the non-DSN ground sites and the flight receiver. To overcome the formidable problem of modeling the motion of a dynamically complex low-orbiting platform, in 1985 we proposed a non-dynamic or kinematic tracking strategy made possible for the first time by GPS [Yunck and Wu 1986]. The kinematic technique dispenses with motion modeling for the low orbiting satellite (though retaining it for the ground sites and GPS satellites), relying instead on smoothed

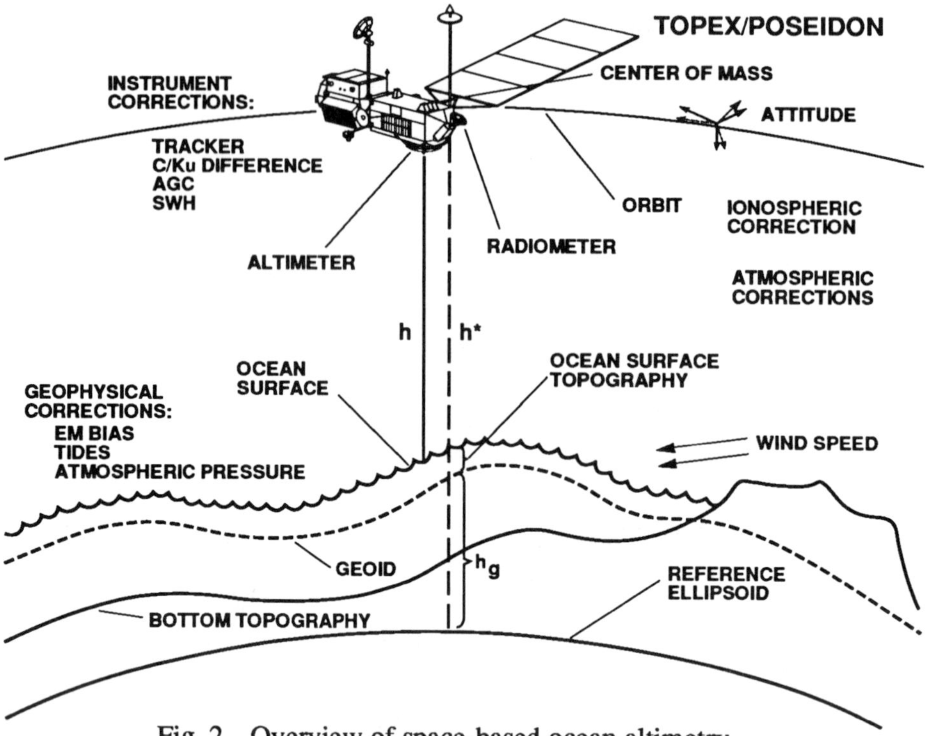

Fig. 2. Overview of space-based ocean altimetry.

geometric positioning. Exhaustive computer studies show that kinematic tracking can achieve sub-decimeter accuracy regardless of orbiter dynamics—as readily for the dynamically confounding Space Shuttle or Space Station as for TOPEX/POSEIDON.

POD in Earth Science

We can illustrate the value of precise orbit determination to ocean altimetry with an example from the study of basin scale ocean circulation. Ocean gyres are key features of ocean circulation that are believed to wobble on a scale of thousands of kilometers with an amplitude of a few centimeters. Quantifying this variability is essential to understanding ocean circulation dynamics and the ocean-atmosphere interaction. Such observations are all but impossible with current altimetric data and TOPEX/POSEIDON will just begin to achieve the needed accuracy. Centimeter POD will enable future altimetry missions to improve observations of basin-scale ocean variability and extend TOPEX/POSEIDON results to interannual and even decade periods, leading to a better understanding of the interplay between the oceans and atmosphere.

Centimeter accuracy will also be valuable in calibrating the biases in orbiting altimeters. GPS-determined altitude, for example, can be compared against altimetry over regions of known elevation, such as lakes and deserts. Where the elevation is not known, a GPS receiver placed on the surface will provide a platform-to-ground baseline that can be determined continuously. This technique, using receivers on ocean platforms, is now being developed for calibrating the TOPEX/POSEIDON altimeter. A more novel application of decimeter tracking is in the construction of interferometric SAR images with data from a single SAR element. In a technique pioneered by Goldstein with data from Seasat [Goldstein, Zebker and Werner 1988], SAR images from repeat ground tracks can be combined to provide 3-dimensional information—so long as the relative platform positions between passes can be determined to much less than a wavelength. This can be done either with a costly and time-consuming fringe-searching process (as was done with Seasat) or directly with decimeter position knowledge. Only the latter is practical for operations.

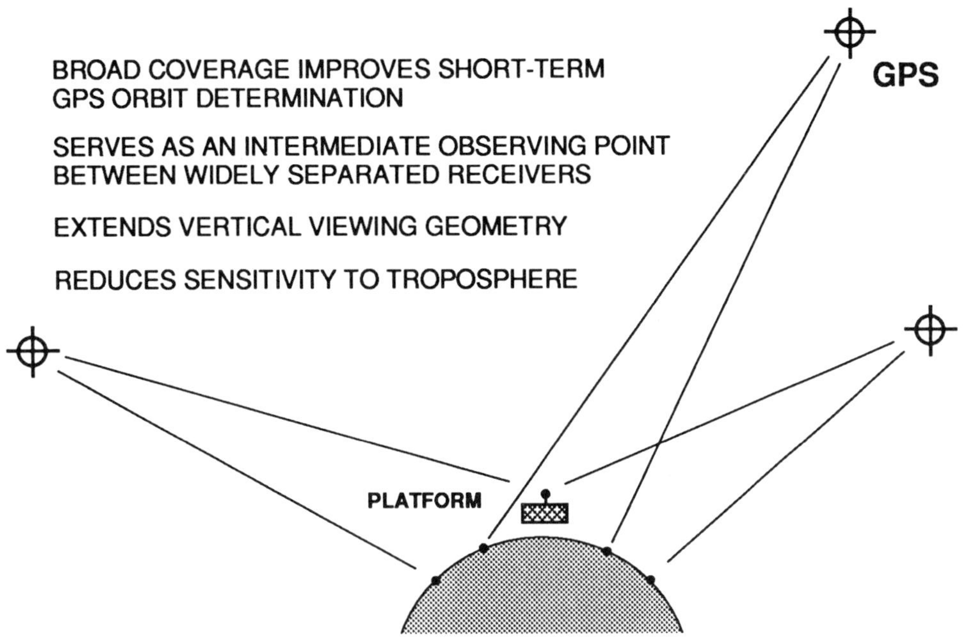

Fig. 3. Several ways an orbiting GPS receiver can improve earth baseline measurements.

Direct Science from GPS

The detailed error studies for TOPEX/POSEIDON POD yielded an unexpected insight that, in a sense, completed an evolutionary circle: Orbiting GPS receivers can markedly improve the measurement of earth baselines and point positions over short time spans. They do this in several ways: by strengthening GPS orbit determination; serving as intermediate observing points between widely separated ground sites having poor common GPS visibility; extending the vertical viewing geometry for all baselines; and desensitizing the solution to media effects by introducing measurements with no intervening atmosphere (Fig. 3). It became apparent through these studies that even without a POD task to perform, suitably designed orbiting GPS receivers would be a valuable addition to any GPS-based geodetic system. The precise positioning capabilities of GPS can thus serve at least 3 earth science interests: ocean altimetry, SAR interferometry, and space geodesy. With care in system design, however, we can also acquire valuable earth science data of a quite different kind.

The orbiting receiver's field of view, which will be a zenith-directed hemisphere on TOPEX/POSEIDON, can be extended to the earth limb both to improve observing geometry (and hence tracking accuracy) and to observe GPS signal occultation by the atmosphere and ionosphere. Radio occultation experiments have a more than 20-year history in planetary science. Variations in signal phase and amplitude during planetary occultations of spacecraft are used to construct profiles of density, pressure, and temperature, and to analyze the composition, turbulence, and other properties of planetary atmospheres [Fjeldbo, Kliore and Eshleman 1971]. In fact, occultation techniques have now been applied to every planet in the solar system, outside of our own, except Pluto, and to numerous planetary moons. Until the advent of GPS, however, we have not had the comprehensive satellite-to-satellite links needed to apply these techniques usefully to Earth. The abundant raypaths from 24 GPS satellites to an assortment of low orbiting platforms will soon give us that opportunity. Lateral sampling of the atmosphere and ionosphere will extend the earth coverage of remote platforms and yield valuable data on geophysical processes ranging from decade and longer changes in global climate to evanescent atmospheric waves. We shall now discuss each of these applications in greater detail.

GPS-BASED POD & PRECISE GEODESY

GPS Principles of Operation

The GPS satellites transmit carriers at two L-band frequencies, 1575 and 1227 MHz. Each carrier is phase modulated by a precise ranging code consisting of pseudorandom bit sequences at 10.23 Mbs. The transmit time, as kept by the clock onboard each GPS satellite, is precisely known for each bit in the sequence. A GPS receiver identifies the incoming code bits and measures their arrival time, as kept by the receiver clock. The difference between the known transmit time and observed arrival time is a measure of the range between the satellite and receiver plus the time offset between transmitter and receiver clocks, a quantity known as "pseudorange". A receiver measuring pseudorange to four satellites can instantaneously determine its three components of position and its time offset from GPS time, typically with an accuracy of 10-15 m. Many receivers can also measure and keep continuous count of carrier phase with a precision of better than 0.5% of a wavelenth (<1 mm) with 1 sec of observation. Continuous phase can be used to construct a record of position change with centimeter precision.

POD & Precise Geodesy

Positioning error drops markedly in relative or differential applications in which a receiver location is determined with respect to one or more known reference sites observing GPS at the same time. Virtually all clock errors are removed by differencing or direct solution and the effect of GPS orbit error is reduced by common mode cancellation. Instantaneous relative position error, typically 1-2 m, can be improved by smoothing observations over extended intervals (reducing random error) and solving for key geometric and other parameters (reducing systematic error). This is simply a restatement of the fiducial concept. Typical geometric parameters include all GPS orbits, the user position (or orbit), and all ground sites except a minimal reference set. Additional dynamic and observational parameters, such as tropospheric zenith delay and solar radiation pressure, may be adjusted as well. The technique can accommodate any number of "user" receivers on the ground or in orbit; indeed, the more users, the stronger the global solution. Thus a single system provides the mutually reinforcing functions of POD and precise geodesy. Direct and differential positioning of an earth satellite are depicted in Fig 4.

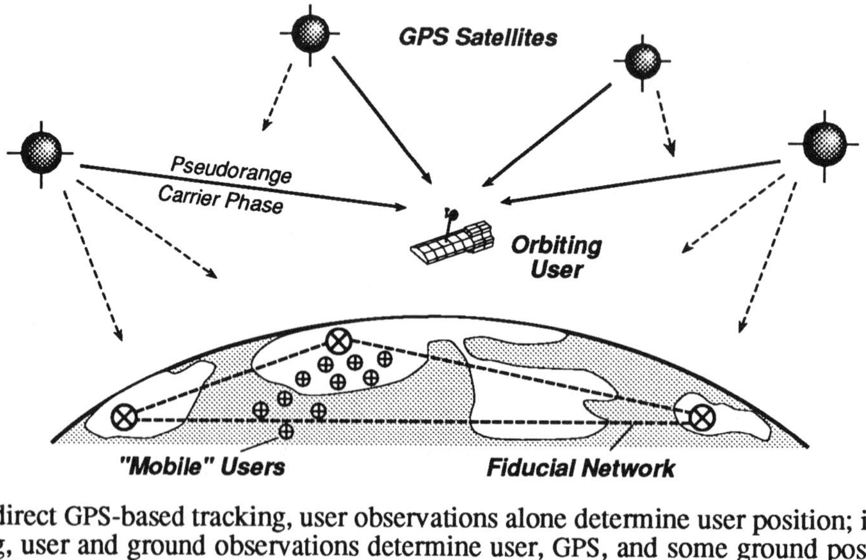

Fig. 4. In direct GPS-based tracking, user observations alone determine user position; in differential tracking, user and ground observations determine user, GPS, and some ground positions with respect to a fixed ground reference network.

POD for low earth orbiters with all tracking systems except GPS (and, when operational, the Soviet Union's GPS-like GLONASS system) must rely on one basic technique: dynamic orbit fitting. GPS, uniquely, offers a versatile alternative in the form of precise geometric positioning combined with kinematic smoothing, as well as a synthesis of the kinematic and dynamic techniques which is more powerful than both.

Dynamic Orbit Fitting

Long arc orbit fitting has classically been dynamic: A model orbit, constructed from an approximate initial state by integrating the equations of motion, is fit to the tracking measurements by least squares adjustment of a limited set of parameters, typically the satellite epoch state (position & velocity) and a few others. Random measurement noise is thereby smoothed against an orbit model. The quality of the dynamic solution is critically dependent on the quality of the force models used to generate the orbit model—an accurate solution requires accurate models. In applications seeking few-decimeter or better POD, such as Lageos, Seasat, and TOPEX/POSEIDON, extraordinary effort has been expended refining models for gravity, drag, radiation pressure, and the like. For Lageos, a dense sphere at an altitude of more than 5000 km, decimeter modeling is now readily achieved. For TOPEX/POSEIDON, at 1334 km where gravity error is worse, decimeter modeling may be within reach, but only with a good deal of model refinement. For future missions, such as the large EOS polar orbiters proposed to fly at 705 km, increased drag and gravity error, moving instruments, and uncertain center of mass will make decimeter modeling exceptionally difficult; gravity error alone may initially be at the meter level. And for the sprawling space station at 400 km, with severe drag, roving astronauts, and so on, dynamic POD is far out of reach. This diversity of platforms for which sub-decimeter orbit accuracy is being sought is illustrated in Fig. 5.

Kinematic POD with GPS

A fundamentally different approach to precise tracking is made possible by two distinctive GPS features: continuous coverage by four or more satellites, providing uninterrupted *geometric* position determination; and the combination of simultaneous phase and pseudorange data types, enabling long term *kinematic* smoothing [Yunck and Wu 1986]. Because this kinematic technique does not use the laws of motion to infer position, knowledge of the changing platform center of mass is not needed; instead position is referred to a fixed point on the GPS antenna, which can be calibrated with sub-millimeter accuracy with respect to antenna phase response. Kinematic tracking is readily adapted to real time position determination with a simple recursive algorithm. Without the ground reference data to improve GPS orbits, however, onboard real time accuracy will be limited by GPS ephemeris error to about 1 m.

Reduced Dynamic Tracking

The kinematic technique discards all platform dynamic information in order to eliminate dynamic modeling error. Often, however, there will be some dynamic information which, if properly weighted, will improve the kinematic result. A simple Kalman filter formulation developed in 1986 combines dynamic and kinematic information in a completely general way, allowing arbitrary relative weighting [Wu, Yunck and Thornton 1990]. In this technique a classical dynamic formulation is augmented with process noise added to the force model, which adds a kinematic component to the solution. Increasing the sigma on the process noise increases the kinematic component of the solution. With sufficiently large sigma the solution becomes purely kinematic; with zero sigma it is purely dynamic. Properly weighted, this reduced dynamic strategy must always equal or surpass both the dynamic and

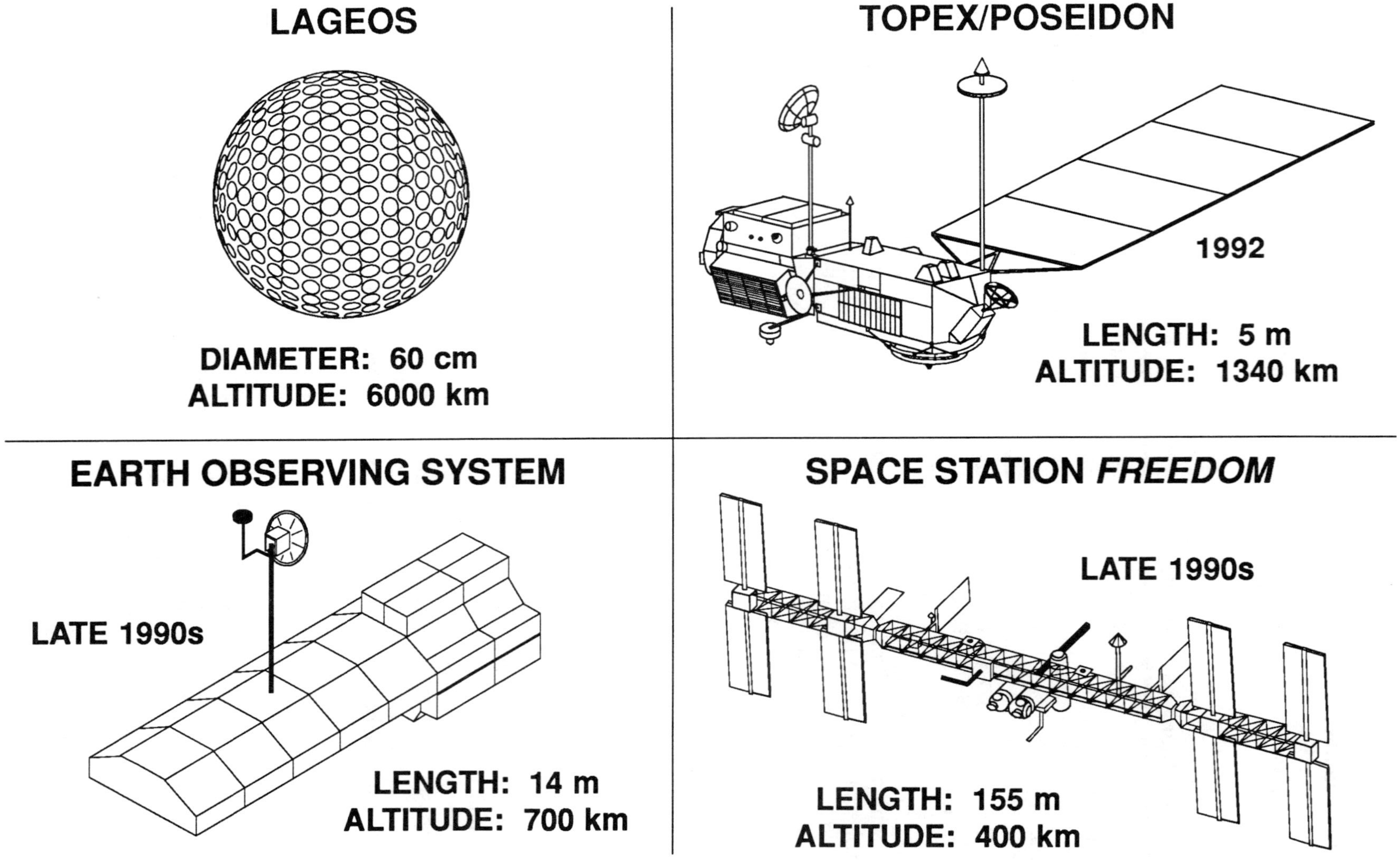

Fig. 5. Four earth satellites that either do or will require precise orbit determination, suggesting a trend toward larger, lower, and more dynamically complex platforms.

kinematic techniques used separately. For dynamically unpredictable platforms like the shuttle or space station, dynamic information will be too weak to offer any real improvement. For quieter craft like the EOS polar orbiters, de-weighted dynamics will be helpful. Because the dynamic characteristics for these platforms are not yet well known, the error analyses presented in the next section employ the more conservative kinematic approach.

We have previously presented analyses indicating that both the kinematic and reduced dynamic techniques could achieve altitude accuracies of a decimeter or better with the GPS configuration to be used for TOPEX/POSEIDON [Yunck et al 1990]. For EOS and the space station we now propose an extended configuration that could bring performance near 1 cm in all components. Key changes include extension of the flight receiver field of view down to the earth limb; increased flight receiver tracking capacity from five to twelve satellites; and a ground network expanded from six to ten sites. Performance will also benefit from improvement in our knowledge of the DSN reference sites from the 5 cm assumed for TOPEX/POSEIDON to about 2 cm by the mid-1990s, and from the expansion of the GPS constellation to 24 satellites. Further improvement is gained by extending the solution arc to 8 hours from the 2-4 hours used in most TOPEX/POSEIDON studies. POD and ground baseline error analyses for this configuration are presented below.

POD and Geodetic Error Analysis

Simulation and covariance studies have been carried out for kinematic tracking of EOS and the Space Station *Freedom*, as well as for ground site geocentric position solutions and both regional and global baseline solutions. All studies assumed a 10-site ground network, with the three DSN sites defining a fixed reference frame and seven supplementary sites that were adjusted in each solution. The regional geodetic studies supplemented the global network with several North and Central American sites. To assess the role of orbiting GPS receivers in ground-based geodesy, all geodetic studies were performed both with and without orbiter data. Receiver performance assumptions used in the studies were consistent with the current performance of JPL's Rogue GPS receiver [Thomas 1988].

Predicted orbit accuracy does not differ significantly between the polar and manned platforms, despite their differing orbits and dynamics. (The space station orbit will be inclined at 28.5°, the polar platforms at 98°.) Typical 2-, 4-, and 8-hour results are shown in Fig 6. Predicted rms accuracies over the arcs are about 3 cm (each component) in 2 hrs falling to 1.3 cm in 8 hrs. Because of the limiting effects of troposphere, fiducial error, and phase measurement error (which is not reduced in kinematic smoothing), there is little reduction beyond 8 hrs.

Figure 7 shows the predicted absolute position errors for several adjusted sites of the network using a 2-hr data arc under two assumptions: no orbiting receivers and two orbiting receivers (1 EOS, 1 space station). The orbiters provide a factor of 2-to-4 improvement, bringing performance down to the scientifically useful 1-3 cm level. These results are for point positions in the frame defined by the three NASA DSN sites. Figure 8 shows a similar comparison for long baselines (4000 to 12500 km) between various of the adjusted sites. The error decreases with increasing arc length. At 8 hours (not shown), all baseline and point position errors are less than 1.5 cm. If fewer ground sites are assumed, the advantage with orbiting receivers is more dramatic.

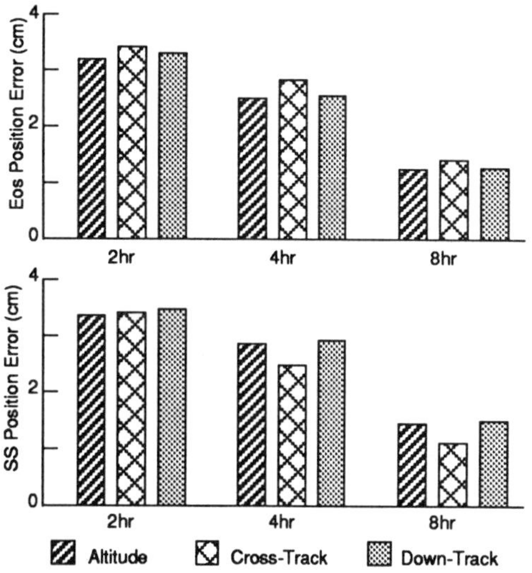

Fig. 6. Results of a covariance analysis for GPS-based precise orbit determination on EOS and Space Station *Freedom*.

A study for regional geodesy examined baselines ranging from 250 to nearly 5000 km in North and Central America and the Caribbean. Figure 9 shows the predicted errors for the N-S and vertical components as a function of baseline length, using a 4-hr data arc both with and without two orbiters. The orbiter advantage increases with baseline length. Significantly, the error in the vertical, which is always the most difficult to recover with upward-looking GPS measurements, falls below 1 cm for even the longest baselines. When the arc length is extended to 8 hrs, the accuracy of all components improves to 5 mm or better.

Geodynamic Science

Ground-based GPS geodesy has become a mainstay in geodynamic science and its use is rapidly expanding. A principal science application is the measurement of crustal plate motion and deformation. Key geophysical questions center on the complex crustal dynamics in regional deformation zones and at plate boundaries. Large scale plate motions are fairly well understood from the geological record, although a good deal of work is going into confirming and refining global tectonic models. Plans are currently being drawn up to apply GPS techniques to the observation of sea floor spreading by acoustically linking GPS receivers mounted on rigid ocean platforms to transponders anchored to the sea floor. The trend in ground-based geodetic technology is toward a permanent global observing network, perhaps a dozen sites, to provide accurate tracking of GPS satellites, supplemented by high density networks in regions of high tectonic activity such as Southern California and Chile. The network will operate semi-autonomously, providing an unbroken record of global and regional deformation.

The advent of orbiting receivers in the 1990s will improve both the accuracy and time resolution of geodetic measurements. Sub-centimeter accuracy over

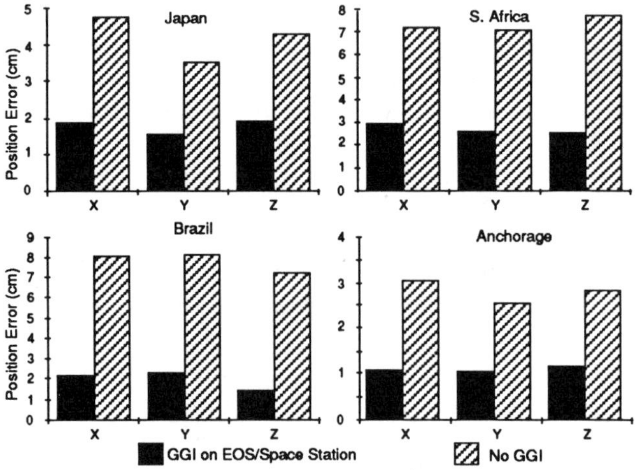

Fig. 7. Covariance analysis for global point positioning with and without GPS data from one EOS platform and the Space Station (2-hr data arc).

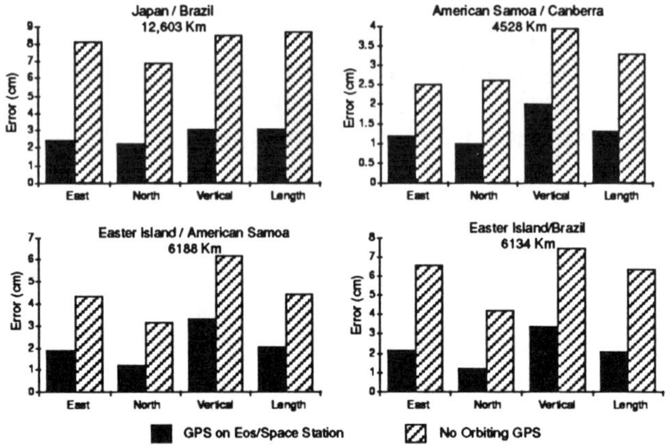

Fig. 8. Covariance analysis for long-baseline geodesy with and without GPS data from one EOS platform and the Space Station (2-hr data arc).

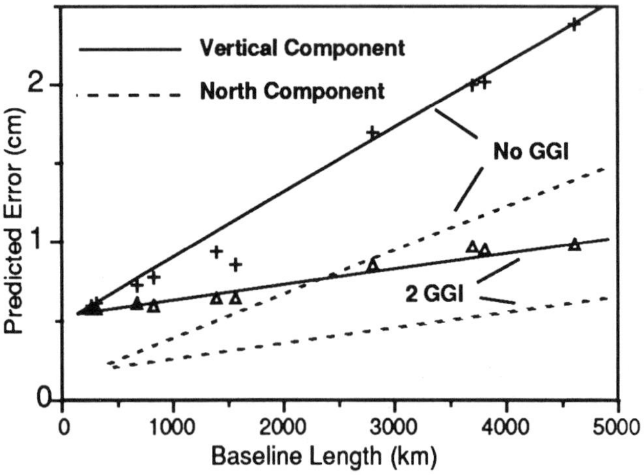

Fig. 9. Covariance analysis for regional geodesy with and without GPS data from two EOS platforms

distances up to 5000 km will be achievable with a few hours of observation. This will permit direct measurement of the rapid redistribution of strain following large earthquakes, information that is critical to understanding the earthquake process and that at present is difficult to obtain. Preliminary studies indicate that GPS will also be valuable in observing polar motion, the earth rotation rate, the location of the geocenter, and the variations of these following large earthquakes [Freedman 1990; Malla and Wu 1989; Chao and Gross 1987].

Gravity Modeling

Receivers on conventional orbiters like TOPEX/POSEIDON and on drag-compensated low orbiters like Gravity Probe B (GP-B) will help improve models of the earth's gravity field and eventually may reveal gravitational time variations due to such processes as post glacial rebound, plate motion, hydrologic phenomena, and the changing distribution of atmospheric mass. All of these can contribute to our understanding of the structure and dynamics of the earth. At its relatively high altitude of 1300 km, TOPEX/POSEIDON can offer rather little in the way of gravity improvement, except at the longest wavelengths. At very low altitudes—altitudes that require drag compensation to reveal gravity effects—an orbiter is far more sensitive to the poorly known short wavelenth gravity components. GP-B, flying in a drag-compensated 600 km polar orbit, will offer a significant advantage over TOPEX/POSEIDON in gravity recovery; Aristoteles, which will fly as low as 200 km, will be better still. The gravity capabilities for all of these missions are still under study. Figure 10 shows the result of a preliminary estimate of the gravity and geoid recovery with GPS data from TOPEX/POSEIDON as a function of the harmonic number. Note how sharply the predicted accuracy falls off with decreasing wavelength.

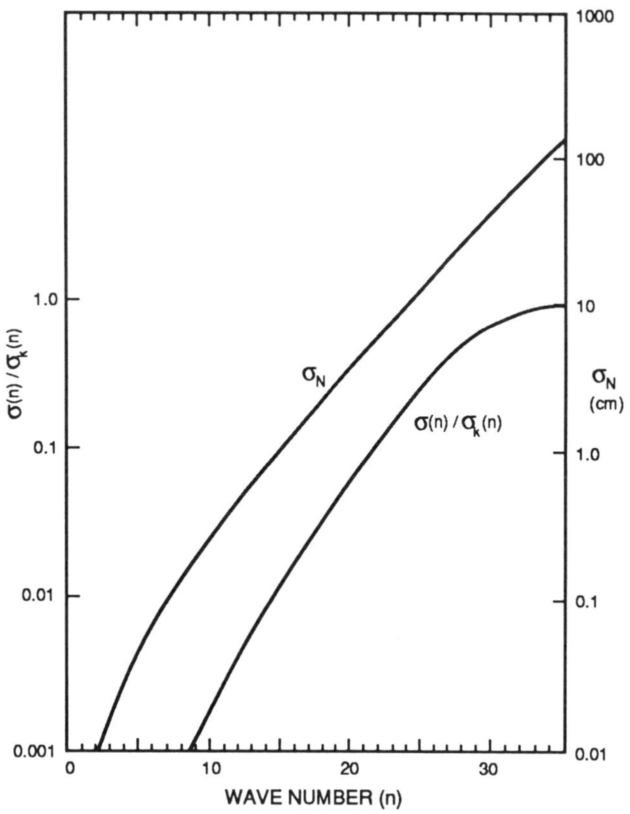

Fig. 10. Estimated accuracy of Earth gravity harmonics recovered from TOPEX/POSEIDON, shown as a function of wave number.

GPS OCCULTATION SCIENCE

In contrast to GPS geodesy, now in worldwide use, science through GPS signal occultation has as yet been little explored. When we recently investigated some of the potential science payoffs of GPS occultation measurements, three particularly promising areas came to light: global atmospheric temperature profiling, acoustic gravity wave analysis, and tomographic mapping of the ionosphere. All of these areas are now under intensive study for possible inclusion as science experiments aboard NASA's two EOS platforms and Space Station *Freedom*. In the following discussion we assume those platforms, flying concurrently, as our bases for occultation measurements. For ionospheric mapping it is also helpful to include measurements from the ground and from down-looking platform instruments.

When a radio signal is occulted by the atmosphere its phase and amplitude are perturbed in a manner related to the vertical refractivity profile, N(h), where h is altitude [Fjeldbo, Kliore and Eshleman 1971]. Phase perturbations reveal refractivity, from which we can determine density, pressure, and temperature. The analysis requires computing the ray asymptote miss distance (a) as a function of the ray bending angle (α), as illustrated in Fig 11. These are given by the spacecraft ephemeris and the Doppler frequency observed on the link. The refractive index distribution is found from a(α) by solving an Abelian integral equation. With GPS L-band data we must form the dual frequency combination of carrier phase to remove the effect of the earth's ionosphere. True atmospheric refractivity profiles can then be computed and used to analyze the thermal structure of the stratosphere and troposphere, below about 50 km. The process requires 3 steps: First we compute the molecular density profile from each refractivity profile; next we compute the vertical pressure distribution by assuming hydrostatic equilibrium, applying our knowledge of the composition (thus the distribution of molecular weights) of the atmosphere, and integrating the air density from the top of the atmosphere downward to the surface; finally we form the temperature profile from the number density and pressure profiles using the equation of state. The GPS occultation experiment is illustrated schematically in Fig. 12. A hypothetical occultation signal is shown in Fig. 13. Note that at Earth grazing the total phase delay is several thousand cycles, a quantity that can be measured to better than 0.01 cycle.

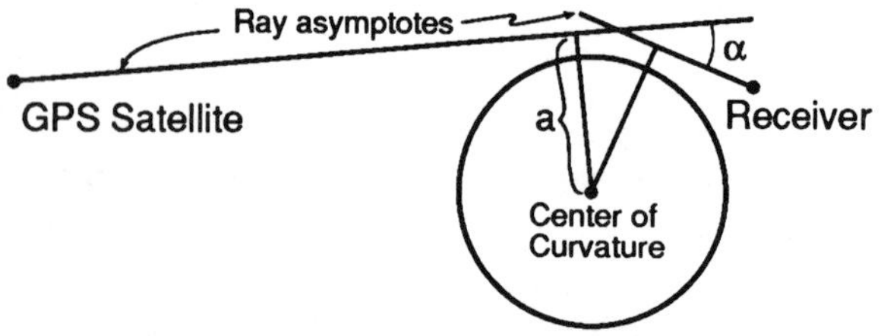

Fig. 11. Ray asymptotes at transmitter and receiver for a GPS atmospheric occultation.

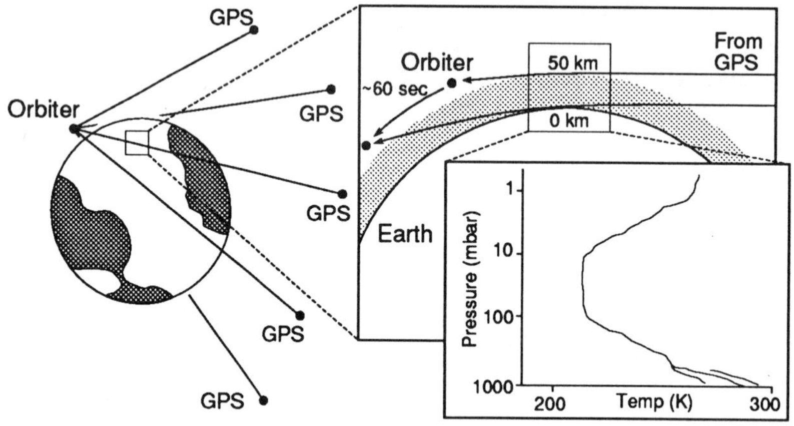

Fig. 12. Schematic view of atmospheric science by GPS signal occultation.

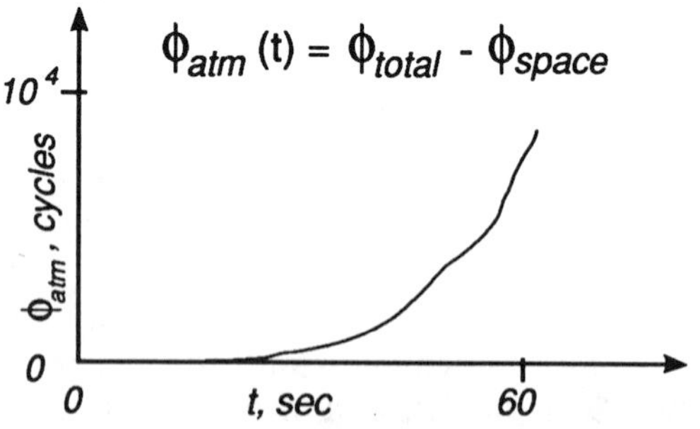

Fig. 13. The magnitude of the phase delay due to atmospheric refraction shown as a function of time for a setting GPS satellite.

Vertical resolution, determined by the limits of geometric refraction, will range from ~0.5 km near the surface to ~1 km near the top of the stratosphere. Temperature accuracy will depend on several factors. In the middle and lower atmosphere random measurement error is a small fraction of total refractivity; systematic errors in modeling platform motion and in removing multipath effects and Doppler bias will dominate. Precise navigation during the observation is critical. At the tropopause (about 12 km), temperature accuracy will be 0.1-0.4 K, with the random component <0.1 K. The error will grow gradually with altitude until loss of refractivity in the thinning atmosphere causes random error to dominate. The major sources of random error will be short term phase instability in the GPS satellite cesium oscillators and thermal noise in the receivers. (Receiver oscillator instability will be differenced out.) The GPS oscillators will contribute an error comparable to or less than that from thermal noise.

We incur a thermal noise penalty of about a factor of 3 from the dual frequency ionospheric calibration. This penalty can be reduced by smoothing the ionospheric correction. The final thermal noise level will depend on the degree of ionospheric smoothing we can achieve, which in turn will depend on the behavior of the ionosphere. Figure 14 gives the predicted temperature accuracy envelope bounded by various uncertainties. This performance surpasses the published EOS science and observational requirements for temperature in the troposphere (1 K required/0.5 K desired), middle atmosphere (2 K/1 K) and upper atmosphere (10 K/5 K), and compares favorably with accuracies quoted for more specialized atmospheric sensors. The altitude resolution is in fact considerably better than that typically achieved more with traditional downlooking instruments.

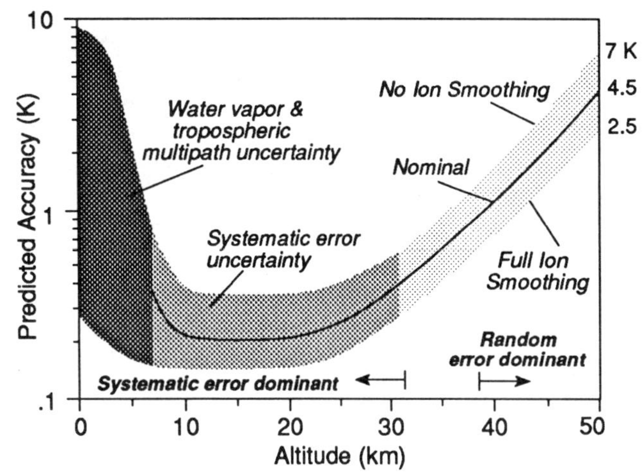

Fig. 14. Predicted accuracy of atmospheric temperature measurements from proposed space-based occultation experiments, with areas of uncertainty highlighted.

We know that the long term climate on earth is governed by a balance between the absorbed solar radiation and the emitted infrared radiation. Models of the terrestrial atmosphere and surface indicate that observed increases in the concentrations of carbon dioxide, methane, chlorofluorocarbons (CFCs), and other trace gases should produce a greenhouse effect at the surface, leading to a significant increase in average global temperature and a consequent melting of polar ice and a rise in ocean levels [Ramanathan 1988]. In the stratosphere, these models predict a lower air temperature because of increased radiative cooling. A doubling of the carbon dioxide concentration may decrease the temperature in the stratosphere by as much as 10 K. The stratospheric temperature may also drop because of reduced absorption of solar radiation in the ozone layer brought about by ozone depletion resulting from widespread use of CFCs. Recent claims to the contrary notwithstanding, we cannot yet tell whether the predicted greenhouse warming has started, or whether recently observed changes in atmospheric temperature and composition herald a long term trend. Extended global coverage provided by GPS occultation measurements, and other orbiting atmospheric instruments, will provide a comprehensive picture of atmospheric thermal structure and composition, and will enable the earliest possible detection of the onset of fundamental (and potentially devastating) climatic changes.

Ionospheric Tomography

Irregular structures of different sizes are generated in the ionosphere as a result of its electromagnetic, corpuscular, and dynamic coupling with the magnetosphere, and its dynamic and aeronomic interactions with the thermosphere and middle atmosphere. The ionosphere, in short, is a complex, highly mutable matrix containing an assortment of transient, inhomogeneous features. To begin to understand it we need an accurate physical description of its different features and the ability to follow their evolution in time and space. Over the years, many observational techniques have been devised for this purpose. One of the most important of these is the measurement of total electron content (TEC) along a path of given cross-section.

With a simple linear combination of dual frequency GPS data we can continuously measure TEC along all GPS-to-receiver raypaths. Pseudorange by itself will give absolute TEC with an accuracy of about 10^{16} electrons/m^2 (16 cm delay accuracy at L1) in 1 sec. Smoothing pseudorange against precise carrier phase will improve this to better than 10^{15} e/m^2 (<2 cm), or about 0.1% of the daytime zenith peak and 0.5% of the typical nighttime minimum. Continuous carrier phase measurements by themselves will give TEC *change* with a precision of better than 10^{14} e/m^2 (<2 mm). Down-looking TEC measurements will strengthen ionospheric sampling. The global coverage offered by the the combination of orbiting platforms and GPS ground sites will provide an unprecedented examination of ionospheric structure and short term variability.

Ground-based multistation TEC data have often been used to study the horizontal variation of ionospheric structures. TEC data cannot, however, provide information on ionospheric variation along the path of observation using conventional analysis techniques. To obtain such information, our co-investigators at the University of Illinois recently introduced TEC analysis by computerized tomography or CT [Austen, Franke and Liu 1988]. CT is a technique whereby an image of an object is constructed from a set of projections, or integrated densities, taken along many lines through the object and is widely used in medicine to produce cross-sectional X-ray images. More recently, 3-dimensional acoustic and seismic tomography have been applied to the oceans and solid earth. In ionospheric tomography, the measurement of TEC along a raypath is the analogue of the integrated density registered by an X-ray. TEC measurements taken along many raypaths can be used to construct maps of electron distribution in 2 or 3 dimensions.

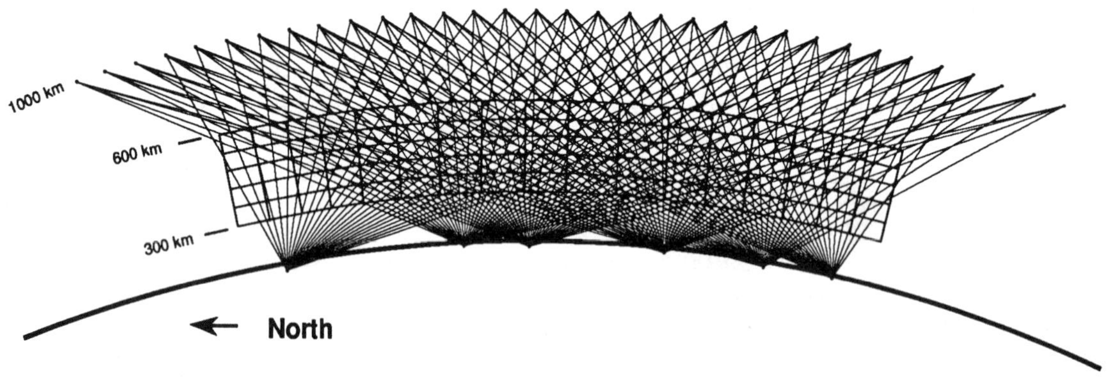

Fig. 15. Typical placement of pixels and rays for recent ground-based ionospheric tomography experiments conducted by colleagues at the Univ. of Illinois [Austen et al, 1988].

The Illinois group has shown that with TEC data from an orbiting satellite collected by a linear array of ground receivers, it is possible to resolve features of a few tens of kilometers [Austen, Franke and Liu 1988]. Figure 15 shows the typical observing geometry used in these satellite-to-ground experiments. A major limitation on reconstruction with ground observations is that the vertical profile of the background ionosphere cannot be recovered. For that, horizontal observing paths through the structure are required—paths that will be abundantly provided by orbiting receivers. Figure 16 illustrates qualitatively the far more extensive ionospheric coverage we can obtain from just a single orbiting platform. GPS receivers and down-looking instruments on three platforms, supplemented by ten (and possibly many more) ground sites, will provide an immensely rich data set for both regional and global ionospheric investigations.

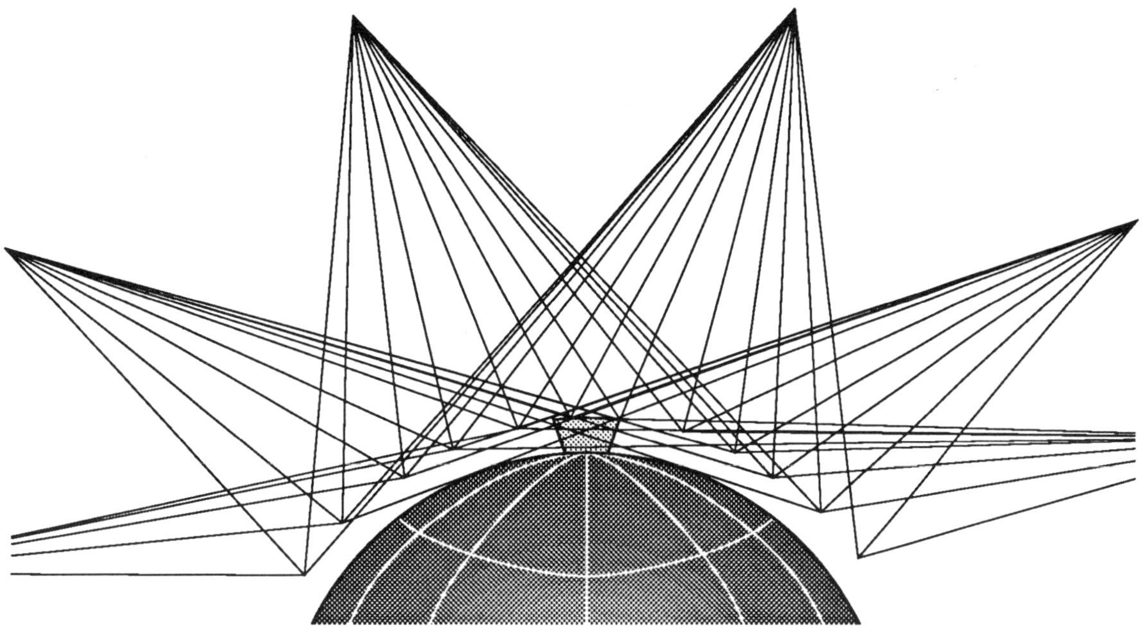

Fig. 16. Expanded ionospheric coverage offered by one orbiting GPS receiver.

Acoustic Gravity Wave Detection

Ionospheric waves called traveling ionospheric disturbances (TIDs) are the manifestation of acoustic gravity waves (AGWs) propagating in and intersecting with the ionosphere. AGWs are present at all levels of the atmosphere and have been detected by a variety of experimental techniques from the earth's surface to the ionosphere [Hines 1960; Yeh and Liu 1974]. AGWs can be caused by unusual events such as volcanic eruptions, earthquakes, and explosions in the atmosphere. A magnitude 7 earthquake, for example, will create a surface uplift of several centimeters over several hundred kilometers, launching an atmospheric gravity wave. We have estimated that in the ionosphere the gravity wave can temporarily increase TEC by 5×10^{15} e/m^2, or 8 cm of delay at 1.6 GHz—easily observable with millimeter-precision GPS carrier phase. In most cases, however, atmospheric waves have their origin in meteorological and geophysical events such as strong wind shear, distortion in jet streams, convective storms and fronts, and auroral-related high latitude events.

Once generated, gravity waves carry energy and momentum from the source to other parts of the atmosphere. As they ascend through the thinning medium the particle motions are amplified through conservation of energy. The wave associated velocity increases with height until nonlinear effects, such as resonant interactions or wave breaking, take place. When this happens, energy and momentum carried by the wave may be deposited to the background atmosphere. It is currently believed, for example, that gravity waves play a key role in determining the general large-scale circulation in the middle atmosphere [Geller 1983]. In addition, a principal means by which the lower atmosphere couples with the thermosphere is through the dissipation of gravity waves, resulting in momentum and heat flux divergences and enhanced turbulent mixing. At high latitudes, energy input from the magnetosphere generates AGWs that propagate around the globe in the thermosphere [Hunsucker 1982], providing a possible channel for coupling of energy from high to low latitudes.

Analysis of ionospheric waves detected with GPS carrier phase measurements can yield a wealth of information about the processes giving rise to the waves and about the flow of energy within and among different earth systems. In the future, when suitable receivers are

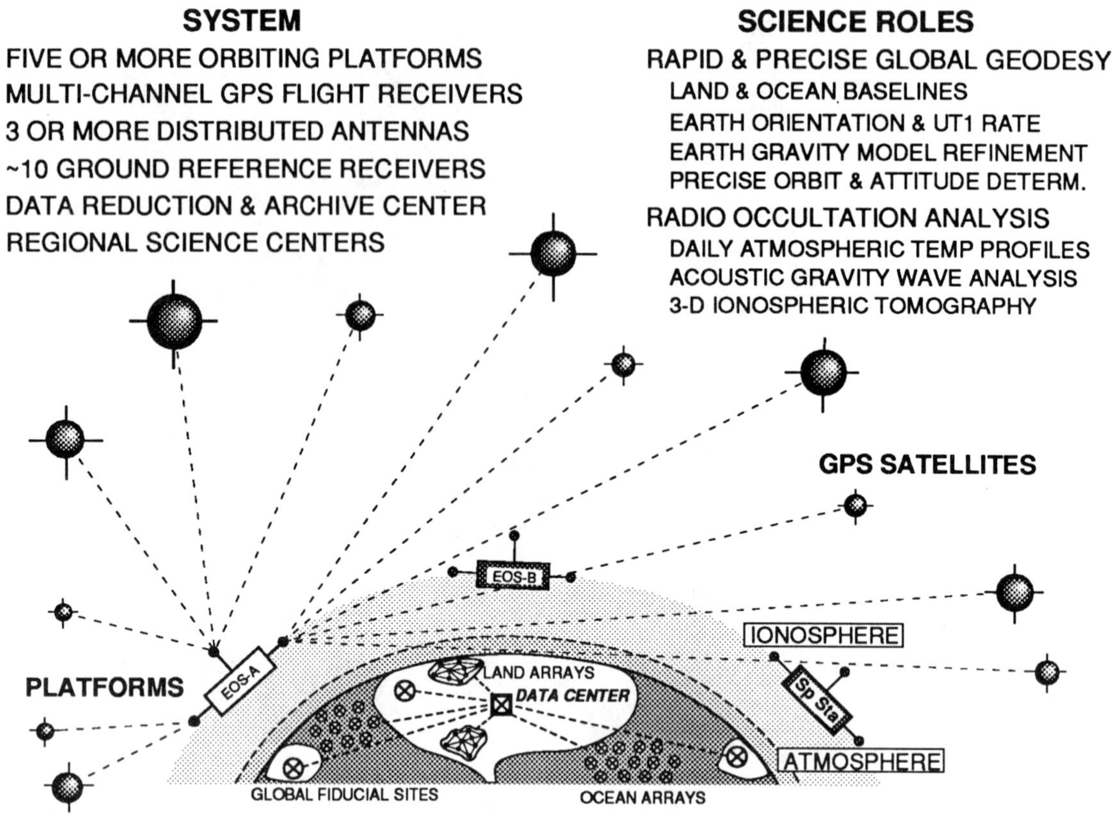

Fig. 17. Schematic illustration of the unified *GPS Earth Observatory*, showing orbiting and ground components and a summary of science objectives.

in orbit, we will attempt to detect and analyze individual wave events, identify the relation (or transfer function) between source events and the resulting gravity waves in the atmosphere, and estimate the energy and momentum input to the atmosphere from gravity waves. Through such studies we hope to understand more clearly the generation and propagation of gravity waves due to all important sources, including magnetospheric auroral sources from above, and weather, orographic (winds over mountains), volcanic, seismic, and other sources from below; the different influences of these waves on the background atmosphere; and their role in the transport of energy throughout the earth system.

GGI & THE GPS EARTH OBSERVATORY

We've seen that an infrastructure comprising a small fleet of orbiting receivers, a global GPS reference network, communication links, analysis centers, and special purpose GPS arrays can serve a diversity of earth science interests—embracing the ionosphere to its outer reaches, the atmosphere and climate, the many forms of ocean variability, and the evolving solid earth. Properly developed, this global instrument can become a central resource in the emerging study of the earth as a dynamic union of interdependent systems. The time for such work is at hand: The International Geosphere-Biosphere Program, the Global Change Initiative, NASA's Mission to Planet Earth, and innumerable smaller collaborations worldwide are now focusing on the complex environmental crises that threaten our planet.

As part of this mobilization, we have proposed that the disparate science initiatives outlined here be assembled into a unified enterprise known as the GPS Earth Observatory, sketched in Fig. 17. A key element of this proposal is a multifunction flight receiver called the GPS Geoscience Instrument (GGI), now in definition phase studies for both NASA EOS platforms and

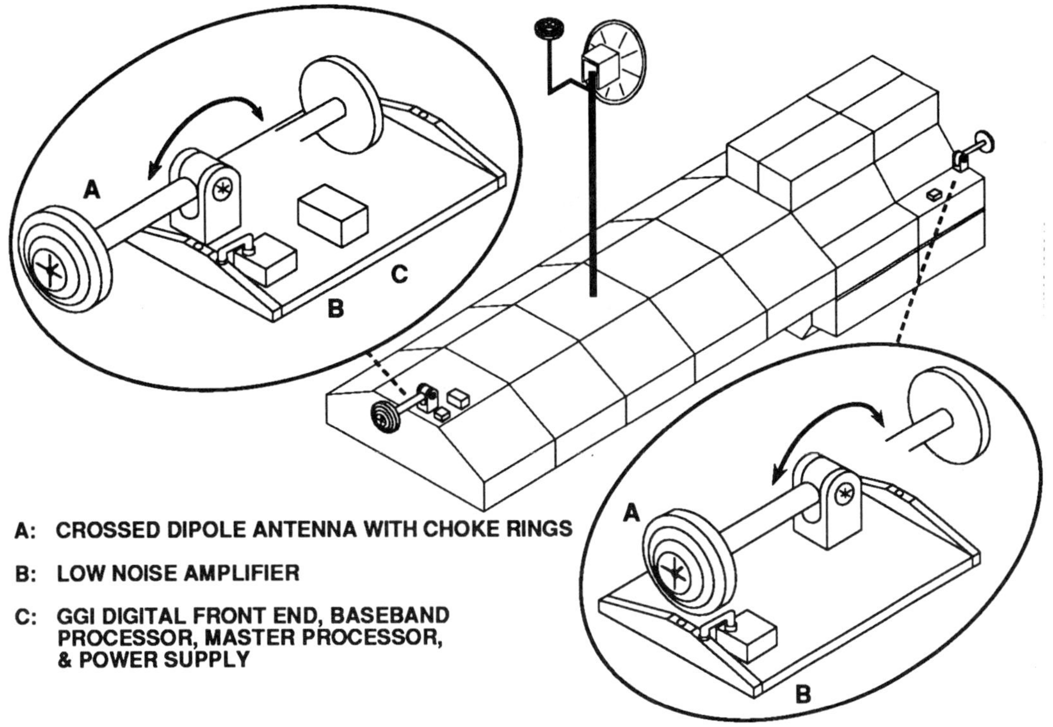

Fig. 18. Strawman layout for GGI on one Earth Observing System polar platform.

Space Station *Freedom*. Although it has its roots in the GPS Demonstration Receiver aboard TOPEX/POSEIDON, the GGI design differs in several important respects. For example, tracking capacity will grow from six to 18 satellites, data rates will increase sharply, and at least three antennas viewing the full sky down to the earth limb will be used. A nominal layout for GGI on the EOS platforms is shown in Fig. 18.

The Observatory will not be limited to the platforms shown in Fig. 17; indeed, data from any high quality receiver, orbiting or ground, will enhance the science. This effort is therefore open to all interested GPS investigators who wish to participate and to include their own institutions among the Science Centers. Our goal is to make as much of the global GPS data set as possible available worldwide with no delay. There is of course some uncertainty in the timeline for all this; schedules for EOS and the Space Station, to take two prominent examples, have proven fragile. There is little doubt, however, that the effort will proceed. Today, at least a half-dozen GPS flight receivers have been or are being built; permanent regional and global networks are partly deployed and arebeing rapidly expanded; communication links, data archives, and distribution systems are being upgraded; and many science investigations are in full swing. The broad multinational programs connected with the IGBP, the Global Change Initiative, and Mission to Planet Earth should provide the impetus and resources that will bring these elements together to effectively serve earth science.

DISCUSSION & CONCLUSIONS

Ground-based observations are fundamental both as reference data in GPS-based POD and as primary data in geodynamics and ionospheric science. With the rapid advance of GPS receiver technology, we can expect within a decade to have geodetic quality receivers costing under $10,000. The processing functions for tracking multiple satellites will be concentrated on a single chip. Receivers with ten or twelve dual frequency channels will soon become standard. Indeed, an argument can be made for providing up to two dozen channels in a dual GPS/GLONASS receiver. Such an instrument would ensure continued full strength operation should one or the other satellite system become unavailable. Moreover, the greater data available when both systems are operating will improve geometric strength and provide reductions in the error from data noise and multipath—possibly critical in the pursuit of millimeter geodesy.

Perhaps more significant may be a reduction in the error from tropospheric delay. Imperfect calibration of tropospheric delay is one of the limiting errors in baseline measurements of more than a few tens of kilometers. Recent studies have shown that one of the most effective methods of calibrating the troposphere is to solve for both the wet and dry components with GPS observations themselves using carefully tuned stochastic models. This requires comprehensive sampling of the visible sky. Doubling the number of signal sources will double the sky sampling density and strengthen the tropospheric solution, possibly improving baseline results well beyond what might be expected from the simple increase in data volume. Greater sampling density may be particularly helpful when azimuthal asymmetry is incorporated into the troposphere models.

Selective availability and anti-spoof—the intentional degradation of system accuracy and encryption of the signal—are a continuing concern to civil GPS users. Happily, selective availability (SA), which is expected to operate continuously, does not affect the ability of GPS receivers to track the satellites and acquire quality data. The positioning applications described here—both POD and ground geodesy—are inherently differential, employing precisely known reference networks. This permits the virtual elimination of SA effects in data analysis. The occultation applications are more problematical. Satellite differencing schemes are possible but not desirable; without such schemes, the potential degradation, if any, will depend upon the SA levels in effect. We expect that information to remove SA degradation will be made available to general users after a respectful waiting period, perhaps a few weeks. For most scientific applications, this should be satisfactory.

Unhappily, anti-spoof (AS) will defeat any code-tracking receiver not privileged with the decrypting keys. Although it is widely rumored that AS will be active only infrequently, the Department of Defense (DoD) advises users not to depend on it. An alternative is offered by codeless receivers, which are unaffected by AS encryption. Although codeless receivers sacrifice measurement precision, in many uses data noise is not the major error and performance approaching the levels quoted here can be achieved. (EOS and the Space Station have stringent real time GPS requirements which may compel them to carry code-tracking receivers equipped to remove SA and AS.) It is possible, therefore, to pursue the range of GPS science presented here, in the full presence of SA and AS, without access to the classified keys—though perhaps with some inconvenience. There is one more prospect to consider: DoD has stated that the GPS performance available to civil users will be no worse than the performance generally available by other means. With an open GLONASS in the offing, and various commercial satellite positioning systems in operation, that could in time mean no SA or AS at all.

GPS has already begun to transform a number of areas of earth science. This account of the emerging GPS role has been necessarily limited, but it touches many of the areas that show promise today. A decade from now we may be surprised by the number of developments unforeseen. With deployment of the operational satellites now well underway, we can look forward to a productive decade of GPS geoscience.

ACKNOWLEDGEMENT

Among those who contributed to these studies are C. Liu at the Univ. of Illinois; B. Hager at MIT; and S. Bassiri, W. Bertiger, G. Hajj, S. Lichten, G. Lindal, U. Lindqwister, S. Wu, J. Wu, and L. Young at JPL. Much of the work described in this paper was carried out by the Jet Propulsion Laboratory, California Institute of Technology, under contract with the National Aeronautics and Space Administration.

REFERENCES

Austen, J., S. Franke and C. Liu (1988), Ionospheric imaging using computerized tomography, *Radio Science, 23*, 299-307.

Blewitt, G (1989). Carrier phase ambiguity resolution for the global positioning system applied to geodetic baselines up to 2000 km, *J. Geophys. Res., 94*, 10187-10203.

Bosworth, J. (1990), Solid Earth Sciences Program Plan, (NASA document, in preparation).

Chao, B. and R. Gross (1987), Changes in the earth's rotation and low-degree gravitational field induced by earthquakes, *Geoph. J. R. Astr. Soc., 91*, 569-596.

Dong, D. and Y. Bock (1989), GPS network analysis with phase ambiguity resolution applied to crustal deformation studies in California, *J. Geophys. Res., 94*, 3949-3966.

Fjeldbo, G., A. Kliore and V. Eshleman (1971), The neutral atmosphere of Venus as studied with the Mariner 5 radio occultation experiments, *Astron. J., 76*, 123-140.

Freedman, A. (1990), Determination of Earth orientation using the Global Positioning System, *TDA Progress Report* (JPL internal document, in press).

Geller, M. (1983), Dynamics of the middle atmosphere, *Space Sci. Rev., 34*, 359.

Goldstein, R., H, Zebker and C. Werner (1988), Satellite radar interferometry: two-dimensional phase unwrapping, *Radio Science, 23*, 713-720.

Hines, C. (1960), Internal atmospheric gravity waves at ionospheric heights, *Can. J. Phys., 38*, 1441-1481.

Hunsucker, R. (1982), Atmospheric gravity waves generated in the high-latitude ionosphere: a review, *Rev. Geophys. Space Phys., 20*, 293-315.

Lichten, S. and W. Bertiger (1989), Demonstration of sub-meter GPS orbit determination and 1.5 parts in 108 three-dimensional baseline accuracy, *Bull. Geod., 63*, 167-189.

Malla, R. and S. Wu (1989), GPS inferred geocentric reference frame for satellite positioning and navigation, *Bull. Geod., 63*, 263-279.

Mueller I. and S. Zerbini, eds. (1989) *The Interdisciplinary Role of Space Geodesy*, Springer-Verlag, West Germany.

Ramanathan, V. (1988), The greenhouse theory of climate change: a test by an inadvertant global experiment, *Science, 240*, 293.

Thomas, J. (1988), *Functional Description of Signal Processing in the Rogue GPS Receiver*, JPL Pub. 88-15, Jet Propulsion Laboratory.

Thornton, C., J. Fanselow and N. Renzetti (1986), GPS-based geodetic measurement systems, *Space Geodesy and Geodynamics*, A. Anderson and A. Cazenave, eds., Academic Press, 197-218.

Wu, S., T. Yunck and C. Thornton (1990), A reduced dynamic technique for precise orbit determination of low earth satellites, *J. Guidance, Control, & Dyn.* (to appear).

Yeh, K. and C. Liu (1974), Acoustic gravity waves in the upper atmosphere, *Rev. Geophys. Space Phys., 12*, 193-216.

Yunck, T. and S. Wu (1986), Non-dynamic decimeter tracking of earth satellites using the Global Positioning System, AIAA paper 86-0404, Reno, NV.

Yunck, T., S. Wu, J. Wu and C. Thornton (1990), Precise tracking of remote sensing satellites with the Global Positioning System, *IEEE Trans. Geosci. & Rem. Sen., GE-28*, (in press).

MAPPING THE EARTH'S GRAVITY FIELD WITH ORBITING GPS RECEIVERS

Oscar L. Colombo
Astronomy Program
University of Maryland
NASA Goddard Space Flight Center, Code 626,
Greenbelt, Maryland 20771, U.S.A.

INTRODUCTION

The orbit of an artificial satellite is shaped primarily by the force of terrestrial gravitation. If the earth were a perfect and homogeneous sphere, the orbit would be a simple Keplerian ellipse. The planet is closer to a flattened ellipsoid than to a sphere; the pull of its equatorial bulge causes the orbit to turn within its plane at a constant rate (precession of the perigee), and also slowly rotates the plane of the orbit about the earth's axis (precession of the line of nodes). But the actual shape and mass distribution is much more complex than for a flattened, homogeneous ellipsoid. Departures from this simpler body cause irregularities (anomalies) in the gravity field that in turn produce "wiggles" or perturbations in the orbit. These departures from a precessing ellipse about a flattened ellipsoid can be sensed with tracking systems that measure the instantaneous distance or velocity of the satellite along the line of sight to the tracker.

The tracking of artificial satellites has been used to map the earth's gravity field since the very early days of space exploration, in the late Fifties and early Sixties. One of the first interesting discoveries was the existence of a small difference between the northern and the southern hemispheres, making the earth slightly pear-shaped (Eckels et al.,1959). The first comprehensive text on the new subject of satellite geodesy was that of Kaula (1966). A recent publication by Lambek (1987) introduces the basic concepts of that discipline and explains its main geophysical applications. The present state of knowledge of the gravity field, the main measurement techniques--both terrestrial and spatial, the geophysical and oceanographic applications of such knowledge, and the scientific priorities for future research in this area, are reviewed in a recent report written by an international and multi-disciplinary group of scientists (Mueller and Zerbini, editors, 1989).

In satellite geodesy, tracking data are used to estimate the parameters of a mathematical representation of the earth gravitational potential (geopotential), to map gravity at or near the earth's surface. The estimated parameters and the corresponding mathematical representation constitute a *gravity field model.* The orbital perturbations due to surface anomalies tend to become smaller the smaller the dimensions of those anomalies, or the shorter their spatial wavelength. This decrease in sensitivity becomes more pronounced with increasing altitude. Global (world-wide) representations of the field have usually the form of *normalized spherical harmonic expansions* (Heiskanen and Moritz, Ch. 1, Par. 14, 1967), truncated, ideally, at a degree corresponding to a wavelength short enough for the related orbital perturbations to be small compared to the accuracy of the data. This wavelength is a function of the height of the satellite. As more and better data have become available, the resolution of the models has increased to degree 55, or about 400 km for the size of the smallest detail (Marsh et al., 1988).

Tracking measures the distance or the relative velocity along the line of sight from tracker to spacecraft. Either the return time of radar pulses, or the Doppler frequency shift in a transmitted signal are measured and converted to distance or velocity, after suitable corrections for refraction and other systematic effects. With the tracking systems used so far to map the terrestrial gravity field, the situation is as shown in Figure 1(a). The tracking

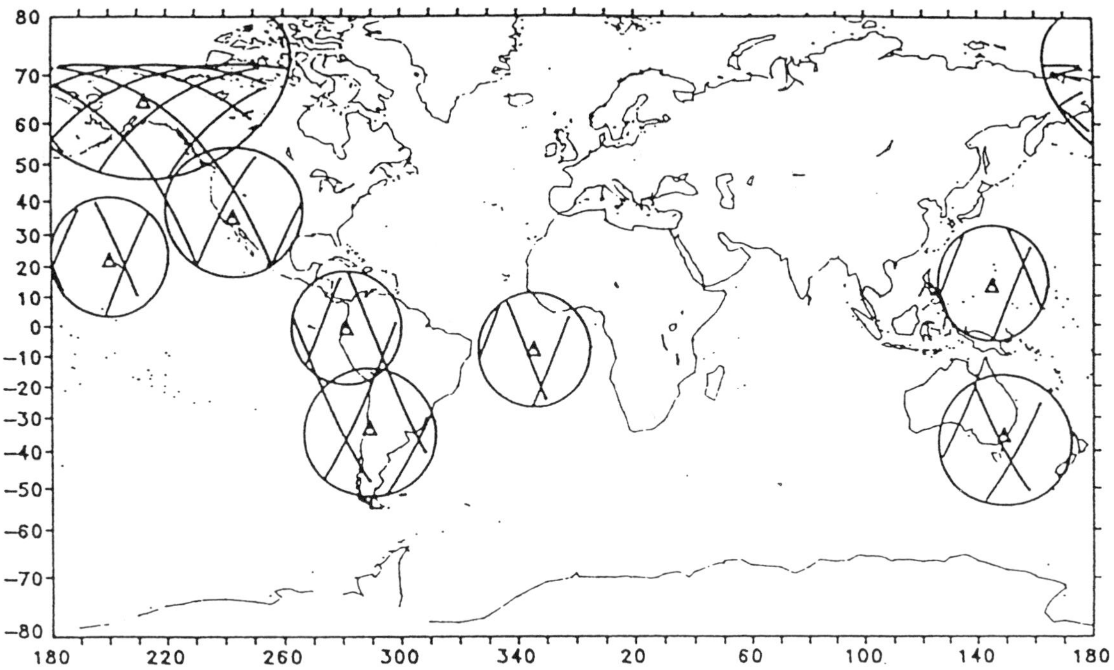

Fig. 1(a). Tracking coverage for a low orbiter from conventional ground tracking stations. The circles shown on the map are regions, centered on the stations, which the spacecraft must overfly for tracking to be possible. This makes coverage spotty, as well as concentrates it on parts of the world where stations are most numerous: North America and Western Europe.

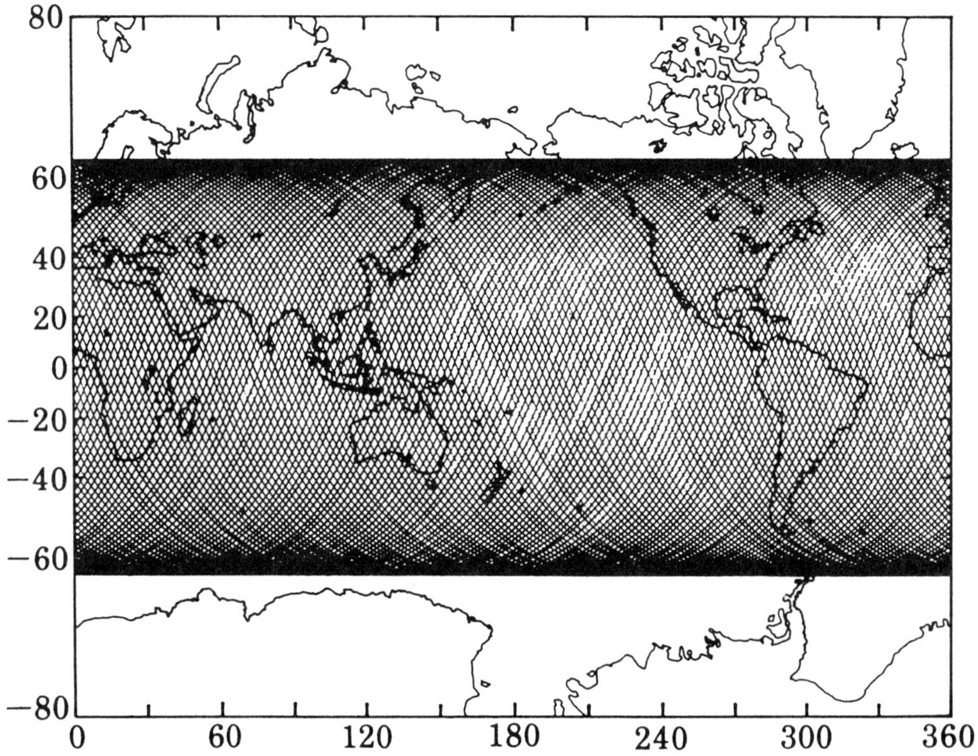

Fig. 1(b). Tracking coverage possible when the lower orbiter carries its own GPS receiver: the coverage band shown is identical to the ground track (in this case, that of the oceanographic satellite TOPEX). With the full GPS constellation (21-24 spacecraft), 4 satellites are in sight almost everywhere simultaneously, providing continuous 3-dimensional tracking.

is done from ground stations, and there is line-of-sight visibility only when the spacecraft is passing over points within the circles that surround each station. Tracking is only possible at those times, which means that coverage, particularly over land, tends to be spotty, and concentrated on certain areas (North America and Western Europe). The situation is helped over the oceans by the availability of satellite altimetry, which can be regarded as tracking from the spacecraft to the mean sea surface. (This has led to the fruitful idea of using altimetry to chart simultaneously the ocean geoid and the stationary sea surface topography in the manner proposed by Wagner, 1986). Not only is coverage constrained by the inability to track a satellite below the horizon of a station; another limitation occurs because orbit perturbations can be detected only along the line of sight. To observe the perturbations in three dimensions, simultaneous tracking is needed from at least three stations. This is rarely done, because such stations are set up at fixed locations on the earth's surface, and are relatively few in number.

The situation just described is in the process of changing because of the introduction of the multi-spacecraft Global Positioning System (GPS). A lower satellite with a suitable receiver can be tracked with several GPS satellites at the same time, and this can be done continuously around the globe once the full constellation becomes available.

There are two main reasons for this: the large number of GPS satellites (up to 24, eventually), and the very high orbits (20000 km), from the vantage of which lower spacecraft can be visible over ground circles more than 6000 km in radius. So GPS can provide both around the world coverage along the ground track, as suggested in Figure 1(b), and simultaneous tracking in three dimensions (to cover the whole world, the lower satellite must be in a polar orbit, as its latitude is confined to between plus and minus the orbit inclination). Using carrier-phase and pseudo-range simultaneously to resolve the ambiguity in the phase, one can have so-called carrier-range, a full one-way range of cm accuracy (after correcting for the ionosphere with both the L1 and L2 frequency signals). However, using a one-way ranging system presents the problem of clock stability. This requires additional data from other receivers, to either operate in a differential mode, canceling out clock errors, or to estimate clock corrections. This could be done if a global network of terrestrial GPS stations operated at the same time as the receiver in the satellite. Fortunately, plans for the implementation of such a network are now well under way (see description of CIGNET in Mader, 1988), with promise of more to come in the form of a "global strain network" of hundreds of GPS receivers placed permanently around the world (NASA Coolfont Meeting, 1989, report in preparation).

Advanced GPS receivers for space applications are under development (e.g., Carson et al., 1988). One example is the oceanographic satellite TOPEX (Melbourne and Davis, 1987), scheduled for 1992. As pointed out by Yunck et al. (1985), a variety of other applications (orbit determination, ground station positioning, etc.) exists for the same data discussed here in relation to gravity field modelling.

As explained later, preliminary studies at several research centers have shown that a few months of data from just one orbiting GPS receiver could give a field model much more accurate than the best available today.

FUTURE MISSIONS THAT MAY CARRY GPS RECEIVERS

There are several missions now under consideration by various space and government agencies in the U.S.A. and Europe that may have GPS receivers on their spacecraft. Those listed in Table 1 are earth-observing missions, so their orbits have high inclinations (for large global coverage) and are nearly circular (to minimize variations in height).

With the exception of the ARISTOTELES satellite, orbit heights range from 600 km, below which air drag becomes quickly a major problem, to 1300 km, where the penetrating radiation of the van Allen belt causes solar panels and electronics to deteriorate quickly unless costly techniques are used to "harden" them. ARISTOTELES is to be drag-compensated by the very frequent thrusting of small rocket engines, and is projected to have an altitude of 200 km.

Given a certain accuracy, ground coverage, tracking data rate, and mission length, the maximum resolution of a gravity field model obtained from that tracking depends on altitude. In the case of a single low orbiter tracked constantly with GPS satellites for a period of one year, preliminary studies indicate that the shorter resolvable wavelength is approximately equal to the orbit height. Accordingly, the more usual measure of resolution, half the wavelength, ranges from 300 km to 600 km for most missions, which are expected to last more than one year, while 6-month long ARISTOTELES will probably give a resolution close to 150 km from its GPS data alone.

In addition to the missions listed in Table 1, the Space Shuttle will also be equipped with a GPS receiver; combining GPS and inertial navigation to measure gravity in orbit is the objective of the STAGE project (Upadhyay and Jekeli, 1988)

Table 1. Missions that may carry a GPS receiver.

Name	Date	Comments
GEOSAT-B	1991	Altimeter, successor to GEOSAT, same orbit (**800 km** height).
TOPEX	1992	USA and French Altimeters and other oceanographic instruments (**1300 km**) Receiver in development.
GRAVITY PROBE-B (GP-B)	1995	Stanford General Relativity Gyroscopic Experiment (**600 km**)
ARISTOTELES	1995	ESA Gravity Gradiometer (**200 km**) 6-month mission.

NOTE: Gravity resolution at the Earth's surface comparable to height/2.

DATA ANALYSIS: MAIN CHARACTERISTICS

General

Differential GPS is likely to be the mode of operation used in precise space applications, such as orbit determination and gravity field modelling (Yunck et al., op. cit.). Simultaneous measurements can be double-differenced to eliminate clock errors when more than one GPS satellite is tracked simultaneously from the lower orbiter and at least one other receiver. While some of the additional receivers needed may be in other spacecraft, the missions listed in Table 1 will probably rely only on ground receivers. Depending on the orbital inclination of the lower satellite, between 6 and 12 ground receivers are needed around the world to form double differences everywhere along the orbit. Some gaps in the coverage may occur during the earlier missions, before the full constellation is available.

An important characteristic of gravity field modelling with GPS is that, on account of the great accuracy, quantity, and geometric strength of the data, it is possible to sense much smaller orbital perturbations than with other forms of tracking, except laser. Because of the more or less complete geographic coverage allowed by GPS, virtually all the field can be modelled from *short-term orbital perturbations* (frequencies higher than one orbital period), without particular reliance on the much larger *long-term perturbations* (typically, frequencies between once every ten days, and once per revolution). Long-term perturbations have been very important in field modelling until now, given the sparseness and overall lower accuracy of the data available. The main problem with these perturbations is that, for a single low orbiter, they do not convey enough information to separate all potential coefficients in the spherical harmonic expansion of the field, from each other. In order to resolve all such coefficients in a model, it has been necessary to use data from many low satellites in different orbits (primarily varying in inclination), with different sensitivities to the various coefficients. With GPS, as suggested by the studies described later on, it should be possible to obtain a full and very accurate field model from the data of a single low satellite, if this is intensively tracked for several months.

The motion of a space vehicle is so smooth and predictable that it is not necessary to determine each position quite independent from the rest of the trajectory. Successive positions are strongly related to each other, as the accelerations involved (mostly gravity, plus some radiation pressure and air drag) change very slowly. For this reason, the orbit parameters (initial conditions, radiation pressure, y-biases, drag coefficients, etc.) may be estimated from occasional tracking, as is the case with laser ranging and Doppler systems. Terrestrial vehicles, with much less predictable motions, require observing a minimum of four GPS satellites at all times. A low orbiter can, in principle, be tracked with less. However, the simultaneous observation of several GPS satellites should improve results, because of the stronger geometry. Using GPS data to track a satellite is a "kinematic" application (positioning a moving receiver) that has much in common with "static" applications (positioning fixed ground stations). While the relative velocity between a GPS satellite and a lower orbiter can be 3 times higher than between a ground station and a GPS satellite (i.e., high Doppler shift), actual changes in the measurements are easy to predict (fairly constant accelerations of the order of 1 **g**) from the immediate past history of measurements, or from a rather coarse knowledge of the orbits involved. Consequently, GPS satellites can be acquired, and lock maintained, using a narrower band-width receiver with less measurement noise than in most kinematic applications. Moreover, inertial navigation (accelerometers and gyroscopes) may be used to enhance the GPS navigation and to help separate gravity information from the effect of other (non-gravitational) forces, in much the same way as in terrestrial vehicles (cf. Upadhyay and Jekeli, op. cit.). Also, because most satellites move above the main layers of the ionosphere, the corresponding refraction is considerably reduced, and for higher orbits (above 600 km) it may be possible to use only the more precise L1-band signal without resorting to the accuracy-degrading ionospheric correction. The problem of tropospheric refraction is, of course, absent from the low orbiter/GPS data (although it will appear in the ground stations/GPS legs of any double differences involving terrestrial receivers, the use of which is explained below). Multipath should be less of a problem than in most other GPS applications, because the only source of unwanted reflections near the orbiting receiver is the low orbiter itself. By carefully positioning the antenna, shielding it from the rest of the spacecraft with an appropriate ground-pad, and using the fairly predictable attitude and geometry of the satellite to estimate the time and nature of any serious multipath that may occur, it should be possible to greatly reduce the occurrence of the problem, and to identify and, if necessary, edit out any data that may be occasionally affected by it.

Analysis Techniques

Non-sequential (batch) adjustment. For satellite altitudes of more than several hundred kilometers, it is possible to model the gravity field with GPS data using current techniques for the analysis of satellite tracking. These techniques are based on the numerical integration of the equations of motion to calculate the orbits, and of the linearized equations of motion to get the sensitivities or "partials" corresponding to the gravity parameters The parameters most often used are the *coefficients of the spherical harmonic expansion of the gravity potential (fully normalized)*, truncated at a sufficiently high degree and order, which depends on tracking accuracy and coverage, orbit height, etc. (In what follows, the letter n will indicate the spherical harmonic degree.) The number of spherical harmonic coefficients can be quite large, and they appear in all observation equations. This tends to rule out the use of recursive estimation schemes, with that type of models, in favor of batch estimation. The number of field parameters is, approximately, the square of n = Nmax, the maximum degree in the truncated expansion. As suggested by preliminary studies, if that degree corresponds to the shortest wavelength that can be resolved, and if HS is the height of the spacecraft, then, since
wavelength = earth's circumference/harmonic degree , it is *Nmax = 40000 km/HS*. For the values of HS between 600 km and 1300 km shown in Table 1, this would range from Nmax = 35 to Nmax = 65, so the number of the non-sequential field parameters can be close to 5000. In the special case of ARISTOTELES (HS = 200 km), the number of spherical harmonic coefficients is in the tens of thousands. However, it is possible to use first data from the gravity gradiometer on board to estimate the field parameters above degree 30 by a different technique (see, for example, Rummel and Colombo, 1985), employing the GPS data to model only the field below n = 30.

With so many parameters present in all observation equations, it becomes impractical, without faster computers, to implement the *sequential* adjustment techniques used often to analyze GPS data, such as Kalman filtering-smoothing. For this reason, *batch* estimation procedures have been employed in gravity field modelling until now (where the full system of normal equations is explicitly set up and solved using all data available and taking little or no advantage of any special structure in the adjustment matrices) These procedures may be efficient enough to analyze the GPS data, provided that there are few nuisance unknowns present, like clock parameters, unresolved ambiguities, etc. The clock parameters can be eliminated by double differencing the data, the remaining unknowns (gravity parameters, orbit initial conditions and other quantities for each satellite involved, as well as non-fiducial stations' coordinates, etc.) can be solved in a simultaneous adjustment. This sort of "grand" adjustment is common in satellite geodesy, and a simulation study recently carried out at NASA's Goddard Space Flight Center by a group that included the author, has shown that it can be a practical way of processing GPS data to map the gravity field. The efficiency of the batch procedure may be improved somewhat by using special methods to solve the normal equations taking advantage of some sparseness in them, since the partials of many nuisance unknowns can be orthogonal to each other.

Alternatively, the grand adjustment could be broken into two steps: first, using only the data from the ground receivers, both orbits and clock corrections for the GPS satellites may be determined. This could be done most efficiently by a sequential adjustment technique. The resulting GPS clock corrections could be applied to the data from the receiver in the orbiter; these data then could be single-differenced between GPS satellites to eliminate the orbiter's clock. In the second step, these single differences could be used to find the gravitational parameters, and simultaneously (re)estimate all the orbits, etc., in a batch adjustment.

The advantage of using double differences over the two-step approach is that, with double differences, undesirable effects other than clock errors cancel out or may be significantly reduced. One example is the error in the known position of the phase center of a transmitting antenna relative to the center of mass of its GPS satellite. On the other hand, estimating first the GPS satellites orbits and clock corrections with ground station data only, and using exclusively orbiter data in the final adjustment, has the advantage that there is no need to have exactly simultaneous ground and orbiter measurements, thus avoiding questions of coordination between ground and space operations (which may be run by different organizations with different primary goals).

Sequential adjustment. Instead of representing the field globally, with a spherical harmonic expansion that requires a batch adjustment, it is possible to represent the field locally. In particular, the accelerations at each observation time may be describe by a few parameters. These parameters could be the three components of the anomalous gravity acceleration vector at that point in the orbit, provided that at least four GPS satellites are simultaneously within sight (Wu et al., 1989), or could be the coefficients of low degree polynomials or other slow-varying functions of time, entering in several consecutive observation equations, to increase redundancy in the estimator. In addition, all other parameters (orbits, station coordinates, biases, refraction, etc.) have to be included. Non-gravitational accelerations (drag, electromagnetic radiation pressure) can be modelled in the same way as in standard orbit determination techniques, by adjusting a few force scale factors per arc. With this approach, so far untried in any extensive way, the gravitational unknowns would appear in a sequential manner, so their sequential estimation could be possible along an orbit arc several revolutions long. An appropriate a priori covariance function for the various parameters (based on what is known about the power spectrum of the field and about the a priori accuracies of the non-gravitational force models) could be used to constrain this overparametrized adjustment. Later, the results from consecutive arcs would be combined to obtain a final solution that incorporates all the available data; this can be done quite efficiently by taking advantage of the great sparseness of the overall normal matrix (Wu et al, op. cit.) Once the gravitational accelerations at various points have been estimated, there remains the task of converting them to a form more appropriate for calculating the field at the earth's surface and above it. This may be done by least-squares fitting a set of spherical harmonics to the estimated accelerations, thus getting the same kind of field model as with batch processing.

Local adjustment. Representing the gravity field over a limited region (usually bound by two meridians and two parallels) is an alternative to mapping this field over the whole world. The use of a *local* representation is discussed briefly in Jekeli (1988). A potential advantage of this approach is that finer details may be resolved, with the same or a smaller number of parameters, than in a *global* map. One example of local representation is a rectangular array of blocks, with gravity anomalies, geoid undulations, or surface densities averaged over each block. (Other examples: square arrays of point masses, two-dimensional Fourier series).

Strong, localized features due to mountain ranges or oceanic trenches may be smoothed out in a global spherical harmonics representation, or appear badly distorted by "ringing", because of the way the degree of truncation is chosen . Usually, this is the degree where the global r.m.s of the gravitational spherical harmonics sink below the noise (unless practical computing limitations dictate a further truncation). However, a local feature may have considerable *local r.m.s. value,* without contributing much to the *global r.m.s value* of the field (averaged over a much larger area). Local representations may be useful in very low orbit missions (like ARISTOTELES), where an extremely large number of spherical harmonic parameters may be necessary to map the finest details.

Problems in forming double differences. One peculiar characteristic of having an orbiting receiver is that its motion, relative to the earth stations, can exceed 8 km per second, while it can be as much as 12 km per second relative to the GPS spacecraft. This complicates considerably the process of choosing the ground stations and the GPS satellites whose observations will be used to form double differences. As shown in Figure 2, one could think in terms of a network with one station riding on the low spacecraft, so the network changes as this station moves. The main requirement in forming double differences is that the orbiter and at least one terrestrial receiver must observe simultaneously the same set of (at least two) GPS satellites. If several satellites are in that set, one may also try to select those that give good geometry, by enforcing a minimum elevation angle, a minimum Geometric Dispersion Of Precision (GDOP), etc. To ensure uninterrupted tracking around the world, the choice of ground receivers and GPS satellites must change often: as the lower spacecraft moves, GPS satellites rise and set on its horizon. Typically, a change will occur at intervals ranging from some 5 to 20 minutes. Occasionally, if low elevation GPS satellites are included, or if its possible to carry antennae both below and above the low spacecraft, the same configuration may be kept for more than half an hour (the orbiter's horizontal plane is higher than the earth's limb, so one can look down as well as up).

Every time a new ground receiver or a new GPS satellite is chosen, new carrier-phase ambiguity parameters are introduced. If the configuration is maintained long enough, those may be resolved by simultaneously observing phase and pseudo-range, and averaging their ionosphere-corrected differences. For this to work, one should average several minutes' worth of data, so the uncertainty of the result is a small fraction of the carrier wavelength. There should not be too many unresolved ambiguity parameters, which have to be included in the overall batch adjustment. So the chosen configurations should last as long as possible, to maximize the number of resolved ambiguities. A further complication is that the number of ground receivers available may be small, and they may not be ideally distributed. Altogether, it is no trivial task to implement a ground stations/GPS satellites selection procedure that:

(a) permits continuous, double differenced tracking coverage of the lower orbiter,

(b) maintains good geometry, and

(c) minimizes the number of unresolved ambiguities.

It is always possible to track the various GPS satellites both from the low orbiter and from the ground stations for longer periods than needed to form the scheduled sets of double differences. In this way, one could have a longer time-interval in which to resolve the ambiguities. This has the drawback that more data from the space-borne receiver will have to be collected than is strictly needed for the double differences, which requires more storage and pre-processing in the spacecraft computer, as well as extra use of the down-link communications channel.

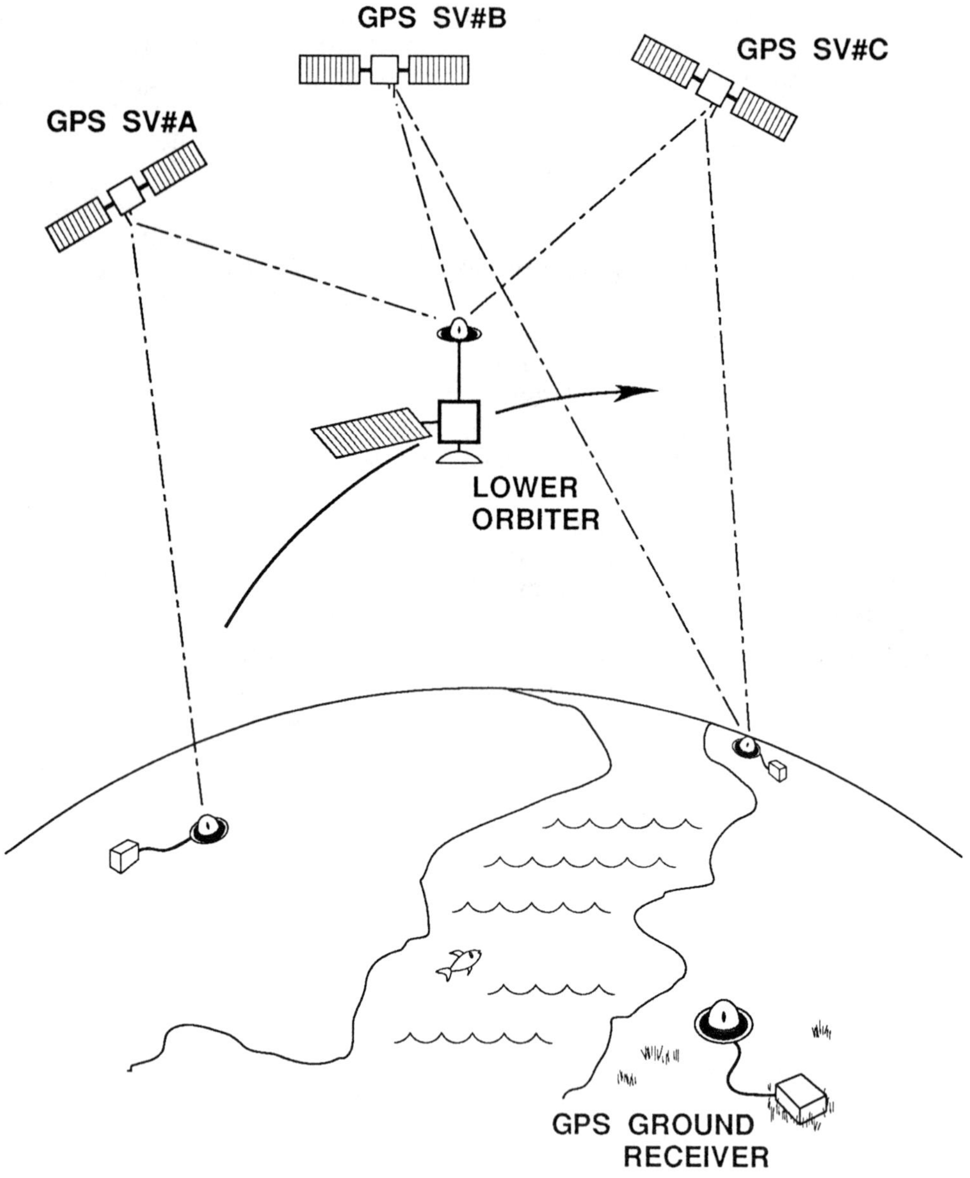

Fig. 2. Picture of ground station receivers, GPS spacecraft, and lower orbiter with GPS receiver. The broken lines are one-way ranges differenced to eliminate clocks (at least four GPS satellites should be visible from one ground station and the low orbiter simulatneously for 3-D orbit observation). Spacecraft moves at about 7 km/sec relative to earth; this complicates the selection of GPS spacecraft and receivers to form double differences while resolving most ambiguities in the carrier phase with help from precise pseudorange.

SUMMARY OF SOME GLOBAL MISSION ANALYSES

Preliminary Error Analyses

Cylindrical earth approach. Earlier work was centered at the Jet Propulsion Laboratory (Melbourne and Tapley (1983), Yunck et al. (1985)) as part of an overall assessment of the value of including a GPS receiver in the TOPEX/POSEIDON spacecraft (see Table 1). Later, these calculations have been repeated to estimate the possible contribution of a similar receiver on the Gravity Probe-B (GP-B) spacecraft (also in Table 1). This work is representative of this type of study in that in it a simplified first order perturbation theory has been used to find the variance-covariance matrix of the adjusted potential coefficients. In this specific case, the earth was treated as an infinite cylinder; the harmonic functions considered were those that are constant along the length, and are simple Fourier series around the circumference (the cylindrical analogues of sectorials on a sphere). With these simplifications, the corresponding orbital perturbations are also Fourier series if the orbit is circular and its plane normal to the axis of the cylinder. This leads to a system of normal equations that is diagonal, so it is trivial to invert its matrix to obtain the variances of the (uncorrelated) gravity parameters. In this study, the distribution of all satellites was in one common plane perpendicular to the cylinder, with three GPS spacecraft spaced 120 degrees from each other and co-rotating with the low orbiter, and another three equally spaced and counter-rotating. Six ground receivers were assumed placed at 60 degree intervals along the circumference of the cylindrical earth.

Flat earth approach. A second approach has been used by J.V. Breakwell and M. Tapley (Everitt et al., 1989), from the GP-B Group at Stanford University, based on a mathematical treatment of short wave orbit perturbations by Breakwell (1979) which involves a flat-earth approximation. In this treatment, the gravity field potential and accelerations are two-dimensional Fourier series, and so are the orbital perturbations, which are treated as linear transforms of the anomalous potential (assuming a continuous geographical coverage with very closely spaced, parallel ground-tracks, although the effect of a coarser coverage is approximately accounted for). In turn, this leads to a diagonal normal matrix, as in the cylindrical earth approach, whose inversion is trivial.

Spherical, rotating earth approach. A third set of studies has been done at NASA's Goddard Space Flight Center (Smith et al., 1988), for GP-B. The orbit dynamics were approximated to first order, assuming the earth to be nominally spherical and rotating about its axis. The perturbation theory used was based on Hill's linearized equations of motion (Kaplan, 1976, Colombo, 1984). In this theory, valid for a nearly circular orbit, the perturbations are given in three orthogonal directions: radial, across (the orbit plane) and along (the orbit). These directions define a rotating Cartesian system, which turns about its across axis once per orbital revolution. The solution of Hills' equations is resonant, with critical frequencies at zero and once per revolution. In the chosen coordinates, the linearized dynamics are those of a simple harmonic oscillator in the across direction, and those of a fourth-order time-invariant oscillator in the radial and along; the two states of the across oscillator are completely decoupled from the four of the radial and along one. As shown in (Colombo, 1984 and 1986), the partials of the spherical harmonic potential coefficients (except for some secular resonant effects) are trigonometric series; assuming an orbit with a repeating ground-track, these become true Fourier series, many of which are orthogonal over a complete orbit repeat (10 days in this study), so the corresponding normal equations are very sparse (block-diagonal, with appropriate ordering of the unknowns) and their matrix can be inverted very quickly (even when the dimension may be quite large) to obtain the variance-covariance matrix of the estimated coefficients. In these studies, the distribution of GPS satellites was supposed to be such that a minimum of four were visible at all times, while ground receivers were so located that three or more double differences could always be formed, and the instantaneous position of the low orbiter totally determined at every point. In this study, as in the others described in this paragraph, observations were supposed to be made at the rate of once per second of both carrier-phase and pseudo-range at the L1 and L2 frequencies. Moreover, the accuracy of the ionosphere-corrected carrier-phase double differences was 2 cm, and the average Dilution of Precision factor for each coordinate was 3 (a value based on a study by Martin and McCarthy, 1987); unresolved ambiguities were treated as step-wise process noise with a triangular covariance function having a correlation length of 15 minutes (estimated average time a GPS satellite would remain in the tracking configuration). Only gravity was supposed to drive the orbit perturbations, which is a reasonable assumption in the case of GP-B, as this will be a drag-free spacecraft. Two different situations were considered: (a) the "best case", where the orbits of the GPS spacecraft are perfectly known, and all the ambiguities are resolved (their a priori variance is zero); (b) the "worst case", where no ambiguity has been resolved exactly (a priori variance is 10 cm, or half a wavelength), and the orbits of the GPS satellites are poorly known, so adjusting them may severely weaken the information content of the data relevant to all three directions: radial, across and along. Because of inter-orbital modulation of the tracking signal, this weakening is most likely to affect the signal near the twice-daily orbital frequency (once per revolution) of the GPS spacecraft, and at pairs of plus/minus the same frequency, centered at each of the (first six) harmonics of the GP-B frequency (once every ninety minutes). Since the orbits were supposed to be estimated in daily arcs, the loss of information was extended to a band of plus/minus one cycle per day about each of the frequencies just mentioned. The conservative assumption was made that the loss of information at the suspect frequencies was, in fact, complete. Assumptions regarding data rates, accuracies, etc., were similar to those made for the two previous techniques.

The results and their implications. When the results of the three methods are compared (after approximately equivalencing cylindrical and flat earth ones to those of the spherical earth approach), they are all in reasonably good agreement (same order of magnitude), as shown in Figure 3 (based on Everitt et al., 1989).

Figure 4 shows plots similar to those of Fig. 3 obtained using the spherical earth method for three different low orbiters: one at the altitude of TOPEX (1300 km), the other at that of GP-B (600 km), and the third at that of ARISTOTELES (200 km). In all three cases the spacecraft is supposed to be drag-free, so the estimates are somewhat optimistic for TOPEX and, particularly, for ARISTOTELES. It is clear from the figure that the horizontal resolution is nearly half the orbit height, assuming the same GPS tracking and coverage as for GP-B in Fig. 3 under the "best" case conditions (i.e, perfect GPS orbits and all ambiguities resolved). Resolution here is the degree Nmax at which the accuracy of a coefficient of that degree equals, on average, the signal spectral power calculated according to Kaula's rule of thumb: *Mean Power per Coefficient* = $10^{-5}/n^2$ where n is the harmonic degree.

Figure 5 compares the expected accuracy for a GP-B field model by harmonic degree with that for the recent GEM-T1 model (Marsh et al., op. cit.), suggesting that the GPS data may bring more than two orders of magnitude improvement at lower degrees. Also shown is the plot of likely accuracies for a future field model incorporating smoothed satellite altimetry, based on studies conducted at Goddard Space Flight Center (S. Klosko, private communication).

Figure 6 compares the spectra of the errors in the model that may be obtained by tracking GP-B with GPS, with the spherical harmonic expansions of several signals of oceanographic and geophysical interest, all equivalenced to dimensionless potential coefficients. Shown, for example, are the mean dynamic sea surface topography (based on a spherical harmonic analysis of the Levitus (1986) global chart); the M2 ocean tide loading potential (based on the analysis of the Schwiderski (1979) model); and the effect on the

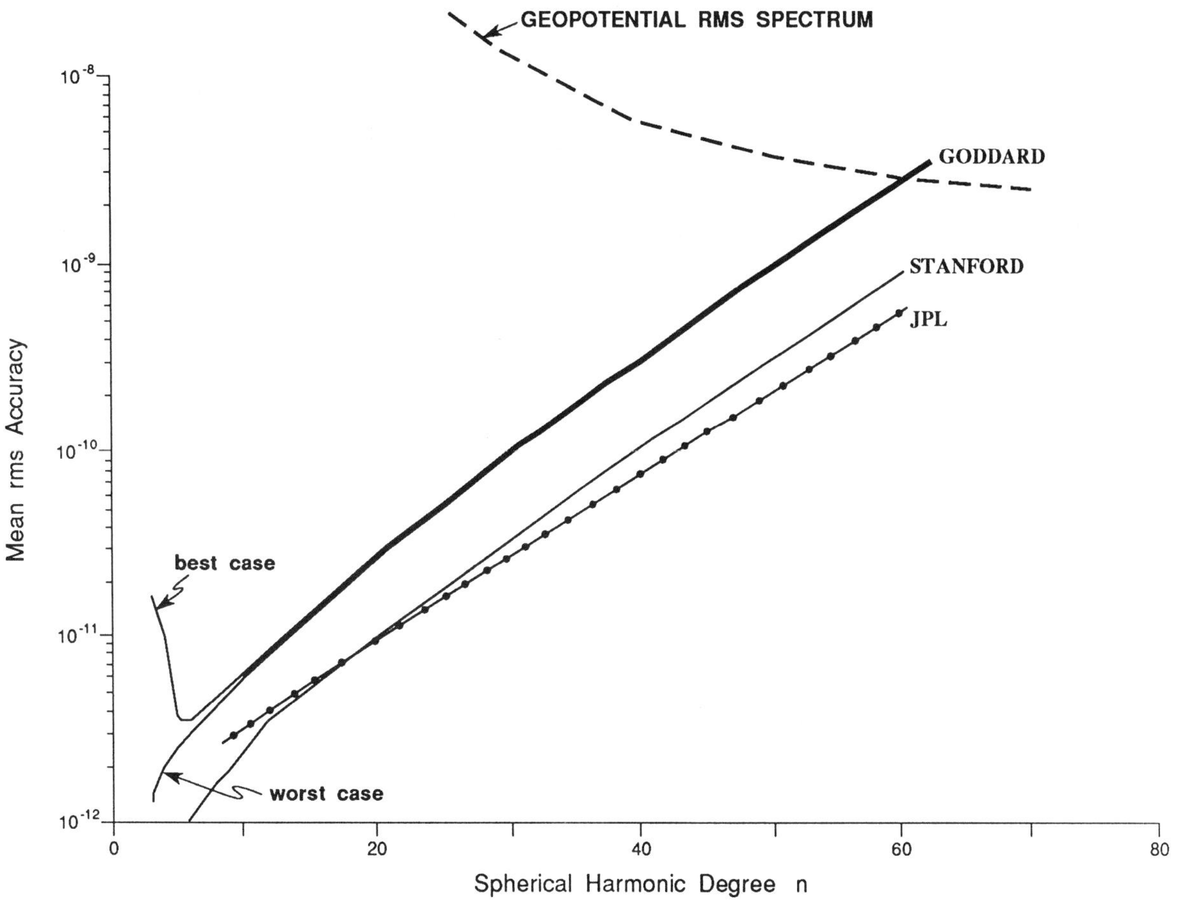

Fig. 3. Plots of cumulative geoid undulation (r.m.s.) errors per degree for estimates based on 2 years of GP-B/GPS data according to: Flat Earth approach (Everitt et al., 1988, Stanford University); cylindrical earth (JPL); spherical, rotating Earth (Smith, Colombo, et al., 1988). Notice reasonable agreement amongst the three methods.

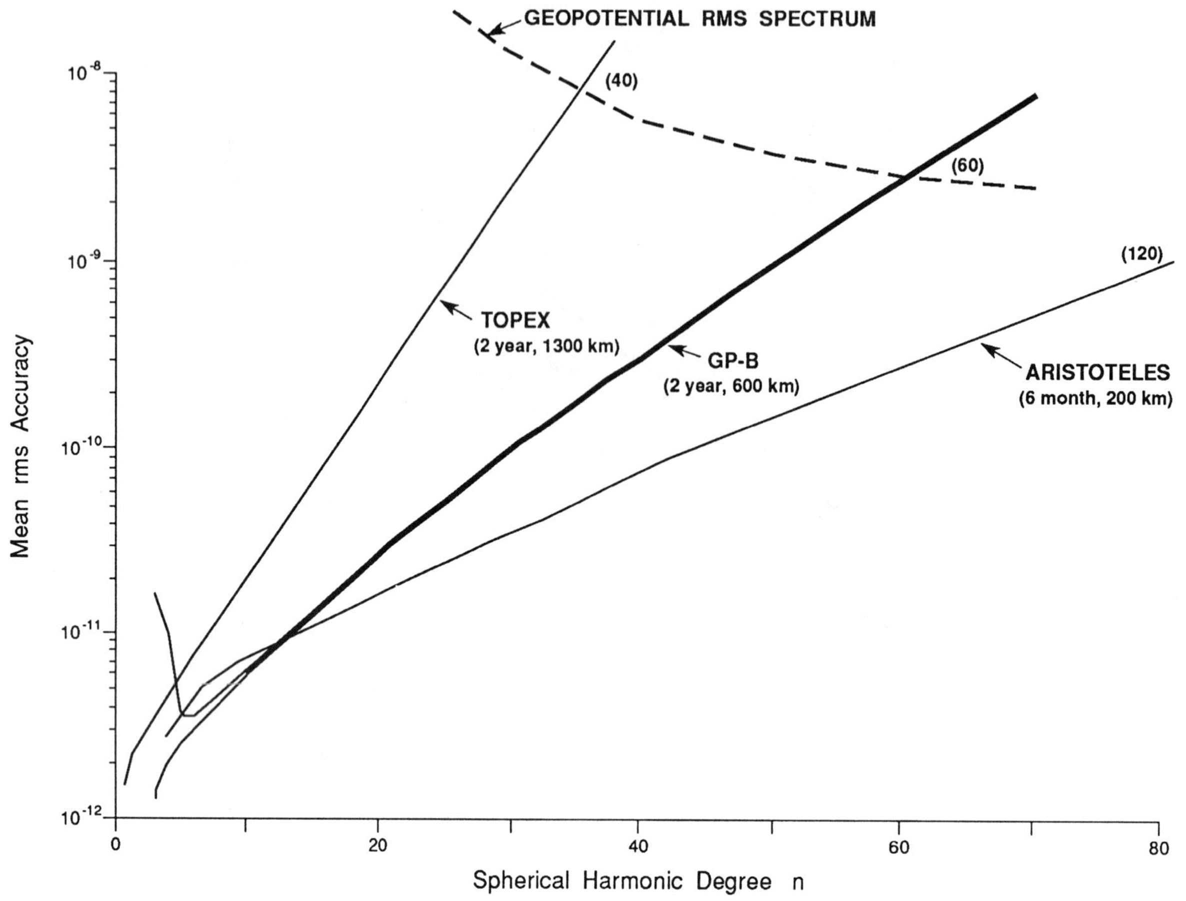

Fig. 4. Potential coefficient acuracies (mean r.m.s. per coefficient per degree n) for three satellites: TOPEX (1300 km altitude, 2 year), GP-B (600 km, 2 year), and ARISTOTELES (200 km, 6 month) . Notice that maximum degree resolved (in brackets) according to geopotential signal spectrum (broken line) corresponds to a half wavelength roughly equal to satellite height.

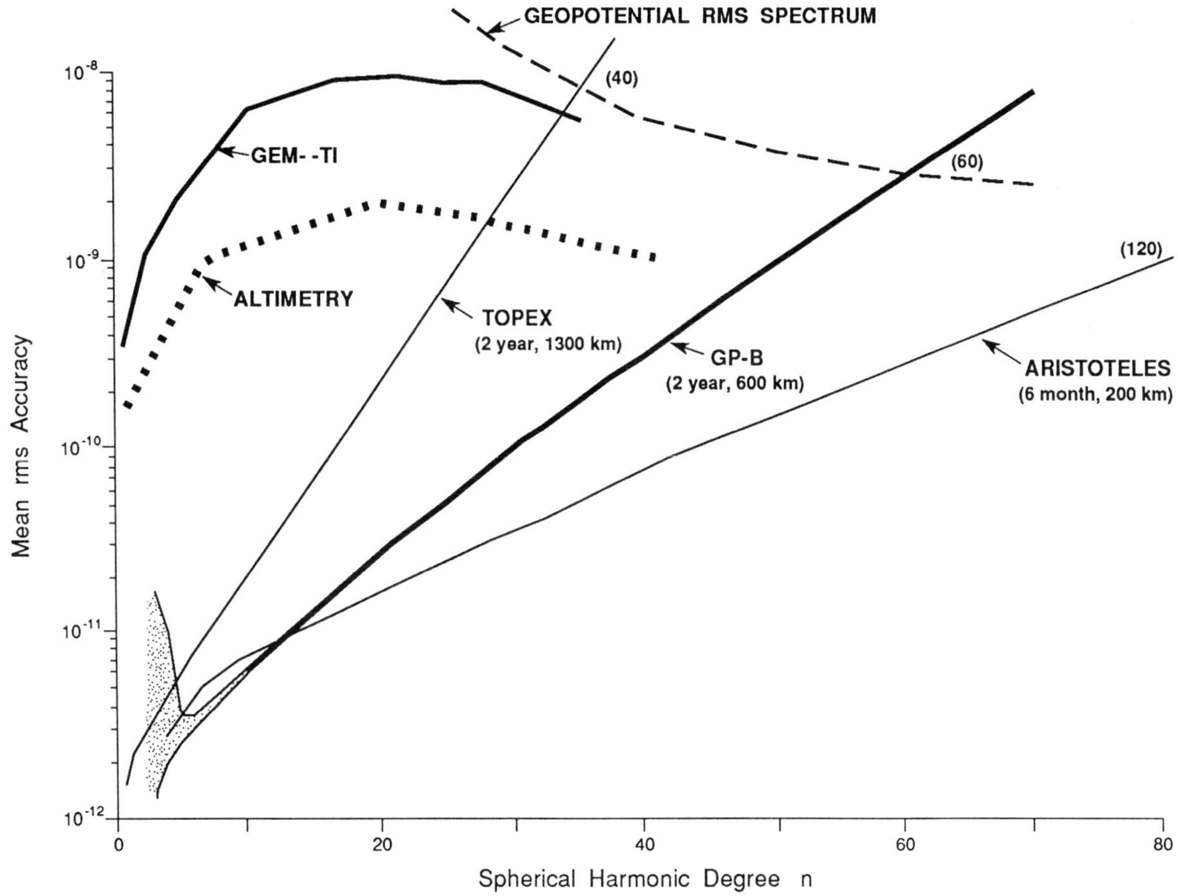

Fig. 5. Potential coefficient accuracies (mean r.m.s per coeff. per degree n) resulting from GPS tracking data of: TOPEX, GP-B, and ARISTOTELES compared with similar accuracies for GEM-T1 model (Marsh et al., 1988, solid line), and with that expected of a model derived from conventional tracking plus satellite altimetry.

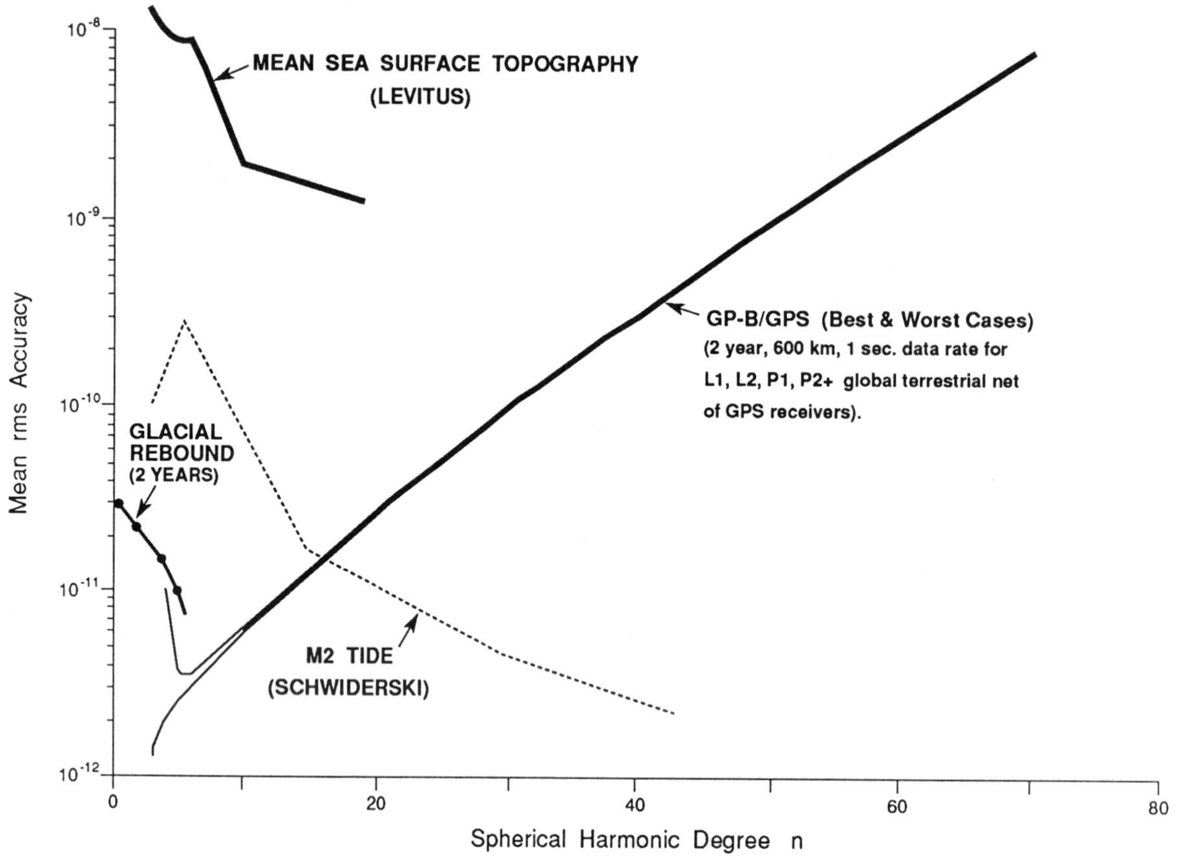

Fig. 6. Potential coefficient accuracies (mean r.m.s. per coeff. per degree n) for GP-B/GPS compared to equivalent spherical harmonic coefficients (dimensionless) of several signals of scientific interest: mean dynamic sea surface topography; loading potential asociated with M2 tide; load. pot ential of glacial rebound.

field of two years of accumulated glacial rebound (for more details on the various geophysical mechanisms behind temporal gravity variations, as well as on their magnitudes, see Ch. 2 and 9 of (Mueller and Zerbini, op. cit.)). The extremely accurate low harmonics predicted for the GPS-derived model suggest that with them the mean sea surface topography should be quite separable from the geoid at long and intermediate wavelengths, while some of the time-varying phenomena, such as tides, could be charted globally, to 10% or better in some cases, at long wavelengths. However, one must keep in mind that these are "noise only" results, and that systematic effects not considered here (due to such things as periodic deformations of the spacecraft to which GPS antennae are attached) can easily obscure these very weak signals. To understand why, one should realize that at low degrees the acceleration associated with a particular harmonic is of the same order of magnitude as the product: (normal gravity **g**) x (size of dimensionless coefficient), so it ranges, according to the figure, between 10^{-9} and 10^{-12} of **g.** Radiation pressure is of the order of 10^{-7} to 10^{-9} of **g,** depending on the satellite area/mass ratio, while drag may vary from 10^{-5} to 10^{-9} of **g**, depending on that ratio and on height (GP-B is expected to have these forces compensated to 10^{-9} of **g** or better). The orbit perturbations due to these time-varying forces are small (1 mm or less), except for those associated with the coefficients of a few resonant spherical harmonic orders, zonals in particular. (These cause very long period perturbations which may build up very gradually to more than one meter in amplitude for some of the tides, but the corresponding spherical harmonics cannot be resolved adequately with data from a single spacecraft. Several satellites at different orbital inclinations are needed.) One may expect to see some improved estimates of tides, but probably not of glacial rebound, because of possible systematic errors. Conversely, if the tidal maps available at the time of these missions are reliable enough (after they are improved through the use of satellite altimetry), one could compare them to the corresponding GPS tidal estimates to verify the accuracy of the latter at very long spatial wavelengths.

A Detailed Simulation Study

To further investigate the potential value of GPS data in mapping the gravity field of the earth, a group associated with Goddard Space Flight Center, including E. Pavlis of the University of Maryland, W.F. Eddy of ST Systems Co., and the author, carried out a simulation of a GPS/lower orbiter tracking data set, and then analyzed it to recover the field parameters used in the simulation, as well as the statistics provided by their variance-covariance matrix.

The spacecraft considered was GP-B, which will carry the Stanford Gyroscopic Relativity Experiment some time in the mid 1990's. It is made almost free from non-gravitational forces, like drag, by the action of small thrusters that act so as to keep a proof mass inside the satellite in continuous free-fall. It is the purely gravitational orbit of this proof mass that is indirectly tracked with GPS, because the position of the antenna relative to this mass is known very accurately (the proof mass is a nearly perfect sphere inside a spherical cavity in a block of fused quartz attached rigidly to the spacecraft, with a gap of 0.1 mm between the sphere and the enclosing cavity wall). There is great scientific interest on this mission, which aims at testing General Relativity by observing the "magnetic" aspect of gravity (associated with gravitational waves and other so far unconfirmed predictions of that theory).This is to be done by measuring the drift of extremely stable gyroscopes in the spacecraft relative to fixed stars observed with on-board telescopes. The gyroscopes and other high precision components will be kept in a cryogenic container of liquid helium. As the helium boils due to internal heating, it will be used as propellant for the drag-compensating thrusters. In addition to the relativity experiment, the spacecraft will have a GPS receiver and also corner cube reflectors for laser ranging from ground stations. These additions make it of particular interest in geodesy; given its polar orbit and low altitude (600 km), the gravity field can be studied in greater detail than with any other spacecraft except ARISTOTELES. It could be also of value in defining a geodetic reference coordinate frame and observing changes in earth rotation, as well as making a continuous connection of the GPS and Satellite Laser Ranging (SLR) reference frames during the length of the mission (up to two years).

The simulation included the 21 primary satellites of the final GPS constellation, GP-B, and 22 ground receivers distributed mostly at SLR and Very Long Baseline Interferometry (VLBI) sites around the world. Tracking data were supposed to be measured simultaneously at all receivers, including the one on GP-B, the data being double-differenced with the help of the ground receivers. One particularly laborious task was that of scheduling the observing times of GPS satellites, both from GP-B and from ground receivers, so as to form double differences while meeting all the conditions outlined in a previous paragraph. This was done as well as possible in the short time available to complete the work. As a result, uninterrupted tracking was achieved over the whole duration of the simulated mission, and three-dimensional tracking (four satellites in common view) over three quarters of that time. Previous to the simulation, Goddard's main orbit analysis program, GEODYN (Martin et al., 1987), was expanded to include all features required for the complete analysis of real GPS data; the simulation study provided a further check that the program was correct (it has been tested also with actual ground receiver data).

The data were considered to be normal points (i.e., average values), each equivalent to 40 raw observations made at the rate of one per second. The raw carrier-phase double differences were supposed to have an accuracy of 2 cm after correcting for the ionosphere, but before being compressed into normal points. Pseudorange was supposed to be available simultaneously, and to be accurate enough to resolve the ambiguities in the corresponding phase measurements within six minutes. If a ground station or a GPS satellite was part of the double differences for less than six minutes, the corresponding ambiguities were treated as unresolved, and included among the variables to be adjusted.

Besides the potential coefficients, the set of unknowns included the initial position and velocity vectors of all 22 spacecraft, solar radiation pressure and y-bias parameters for each GPS satellite, unresolved range ambiguities, tropospheric parameters for each ground station (with a new parameter introduced every four hours), the coordinates of all stations (except for two latitudes and one longitude), and polar motion and length of day corrections. Most of these unknowns were totally unconstrained, or were given light a priori weights, equal to the inverses of several times their expected accuracies.

The total time interval was ten days, divided into five two-day intervals, the beginning of each being the epoch time for a new set of two-day orbit arcs for all satellites. The accuracy estimates obtained for the potential coefficients were then scaled to one year by dividing by the square root of 36.5 (365 days divided by 10). The initial orbital elements of GP-B were chosen so as to ensure an accurate repeat of the ground track after ten sidereal days and 149 orbital revolutions. This orbit gives the finest sampling of the field possible in ten days, making sure that it is dense enough to estimate all coefficients to the limit of resolution shown in Fig. 4. Altogether, there were some 50000 double differences, ionosphere-corrected, obtained from a total of 160000 simulated phase measurements (normal points); and from these data, 6603 unknowns were solved to obtain a spherical harmonic gravity field model complete to degree and order 55 . Computations were carried out in the computer CYBER 205 at Goddard. The solution of the normal equations by Choleski decomposition and the inversion of the normal matrix were done using the program SOLVE; altogether, simulating the observations, then forming the normal equations, and obtaining the final results, required about 8 hours of computer time. To verify those results, Gaussian white noise (quasi-random numbers) was added to the simulated double differences; after the solution, the a posteriori residuals and the errors in the recovered coefficients were compared with their estimated standard deviations in a series of statistical tests. These revealed no inconsistencies between the simulated data and the results. Figure

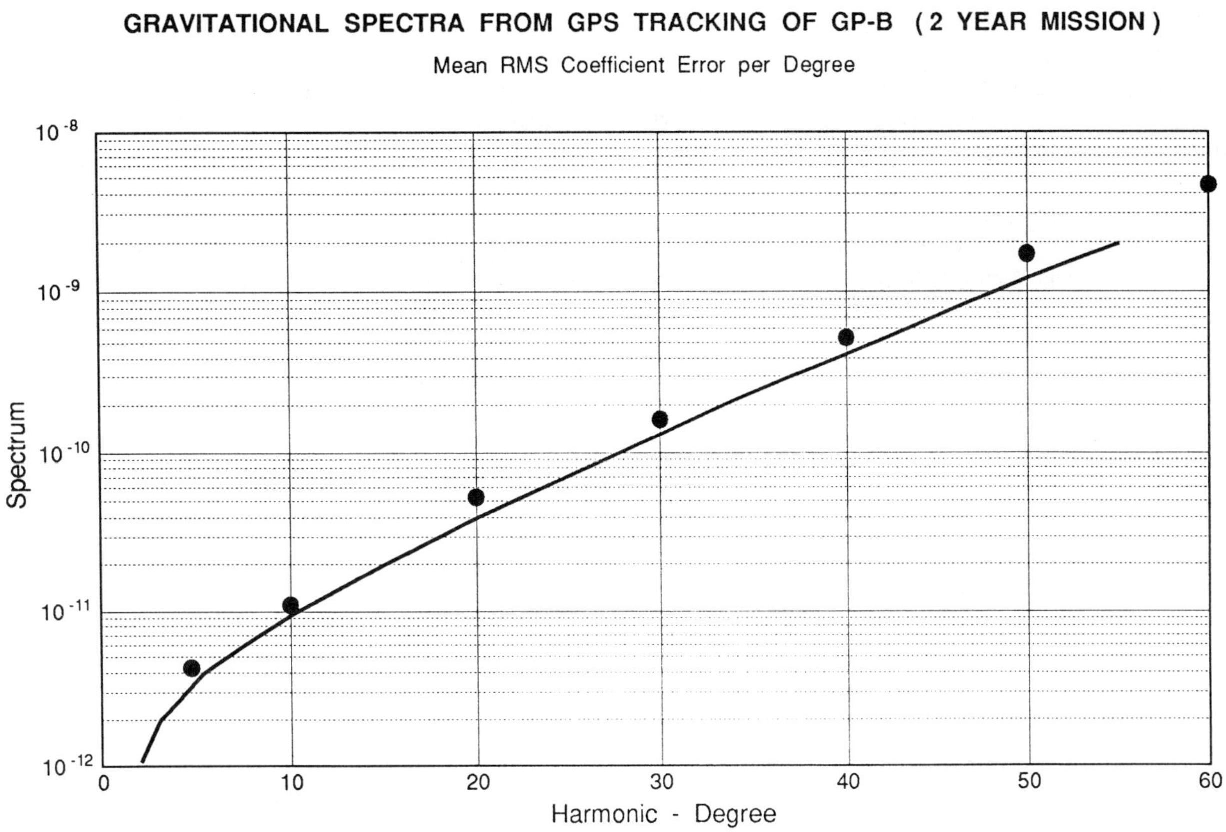

Fig. 7. Comparison of potential coefficients accuracies (mean r.m.s. per coeff. per degree n) predicted by two methods: complete simulation of data and analysis using numerical integration of orbits and partials (with GEODYN)--*continuous line*; analytical method for a spherical, rotating earth and a circular reference orbit--*black circles*.

7 shows a plot of the mean r.m.s. error per coefficient per degree according to the variance-covariance matrix obtained with SOLVE. The black circles superimposed are analytical, spherical, rotating earth results from the "best case" plot for GP-B in Fig.3, scaled upwards so as to correspond to one year worth of data instead of two. The agreement is quite obvious, and also shows the power of the simple analytical technique used to do the preliminary study. That the "best case" turned out to be validated suggests that the simultaneous adjustment of GPS orbits and potential coefficients does not result in any serious loss of information on the gravity field. A full report on this simulation is in the process of being written. (Another simulation has been completed recently, this time for the TOPEX mission; the results will be presented by E.C. Pavlis and D.E. Smith at the 1989 Fall Meeting of the American Geophysical Society.)

The adjustment showed some numerical instability; this was traced to the presence of resonant coefficients (orders 15, 30 and 45, as well as the zonals). The partials for these coefficients are dominated by almost identical resonant perturbations, so the corresponding columns of the matrix of observation equations are nearly proportional to each other, resulting in a normal matrix that can be numerically rank-deficient. This problem is peculiar to the spherical harmonic representation used here, and to the fact that essentially only one satellite, the low orbiter, is sensing the anomalous field, the GPS spacecraft themselves contributing very little additional information, because of their very great height (this is the reason why, until now, all field models have been obtained with tracking data from many satellites--much lower than the GPS ones--and at different inclinations, to have extra redundancy to separate the resonant coefficients). A number of preliminary solutions showed that the severity of the problem increased with the size of the field; it is unlikely to be a serious issue in the case of TOPEX (except for the zonals), because both the maximum degree and order are not likely to exceed 40 (see Fig.4). The problem was obviated by solving only for enough resonant coefficients as were needed to absorb their main overall effect on the data, while keeping the rest of them fixed to their a priori values. In a real solution, it should be possible to deal with this problem much more adequately by a simple transformation that removes the effect of resonance from the partials (and, thence, from the normal equations) while introducing a few extra nuisance parameters to account for this effect in the data. (The idea has been investigated by the author, and will be presented in detail elsewhere.)

One of the most important things learned from this work is that the large amounts of data to be processed (even after condensing into normal points) can be handled by conventional space-geodetic techniques with a reasonable amount of effort using existing computing machinery (the CYBER 205 is only a first generation commercial supercomputer, and is now in the process of being phased out at Goddard). Faster machines, as well as somewhat slower but much cheaper ones, are already available. New approaches, intrinsically more efficient, such as that proposed by Wu et al (op. cit.), may prove even more powerful than current techniques, though this still has to be demonstrated in practice.

CONCLUSIONS

It seems quite likely that, in the coming ten years, there will be a number of spacecraft carrying advanced GPS receivers at a time when a large GPS constellation and a global net of ground receivers also become available. At present, the independent results of several researchers consistently indicate that the combination of these factors will bring about a number of new applications for space techniques, with remarkable effect on geodesy and on the earth sciences. Part of that effect will be a great improvement in our knowledge of the gravitational field of the earth at long and intermediate wavelengths, perhaps including some temporal variations. (There is also the possibility that, in combination with inertial navigation, precise ship- and airborne GPS positioning may help chart gravity at very short wavelengths over large areas of the world.)

Analysis techniques, some quite new, some adaptations of older ones, are being developed and tested in preparation for the day when low orbiter data are available. So far, error analyses and simulations show good agreement concerning the accuracy of the field models to be obtained from such data. Equally important, it has been shown that the large number of measurements expected can be processed using a reasonable amount of computing resources that are available today.

GPS has become a pervasive tool for doing very precise geodesy both locally and on a world-wide basis. In a few years it has transformed the whole field of geodetic science and of surveying, deeply changing the way its practitioners think, work, and what they are able to do. Regardless of what the future of this particular satellite system may be, it has enabled geodesists to demonstrate the power of space geodetic techniques based on radio tracking. The lessons now being learned with GPS should remain useful for a long time.

Acknowledgement. Support for this work has been provided by NASA under grant NAG-5245.

REFERENCES

Colombo, O.L., 1984,The global mapping of gravity with two satellites.Netherlands Geodetic Commission, *Publications in Geodesy (New Series)*, Vol. 7(3), Delft, Holland.

Colombo, O.L.,1986, Notes on the mapping of the gravity field using satellite data. In *Mathematical and Numerical Techniques in Physical Geodesy*. Geodetic Lecture Notes in Earth Science Series, Vol. 7 (H. Sunkel editor), Springer-Verlag, N.York.

Breakwell, J.V., 1979, Satellite determination of short wavelength gravity variations. *J. Astron. Sci.*, Vol 27(4), pp. 329-344.

Eckels, A., O'Keffe, J.A., and Squires, R.K., 1959, VANGUARD measurements give pear-shaped component of earth's figure. *Science*, Vol 129.

Everitt, C.W.F., Breakwell, J.V., Tapley, M., DeBra, D.B., Parkinson, B.W., Smith, D.E., Colombo, O.L., Pavlis, E.C., Tapley, B.D., Nerem, R.S., Yuan, N.D., 1989, Gravity Probe-B as a geodesy mission and its application for TOPEX. *CSTG Bull.*, 11, pp. 55-67.

Heiskanen, W.A., and Moritz, H., 1967, *Physical Geodesy*. W.H. Freedman and Co., San Francisco.

Jekeli, C., 1988, Local gravity disturbance from multiple-high-single-low satellite-satellite tracking. *Proceedings Chapman Conference on Progress in the Determination of the Earth's Gravity Field*, pp 164-167, Ft. Lauderdale, Florida.

Kaula, W.M., 1966, *Theory of Satellite Geodesy*. Blaisdell Publishing Co., Waltham, Massachusetts.

Lambeck, K., 1988, *Geophysical Geodesy--The Slow Deformations of the Earth*. Clarendon Press, Oxford, England.

Levitus, S., 1982, Climatological atlas of the world oceans. NOAA Professional Paper 13, U.S.A.

Mader, G., 1988, Description of CIGNET stations. *GPS Bull.*, Vol 1, CSTG GPS Sub-commission, Maryland.

Marsh, J.G., Lerch, F.J., Putney, B., Christodoulidis, D.C., Felsentreger, T.L., Sanchez, B.V., Smith, D.E., Klosko, S.M., Martin, T.V., Pavlis, E.C., Robbins, J.W., Williamson, R.G., Colombo, O.L., Chandler, N.L., Rachlin, K.E., Patel, G.B., Bhati, S., and Chinn, D.S., 1988, A new gravity field model for the earth from satellite tracking data: GEM-T1. *J. Geophys. Res.*, Vol . 93 (B6), pp. 6169-6215.

Martin, C.F., and McCarthy, J.J., 198 , Error analysis of anomalous gravitational

acceleration from GPS tracking by the STS. Contractor Report, EG&G Lanham, Maryland.
Martin, T.W., Eddy, W.F., Rowlands, D.D., and Pavlis, D.E., GEODYN-II system description. Contractor Report, EG&G, Lanham, Maryland.
Melbourne, W.G., and Tapley, B.D., 1983, The geopotential information in GPS carrier-phase tracking from a low-orbiting gravimetric mission. *EOS Trans. AGU*, Vol. 64(45), page 680.
Melbourne, W.G., and Davis, E.S., 1987, GPS-based precise orbit determination: a TOPEX flight experiment. AAS/AIAAS *Proceedings Astrodynamics Specialists Conference*, Kalispell, Montana.
Mueller, I.I., and Zerbini, S. (editors), 1989, The interdisciplinary role of space geodesy. *Proceedings of an International Workshop held at the Ettore Majorana Center for Scientific Culture in Erice, Italy*. Springer-Verlag, New York.
Rummel, R., and Colombo, O.L., Gravity field determination from satellite gravity gradiometry. Bull. Geod., 59, pp. 233-246.
Schwiderski, E.W., 1979, Global ocean tides, part II: the semidiurnal principal tide (M2); Atlas of tides charts and maps. Report NSWC TR 79-414, Washington D.C.
Smith, D.E., Lerch, F.J., Colombo, O.L., and Everitt, C.W.F., 1988, Gravity field information from Gravity Probe-B. *Proc. Chapman Conf. Progress in Det. of Earth's Gravity Field*, pp. 159-163, Ft. Lauderdale, Florida.
Upadhyay, T., and Jekeli, C., 1988, Satellite-to-satellite tracking experiment for global gravity field mapping. *Proc. Chapman Conf. Progress in Det. of Earth's Gravity Field*, pp. 168-171, Ft. Lauderdale, Florida.
Wagner, C.A., 1986, Accuracy estimate of geoid and ocean topography recovered jointly from satellite altimetry. *J. Geophys. Res.*, Vol. 91(B1), pp. 453-461.
Wu, S.C., Wu, J.T., and Bertiger, W.I., 1989, An efficient technique for gravity recovery using a low satellite.*Proc. Fifth Int. Geodetic Symp. on Satellite Positioning*, Las Cruces, New Mexico.
Yunck, T.P., Melbourne, W.G., and Thorton, C.L., 1985, GPS-based satellite tracking system for precise positioning. *IEEE Trans. Geosci. & Remote Sensing*, Vol GE-23 (4), pp. 450-457.

THE DORIS SATELLITE RADIO TRACKING SYSTEM: STATUS AND PLANS

P. Willis, C. Boucher, M. Kasser
Institut Géographique National
R. Biancale, A. Cazenave
Groupe de Recherche en Géodésie Spatiale
M. Dorrer, F. Nouel
Centre National d'Etudes Spatiales

INTRODUCTION: THE CONCEPTS OF THE DORIS SYSTEM

The goal of this paper is to present the status of the French DORIS system, the on-going developments and the plans for the future. The DORIS system is being developped by the combined efforts of three French groups: the **CNES** (Centre National d'Etudes Spatiales), the **GRGS** (Groupe de Recherche en Géodésie Spatiale) and the **IGN** (Institut Géographique National). DORIS is an acronym for Doppler Orbitography and Radiopositioning Integrated by satellite. Its main purpose is the precise orbit determination. Nevertheless, this system will also have significant positioning applications for geodesy and geophysics. This system is being built to be the nominal French system for orbit determination on the joint US-French oceanographic project **TOPEX/POSEIDON** planned for 1992.

The space segment

DORIS will fly on low altitude orbiting satellites. In order to prepare the TOPEX/POSEIDON mission, a feasability test will be performed on the near-to-come SPOT2 satellite (altitude of 830 km). DORIS is an dual frequency uplink DOPPLER system (same philosophy as for ARGOS). The two nominal frequencies are: 2.03625 GHz and 401.25 MHz. The on-board electronics contain only a Doppler receiver that performs the two Doppler counts on these two nominal frequencies. The satellite also contains an omni directional antenna and an ultra stable oscillator (BVA quartz) that allows stabilitiy of 5×10^{-13} for periods of 10 to 100 seconds. All the Doppler data collected on board of the satellite are retransmitted to the control station in Toulouse.

The SPOT 2 flight is assumed to take place in fall 1989. The DORIS system will also be a payload on the next SPOT mission: SPOT 3, SPOT 4 (under consideration).

The control segment

In order to determine precisely the orbit of the satellite a permanent dense network of 50 permanent tracking beacons is being installed by the IGN. At present, 35 beacons are already on site thanks to a large collaboration of various institutions in the world. These beacons are under the responsability of the IGN/SMB (beacon installation and maintenance center). They are adapted to laboratory environment and exist in a rack form. The criteria used to select the tracking sites were the following: equally spaced on the surface of the earth (in order to continuously track the satellite), having precise coordinates (VLBI, SLR, LLR sites belonging to the IERS terrestrial frame).

The user segment

In parallel, mobile units of the DORIS beacons are being developped by the CNES for geodetic and geophysical applications. The electronics will be almost the same as for the tracking units but the weight and power requierements will be compatible with field operations. These mobile units could be left unattended on sites for months in case of geophysical applications (active zones, volcanos, ice shields,...).

REQUIEREMENTS FOR PRECISE ORBIT DETERMINATION

The DORIS reference system

One of the key improvement of DORIS is the dense network of the 50 permanent tracking stations. As far as possible these stations have been collocated with the IERS terrestrial system sites (**VLBI**, **SLR**, **LLR** or related sites (**GPS**, **TRANSIT**)). For these stations, the estimated coordinates accuracy will be at a few centimeter level. For the other stations, mainly isolated islands, a TRANSIT/NNSS positioning was performed and will provide us with meter level a priori coordinates. As soon as the DORIS data will be avaliable, new coordinates will be determined for these sites. The realization of the DORIS reference frame and its refinements will be done by IGN.

A new dedicated gravity field

As DORIS will fly on low altitude satellites (830 km for SPOT 2, 1300 km for TOPEX/POSEIDON), the precision of the gravity model will be crucial for the quality of the results. In preparation of DORIS, a new model GRIM 4 is being computed in collaboration between the German and French groups **DGFI** and **GRGS**. A large set of satellite observations will be used (camera, Doppler, Laser) over 25 years since 1962. The unreduced output of GRIM 4 will concern:

1. sperical harmonic coefficients of the earth's gravity field up to degree 40 plus resonances up to degree 50 and the time variation of some zonal terms.
2. spherical harmonic coefficients of 11 ocean tide waves (those of Schwiderski's model) up to degree 6.3 station coordinates plus their tectonic drifts if needed.

This model will allow to use DORIS data during the first 6 months of the SPOT 2 mission. These 6 months of data will be incorporated in the a priori GRIM 4 solution and will provide a more accurate dedicated gravity field.

Orbit determination

To obtain the ultimate accuracy from the DORIS data, a new orbit determination software, called ZOOM, based on a classical numerical integrator technique, is being developped at CNES. A special care has been taken for the inertial reference frame (IAU and IUGG recommendations) and for the terrestrial reference frame (IERS). The expected accuracy of the orbit determination is 10 cm in the radial component and fullfills the TOPEX/POSEIDON requierements for this oceanographic mission.
In the context of the TOPEX/POSEIDON project, comparaisons have been conducted with regard to the NASA software (GEODYN) and with the University of Texas software (UTOPIA) to calibrate the force models. Those comparaisons showed agreement at a few millimeters level.

GEODETIC AND GEOPHYSICAL APPLICATIONS

In order to prepare the SPOT 2 launch, simulations studies were conducted at the IGN to test the positioning capabilities of DORIS. In these simulations, we used an error budget, coming from CNES, and we let this noise propagate in the DORIS measurements. The performance showned here should be more realistic than those coming from a standard variance-covariance analysis (as long as the error budget is realistic). In all these studies, we will try to find out how long it takes to stabilize the results on the coordinates (period of convergence) and also, when the solution is stabilized, what is the difference between the estimated values of the position coordinates and their theoretical values.

Absolute positioning

Figure 3 shows the results for absolute positioning obtained for one particular case of the simulated noise. We can see that the convergence is obtained after a few days (basically one or two weeks). The disagreement between the estimated and the theoretical position is at the 10 cm level. Those simulations have shown that DORIS should allow precise absolute positioning:

8 cm (1σ)	for less than 40 passes

This high absolute accuracy (in a global reference frame, referred to the center of mass of the earth) should allow several important geodetic applications: connection of national geodetic datums to a global reference frame (IERS terrestrial reference frame), connection of isolated islands, precise absolute positioning for military applications (radars,...), precise absolute positioning for one way time transfer (eg GPS).

Relative positioning

A large part of the absolute errors still comes from the orbit position and then cancells in a relative mode. Figure 4 shows the results obtained for relative positioning for a baseline of 100 km. Different length of baselines have been simulated and led us to the folowing expected performance of DORIS for relative positioning:

2 cm + 0.1 ppm (1σ)	for less than 30 passes

Several applications are foreseen: realization of national geodetic datum, connection for islands, tide gauges fixing, collocation with other systems (VLBI, SLR, LLR).

Permanent positioning

DORIS is a centralized system: no data is recorded on the field, all the data are directly processed and archived at Toulouse. There is no need of human intervention. The mobile DORIS beacon could be left in remote areas were deformations are to occur. The computation could be done on a permanent basis (dayly, weekly, monthly basis).

Several applications are foreseen for geophysics: geophysical monitoring, active engineering (dams monitoring,...), glaciology (Antartica, large glaciers,...).

PLANS FOR THE FUTURE

We summarize here the future key steps of the DORIS project for the following years:

1. launch of the SPOT 2 satellite (November 1989): First DORIS data avaliable.
2. Pilot experiments: In order to test, on a field basis, the positioning capabilities of the DORIS system (absolute, relative and permanent), several test experiments will be conducted for geodetic and geophysical purposes. These tests are still under discussion with collaborative institutes and will probably be the following: **Djibouti** (active tectonic area), **Hawaï** (volcanic zone), **Japan** (geodetic test of precision), **France** (landslide).
3. GRIM4: new dedicated gravity field.
4. ZOOM software: refined DORIS terrestrial system for precise orbit determination.
5. Launch of the TOPEX/POSEIDON satellite (mid 1992): DORIS will be the French nominal system for this US-France joined project.
6. Launch of the SPOT 3 satellite (1992).

REFERENCES

De Moegen, N., Boucher, C. (1985), Positioning accuracy of the DORIS/SPOT system, Satrape meeting, Paris, IGN/SGN CC/G n° 505.

Dorrer, M. (1989), Cours international de technologie spatiale, Toulouse.

Imbert, G. (1988), Etat du réseau orbitographique DORIS, IGN/SMB.

Le Feurmou, L. (1988), Projet DORIS: Etude du positionnement permanent, mémoire de diplome d'ingénieur ESGT.

La Brune, Y., Nouel, F., Jayles, C. (1986), Precise orbit determination with DORIS system and the associated ZOOM software, 2nd international symposium on spacecraft flight dynamic, Darmstaadt (RFA), ESA/SP n°255.

The DORIS system : Block diagram

Figure 1

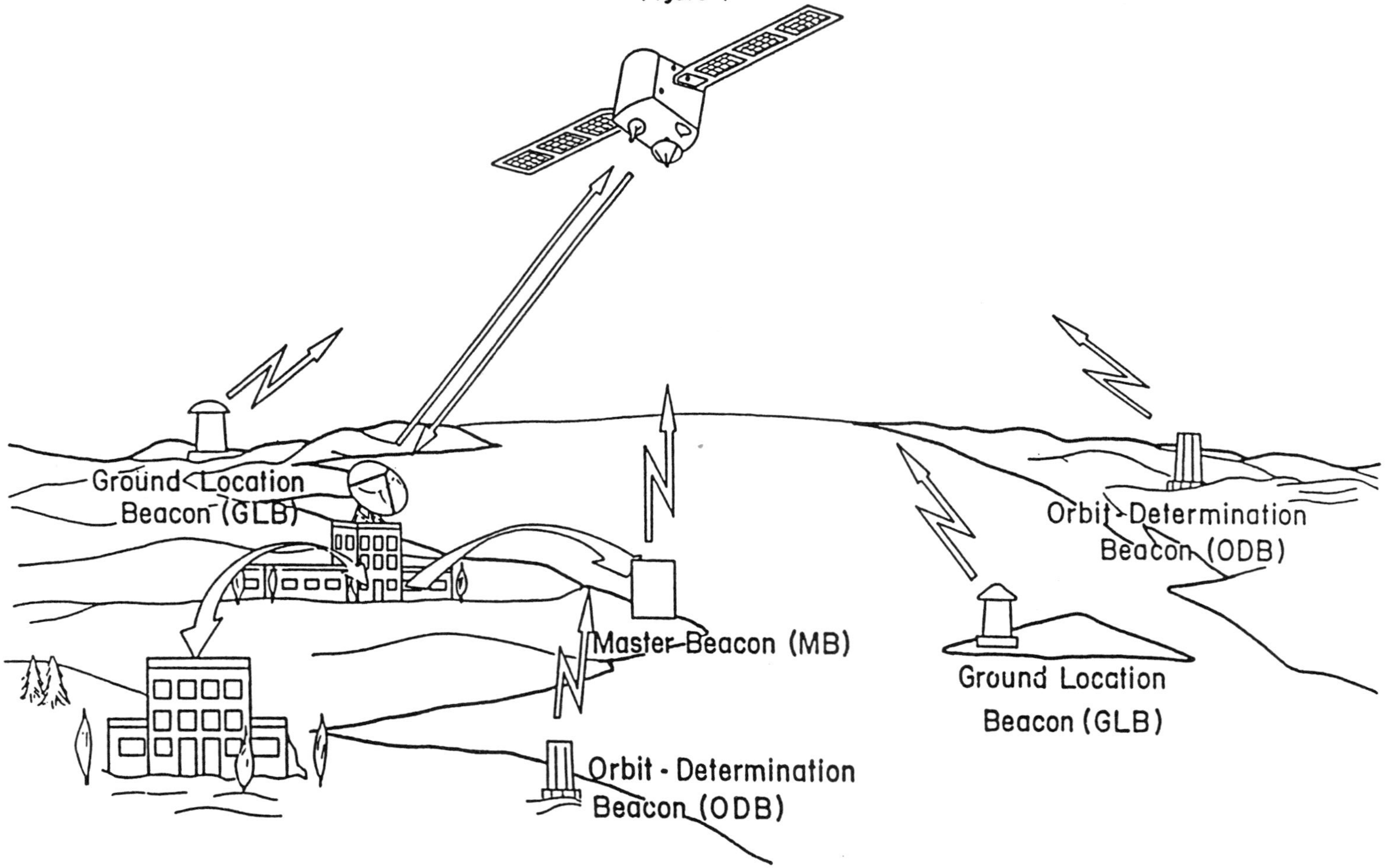

The DORIS tracking network

Figure 2

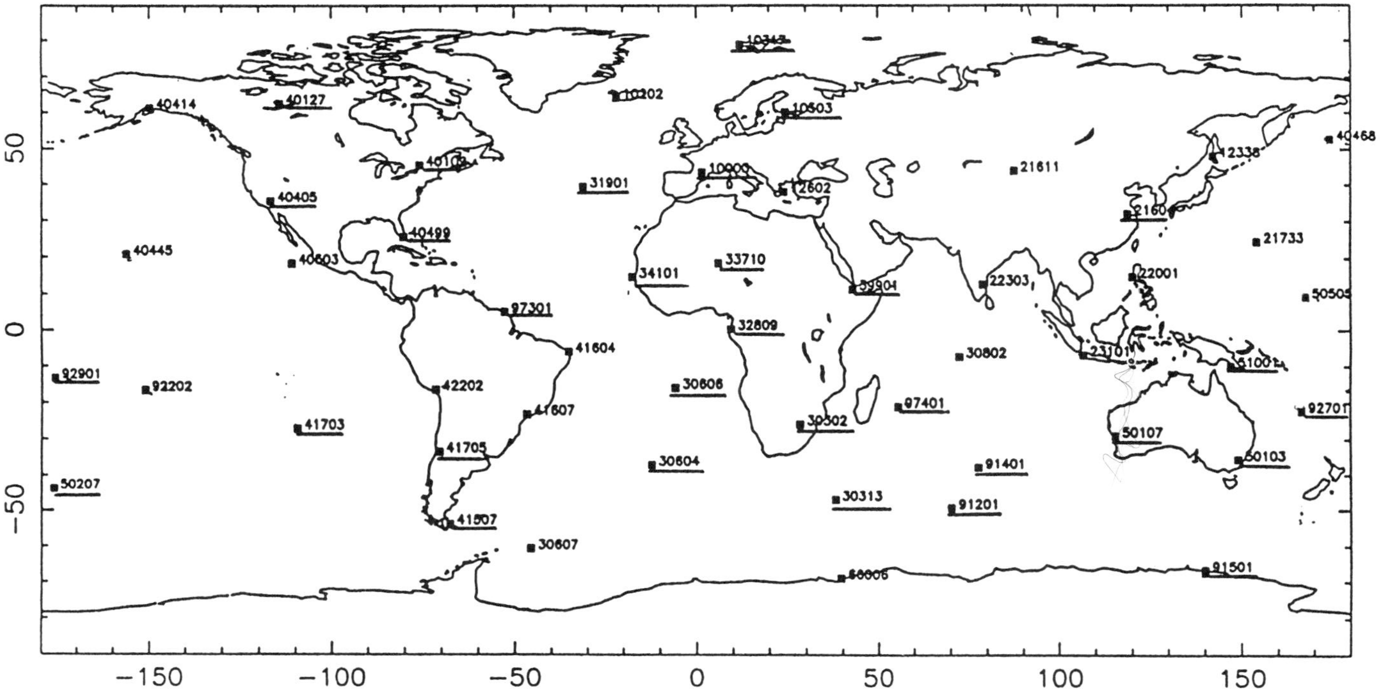

EDITION : 07/02/89

■ : installed beacon ——
■ : planned beacon

Simulations of the DORIS absolute positioning

Figure 3

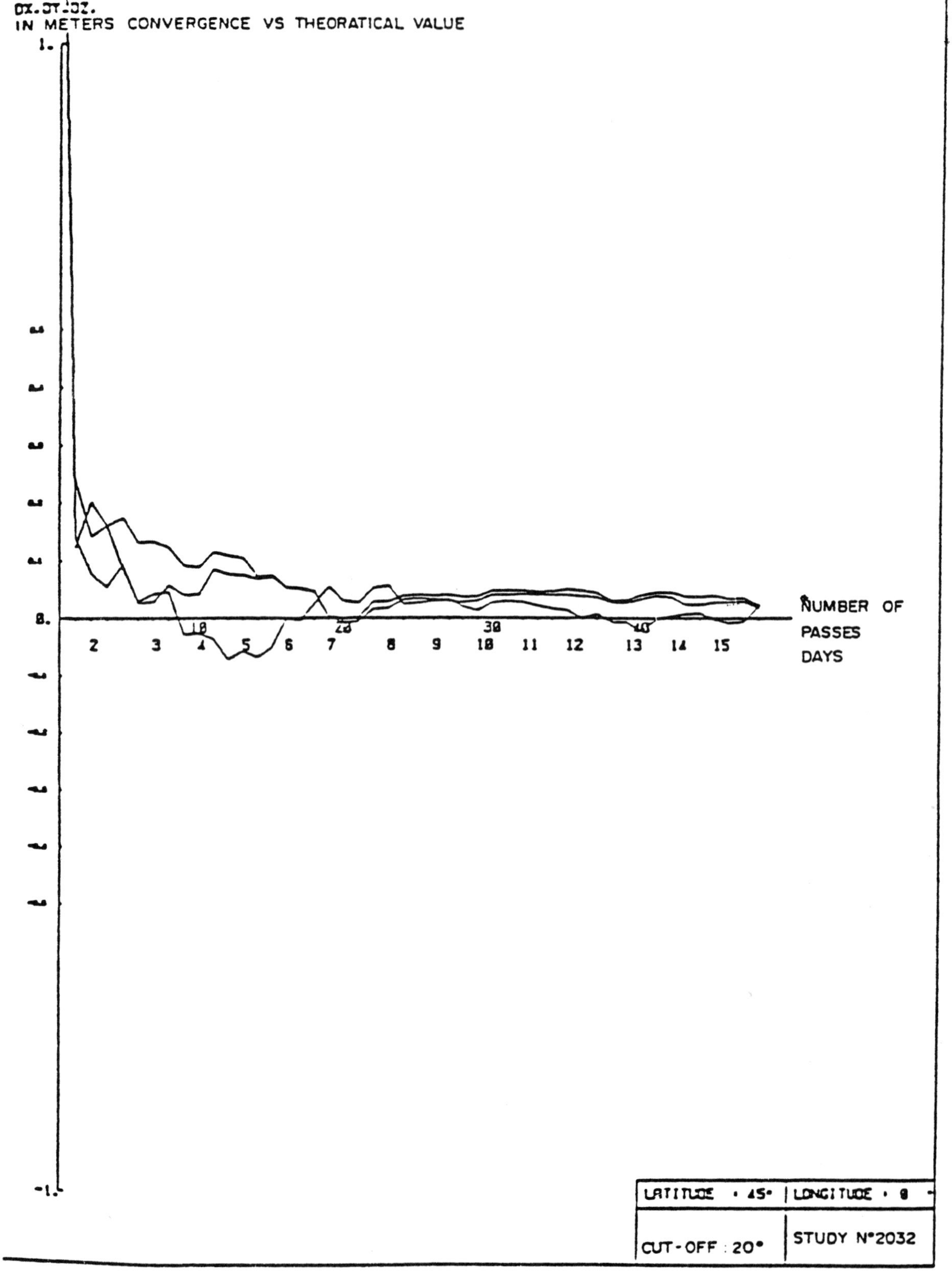

Simulations of the DORIS relative positioning

Figure 4

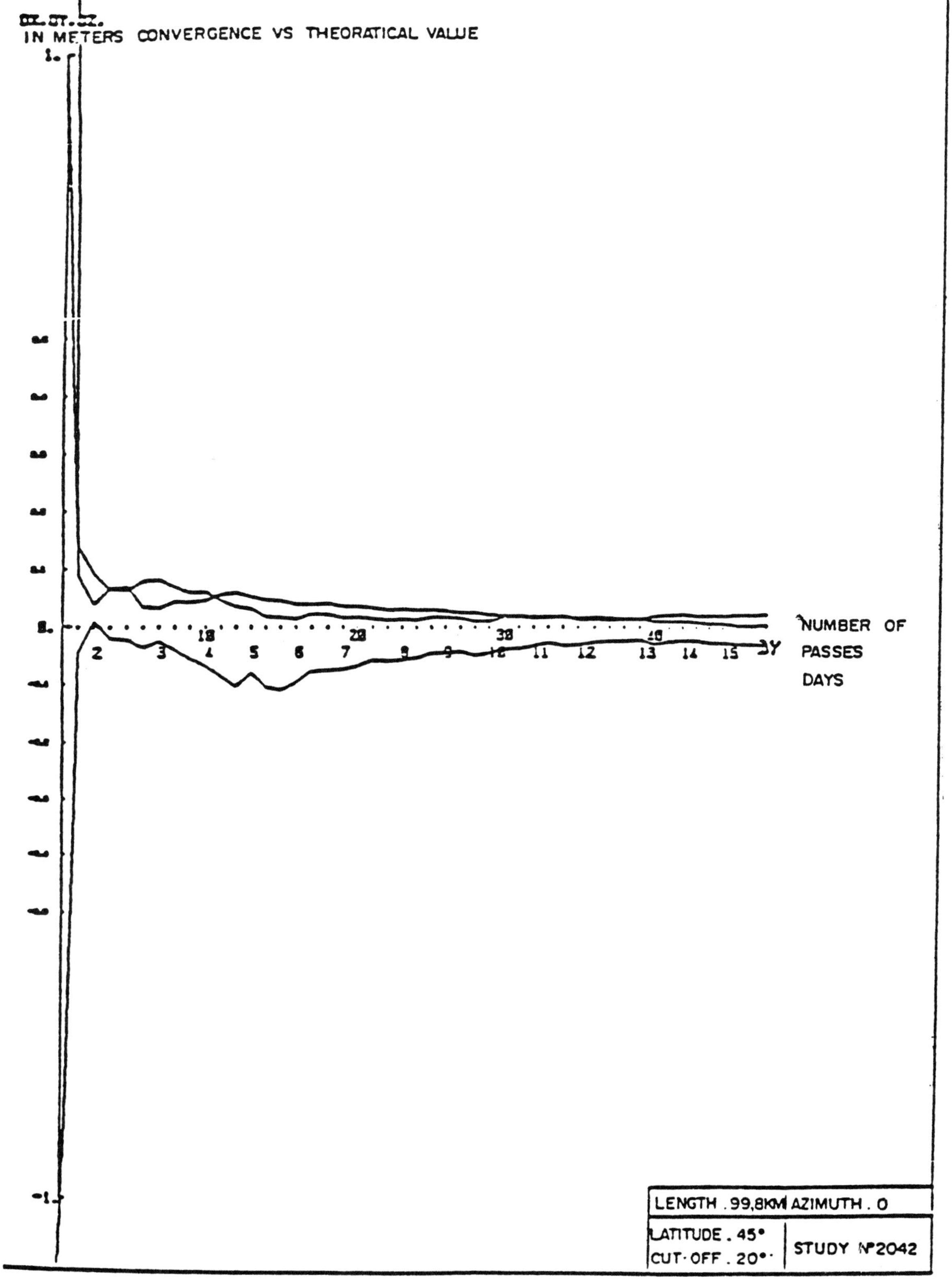

PRARE — STATUS AND PROSPECTS

Herbert Wilmes, Christoph Reigber
Deutsches Geodätisches Forschungsinstitut, DGFI, Abt.1, München
Federal Republic of Germany

INTRODUCTION

The Precise Range and Range Rate Equipment, PRARE, was proposed as additional experiment (Hartl, Reigber 1982) on ESA's First Remote Sensing Satellite ERS-1, launch date in September 1990. The development of the PRARE system, a national experiment of the Federal Republic of Germany under grant of the German Ministry of Research and Technology (BMFT) is being performed by the institutions:

- INS: Institut für Navigation, Universtät Stuttgart (system development)
- Kayser Threde GmbH, München (industrial contractor space segment)
- Dornier GmbH, Friedrichshafen (industrial contractor ground stations)
- DGFI: Deutsches Geodätisches Forschungsinstitut, Abt. 1, München (master station installation and operation, tracking data processing, tracking network coordination / installation)

THE PRARE SYSTEM

PRARE is a new satellite tracking system with

- a compact versatile space segment, an independent selfcontained hardware unit with a mass of 17 kg, size 400 x 240 x 180 mm^3 and a power consumption of 39 VA in operation and 8 VA in stand-by mode respectively,

- a ground station network of small automated transportable ground stations and

- a control segment consisting of a master and dumping station, a time control unit and preprocessing computer in connection with the processing and archiving facility.

The major advantages of this concept of spaceborne microwave tracking are
- high precision two-way ranging and two-way doppler measurement relative to a ground station network with the additional measurement of ionospheric delay parameters and the determination of tropospheric refraction parameters,
- all-weather tracking capability,
- almost unattended operation of the ground stations,

- collection of the measurement data in the space segment and transfer to the master station for central processing at a minimum time delay,
- simultaneous tracking of up to 4 ground stations which enables the processing of the tracking data in geometric mode,
- capability to view and store the tracking data at the individual ground station using a standard PC-computer.

The Measurement Principle

In the following the PRARE measurement principle is briefly explained:
Two continuous signals are emitted from the space segment to earth, one of which is in the S-band (2.2 GHz), the other in the X-band (8.5 GHz). Both signals are modulated with a PN-code (pseudo random noise, 1 MChips/s for the S-band and 10 MChips/s for the X-band) used for the distance measurement and containing data signals ("broadcast information") for the ground station operation (prediction of visibility etc.). The time delay in the reception of the two simultaneously emitted signals is measured in the ground stations with an accuracy better than 1 ns and the result is retransmitted and stored in the onboard memory for later ionospheric correction of the data. In the ground station the received X-band signal is transposed to 7.2 GHz, coherently modulated with the regenerated PN code (or with one of three orthogonal copies for code multiplexing) including the two-frequency delay data, housekeeping and meteorological ground data and is retransmitted to the space segment.

In the space segment the PN code is fed into a correlator to determine on board the two-way signal delay which is a measure for the two-way slant range between the satellite and the ground station. In addition to this, the received carrier frequency is evaluated in a doppler counter to derive the relative velocity of the spacecraft relative to the ground station. This two-way tracking principle secures highest measurement precision because the signal generation and the (time delay and doppler) measurement are based upon the same ultrastable quartz frequency generator (USO) with its system-inherent high short-term stability. Both, the measurements and the housekeeping data from spacecraft and ground stations are stored in the onboard memory.

Four independent correlators and four doppler counters in the space segment allow simultaneous measurements to up to four ground stations in a code multiplex mode.

The information collected in the space segment will be retransmitted to the master and control station during every agreeable pass. By this procedure the global dataset of tracking data is made available at least two or three times per day.

Accuracies

The PRARE measurement accuracy can be estimated on the basis of the hardware limitations and the impact of refraction model uncertainties:

- Noise values (referring to 1 measurement per second for ranging and 30 seconds integration for doppler, one-way distances and velocities at 90^0 elevation)
  ± 1.7 cm for X-band ranging
  ± 30 cm for S-band ranging
  ± 0.1 mm/s for X-band doppler

- bias values
  < 1 cm for X-band ranging
  < 3 cm for S-band ranging

The resulting overall ranging error (RSS) after corrections applied including refraction influences can be assessed as 3 ... 7 cm.

THE REALISATION OF PRARE ON ERS-1

PRARE will be flown for the first time on the first European Remote Sensing Satellite ERS-1 at the end of 1990 or early in 1991. Besides the PRARE system the satellite is equipped with (Paci, Fea 1989)

- an Active Microwave Instrument (AMI) operating either as Synthetic Aperture Radar (SAR) or as Wind and Wave Scatterometer
- a Radar Altimeter (RA)
- an Along Track Scanning Radiometer (ATSR)
- a Laser Retroreflector (LRR)

The PRARE master station will be installed at the DLR in Oberpfaffenhofen interfacing closely to the German Processing and Archiving Facility (D-PAF) where the processing of ERS-1 tracking data will be executed. This ensures a central availability of the tracking data at the processing centre, short turn-around times for the generation of orbital products and retransmission of prediction information.

The objective of the PRARE experiment is to prove the operability of the system and the measurement performance under space conditions. In addition to these system-related aims PRARE complements ERS-1's primary tracking system SLR (Satellite Laser Ranging) with range and doppler observations from its globally distributed tracking network with emphasis on remote locations and areas with disadvantageous weather conditions like high cloud coverage. By this, the PRARE tracking system shall help to improve the gravity field determination and to meet the requirements posed by Radar Altimetry and SAR data processing as concerns decimetre orbit accuracy in the radial component.

PRARE Space Segment

Responsible organisations for the realisation of the space segment are the Institut für Navigation, Stuttgart together with the industrial contractor, Kayser Threde GmbH. As shown on the bar chart in Figure 1, the development and manufacturing of the space segment hardware, the Engineering Model (EM) and two Flight Models (FM) has been completed. The flight hardware has been integrated into the satellite platform and tests have been finished by the end of 1988. The space segment software development and tests are presently being completed.

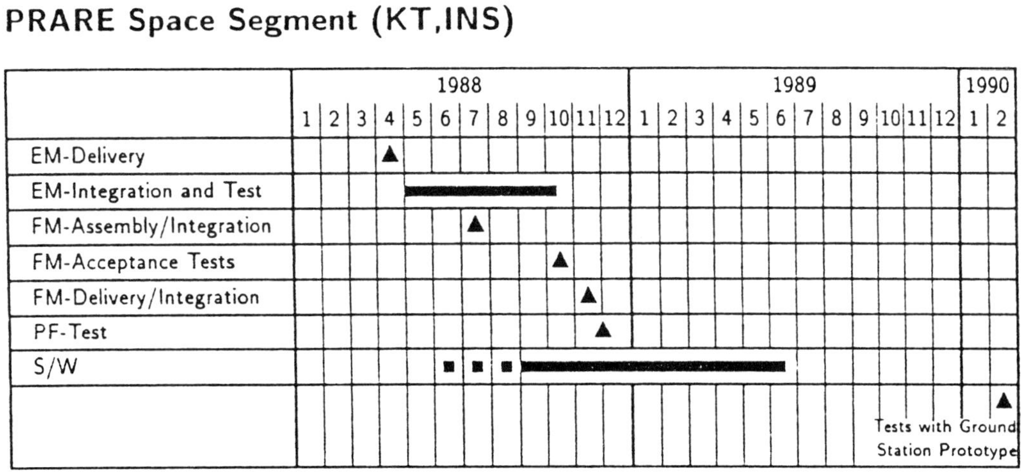

Figure 1. PRARE Space Segment Development

PRARE Control Segment

The tracking information which is collected in the PRARE space segment and stored in the onboard memory is transmitted to the master control station whenever there is a contact between the S/C and the master station. In the master station all incoming data are preprocessed, precision processed, archived and finally disseminated to the users. The PRARE system time is compared with UTC and updates of the broadcast data are sent to the onboard memory.

The PRARE master control station is being developed jointly by the DGFI, INS and DLR. The time system and control station hardware will be installed in 1990. At present the main efforts are concentrated on the software development.

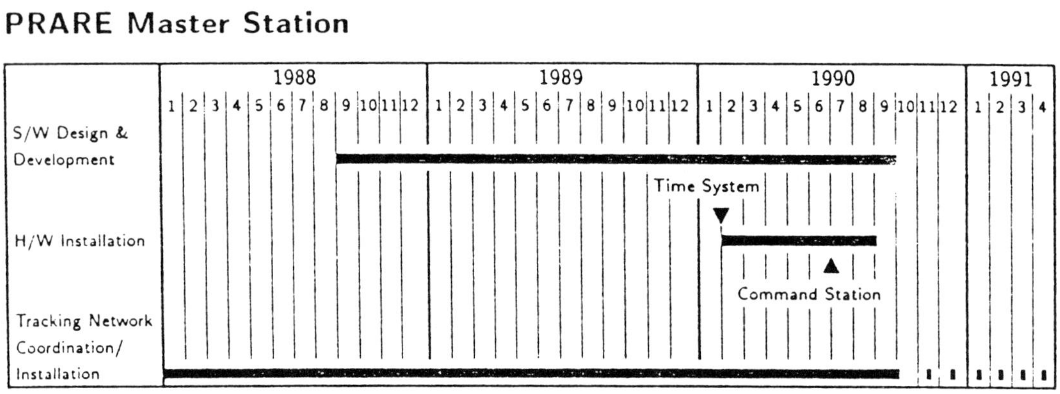

Figure 2. PRARE Master Control Station Development

PRARE Ground Stations

A redesign and the preparation of a small series production at agreeable costs for the individual station was started by the company Dornier in April 1989. To enable easy transportability and handling this concept foresees the modular design of the hardware in three major units (Dornier 1989):

- **antenna unit** comprising the 60 cm diameter offset antenna, the mechanical antenna pointing unit and the frontend electronics
- **electronics unit** with the S- and X-band receiver and X-band transmitter, time interval counter and the station processor
- **monitor and test computer,** an optional feature for the station set-up, for the monitoring and storage of measurement data and to provide access to detailed status and test information.

The prototype is going to be presented in April 1990. To enable the series production starting early 1990, orders for ground stations have to be placed by December 1989. Coordination of the network and positioning of the sites is an ongoing activity.

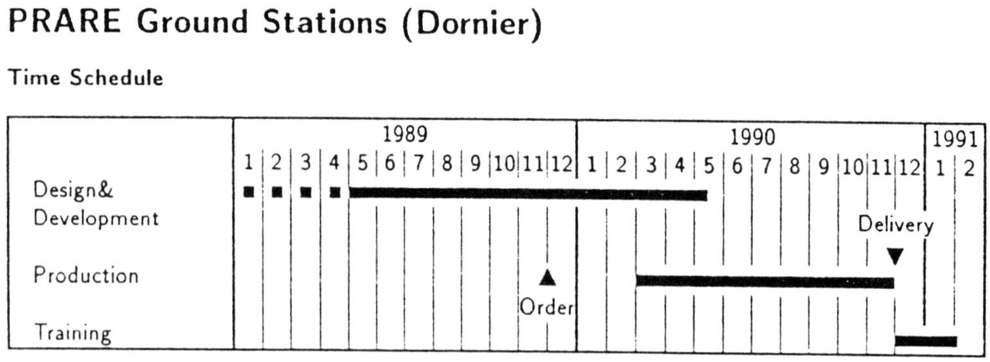

Figure 3. PRARE Ground Station Development

The present - still preliminary - status of the PRARE network composition is visualised in Figure 4. The stations Oberpfaffenhofen, Stuttgart, Wettzell, Trondheim and Nye Alesund in Europe, Hobart in Australia and the Antarctic station Esperanza have been selected definitely. All of the final European stations and a few stations on the American continent have been surveyed by GPS in 1988/1989.

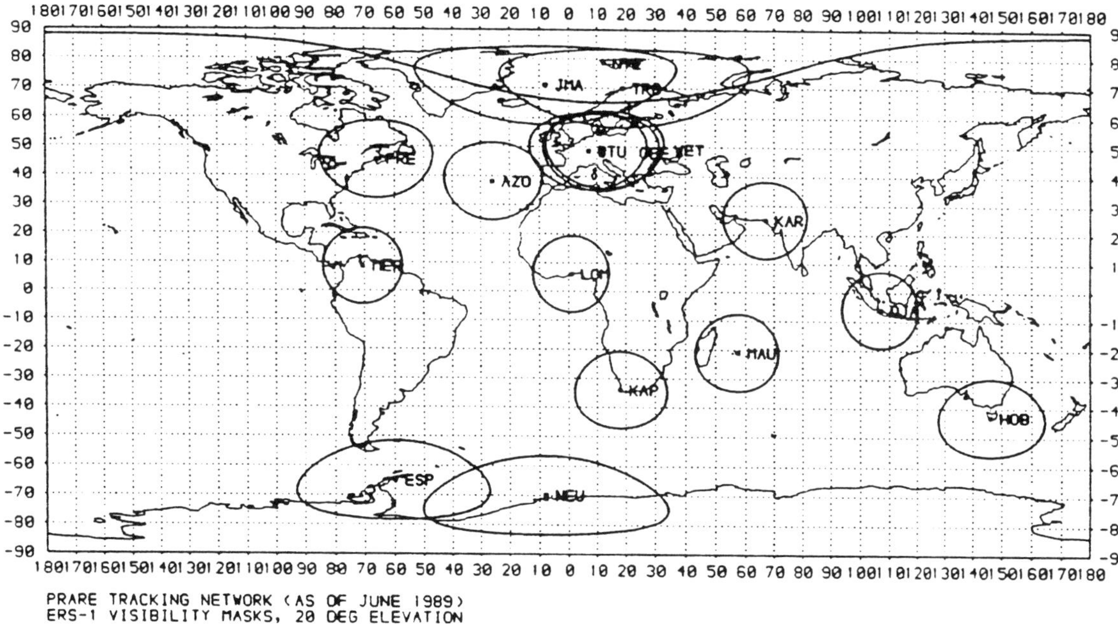

Figure 4. PRARE Tracking Network Status (ERS-1 Visibility Masks for 20^0 Elevation)

FUTURE APPLICATIONS OF PRARE

The system features as described in the previous sections enable the application of the PRARE system in a wide range of space missions. These possible applications range from the fields of solid earth physics, ocean / ice / land remote sensing to planetology and spacecraft control. There is a good chance of missions being realised within about the next decade in the earth observation domain such as: ERS-2, ATMOS, ARISTOTELES, the European Polar Platform and the GPS/ACRE experiment.

PRARE on an Experimental GPS Satellite

In response to an announcement of opportunity by the US Naval Research Laboratory the PRARE investigators proposed the installation of a PRARE space segment on an experimental GPS satellite, ACRE (Advanced Clock / Ranging Experiment)(Hartl, Reigber 1987). The major objectives of the planned GPS, Laser, VLBI and PRARE combination in this mission would be in the fields of

- precise time transfer for monitoring and modelling the spacecraft clock drift
- Intercomparison and combination of GPS, Laser and VLBI / PRARE measurement techniques and reference frames
- operational improvements like (1) short-term availability of precise orbit ephemeris by the space-based data collection, (2) increased orbital accuracies based upon two-way measurements at higher frequencies and reduced ionospheric influences and (3) enhancements for the time dissemination by more accurate and prolonged time synchronisation

PRARE Extended Version - PRAREE

At present tropospheric refraction influences are the limiting error influences for the PRARE system's accuracy. To get rid of this impact and to improve the tracking system performance by at least one order of magnitude, an extended version of the PRARE system, called PRAREE was proposed to be flown on the first European Polar Platform (Reigber, Hartl et al. 1988).

The proposal is to augment the original PRARE system by the following features (c.f Figure 5.):

- an additional 100 MHz subcarrier in the X-band up- and down-link which leads to an increase in the PRN (pseudorandom noise) ranging accuracy and can be used for a direct and unambiguous carrier phase measurement,
- an additional 400 MHz downlink with PRN modulation which allows a more detailed analysis of the ionosphere;
- a two-colour semiconductor continuous wave Laser uplink allows (under the condition of clear visibility to the spacecraft) the direct measurement of atmospheric refraction. The difference in propagation between the two Laser colours is a measure for the dry component of the tropospheric refraction and the signal delay time between the microwave and optical channel determines the wet component,
- additional channels (up to 10) for simultaneous target tracking.

PRAREE is thus an autonomous two-way, 3-microwave-frequency tracking system with its own telemetry, telecommand, data storage, timing and data transmission capability which is complemented by two Laser channels in different colours, modulated by ranging codes.

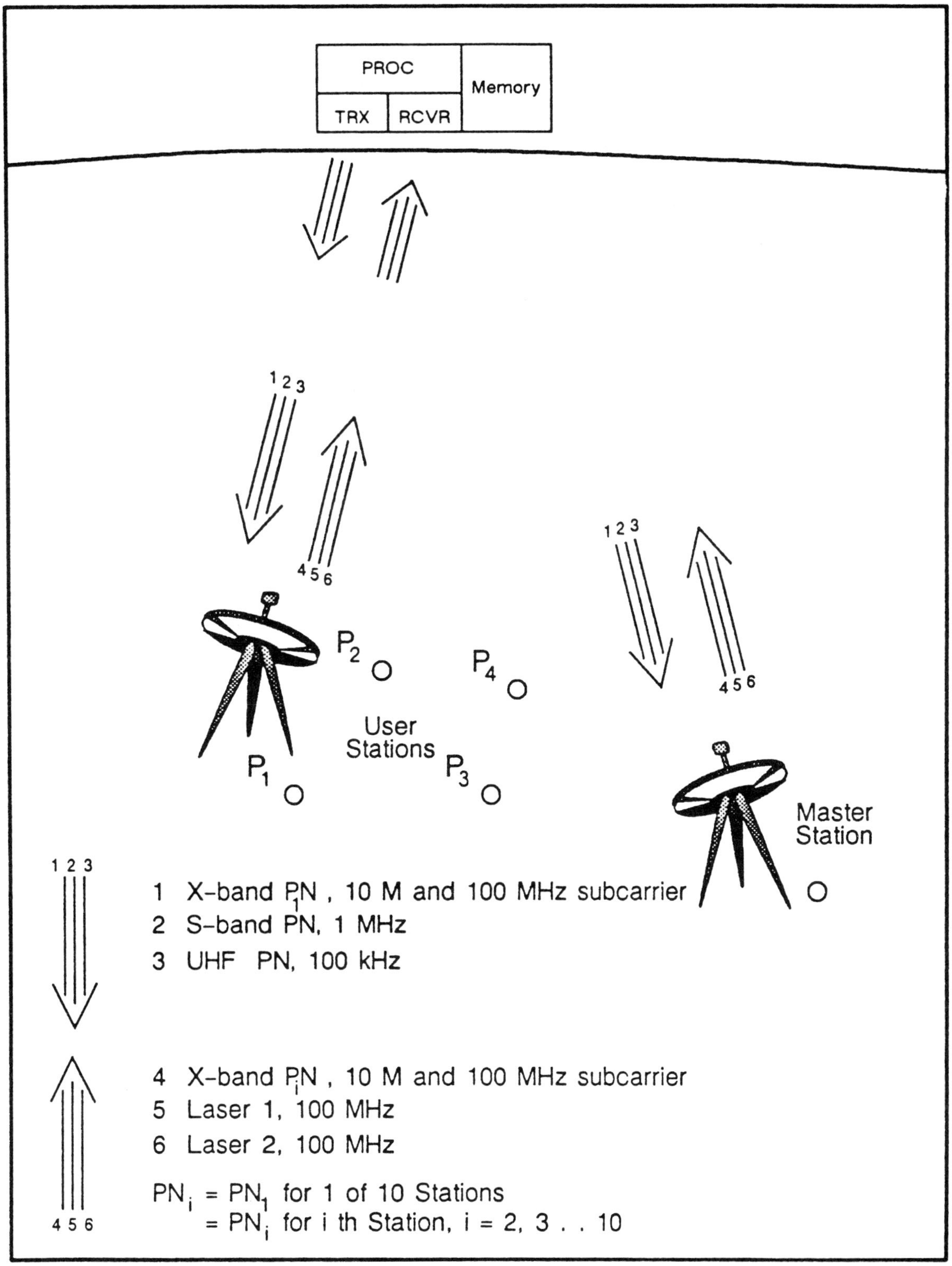

Figure 5. Basic Concept of PRAREE

CONCLUSIONS

PRARE is one of the first space-based two-way tracking systems providing range and doppler observations at the same time and with an accuracy which for the single range is comparable to the single shot accuracy of the 2nd to 3rd Laser generation systems. In its extended version PRAREE could achieve the few mm system accuracy which is needed to satisfy the scientific requirements of the future in geodesy and atmospheric sciences.

Acknowledgement: The project on which the report is based is funded by the German Federal Minister of Research and Technology (BMFT) under grant No. 01 QS 87 157.

REFERENCES

Beard, R., L.: The Advanced Clock / Ranging Experiment (ACRE), International Coordination of Space Techniques for Geodesy and Geodynamics (CSTG), Bulletin No. 11, München, Juni 1989

Dornier GmbH: PRARE Ground Stations, Technical Description, Friedrichshafen, August 1989

Hartl, Ph., Reigber, Ch.: The Precise Range and Range Rate Equipment (PRARE), Experiment Description, Proposal to ESA, 1982

Hartl, Ph., Reigber, Ch.: The Precise Range and Range Rate Equipment PRARE on a GPS-Satellite, Proposal to NATO, December 1987

Paci, G., Fea, M.: The First ESA Remote Sensing Satellite Mission, International Coordination of Space Techniques for Geodesy and Geodynamics (CSTG), Bulletin No. 11, München, Juni 1989

Reigber, Ch., Hartl, Ph., Grassl, H., Kohl, H., Markwitz, W., Schäfer, W., Schwintzer, P., Zschau, J.: Precise Range and Range Rate Equipment PRARE - Extended Version - for Application on the European Polar Platform: PRAREE, Proposal to ESA, July 1988

METHODS OF REALISATION OF COHERENT SIGNALS IN SATELLITE SYSTEMS

M.T.Prilepin

The Soviet Geophysical Committee, Moscow, USSR

Abstract. The development of a satellite system proceeding from the concept of coherent signals allows us to increase the accuracy of absolute and relative measurements of terrestrial coordinates and to improve the accuracy of determination of satellite orbits by reducing errors of initial conditions and by using precisely measured distances between satellites. Two methods of realisation of coherent signals, emitted by satellites, are analysed. In the case of the first variant, the satellites continuously exchange frequencies whose transformation permits the realisation of coherent signals and the prescise measurement of distances between satellites. In the second variant the geostationary satellite is employed to transmit signals from the master ground-based generator to the satellites of the system and to retranslate these signals further to the user.

INTRODUCTION

In the papers (Prilepin, 1986, 1988) has formulated the reasons in favour of the usage of coherent signals when developing specialised satellite systems for geodetic and geophysical applications and modifying the global systems of GPS or GLONASS type functioning now in experimental phase is considered.

The development of a multisatellite system on the concept

of coherent signals allows us to use on board satellites more portable and light-weight frequency generators with less consumed power than those necessary in systems with independent signals. This approach greatly facilitates the ground-based service because to obtain equal precision in the case of coherent signals the requirements to frequency stability are reduced by about three orders. In practice, highquality quartz generators can operate on board satellites instead of the rubidium, cesium or hydrogene generators. The weight, consumed power and volume for cesium generators are 10; 12.5;10 times more than for the quartz generators, respectively; for the hydrogene generators these numbers are 25;15 and 25 (Bartolomew, 1978).

Moreover, if we take into account that in the working phase the satellite has no less than three generators and that the launching of one kilogram of productive load into orbit at about 20 th.km altitude costs about 10 th. dollars, then the real advantages of coherent signals concept become apparent.

The application of the "between-satellite difference"technique with coherent signals allows essential simplification on quicker processing of data since it is not necessary to model the behaviour satellite generators in time, which is indispensable in absolute determinations with independent signals.

If coherent signals are formed by frequency exchange between satellites, then there is a possibility to measure with precision (no less than 0.1 fraction of phase cycle) the distance between satellites, which enables to improve significantly the accuracy of the geometric solution in absolute and relative measurements. This procedure is also valid when determining the satellite orbits from tracking stations, thus providing more precise initial conditions (satellite coordinates and their changes in time) and auxiliary data (the distances between satellites and their changes in time) when integrating equations of satellite motions.

In this paper the author discusses two approaches to the realisation of the concept of coherent signals. The first method is based on mutual exchange of phase information between satellites working in a pair, whereas the second employs the frequency transmitted by the geostationary satellite.

2. OBSERVATION EQUATIONS FOR INDEPENDENT AND COHERENT SIGNALS

The differences between the concepts of coherent and independent signals are most vividly seen in observation equations.

When observing from point P two satellites j and k from emitting frequencies ω_1 and ω_1' in the diapason L_1, the difference of phases $\Delta\varphi_P^{j,k}$ shall depend on distances from point to satellites r_1 and r_2 as follows (Goad, 1985; Fell, 1986) :

$$\Delta\varphi_P^{j,k} = 2\pi N + \delta\varphi = (\omega_1 - \omega_1')(t - t_c) + \omega_1 \frac{r_1}{v_1} - \omega_1' \frac{r_2}{v_2} \qquad (1)$$

where $\delta\varphi$ is the "measured" fraction of the phase cycle;

N is the integer of cycles;

v is the propagation velocity of signals;

t is the instant time;

t_c is the moment of time assumed as initial.

In the case of coherent signals the corresponding formula shall be (Prilepin, 1988; 1989):

$$\Delta\varphi_p^{j,k} = \omega_1^0\left(\frac{\tau_2}{v_2} - \frac{\tau_1}{v_1}\right) + \left(\omega_1^0 - 2\omega_2\right)\frac{\tau_0}{v_0} \qquad (2)$$

where ω_1^0 is the frequency of the first ("master") satellite in diapasone L_1;

ω_2 is the frequency of the second satellite in diapasone L_2;

τ_0 is the distance between satellites;

v_0 is the signal velocity between satellites.

As we can see with coherent signals the phase difference $\Delta\varphi_p^{j,k}$ is independent of the instant time of observations and, therefore, does not depend on the divergence of frequencies of different satellites.

Even with very good generators, for example, the cesium ones, the shift per hour of the frequency is about 0.001 Hz, whereas the short-period (several seconds) fluctuations may reach 1/3 of this value (Bendex, 1986). In accordance with formula (1), the additional phase fluctuations and shifts during the observation period (1.5-3.0h.) may reach several phase cycles (4÷12). All these factors complicate the processing of data by requiring the modelling of the behaviour of each generator of the satellite, the introduction of the additional unknown quantities, which finally results in lower accuracy. As shown by experiments (Beutler at al, 1984; Bock at al, 1985), the errors of absolute determinations may reach 5÷10 m, and one of the primary causes is the frequency shift of the satellite. These problems are eliminated with the use of coherent signals.

3. REALISATION OF COHERENCY BY AN EXCHANGE OF PHASE INFORMATION BETWEEN SATELLITES

The method of realisation of coherent signals in the development of an autonomous two-satellite system was described earlier in two papers (Prilepin, 1986, 1987). In this case the satellites emit the same frequency equal to the difference of frequencies on each satellite.

The problem, as applied to the realisation of coherence for two satellites in the functioning GPS or GLONASS systems, can be solved by means of a more complicated frequency exchange when the second satellite transmits to the first (master) satellite the frequencies ω_1 (L_1 diapasone) and ω_2 (L_2 diapasone), whereas the first satellite transmits to the second satelite the differences of $\omega_1^o - \omega_2^o$, $\omega_1^o - \omega_2$ and $\omega_1 - \omega_2^o$. In this case, from both satellites the same frequencies ω_1^o and ω_2^o to the user are transmitted. Moreover, the second satellite can determine the distance between satellites (Prilepin, 1988). Since the ground-based station conducts measurements on two frequencies in different diapasones (L_1 and L_2), the ionospheric correction can be, therefore, determined. The range between the satellites can be determined at the observation station as well, if the transmission of the differential frequency $\omega_1^o - \omega_2^o$ from both satellites directly to the user is additionally introduced.

Fig. 1 illustrates the exchange of frequencies between satellites in this case.

We shall have the following equations for phase differences during observations of two satellites:

$$(\Delta\varphi_p^{j,k})_{\omega_1^\circ} = \omega_1^\circ\left(\frac{\tau_2}{\upsilon_2} - \frac{\tau_1}{\upsilon_1}\right) + (\omega_1^\circ - 2\omega_2)\frac{\tau_0}{\upsilon_0} \quad (3)$$

$$(\Delta\varphi_p^{j,k})_{\omega_2^\circ} = \omega_2^\circ\left(\frac{\tau_2}{\upsilon_2} - \frac{\tau_1}{\upsilon_1}\right) + (\omega_2^\circ - 2\omega_1)\frac{\tau_0}{\upsilon_0} \quad (4)$$

$$(\Delta\varphi_p^{j,k})_{\Delta\omega_{1,2}^\circ} = (\omega_1^\circ - \omega_2^\circ)\left(\frac{\tau_2}{\upsilon_2} - \frac{\tau_1}{\upsilon_1} + \frac{\tau_0}{\upsilon_0}\right) \quad (5)$$

Formulas (3)-(5) solve the formulated problem of concurrent determination of unknown quantities ($\frac{\tau_2}{\upsilon_2} - \frac{\tau_1}{\upsilon_1}$) and $\frac{\tau_0}{\upsilon_0}$ taking into account ionospheric effects.

The analysed scheme of realisation of coherency of signals on two satellites by an exchange of frequencies between them can be principally expanded to cover the whole constellation in newly developed multisatellite systems and in the now functioning GPS and GLONASS systems.

It is, however, presumed that for geodetic and geophysical measurements in static conditions it is sufficient to have 1÷3 separate appropriately matched pairs of satellites. The final selection of the number of pairs and the character of their orbits is to be determined later after numerical modelling of the problem.

The details that should be taken into consideration when selecting the kind of communication between satellites are as follow: the distance between the satellites should not exceed the operation range of the measuring facilities; the distance between the satellites should be varied within small limits, i.e., the rate of the change τ_0 should be minimal. It is desirable to avoid disconnections of communication between satellites.

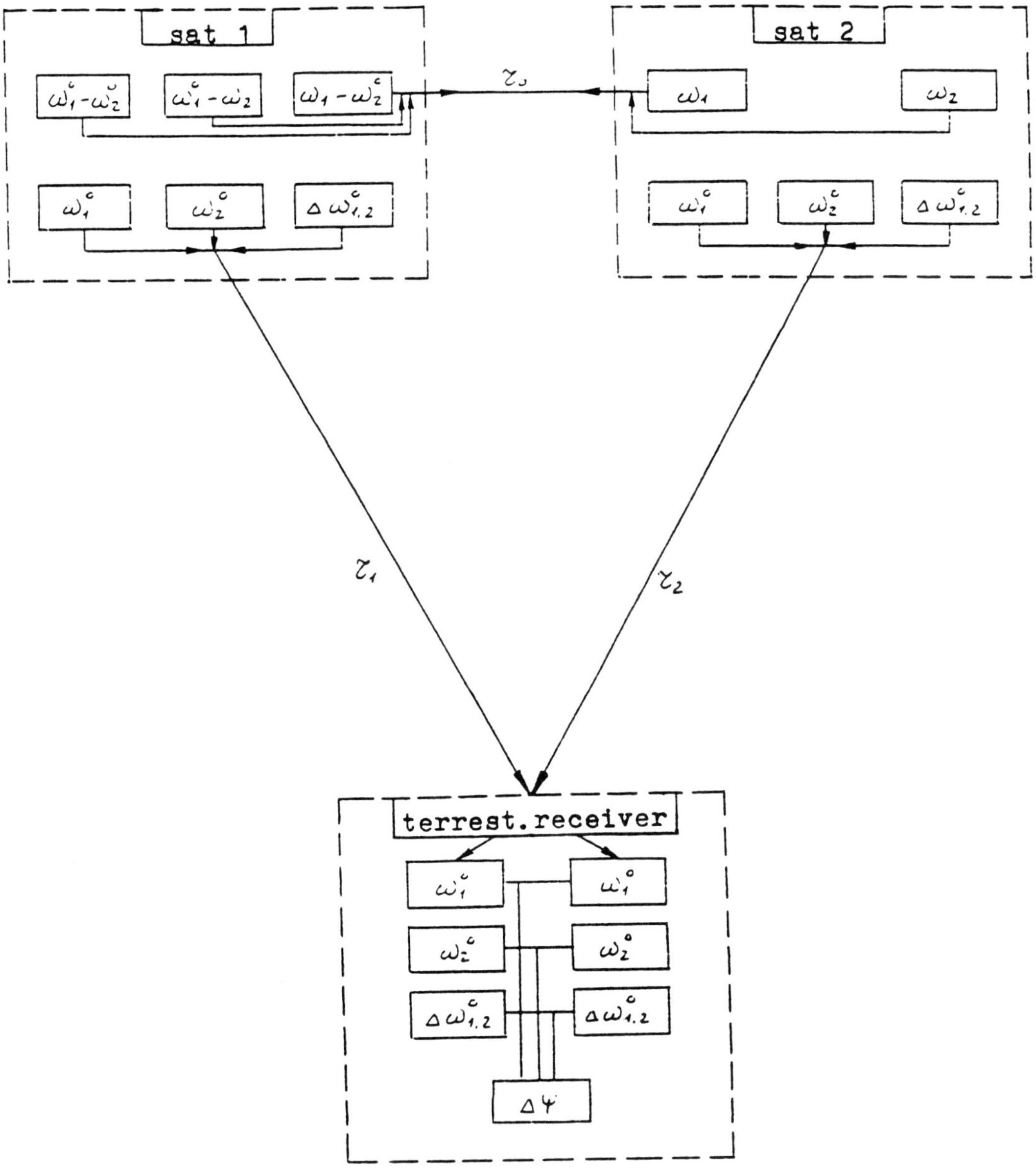

Fig. 1. The scheme of exchange of signals between satellites functioning in a pair.

4. REALISATION OF COHERENCE USING GEOSTATIONARY SATELLITES

Several suggestions are indicated in the literature as regards the utility of geostationary satellites (GS) in the solution of geodetic and navigation tasks (NAVSAT, GEOSTAR).

The navigation STARFIX regional system already showed good results by using GS. The system has good prospects with respect to higher accuracy and to transition from obtaining plane coordinates (X,Y) to space coordinates (Ott at al, 1989) (X,Y,Z).

In our case GS are intended for precise geodetic and geophysical measurements and are used only as translators of signals of the GPS or GLONASS type of systems to realise coherence. Since we discuss the idea of the method, we shall presume that only one GS is used. In this case the energy carrying the measurement signal is directed to the solid-angle encompassing the GPS or GLONASS orbits.

Fig.2 shows the scheme of the geometry of solution of the problem. The data load center transmits a high carrier ν (in the order of 30-40 GHz) on which the carriers ω° and the service information of the GPS or GLONASS systems are imposed as modulating. On GS the signal of the frequency $\nu \pm \omega^{\circ}$ is amplified and retranslated to the satellite systems GPS or GLONASS. On each of the satellites of these systems, in the visibility range of GS, the signal of frequency is detected amplitied and directed to the user. This operation eliminates the difficulties of receipt and transmission on the same frequency.

The phase difference from satellites j is fixed at point P by ground-based receiver:

$$\Delta\Psi_p^j = (\Omega - \omega^c)(t - t_c - \frac{\rho_j}{v_j'} - \frac{\tau_j}{v_j}) \qquad (6)$$

where Ω is the circle frequency of the receiver;

ρ_j is the distance from GS to satellite;

τ_j is the distance between satellite and terrestrial point.

In the case of synchronous observations of two satellites we shall have:

$$\Delta\Psi_p^{j,k} = \omega^c(\frac{\tau_j}{v_j} - \frac{\tau_k}{v_k} + \frac{\rho_j}{v_j'} - \frac{\rho_k}{v_k'}) \qquad (7)$$

The sum of differences of distances determined by (7) is expressed through unknown coordinate corrections in the form symmetrical to $\delta x, \delta y, \delta z$

$$(\tau_j - \tau_k) + (\rho_j - \rho_k) = -\frac{2(x_k - x_j)}{\tau_j^0 + \tau_k^0}\delta x_p - \frac{2(x_k - x_j)}{\rho_j^0 + \rho_k^0}\delta x_{GS} + \\ + \frac{(x_k - x_p^0)^2 - (x_j - x_p^0)^2}{\tau_j^0 + \tau_k^0} + \frac{(x_k - x_{GS}^0)^2 - (x_j - x_{GS}^0)^2}{\rho_j^0 + \rho_k^0} + \\ + F(y) + F(z) \qquad (8)$$

As obvious from equation (8), the coefficients at unknown quantities of the determined point and geostationary satellite are different; therefore, the unknown quantities of the observation program including several satellites can be determined separately.

If the coordinates of the data load center are known, we may, in principle, use another equation which directly ties the coordinates of terrestrial point P and GS (the dashed line in Fig.2). In this case the ground-based receiver should be able to demodulate frequency ω^c.

A combined system including three GS exchanging frequencies, as shown in the previous paragraph, allows us to use one groundbased generator to provide a practically global coverage

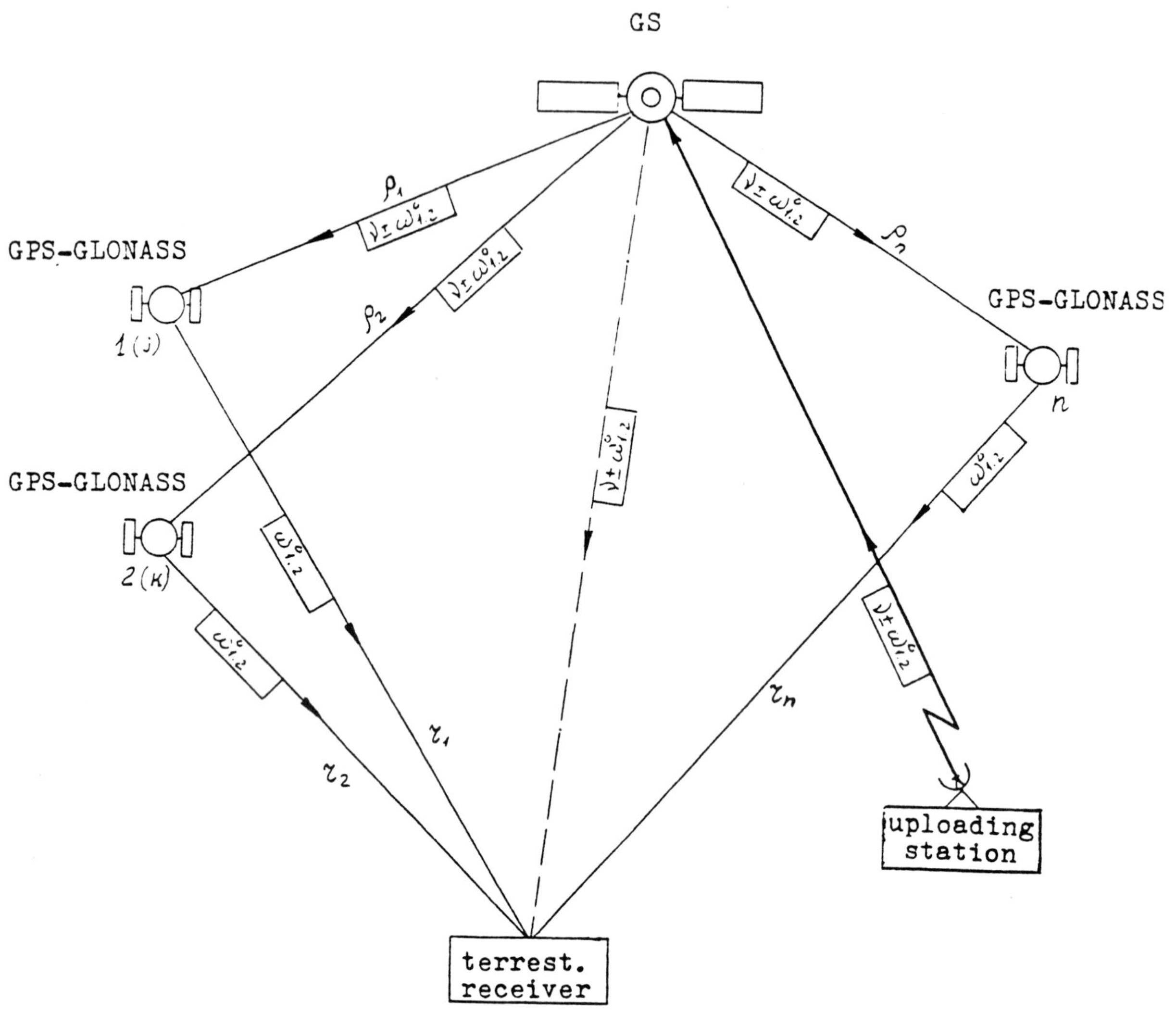

Fig. 2. The scheme of exchange of signals through the geostationary satellite.

except the "polar cap" (about 2% of the surface of the Earth).

CONCLUSION

The method of realisation of coherency of oscillations by frequency exchange between satellites provides higher accuracy of determinations by eliminating the effect of satellite frequency drift and using measurements of distances between satellites. This technique can be profitably applied in the systems of global coverage both in newly developed autonomous systems and when modernising the functioning systems of the GPS or GLONASS type.

Realisation of coherence using one GS is easier when solving regional problems, such as monitoring of crustal deformations for earthquake prediction, because in this case GS can effectively operate as transmitters of measurement data.

The GS coordinates are determined simultaneously with the coordinates of terrestrial points at almost the same level of accuracy, which is important for the solution of a number of scientific and practical problems.

REFERENCES

1. C.A.Bartolomew. Satellite frequency standards. Navigation. v. 25, N 2, 1978.
2. Bendix field engineering corporation. Time and Frequency Services.
3. G.Beutler, D.A.Davidson, R.B.Langley, R.Santerre, P.Vanicek, D.E.Wells. Some theoretical and practical aspects of geodetic positioning using carrier phase difference observations of GPS satellites. Technical Report N 109. Depart.

of surveying Engineering. University of New Brunnswick, 1984.

4. J.Bock, R.I.Abbot, C.C.Counselman III, S,A.Gourevitch, R.N.King. Establishment of three-dimensional geodetic control by interferometry with Global Positioning System. J.G.R., v.90, 7689-7703, 1985.

5. P.J.Fell. Conv. Proceedings of the 4th Int. Geodetic Symposium on Satellite Positioning. Austin, 1986.

6. C.C.Goad, ed. Proceedings of the First Int. Symposium on Precise Positioning with GPS, Rockville, 1985.

7. L.Ott, A.Bogle Starfix: commersial satellite positioning. John E. chance and associates, inc., 1989.

8. M.T. Prilepin. Use of a Satellite System for Studying Crustal Deformations. Abstracts of Papers Presented at the International Symposium on Recent Crustal Movements, Tallinn, September, 1986.

9. M.T. Prilepin. The application of satellite systems for crustal deformation studies. Journal of Geodynamics. v. 10, 1988,

10. M.T.Prilepin. Monitoring of crustal deformations by satellite methods. The paper presented at XIX GA IUGG, Vancouver, 1987.

11. M.T.Prilepin. The satellite geodetic system with coherent signals. Paper presented at the Int. Symposium "Global positioning sistems". Brisbane, Australia, October, 1988.

12. M.T.Prilepin. Improvement of accuracy of single-point determinations. Paper presented at 5th Intern. Geodetic symposium on satellite positioning, Lus Cruces, New Mexico,1989.

Appendix 1

125 Years of International Cooperation in Geodesy

Presidential Address

IAG General Meeting
Edinburgh, Scotland
August 7, 1989

Ivan I. Mueller
Department of Geodetic Science & Surveying
Ohio State University, Columbus, Ohio, USA

Prologue: Why History?

"... The great private problem of man: death as the loss of self. But what is self? It is the sum of everything we remember. Thus, what terrifies us ... is not the loss of the future but the loss of the past. Forgetting is a form of death ever present ... Intentional forgetting is a form of suicide ... Organized forgetting is a form of murder ..."

—Milan Kundera, writer

Background

- Carl Friedrich Gauss 1777-1855
  Director, Göttingen Observatory
- Friedrich Bessel 1784-1846
  Director, Königsberg Observatory
- Johann Jakob Baeyer 1794-1885
  General, Prussian Army

The First 60 Years:
The Beginning

1861 Johann Jakob Baeyer's report
"Entwurf zu einer mitteleuropäischen Gradmessung"
approved by the King of Prussia.

... what one cannot achieve alone, several can achieve together ...

The First 60 Years:

18 Signatories to Central European Arc Measurements, 1862

Austria	The Netherlands
Belgium	Norway
Denmark	Russia for Poland
France	Sweden
8 German states	Switzerland
Italy	

The First 60 Years:

General Conferences

1864 Berlin	1883 Rome	1906 Budapest
1867 Berlin	1886 Berlin	1909 Cambridge
1871 Vienna	1889 Paris	1912 Hamburg
1874 Dresden	1903 Copen-hagen	1917 St. Peters-burg*
1877 Stuttgart		
1880 Munich		

*cancelled

The First 60 Years: Main Events

- 1864 Central European Arc Measurement

  President 1864-74: Dr. Hansen
  Director, Central Bureau 1864-85:
  J.J. Baeyer

- 1867 European Arc Measurement

  Spain & Portugal join, total of 20 states

- President 1874-92: Carlos Ibáñez, Spain

The First 60 Years: Main Events

- 1884 Greenwich meridian and GMT adopted
- 1885 General Baeyer dies
- 1886 International Association of Geodesy (Internationaler Erdmessung) 10-year charter through 1896

  Director, Central Bureau 1886-1917:
  F.R. Helmert, Germany

The First 60 Years: Main Events

- President 1892-1903: Herve Etienne August Alband Faye, France
- 1895 Renewal of IAG charter through 1906, USA and Japan join
- 1896 International Latitude Service, 5 stations
- President 1903-17: Jean Antonin Leon Bassot, France
- 1906 Renewal of IAG charter through 1916

The First 60 Years: Main Events

- 1913 BIH, Bureau International de l'Heure

- **1914-18 World War I**

- 1916 Charter expires
- 1917 Bassot & Helmert die

The First 60 Years: Main Events

- 1917-22 Reduced geodetic association among neutral nations: Denmark, The Netherlands, Norway, Spain, Sweden, Switzerland

  Leaders:

  R. Gautier, Switzerland, and H.G. van de Sande Bakhuyzen, The Netherlands

"The neutral nations maintain the existence of the Association ... for a period that cannot be precisely defined at present ..."

The First 60 Years: Main Events

1919 Brussels Assembly,
1st International Research Council

(from 1931, International Council of Scientific Unions, ICSU)

-Geodesy & Geophysics, IUGG
-Astronomy, IAU
-Pure and Applied Chemistry
-Radio Science

The First 60 Years:
Horizontal Control

- No. of triangulation points increases from 2010 in 1862 to 5546 in 1912
- Meridian arcs: Shetland Islands to Algiers
  N. America 98° W
  Africa 30° E
  S. America Pacific Coast

The First 60 Years:
Horizontal Control

- Parallel arcs: Ireland -Urals 52° N
  Brest - Astrachan 47.5° N
  Across N. America 39° N
- Deflection of the vertical determinations in all participating countries
- 1875 International Meter Convention

The First 60 Years: Gravity

- 1887 Improvement of the Sterneck pendulum apparatus
- 1900 Vienna Gravity System, Helmert
- Earth tide measurements with horizontal pendulums
- 1901 Helmert's gravity formula

The First 60 Years: Gravity

- 1903 1st barometric sea gravity observations, Hecker
- 1906 Kühnen & Furtwängler absolute gravity measurement in Potsdam
- 1906 Eötvös torsion balance
- Isostatic investigations, Hayford & Helmert
- 1909 Potsdam Gravity System

The First 60 Years:
Figure of the Earth

- 1873 Term "geoid" coined
- 1878 Clark's readjustment of arc measurements
- H. Bruns' "Die Figur der Erde"
- 1880 F.R. Helmert
  "Die mathematischen und physikalischen Theorien der Höhere Geodäsie"
- 1909 Hayford's ellipsoid—to be adopted in 1924 as the International Ellipsoid

The First 60 Years:
Geodetic Astronomy

- 1882-86 Küstner latitude observations in Berlin
- 1891-92 S. Chandler, "On the Variation of Latitude"
- 1891-92 Simultaneous latitudes in Honolulu & Berlin
- 1896 International Latitude Service
- 1905 Adjustment of Central European longitude net
- 1911 Wireless time transfer in Paris

1920's: General Activities

- President 1920-33: William Bowie, USA
- General Secretary 1920-46: Georges Perrier, France
- 1922 Rome 1st IUGG General Assembly: Constitution of Section of Geodesy (IAG since 1930) 24 countries, 79 delegates
- 1924 Madrid General Assembly IAG statutes based on 1922 IUGG Assembly

1920's: General Activities

- Bulletin Géodésique 1924-42, 1946-
- IAG Gravity Commission
- International Bibliography of Geodesy 1924- (restart after 1876, 1881 & 1889)
- IAG - IAU World Longitude Net to determine continental drift
- 1927 Prague General Assembly

1920's: Technology

- 1923- Submarine gravity pendulum observations (Vening-Meinesz), ± 5-10 mgal
- 1924 Invar wires for base measurements $\pm 10^{-6}$
- 1929-30 1st spring pendulum and spring balance gravimeters (Ising, Lejay), ± 1-5mgal

1920's: Investigations

- 1924- Isostasy (Bowie, Heiskanen, Vening-Meinesz)
- 1924 International Ellipsoid (Hayford)
- 1928 Gravimetric deflection computations (Vening-Meinesz)
- Normal gravity field (Somigliana, Cassinis, Heiskanen)

1930's: General Activities

- 1930 Stockholm General Assembly:

  IAG Commission on Adjustment of European Triangulation
- 1933 Lisbon General Assembly
- President 1933-46: F.A. Vening-Meinesz, The Netherlands
- 1936 Edinburgh General Assembly 34 countries

1930's: General Activities

- IAG Isostatic Institute in Helsinki (Heiskanen)
- International gravity survey for geoid determination (de Graaff-Hunter)
- Joint Commissions on Variations on Mean Sea Level (IAPSO) and on the Study of the Earth Crust (IASPEI)
- 1939 Washington General Assembly

1930's: Technology

- Bifilar tidal gravimeter, ±0.01 mgal
- Horizontal pendulum, ±0".001
- 1st experiments with marine gravimeters
- Reversible pendulum absolute gravity, Washington & Teddington (Clark, Heyl & Cook), ± 3 mgal

1930's: Investigations

- 1930-36 Gravimetric geoid calculations (Jeffreys, Hirvonen)
- Leveling error theory for (Vignal)
- Corrections to gravity observations on moving platforms (Browne)
- 1936- Boundary value problem (Molodensky, Krassowsky, Michailov)
- 1937 Leveling & refraction (Kukkamäki)

World War II

1939 - 1945

1940's: General Activities

- 1946 Cambridge Extraordinary General Assembly: IUGG reorganization
- 1946 Paris IAG Executive Committee: Reorganization under new sets of IUGG rules
- President 1946-51: Walter D. Lambert, USA
- General Secretary 1946-60: Pierre Tardi, France

1940's: General Activities

- 1946- Bulletin Géodésique Nouvelle Série
- Establishment of IAG Central Bureau, Paris
- 1948 Oslo General Assembly
  28 countries, 102 IAG participants
  New IAG statutes and by-laws

1940's: Technology

- 1946 Free fall absolute gravity, Sevres (Volet), ±10 mgal
- 1945 Shoran triangulation, ±10 m
- 1946 Stellar triangulation (Väisäla)
- 1947 Optical interferometric calibration baselines (Väisäla), $\pm 10^{-7}$
- 1945-47 Solar eclipse techniques (Bonsdorff)
- 1948 Electro-optical Geodimeter (Bergstrand), $\pm 2 \times 10^{-6}$

1940's: Investigations

- Proposal for United European Leveling Network, UELN
- Vertical deflection collection and geoid computations (Bomford, de Graaff- Hunter)
- Gravimetric geoid (Tanni), ±10 m
- Intrinsic geodesy (Marussi)
- 3-dimensional geodesy (Hotine, Levallois, Wolf)

1950's: General Activities

- 1951 Brussels General Assembly
- President 1951-54: Fridolin Baeschlin, Switzerland
- International Gravity Bureau, Paris
- Amendments to by-laws and statutes
- 1954 Rome General Assembly
  31 countries, 235 IAG participants

1950's: General Activities

- President 1954-57: J. de Graaff-Hunter, U.K.
- 1957 Toronto General Assembly
- President 1957-60: Gino Cassinis, Italy
- International Earth Tide Center, Brussels

1950's: Technology

- 1950 Hiran triangulation, ±5 m
- Automatic level
- High-precision spring gravimeters, ±0.02 - 0.05 mgal
- Sea gravimeter (Graf, LaCoste), ±5 mgal

1950's: Technology

- 1957 Markowitz moon camera
- 1956 Microwave Tellurometer (Wadley), $\pm 5 \times 10^{-6}$
- 1957 Sputnik 1 & Vanguard 1 satellites
- 1958 Satellite geodesy starts: Moon-watch teams; Baker-Nunn camera in New Mexico; test of Transit Doppler system

1950's: Investigations

- Equilibrium figure of the Earth (Bullard, Jeffreys)
- Extension of adjustment theory & use of statistics (Tienstra, Bjerhammar, Wolf, ...)
- Molodensky's theory of gravimetric geodesy
- ED50 European datum adjustment

1950's: Investigations

- World Geodetic System proposal (Heiskanen)
- Cape-Cairo meridian arc finished
- Bomford's astrogeodetic geoid, ±1-5 m
- World gravity network (Woollard, Morelli)

1950's: Investigations

- Columbus gravimetric geoid (Heiskanen), ±5 m
- 1958 G. Veis' "Geodetic Applications of Observations of the Moon, Artificial Satellites and Rockets"
- Satellite positioning with cameras: ±100 m absolute, ±20 m relative
- The pear-shaped earth: determination of J_2 and J_3 with satellites (O'Keefe)

1960's: General Activities

- 1960 Helsinki General Assembly
- President 1960-63: Charles A. Whitten, USA
- General Secretary 1960-75: J.J. Levallois, Fr.
- 1963 Berkeley General Assembly
- President 1963-67: Guy Bomford, U.K.

1960's: General Activities

- 1963 International Polar Motion Service, Mizusawa
- 1963 Central Bureau for Satellite Geodesy, Cambridge, Massachusetts
- 1966 Marine Geodesy Symposium, Columbus
- 1967 Lucerne General Assembly
- President 1967-71: Antonio Marussi, Italy

1960's: Technology

- 1963 Free-fall interferometric absolute gravity meter (Faller, Sakuma), ±0.005 - 0.01 mgal
- Mekometer, ±0.1 mm
- 1967 Rise & fall absolute gravity meter (Cook), ±0.1 mgal
- 1968 Transportable absolute gravity meter (Faller), ±0.05 mgal

1960's: Technology

- 1968 Superconducting gravimeter (Goodkind), ±0.1 μgal
- Geodetic satellites: Transit 1B, Echo 1-2, ANNA 1B, BE-B & C, GEOS 1-2, Pageos, DIADEME 1-2, Oscar 14, Timation 2
- 1969 Apollo 11 corner reflector on the Moon, 1st returns at Lick and McDonald Observatories

1960's: Technology

- Ballistic satellite cameras
- Satellite Doppler equipment: Transit vans, 1969 Geoceiver
- Satellite laser tracking ±1 m: 1964 GODLAS, 1965 LARGOS, 1966 SAO laser at New Mexico, 1967 MOBLAS 1, 2 CNES stations, 1968 Athens
- VLBI: Mark I receiver development

1960's: Investigations

- 1966 Wm. Kaula's "Theory of Satellite Geodesy"
- Reformation of geodesy, without hypothesis (Bjerhammar, Hirvonen)
- Gravity field statistical investigations (Kaula, Bjerhammar, Moritz, ...)
- Network reliability studies (Baarda, Bjerhammar)

1960's: Investigations

- Terrestrial refraction studies (Tengstrom)
- 1967 Geodetic Reference System
- Least squares collocation (Krarup, Moritz)
- Recent vertical crustal movements from leveling and gravity (Boulanger)

1960's: Investigations

- Satellite gravity fields: 1962 J_2-J_9, J_{41}, J_{22}; 1969 n = m = 16 (SE II)
- Satellite positioning using cameras/ Doppler: ±20 m absolute, ±5 m relative
- 1969 1st VLBI baseline: Haystack-Greenbank, ±2 m
- Polar motion monitoring starts using Doppler satellites

1970's: General Activities

- 1971 Moscow General Assembly
  Revision of IAG statutes and by-laws
  44 countries, 450 IAG participants
- President 1971-75: Y.D. Boulanger, USSR
- 1971-75 IAG sponsored 20 symposia
- 1975 Grenoble General Assembly
  51 countries, 410 IAG participants

1970's: General Activities

- President 1975-79: T.J. Kukkamäki, Finland
- Secretary General 1975- : Michel Louis, Fr.
- 1975 1st Bomford Prize: Erik W. Grafarend
- 1975- Independent, reviewed editing of Bulletin Géodésique

1970's: Activities

- 1975-79 IAG sponsored 28 symposia
- 1979 Canberra General Assembly
- President 1979-83: Helmut Moritz, Austria
- 1979 1st Levallois Medal: C. H. Whitten
- 1979 Bomford Prize: Fernando Sansò

1970's: General Activities

- 1979 IAG Commission for Africa established
- 1979 International Lithosphere Project
- 1979 CSTG established - International Coordination of Space Techniques for Geodesy and Geodynamics, IAG–COSPAR joint commission

1970's: Technology

- 1975 Inertial survey systems
- Transportable absolute gravimeters, ±0.01 mgal
- Helicopter-borne gravimetry, ±1 mgal
- Terrameter (2 wavelengths), $\pm 10^{-7}$
- Portable Doppler satellite receivers: Magnavox, JMR, etc.

1970's: Technology

- Tranet-2 Doppler system
- Satellite laser tracking: 6 SAO, 8 MOBLAS, Interkosmos stations, Dodaira, Wettzell, Orroral, Grasse, 1st mobile TLRS-1

  Break in refraction & waveform digitizing (±15-20 cm); Nd-Yag 0.2 ns, cavity dump and pulse chop (±10 cm)
- VLBI: Mark II & III developments MV-1 1st mobile system

1970's: Technology

- Lunar laser ranging: Apollo 14–15 & Lunakhod II reflectors on the Moon; McDonald operation, first returns at Maui and Orroral
- Satellites: DIAL, Oscar 19, PEOLE, STARLETTE, Geos-3 1st altimeter, LAGEOS, Navstar 1, Seasat-1.
- 1979 GPS—Global Positioning System—receivers: MITES series

1970's: Investigations

- 1971 IGSN International Gravity Standard Net (±0.1 mgal)
- Recent vertical crustal movements map for Eastern Europe
- Gravimetric earth tide profiles
- Solution of large systems of equations, Meissl
- 4-dimensional geodesy, Mather

1970's: Investigations

- WGS 72, World Geodetic System
- RETRIG Phase II-ED 79, Kobold
- Gravity field covariance models
- Ocean tide models

1970's: Investigations

- Satellite projects: ISAGEX, SAFE exp., EROLD, EODAP '75-79, EDOC 1-2, MEDOC, EROS 1-3, ESA ERG, NASA Crustal Dynamics Project starts 1979
- Satellite & terrestrial gravity fields: n = m = 36, 1366 coefficients (GRIM-2, GEM 10B)
- Satellite Doppler and laser positioning: 1 m absolute; 0.2 m relative

1970's: Investigations

- VLBI baselines:
  1972-79 1st transcontinental experiment
  Haystack-Goldstone ±0.2 m

  1979 1st trans-Atlantic experiment
  Haystack-Onsala
- 1972 Earth rotation parameters:
  UT0 ±0.5 ms from LLR
  Polar motion ±0".02 from Doppler

1980's: General Activities

- 1980 1st Geodesist's Handbook
- 1979-83 IAG sponsored 19 symposia
- 1982 Tokyo General Meeting, 9 symposia
- 1983 Hamburg General Assembly
  Revision of statutes and by-laws
- President 1983-87: Peter V. Angus Leppan, Australia

1980's: General Activities

- 1983 Levallois Medal: Rudolf Sigl
- 1983 Bomford Prize: John Wahr
- 1983-87 IAG sponsored 39 symposia
- 1987 Vancouver General Assembly
  43 countries
- President 1987- : Ivan I. Mueller, USA

1980's: General Activities

- 1987 Levallois Medal: Arne Bjerhammar
- 1987 Bomford Prize: Peter Teunissen
- 1988 International Earth Rotation Service
- 1989 Edinburgh General Meeting, 5 symposia

1980's: Technology

- Sea & airborne gravimetry with improved navigation, to ±1 mgal
- Gravity gradiometer systems tested
- Development of superconducting and other types of satellite gradiometry to $\pm 10^{-4}$ E
- Satellite-to-satellite tracking

1980's: Technology

- Satellite laser tracking:
  TLRS 2-4, MTLRS 1-2, Yarragadee, Metsahovi, Quincy, Simosato, Mazatlan, Matera, Zimmerwald, Graz, Monument Peak, Shanghai, etc.

  Mode locked lasers (±3-4 cm), with GPS timing (±1-2 cm)
- VLBI: Polaris & IRIS network (Westford, GRAS, Onsala, Richmond, Wettzell), MV 2-3, Medicina, etc., additional correlator facilities

1980's: Technology

- Lunar laser (Nd:Yag) ranging: CERGA, MLRS
- GPS: Portable receivers from 10 manufacturers
- Satellites: Navstar 6-13, Nova 1-3, Oscar 30, BULGARIA-1300, METEOR-3, AJISAI, GEO-IK, ETALON 1-2

1980's: Investigations

- Integrated geodesy
- Operational geodesy
- Network optimization studies
- Absolute gravimeter comparisons, Sevres
- North American Datum, NAD-83
- World Geodetic System, WGS-84

1980's: Investigations

- European Datum, ED-87
- United European Leveling Network, UELN-73
- International Absolute Gravity Base Station Network—IAGBN—planning
- Sea surface from satellite altimetry
- Static and kinematic GPS surveys, to 10^{-7}

1980's: Investigations

- Projects
  MERIT-COTES 1980-87
  WEDOC
  WEGENER-MEDLAS 1981–
  ADOS 1981–86
  NASA Crustal Dynamics Project continues
  etc.
- Rates for tectonic plate motions and regional crustal deformation

1980's: Investigations

- Satellite & terrestrial combined gravity fields: n = m = 360 (OSU86F) from n = 2, m = 0 in 1959
- Geoid determinations, regionally ±10 cm from 5 m in 1950's, locally ±1 cm from 1 m
- Space technique positioning: ±0.1 m absolute, ±0.01 m (or 10^{-7}) relative from 100 m & 20 m in 1961
- Earth rotation parameters: 0".001 from 0".02 in 1970's

Epilogue

"... the Association has known many setbacks in its history. It has overcome them all ... because geodesists feel at home in it. The mutual respect that arises out of ... direct contact and ... shared concerns creates ties of friendship that transcend ... jealousies and ideologies ... These relationships are extremely powerful forces for scientific progress ... "

– J.J. Levallois

IAG Honorary Secretary General

Acknowledgements

My friends Heinz Kautzleben, Michel Louis and Wolfgang Torge, members of the IAG Bureau, contributed in many ways to this presentation.

Appendix 2 — List of Participants

Michael AARESTRUP, National Survey & Cadastre (KMS), Svanemollevej 89, DK 2900 COPENHAGEN, Denmark

Dr Ing Jean ABLA, IGCI 01 B.P. 3862, ABIDJAN 01, Ivory Coast

Eng. J. M. AGRIA TORRES, Instituto Geografico Cadastral, Praca da Estrela, 1200 LISBOA, Portugal

Zuheir ALTAMIMI, Institut Geographique National, 2 ave Pasteur, F-94160 SAINT MANDE, France

Dr Martine AMALVICT, Institute de Physique du Globe, 5 rue Rene Descartes, F-67084 STRASBOURG-Cedex, France

Ir Bouderwijn AMBROSIUS, Dept Aerospace Engineering, Delft University of Technology, Kluyverweg 1, 2629 HS DELFT, Netherlands

Mr Giles ANDRE, Mapping & Charting Establishment, Elmwood Avenue, FELTHAM, Middlesex TW13 7AF, United Kingdon,

Prof Peter ANGUS-LEPPAN, School of Surveying, University of New South Wales, P.O.Box 1, KENSINGTON N.S.W. 2033, Australia

Marcio H O AQUINO, Institute of Engineering Surveying & Space Geodesy, University Park, NOTTINGHAM NG7 2RD, United Kingdom

Dr Brent ARCHINAL, Earth Orientation Division, US Naval Observatory, Washington DC 20392, USA

Elisa Felicitas ARIAS, Central Bureau of IERS, 61 avenue de l'Observatoire, F-75014 PARIS, France

Prof Vidal ASHKENAZI, Institute of Engineering Surveying & Space Geodesy, University Park, NOTTINGHAM NG7 2RD, United Kingdom

Dr Wolfgang AUGATH, c/o NLVA-Landesvermessung, Warmbuchenkamp 2, D-3000 HANNOVER 1, Fed Rep Germany

Dr Trevor BAKER, Proudman Oceanographic Laboratory, Bidston Observatory, BIRKENHEAD, Merseyside L43 7RA, United Kingdom

Prof Paolo BALDI, Inst di Scienze della Terra, Viale Ungheria 43, I-33100 UDINE, Italy

Dr Georges BALMINO, CNES-GRGS-BGI, 18 Ave Edouard Belin, F-31055 TOULOUSE CEDEX, France

Dr Francois E BARLIER, CERGA, Avenue Copernic, F-06130 GRASSE, France

Guy BASKER, Institute of Engineering Surveying & Space Geodesy, University Park, NOTTINGHAM NG7 2RD, United Kingdom

Colin BEATTY, Global Surveys Ltd, 2 Ridgacre Lane, Quinton, BIRMINGAM B32 1ES, United Kingdom

Norman BECK, Geodetic Survey of Canada, 615 Booth St, OTTAWA Ont K1A 0E9, Canada

Dr-Ing Matthias BECKER, Institut fur Physikalische Geodasie, Petersenstrasse 13, D-6100 DARMSTADT, Fed Rep Germany

Gian Battista BENCIOLINI, Dipartimento di Ingegneria, Localita Mesiano di Polo, TRENTO, Italy

Dr Alain BERNARD, ONERA, BP 72 - 92322 CHATILLON CEDEX, France

Giovanna BERRINO, Osservatorio Vesuviano, Via A Manzoni 249, I-80123 NAPOLI, Italy

Barbara BETTI, Instituto di Topografia, Politecnico di Milano, Piazza Leonardo da Vinci 32, I-20133 MILANO, Italy

Dr Gerhard BEUTLER, Astronomical Institute, Sidlerstrasse 5, CH-3012 BERNE, Switzerland

Dr Richard BIANCALE, CNES/GRGS, 18 ave Edouard Belin, F-31055 TOULOUSE CEDEX, France

Prof Peter BIRO, Technical University of Budapest, H-1521 BUDAPEST, Hungary

Dr Geoffrey BLEWITT, Jet Propulsion Lab, MS 238/625, 4800 Oak Grove Drive, PASADENA Ca 911909, USA

Prof Denizar BLITZKOW, IAG-USP, Caixa postal 30627, 01051 SAO PAULO SP, Brazil

Dr Yehuda BOCK, MIT 54-510, CAMBRIDGE Ma 02139, USA

Dr Gerd BOEDECKER, Bavarian Academy of Sciences, Marstallplatz 8, D-8000 MUNCHEN, Fed Rep Germany

Leendart BOER, Delft University of Technology, Thijsseweg 11, 2629 JA DELFT, Netherlands

Sklomo BOGNER, Technion, HAIFA-Technion City, Israel

Ronald F BORDLEY, Ordnance Survey, Romsey Road, Maybush, Southampton SO9 4DH, United Kingdom

Serge BOTTON, Institut Geographique National, 2 ave Pasteur, F-94160 SAINT MANDE, France

Claude BOUCHER, Institut Geographique National, 2 ave Pasteur, F-94160 SAINT MANDE, France

Mrs A BRAUN, Springer-Verlag, HEIDELBERG, Fed Rep Germany

Piere BRIOLE, Institut de Physique du Globe, 4 Place Jussieu, F-75252 PARIS CEDEX 05, France

Prof Dr Peter BROSCHE, Obsrvatorium Hoher List, Universitatsternwarte Bonn, D-5568 DAUN, Fed Rep Germany

Dr Ing Frits J J BROUWER, Rijkswaterstaat Survey Dept, Kanaalweg 3-B, PO Box 5023, NL-2600 GA DELFT, Netherlands

Maria A BROVELLI, Instituto di Topografia, Politecnico di Milano, Piazza Leonardo da Vinci 32, I-20133 MILANO, Italy

Dr John BROZENA, Bullard Laboratories, Cambridge University, Madingley Road, CAMBRIDGE CB3 0EZ, United Kingdom

Mrs Carine BRUYNINK, Royal Observatory of Belgium, Ave Circulaire 3, B-1180 BRUSSELS, Belgium

Dr Aleksander BRZEZINSKI, Space Research Centre, Polish Academy of Sciences, Bartycka 18, 00-716 WARSAW, Poland

Dr Beat BUERKI, Inst Geodesy & Photogrammetry, ETH Zurich, CH-8093 ZURICH, Switzerland

Prof DrSc Milan BURSA, Astronomical Institute, Czechoslovak Academy of Sciences, Budecska 6, 120 23 PRAHA 2, Czechoslavakia

Prof Dr Ing James CAMPBELL, Geodetic Institute, University of Bonn, Nussallee 17, D-5300 BONN 1, Fed Rep Germany

Dr Nicole CAPITAINE, Observatoire de Paris, 61 ave de l'Observatoire, F-75014 PARIS, France

Prof Alessandro CAPORALI, Dept Physics, via Marzolo 8, I-35131 PADOVA, Italy

Dr Yves CARISTAN, CEA/RCP, BP 12, 91680 BRUYERES LE CHATEL, France

Dr William E CARTER, National Geodetic Survey, NOAA N/CG114, 11400 Rockville Pike, ROCKVILLE Md 20852, USA

CHANG RueyGang, Dept Surveying & Mapping, Chung Cheng Inst Technology, TASHI 33509, Taiwan

Ing Mohamed CHARFI, Topographic & Cartographic Office, Cite Olympic, TUNIS, Tunisia

Prof Dr CHEN Junyong, Director General, National Bureau of Surveying & Mapping, Baiwanzhuang, BEIJING, China

CHEN Wu, Department of Surveying, The University, NEWCASTLE UPON TYNE NE1 7RU, United Kingdom

CHEN Zongping, Inst Astron Physikal Geodasie, TU Munchen, Hochkalterstr 6 D-8000 Munchen 90, Fed Rep Germany

Ross CHRISTIE, Ordnance Survey, Romsey Road, Maybush, Southampton SO9 4DH, United Kingdom

Prof Adam CHRZANOWSKI, Dept of Surveying Engineering, University of New Brunswick, PO Box 4400, FREDERICTON N.B. E3B 5A3, Canada

Dr Steven COHEN, Goddard Space Flight Center, Code 621, GREENBELT Md 20771, USA

Michael COLES, Measurement Devices Ltd, 3 Old Hall Mews, Colney, NORWICH NR4 7TX, United Kingdom

Dr Oscar L COLOMBO, Goddard Space Flight Center, Code 621, GREENBELT Md 20771, USA

Prof Gennaro CORRADO, Dept Geofisica e Vulcanologia, S Marcellino 10, I-80139 NAPOLI, Italy

Bruno COULON, CEA-RCP, BP 12, F-91680 BRUYERES-LE-CHATEL, France

Michael CRAYMER, University of New Brunswick, Department of Edgineering Surveying, P O Box 4400, FREDERICTON N B E3B 5A3, Canada

Mattia CRESPI, Instituto di Topografia, Politecnico di Milano, Piazza Leonardo da Vinci 32, I-20133 MILANO, Italy

Dr Bruno CRIPPA, Instituto di Topografia, Politecnico di Milano, Piazza Leonardo da Vinci 32, I-20133 MILANO, Italy

Prof P A CROSS, Department of Surveying, The University, NEWCASTLE UPON TYNE NE1 7RU, United Kingdom

Dr Techn Arpad CZOBOR, Inst Geodesy Cartography & Remote Sensing, PB 546, H-1373 BUDAPEST, Hungary

Loic DANIEL, Institut Geographique National, 2 ave Pasteur, F-94160 SAINT MANDE, France

Peter John DARE, Dept of Land Surveying, North East London Polytechnic, Longbridge Road, DAGENHAM Essex RM8 2AS, United Kingdom

Abdullah DAUD, Department of Surveying, The University, NEWCASTLE UPON TYNE NE1 7RU, United Kingdom

Dr Mark DAVISON, Institute of Engineering Surveying & Space Geodesy, University Park, NOTTINGHAM NG7 2RD, United Kingdom

Geoffrey DAY, British Geological Survey, Murchison House, West Mains Road, EDINBURGH EH9 3LA, United Kingdom

Ir Cees DE JONG, Department of Geodesy, Delft University of Technology, Thijsseweg 11, NL-2629 JA DELFT, Netherlands

Dr Clive DE LA FUENTE, 21 Anderson Cresecent, Beeston, NOTTINGHAM NG9 2PS, United Kingdom

D DELIKARAOGLOU, Canada Centre for Surveying, 615 Booth Street , OTTAWA Ont K1A OE9, Canada

Mrs G DELIS, Springer-Verlag, HEIDELBERG, Fed Rep Germany

Luc DEMONIE, National Geographical Inst, Abdij ter Kameren 13, B-1050 BRUSSELS, Belgium

Dr Heiner DENKER, Institut fur Erdmessung, Nienburgerstrasse 6, D-3000 HANNOVER 1, Fed Rep Germany

Dr Jean DICKEY, Jet Propulsion Laboratory, 4800 Oak Grove Drive M/S 238-332, PASADENA Ca 91109, USA

Dr Reinhard DIETRICH, Akad Wissenschaften DDR, Zentralinstitut fur Physik der Erde, Telegrafenberg A17, 1561 POTSDAM, German Dem Rep

Prof Dragutin DJUROVIC, Faculty of Sciences, Studentski trg 16, YU-11000 BELGRADE, Yugoslavia

Dr Alan DODSON, Dept of Civil Engineering, University Park, NOTTINGHAM NG7 2RD, United Kingdom

Dominici DONATELLA, Scienze Mater e della Terra, Facolta di Ingengeria, Universita de Ancona, Via Brecce Bianche, I-60100 ANCONA, Italy

DONG Da-Nan, Dept Earth Atmospheric & Planetary Sciences, MIT Rm 54-510, CAMBRIDGE Ma 02139, USA

Dr Nancy B DOUGLAS, NASA HQ, Code EEG, WASHINGTON DC 20546, USA

Dr John M DOW. European Space Operations Centre, Robert Bosche Strasse 5, D-6100 DARMSTADT, Fed Rep Germany

Dr Herb DRAGERT, Pacific Geoscience Centre, 9860 West Saanich Rd, SIDNEY, B C V8L 4B2, Canada

Dr Ing Hermann DREWES, DGFI/Abt I, Marstallplatz 8, D-8000 MUNCHEN 22, Fed Rep Germany

Adrian DREWETT, Department of Surveying, The University, NEWCASTLE UPON TYNE NE1 7RU, United Kingdom

Asmund H DROTTNING, GRGS/CNES, 18 ave Edouard Belin, F-31055 TOULOUSE Cedex, France

Dr Andrzej DROZYNER, Deutsches Geodatisches Forschungsinstitut, Marstallplatz 8, D-8000 Munchen, Fed Rep Germany

Dr Peter DUNN, STX Inc, 9701 L Philadelphia Court, LANHAM Md 20706, USA

Ing Hubert DUPRAZ, Ecole Polytechnique Federal, Geodesie et Mensuration, GR-Ecublens, CH 1015 LAUSANNE, Switzerland

Richard J EANES, Centre for Space Research, University of Texas at Austin, AUSTIN Tx 78712, USA

Dr Donald H ECKHARDT, A.F.G.L./L.W., HANSCOM A.F.B. Ma 01731, USA

Dr Robert J EDGE, Proudman Oceanographic Lab, Bidston Observatory, BIRKENHEAD, Wirral L43 7RA, United Kingdom

Dr-Ing Walter EHRNSPERGER, Bayer Komm Internat Erdmessumg, Marstallplatz 8, D-8000 MUNCHEN, Fed Rep Germany

Dr Martin EKMAN, National Land Survey, Division of Geodetic Research, S 801 82 GAVLE, Sweden

Dr W ENGEL, Springer-Verlag, HEIDELBERG, Fed Rep Germany

Dr Theodossius ENGELIS, ST Systems Corp, 9701 L Philadelphia Court, LANHAM Md 20706, USA

Bjorn ENGEN, Statens Kartverk, Geodesy Division, N-3500 HONEFOSS, Norway

Dr Erhard ERKER, Bundesamt fur Eich und Vermessungswesen, Schiffamtsgasse 1-3, A-1025 WIEN, Austria

Dr Alan G EVANS, Naval Surface Warfare Center K13, DAHLGREN Va 22448-5000, USA

Major Gen Patrick F FAGAN, Directorate General of Military Survey, Elmwood Avenue, FELTHAM Middlesex TW13 7AH, United Kingdom

Dipl Ing Thomas FAHLBUSCH, Geodatisches Institut, Universitat Karlsruhe (TH), Englerstrasse 7, Postfach 6980, D-7500 KARLSRUHE 1, Fed Rep Germany

James E FALLER, JILA, Campus Box 440, University of Colorado, BOULDER Co 80309, USA

Hassan H FASHIR, Department of Surveying, The University, NEWCASTLE UPON TYNE NE1 7RU, United Kingdom

William FEATHERSTONE, Department of Earth Sciences, Parks Road, OXFORD OX1 3PR, United Kingdom

Dr Martine FEISSEL, Central Bureau of IERS, 61 avenue de l'Observatoire, F-75014 PARIS, France

Dr Istvan FEJES, FOMI Satellite Geodetic Observatory, H-1373 BUDAPEST PF 546, Hungary

Marco FERMI, Telespazio SPA, via Bergamini 50, I-00159 ROME, Italy

Geraint FFOULKES-JONES, Institute of Engineering Surveying & Space Geodesy, University Park, NOTTINGHAM NG7 2RD, United Kingdom

Rene FORSBERG, National Survey & Cadastre, Gamlehave Alle 22, DK-2920 CHARLOTTENLUND, Denmark

Dr Alain FOURGASSIE, Service Hydrograph Oceanograph Marine, B P 426, 29275 BREST CEDEX, France

Prof Dagogo M FUBARA, University of Science & Technology, PORT HARCOURT, Nigeria

J FUJINAWA, National Research Centre for Disaster Prevention, Tennodai 3-1, Tsukuba-shi, IBARAK-KEN 305, Japan

Dr Roman GALAS, Universitatsternwarte Bonn, Auf dem Hugel 71, D-5568 DAUN, Fed Rep Germany

Dr Daniel GAMBIS, Central Bureau IERS, Observatoire de Paris, 61 Ave de l'Observatoire, F-75014 PARIS, France

Dr A GEIGER, Institute of Geodesy & Photogrammetry, ETH-Zurich, CH-8093 ZURICH, Switzerland

Dr Yola GEORGIADOU, Geodetic Research Laboratory, Dept Engineering Surveying, University of New Brunswick, P O Box 4400, FREDERICTON N B E3B 5A3, Canada

Prof Dr Nicola I GEORGIEV, Central Geodesy Laboratory, Bulgarian Academy of Sciences, 1000 SOFIA, 7 noemvri str no 1, Bulgaria

Sara GERRARD, Institute of Engineering Surveying and Space Geodesy, University Park, NOTTINGHAM NG7 2RD, United Kingdom

Prof Dr Carl GERSTENECKER, Institut Phys Geodasie, Petersenstrasse 13, D-6100 DARMSTADT, Fed Rep Germany

Prof Erik W. GRAFAREND, Dept Geodetic Science, Stuttgart University, Keplerstr. 11, D-7000 STUTTGART, Fed Rep Germany

Jean-Philippe GRELOT, Institut Geographique National, 136 bis rue de Grenelle, F-75700 PARIS, France

Dr Richard GROSS, Mail Stop 238-332, Jet Propulsion Lab, 4800 Oak Grove Drive, PASADENA Ca 91109, USA

Prof Erwin GROTEN, Inst Phys Geodesy, TU Darmstadt, Petersenstrasse 13, D-6100 DARMSTADT, Fed Rep Germany

Erich GUBLER, Federal Office of Topography, Seftigenstrasse 264, CH-3084 WABERN, Switzerland

Dr Werner GURTNER, Astronomical Institute, University of Berne, Sidlerstr. 5, CH-3012 BERNE, Switzerland

Roger HAAGMANS, Department of Geodesy, Delft University of Technology, Thijsseweg 11, NL-2629 JA DELFT, Netherlands

Walter HAEN, Univ Bundeswehr Munchen, Werner-Heisenberg-Weg 39, D-8014 NEUBIBERG, Fed Rep Germany

Dr Lars-Ake HALLER, Lantmateriverket, S-80182 GAVLE, Sweden

Paul HANNA, Institute of Engineering Surveying & Space Geodesy, University Park, NOTTINGHAM NG7 2RD, United Kingdom

Prof Ian A HARLEY, University College, Gower Street, LONDON WC1E 6BT, United Kingdom

Dr John C HARRISON, Geodynamics Corporation, 5520 Ekwill St., Suite A, SANTA BARBERA, CA 93111, USA

Bjorn Geirr HARSSON, Statens Kartverk, Geodesy Division, N-3500 HONEFOSS, Norway

Hamdan HASHIM, Department of Surveying, The University, NEWCASTLE UPON TYNE NE1 7RU, United Kingdom

Dr Hermann HAUCK, Institut fur Angewandte Geodasie, Richard Strauss Allee 11, D-6000 FRANKFURT AM MAIN, Fed Rep Germany

Alan HAUGH, B.P. Exploration Co., Britannic House, Moor Lane, LONDON EC2Y 9BU, United Kingdom

Dr-Ing Bernhard HECK, Department of Geodetic Science, University of Stuttgart, Keplerstasse 11, D-7000 STUTTGART, Fed Rep Germany

Dr B HECKMANN, Hessisches Landesvermessungsamt, Schaperstr. 16, D-6200 WIESBADEN, Fed Rep Germany

Prof Dr Gunter W HEIN, Inst Astr & Physical Geodesy, University FAF Munich, Werner-Heisenberg-Weg 39, D-8014 NEUBIBERG, Fed Rep Germany

Kosuke HEKI, Kashima Space Research Centre, Communications Res Lab 893-1, Hirai, Kashima-machi, IBARAKI 314, Japan

Dr Warren HELLER, Analytic Sciences Corp, 55 Walkers Brook Drive, READING Ma 01867, USA

Prof Dr Heinz HENNEBERG, Apartado 6, MARACAIBO, Venezuela

Siegfried HIEBER, European Space Agency, 8-10 rue Mario Nikis, 75738 PARIS CEDEX 15, France

Christopher John HILL, Institute of Engineering Surveying & Space Geodesy, University Park, NOTTINGHAM NG7 2RD, United Kingdom

Dr Roger G HIPKIN, Department of Geophysics, University of Edinburgh, Mayfield Road, EDINBURGH EH9 3JZ, United Kingdom

Prof B HOFMANN-WELLENHOF, Technical University, Rechbauerstrasse 12, A-8010 GRAZ, Austria

Dr Petr HOLOTA, Research Institute of Geodesy, 25066 Zdiby 98, PRAGUE-EAST, Czechoslovakia

Sabine HOURY, GRGS/CNES, 18 avenue Edouarde Belin, F-31400 TOULOUSE, France

Dr James R. HUDDLE, Chief Scientist, Litton Guidance & Control Div., 5500 Canoga Ave, WOODLAND HILLS Ca. 91364, USA

Ing Laurent HUGUENIN, Ecole Polytechnique Federal, Geodesie et Mensuration, GR-Ecublens, CH 1015 LAUSANNE, Switzerland

Dr Ing Johannes IHDE, VEB Geodasie & Kartographie, Forschungscentrum, Gohliserstr 4, GDR-7022 LEIPZIG, German Dem Rep

Dr Jonathan C ILIFFE, Department of Photogtammetry & Surveying, University College, Gower Street, LONDON WC1E 6BT, United Kingdom

Prof David D JACKSON, Department of Earth and Space Science, UCLA, LOS ANGELES Ca 90024-1567, USA

Dr Reiner JAEGER, Geodaetisches Institut, Universitaet Karlsruhe (TH), Englerstr 7, Postfach 6980, D-7500 KARLSRUHE 1, Fed Rep Germany

Cord-Hinrich JAHN, Institut fur Erdmessung, Universitat Hannover, Nienburgerstr 5, D-3000 HANNOVER 1, Fed Rep Germany

Dr Christopher JEKELI, AFGL/LWG, HANSCOM AFB Ma 01731, USA

Michael JENSEN, Shell UK Expl & Production, Shell-Mex House, The Strand, LONDON WC2R 0DX, United Kingdom

Gordon JOHNSTON, Racal Survey (UK) Ltd, Unit 2 Deemouth Centre, South Esplanade East, ABERDEEN AB9 8JZ, United Kingdom

Bo JONSSON, National Land Survey of Sweden, S-80182 GAVLE, Sweden

Prof Dr Istvan JOO, College for Surveying, University of Forestry and Timber, 1-3 Pirosalma Str, H-8000 SZEKESFEHERVAR, Hungary

Dr J KAARIAINEN, Finnish Geodetic Institute, Ilmalankatu 1A, SF-00240 HELSINKI, Finland

Majid KADIR, Faculty of Surveying, University of Technology, Karung Berkunci 791, 80990 JOHOR BAHRU, Malaysia

Prof Johani KAKKURI, Finnish Geodetic Institute, Ilmalankatu 1A, SF-00240 HELSINKI, Finland

Dr Michel KASSER, Institut Geographique National, 2 ave Pasteur, F-94160 SAINT MANDE, France

Prof Dr Heintz KAUTZLEBEN, Zentralinstitut f Physik der Erde, Telegraphenberg, 1561 POTSDAM, German Dem Rep

Paul KEARNEY, NERC Geophysical Equipment Pool, James Clerk Maxwell Building, Mayfield Road, EDINBURGH EH93JZ, United Kingdom

Dr William KEARSLEY, School of Surveying, University of New South Wales, P.O.Box 1, KENSINGTON N.S.W. 2033, Australia

Dr Adam J KERR, Director, International Hydrographic Bureau, Av Pres J F Kennedy, BP 445, 98011 MONACO Cedex, Monaco

Mohamed Ahmed KHALID, Department of Surveying, The University, NEWCASTLE UPON TYNE NE1 7RU, United Kingdom

Mohammad Hassan KHODDAM MOHAMMADI, National Cartographic Centre, Mehraj Street, TEHERAN, Iran

Dr Robert W KING, MIT 54-620, CAMBRIDGE, Mass. 02139, USA

Dipl Ing Ibrahim KINIK, General Command Mapping, Harita Genel Komutanligi, Geodezi Dairezi, CEBECI-ANKARA, Turkey,

Prof Aimo A. KIVINIEMI, Finnish Geodetic Institute, Ilmalamkatu 1A, 00240 HELSINKI, Finland

Dr Ing Gunter KLEIN, Bayer Kom Int Erdmessung, Marstallplatz 8, D-8000 Munchen 22, Fed Rep Germany

Dr Jaroslav KLOKOCNIK, Astronomical Institute, Czech Acad Sci, CS-251 65 Ondrejov Observatory, Czechoslovakia

Steven M KLOSCO, ST Systems Corp, 9701 L Philadelphia Court, LANHAM Md 20706, USA

Per KNUDSEN, National Survey & Cadastre, Gamlehave Alle 22, DK-2920 CHARLOTTENLUND, Denmark

Prof Karl-Rudolf KOCH, Inst Theor Geodasie, Nussallee 17, D-5300 BONN 1, Fed Rep Germany

Johan J KOK, Department of Geodesy, Delft University of Technology, Thijsseweg 11, NL-2629 JA DELFT, Netherlands

Prof Barbara KOLACZEK, Space Research Centre, Polish Academy of Sciences, Bartycka 18, 00-716 WARSAW, Poland

Dr Yoshiteru KONO, Faculty of Science, Kanawawa University, 1-1 Marunouchi, KANAZAWA 920, Japan

Ir Radboud KOOP, Department of Geodesy, Delft University of Technology, Thijsseweg 11, NL-2629 JA DELFT, Netherlands

Wieslaw KOSEK, Space Research Centre, Polish Academy of Sceinces, ul Bartycka 18, 00-716 WARSAW, Poland

Ir Anton KOSTERS, Department of Geodesy, Delft University of Technology, Thijsseweg 11, NL-2629 JA DELFT, Netherlands

Dr Jan KRYNSKI, Dept Surveying Engineering, 2500 University Drive NW, CALGARY Alberta T2N 1N4, Canada

Dr Muneendra KUMAR, 10625 Wayridge Drive, MONTGOMERY VILLAGE Md 20879, USA

Dr Muneendra KUMAR, Defence Mapping Agency, 10625 Wayridge Drive, MONTGOMERY VILLAGE Md 20879, USA

Dr. Richard B LANGLEY, Dept of Surveying Eng, University of New Brunswick, P O Box 4400, FREDERICTON, New Brunswick E3B 5A3, Canada

Jean-Michel LEMOINE, CNES-GRGS-BGI, 18 Ave Edouard Belin, F-31055 TOULOUSE CEDEX, France

Norman LEPPARD, Mapping & Charting Establishment, Elmwood Ave, Feltham, MIDDLESEX TW13 7AE, United Kingdom

Dr LI Yanxing, State Seismoligical Bureau, Yueyahe, Yihaoqiao, Jintang Road, Hedong District, TIANJIN, China

Jacques LIARD, Geological Survey of Canada, 1 Observatory Crescent, OTTAWA Ont K1A 0Y3, Canada

Dr Stephen LICHTEN, Jet Propulsion Lab, Mail Stop 238-640, 4800 Oak Grove Drive, PASADENA Ca 91109, USA

Prof Dr-Ing Klaus LINKWITZ, Inst fur Geodasie im Bauwesen, Universitat Stuttgart, Keplerstrasse 10, D-7000 STUTTGART, Fed Rep Germany

Ing Michel LOUIS, 136 bis rue de Grenelle, F-75700 PARIS, France

Alastair MACDONALD, Ordnance Survey, Romsey Road, Southampton SO9 4DH, United Kingdom

Roderick T MACLEOD, Sagem Australasia Pty Ltd, PO Box 374, 2/122 Euston Road, ALEXANDRIA 2015, Australia

Dr Seiji MANABE, National Astronomical Observatory, MIZUSAWA-shi, Iwate-ken 023, Japan

Ambrogio MANZINO, Instituto di Topografia, Politecnico di Milano, Piazza Leonardo da Vinci 32, I-20133 MILANO, Italy

Prof Claudio MARCHESINI, Istituto di Scienze della Terra, Viale Ungheria 43, I-33100 UDINE, Italy

Prof Iginio MARSON, Inst di Miniere e Goefisica Applicata, Via Valerio 10, I-34127 TRIESTE, Italy

Prof Ignazio G MASSA, Dipartimento di Fisica, Universita di Cagliari, Via Ospedale 72, I-09124 CAGLIARI, Italy

Clive MASTERTON, Photogrammetry & Surveying, University College, Gower Street, LONDON WC1E 6BT, United Kingdom

Dr Takeshi MATSUMOTO, Japan Marine Science & Technology Centre, 2-15 Natsushima-cho, YOKOSUKA 237, Japan

Wayne MCVEIGH, Mapping & Charting Establishment, Elmwood Avenue, FELTHAM, Middlesex TW13 7AF, United Kingdom

P P MEDVEDEV, Academy of Sciences, Soviet Geophysical Committee, Molodezhnaya 3, MOSCOW GSP-1, USSR

Dr William MELBOURNE, Jet Propulsion Laboratory, MS 238-540, 4800 Oak Grove Drive, PASADENA Ca 91109, USA

Prof MENG Jiachun, Institute of Geodesy and Geophysics, Chinese Academy of Sciences, 54 Xu Dong Lu, WUCHANG, Hubei, China

Prof Charles MERRY, Department of Surveying, University of Cape Town, 7700 RONDEBOSCH, South Africa

Federica MIGLIACCIO, Instituto di Topografia, Politecnico di Milano, Piazza Leonardo da Vinci 32, I-20133 MILANO, Italy

MIN Tae-Jung, Director Geodetic Division, National Geographic Institute, Wonchun-dong 111, SUWON-shi Kyonggi-do, Korea

Prof Dr Sjamsir MIRA, Institut Teknologi Bandung, Jl Kanayakan 6, BANDUNG 40135, Indonesia

Prof Hiroo MIZUNO, Department of Earth Sciences, Kagawa University, 1-1 Saiwai-cho, Takamatsu, KAGAWA 760, Japan

Prof H MONTAG, Acad Wissenschaften DDR, Zentralinst Physik der Erd, Telegrafenberg A17, DDR-1561 POTSDAM, German Dem Rep

Dr Terry MOORE, Institute of Engineering Surveying and Space Geodesy, University Park, NOTTINGHAM NG7 2RD, United Kingdom

Prof Carlo MORELLI, Istituto di Miniere e Geofisica Applicato, 1 Piazzale Europa, 34126 TRIESTE, Italy

Dr Peter MORRIS, BP Exploration, Britannic House, Moore Lane, LONDON EC2Y 9BU, United Kingdom

Coronel Luis Rojo MOYA, Instituto Geografico Militar, SANTIAGO, Chile

Prof Ivan I MUELLER, Department of Geodetic Science, Ohio State University, 1958 Neil Ave, COLUMBUS Oh 43210, USA

Prof Luigi MUSSIO, Instituto di Topografia, Politecnico di Milano, Piazza Leonardo da Vinci 32, I-20133 MILANO, Italy

Prof Ichiro NAKAGAWA, Geophysical Institute, Kyoto University, Sakyo-ko, KYOTO 606, Japan

Dr Michael NAPIER, Institute of Engineering Surveying & Space Geodesy, University Park, NOTTINGHAM NG7 2RD, United Kingdom

Ruth E NEILAN, Jet Propulsion Laboratory, MS 238-640, 4800 Oak Grove Drive, PASADENA Ca 91101, USA

R Steve NEREM, Centre for Space Research, University of Texas at Austin, AUSTIN Tx 78712, USA

Simon P NEWBY, Dept of Surveying Engineering, University of New Brunswick, PO Box 4400, FREDERICTON NB E3B 5A3, Canada

Prof Dr Bogdan NEY, Institute of Geodesy and Cartography, ul Jasna 2/4, 00-950 WARSZAWA, Poland

Beatrice NHUN FAT, CNES-GRGS, 18 ave Edouard Belin, F-31055 TOULOUSE CEDEX, France

Vanya ODDEN, Norwegian Hydrographic Service, PO Box 60, N-4001 STAVANGER, Norway

Dr Shuhei OKUBO, Earthquake Research Institute, University of Tokyo, Bunkyo-ku, TOKYO 113, Japan

Mark OLDHAM, School of Physics, The University, NEWCASTLE UPON TYNE NE1 7RU, United Kingdom

Prof Dr J Pereira OSORIO, Observatorio Astronomico, Monte da Virgem, 4400 VILA NOVA de GAIA, Portugal

Prof Dr H PAPO, Technion, Dept of Civil Engineering, HAIFA 32000, Israel

Prof Paul PAQUET, Royal Observatory of Belgium, Avenue Circulaire 3, B-1180 BRUSSELS, Belgium

Prof Dr Teuvo PARM, Finnish Geodetic Institute, Ilmalankatu 1A, SF-00240 HELSINKI, Finland

Prof Dr Ing Hans PELZER, Geodatisches Institut, Universitat Hannover, Nienburgerstr 1, D-3000 HANNOVER 1, Fed Rep Germany

Dr Bartolomeo PERNICE, Agenzia Spaziale Italiana, Centro di Geodesia Spaziale, cp 155, I-75100 MATERA, Italy

Dr C. Frederick PETERS, GE Aerospace, 1017 Derwydd Lane, BERWYN Pa 19312-2008, USA

Ingmar PETERSON, National Land Survey, Geodetic Research Division, S-80182 GAVLE, Sweden

Ing Gerard PETIT, Institut Geographique National, 2 ave Pasteur, F-94160 SAINT MANDE, France

Heert PIERSMA, Dept Aerospace Engineering, Delft University of Technology, Kluyverweg 1, NL-2629 HS DELFT, Netherlands

Knud PODER, National Survey & Cadastre, Gamlehave Alle 22, DK-2920 CHARLOTTENLUND, Denmark

Christian POITEVIN, Centre de Geophysique Interne, Observatoire Royal de Belgique, ave Circulaire 3, B-1180 BRUXELLES, Belgium

Dr Josef POPELAR, Geophysics Division, Geological Survey of Canada, 1 Observatory Crescent, OTTAWA K1K 0Y3, Canada

Major James PRAIN, Military Survey (Systems & Techniques Unit), Elmwood Ave, FELTHAM Middx TW13 7AE, United Kingdom

Prof Mikhial PRILEPIN, Academy of Sciences, Soviet Geophysical Committee, Molodezhnaya 3, MOSCOW GSP-1, USSR

Fabio RADICIONI, Scienze dei Materiali e della Terra, Facolta di Ingengeria, Universita de Ancona, Via Brecce Bianche, I-60100 ANCONA, Italy

Prof Jacub RAIS, Bakosurtanal, Jalan Raya Bogor Km 46, Cibinong, BOGOR, Indonesia

Peter RANDS, Department of Surveying, The University, NEWCASTLE UPON TYNE NE1 7RU, United Kingdom

D RAVANBAKHSH, National Cartographic Centre, TEHERAN, Iran

Martin RAYSON, Department of Surveying, The University, NEWCASTLE UPON TYNE NE1 7RU, United Kingdom

Prof Dr Christoph REIGBER, Deutsches Geodatisches Forschunginstitut, Abt. I, Marstallplatz 8, D-8000 MUNCHEN 22, Fed Rep Germany

Prof Ewald REINHART, Institut fur Andgewandte Geodasie, Richard-Strauss-Allee 11, D-6000 FRANKFURT am MAIN 70, Fed Rep Germany

Dr Ole REMMER, National Survey & Cadastre, Gamle Have Alle 22, DK-2920 CHARLOTTENLUND, Denmark

Dr Nicholas A RENZETTI, Jet Propulsion Laboratory, 4800 Oak Grove Drive, PASADENA, California 91109, USA

Dr Peter RICHARDUS, 10 Heimanslaan, 6705 AG WAGENINGEN, Netherlands
Dr Bernd RICHTER, Institut fur Andgewandte Geodasie, Richard-Strauss-Allee 11, D-6000 FRANKFURT am MAIN 70, Fed Rep Germany
George F RITCHIE, Positioning Resources Ltd, 64 Commerce Street, ABERDEEN AB2 1BP, United Kingdom
Dr Chris RIZOS, School of Surveying, University of New South Wales, P O Box 1, KENSINGTON 2033, Australia
Dipl-Ing Rudiger H RODER, Institut fur Erdmessung, Universitat Hannover, Nienburger Str 6, D-3000 HANNOVER 1, Fed Rep Germany
Henry H M ROGERS, Geodesy Division, Mapping & Charting Establishment, Elmwood Avenue, FELTHAM, Middlesex TW13 7AF, United Kingdom
Prof Reiner RUMMEL, Department of Geodesy, Delft University of Technology, Thijsseweg 11, DELFT, Netherlands
Dr S John SABURI, Dept of Land Surveying, Ardhi Institute, P O Box 35176, DAR-ES-SALAAM, Tanzania
Prof Fausto SACERDOTE, Dipartimento di Matematica, via Buonarroti 2, I-56100 PISA, Italy
Thada SAI-SANGUAN, Survey Division, Royal Irrigation Dept, 78 Tiwanon Road, PAKKRED, NONTHABURI, Thailand
Dr Akihiko SAKUMA, Bureau Int Poids et Mesures, Pavillon de Breteuil, F-92312 SEVRES Cedex, France
Mahmoud Mohamed SALAH, Nat Res Inst Astronomy & Geophysics, Helwan, CAIRO, Egypt
Dr David A SALSTEIN, Atmospheric and Environmental Res Inc, 840 Memorial Drive, CAMBRIDGE Ma 02139, USA
Fernando SANSO, Instituto di Topografia, Fotogrammetria e Geofisica, Politecnico di Milano, Piazza Leonardo da Vinci 32, 20133 MILANO, Italy
Michel SARRAILH, CNES-GRGS-BGI, 18 Ave Edouard Belin, F-31055 TOULOUSE CEDEX, France
Dr R G S SASTRY, Earth Sciences Department, University of Roorkee, ROORKEE 247667, India
Dr Mikio SATOMURA, Faculty of Liberal Arts, Shizuoka University, Ohya 836, SHIZUOKA 422, Japan
Prof Dr Ing Burkhard SCHAFFRIN, Geodetic Institute, Stuttgart University, Keplerstrasse 11, D-7000 STUTTGART 10, Fed Rep Germany
Prof Dr-Ing Gunter SCHMITT, Geodaetisches Institut, Universitaet Karlsruhe (TH), Englerstr 7, Postfach 6980, D-7500 KARLSRUHE 1, Fed Rep Germany
Ir Ernst J O SCHRAMA, Department of Geodesy, Delft University of Technology, Thijsseweg 11, NL-2629 JA DELFT, Netherlands
Dr Bob E SCHUTZ, Centre for Space Research, University of Texas at Austin, AUSTIN Tx 78712, USA
Dr Klaus-Peter SCHWARZ, Department of Surveying Engineering, University of Calgary, CALGARY Alta T2N IN4, Canada
Prof Dr Gunter SEEBER, Institut fur Erdmessung, Universitat Hannover, Nienburgerstr 5, D-3000 HANNOVER, Fed Rep Germany
Prof Dr Hermann SEEGER, Inst Angewandte Geodasie, Richard Strauss Allee 11, D-6000 FRANKFURT am MAIN 70, Fed Rep Germany
Prof Naosuke SEKIGUCHI, National Astronomical Observatory, Mitaka, TOKYO 181, Japan

Dr Paul SELLERS, Department of Surveying, The University, NEWCASTLE UPON TYNE NE1 7RU, United Kingdom

Dr C K SHUM, Centre for Space Research, University of Texas at Austin, AUSTIN Tx 78712, USA

Michael G. SIDERIS, Dept of Surveying Engineering, University of Calgary, 2500 University Drive NW, CALGARY Alberta T2N 1N4, Canada

Antonio SILVA, Institute of Engineering Surveying & Space Geodesy, University Park, NOTTINGHAM NG7 2RD, United Kingdom

Prof Lars Erik SJOBERG, The Royal Institute of Technology, Department of Geodesy, STOCKHOLM, Sweden

Ingeborg SMEETS, Astronomisch Institut, Katholieke Universiteit Leuven, Celestijnenlaan 200 B, B-3030 HEVERLEE, Belgium

Jeroen SMIT, Department of Geodesy, NL-2629 JA Delft University of Technology, Thijsseweg 11, DELFT, Netherlands

Dr D E SMITH, Geodynamics Branch, Goddard Space Flight Centre, GREENBELT Md 20771, USA

James R SMITH, Dept of Civil Engineering, Portsmouth Polytechnic, Burnaby Road, PORTSMOUTH, Hants, United Kingdom

Prof Dr Jozsef SOMOGYI, PO Box 5, Geodetic and Geophys Res Inst, Hungarian Acad of Sciences, H-9400 SOPRON, Hungary

Dr Fred SPILHAUS, American Geophysical Union, 2000 Florida Ave NW, WASHINGTON DC 20009, USA

Michael P STEWART, Department of Geophysics, University of Edinburgh, James Clerk Maxwell Building, Mayfield Road, EDINBURGH EH9 3JZ, United Kingdom

Dr Elke STOCKER-MEIER, Inst Theor Geodasie, Universitat Bonn, Nussallee 17, D-5300 BONN 1, Fed Rep Germany

Theresa STRANGE, Department of Earth Sciences, Oxford University, Parks Road, OXFORD OX1 3PR, United Kingdom

Dr William E STRANGE, National Geodetic Survey N/CG1X2, 11400 Rockville Pike, ROCKVILLE Md 20852, USA

John STREICKER, Dept of Surveying Engineering, University of New Brunswick, PO Box 4400, FREDERICTON NB, Canada

Yuji SUGIMOTO, Kahima Space Research Centre, Communications Research Lab, 893-1 Hirai, KASHIMA, Ibaraki 314, Japan

Philip SUMMERFIELD, Institute of Engineering Surveying & Space Geodesy, University Park, NOTTINGHAM NG7 2RD, United Kingdom

Prof H SUNKEL, Technical University Graz, Department of Mathematical Geodesy, Technikerstrasse 4, A-8010 GRAZ, Austria

Dr Shuzo TAKEMOTO, Disaster Prevention Research Instutute, Kyoto University, Gokasho, Uji, KYOTO 611, Japan

Dr Minoru TANAKA, Crustal Dynamics Department, Geographical Survey Institute, Kitasato 1, TSUKUBA SCIENCE CITY 305, Japan

Dr Torao TANAKA, Disaster Prevention Research Institute, Kyoto University, Gokasho, Uji, KYOTO 611, Japan

Prof James G TANNER, University of Western Ontario, RR#1, UNION Ont NO1 2L0, Canada

Prof TAO Benzao, Head of Geodesy Department, Wuhan Technical University, 39 Luo Yu Lu, WUHAN 430070, China

Mark TAPLEY, GP-B Hansen Labs, Stanford University, STANFORD Ca 94305-4085, USA
Prof Ali Abd-el-Azim TEALEB, National Research Institute of Astronomy and Geophysics, Helwan, CAIRO, Egypt
John TEMPLE, Institute of Engineering Surveying & Space Geodesy, University Park, NOTTINGHAM NG7 2RD, United Kingdom
Dr-Ing Peter TEUNISSEN, Department of Geodesy, Univesity of Delft, Thijsseweg 11, NL-2629 JA DELFT, Netherlands
Somphorn THUPTHONG, Survey Division, Royal Irrigation Department, 78 Tiwanon Road, PAKKRED, Nonthaburi, Thailand
Prof Wolfgang TORGE, Institut fur Erdmessung, Universitat Hannover, Nienburgerstr 6, D-3000 HANNOVER, Fed Rep Germany
Prof Dr Kennert TORLEGARD, Photogrammetry, KTH, S-10044 STOCKHOLM, Sweden
Prof Beniamino TORO, Dept di Scienze della Terra, P Aldo Moro 5, I-00185 ROMA, Italy
Mark H TORRENCE, ST Systems Corp, 9701 L Philadelphia Court, LANHAM Md 20706, USA
Michael TULLY, GEKO A/S, Kjorbokollen, N-1300 SANDVIKA, Norway
Terrence T TURNOR, Mapping & Charting Establishment, Elmwood Avenue, FELTHAM, Middlesex TW13 7AF, United Kingdom
Nicole VALES, CNES-GRGS-BGI, 18 Ave Edouard Belin, F-31055 TOULOUSE CEDEX, France
Herbert D VALLIANT, LaCoste & Romberg Gravity Meters Inc, 4807 Spicewood Springs Road, Bldg 2, AUSTIN Tx 78759, USA
Robbert VAN ETTEN, Delft University of Technology, Thijsseweg 11, NL-2629 JA DELFT, Netherlands
Ir Martin VAN GELDEREN, Department of Geodesy, Delft University of Technology, Thijsseweg 11, NL-2629 JA DELFT, Netherlands
Dr Herman VAN GYSEN, Dept of Surveying & Mapping, University of Natal, King George V Ave, 4001 DURBAN, South Africa
Prof Jan VAN MIERLO, Geodaetisches Institut, Universitaet Karlsruhe (TH), Englerstr 7, Postfach 6980, D-7500 KARLSRUHE 1, Fed Rep Germany
Dr Petr VANICEK, Dept of Surveying Engineering, University of New Brunswick, P O Box 4400, FREDERICTON NB E3B 5A3, Canada
Margarita VEI, Inst Astron Physikal Geodasie, TU Munchen, Hochkalterstr 6 D-8000 Munchen 90, Fed Rep Germany
Prof George VEIS, National Technical University, 9 Herroon Polytechniou, Zographou, ATHENS 157 73, Greece
Martin VERMEER, National Survey & Cadastre, Gamlehave Alle 22, DK-2920 CHARLOTTENLUND, Denmark
Dr Grazia VERRONE, Department of Physics, University of Bari, via Amendola 173, I-70126 BARI, Italy
Prof Raimondo O VICENTE, Faculty of Sciences, R Mestre Aviz 30 R/C, 1495 LISBOA, Portugal
Dr Ing Pavel VYSKOCIL, VUGTK, CS-25066 ZDIBY 98, Czechoslovakia
Prof John WAHR, Department of Physics, University of Colorado, Campus Box 390, BOULDER CO 80309, USA
Dr Attallah M WASSEF, University of Toronto, Centre for Surveying Science, Erindale Campus, MISSISSAUGA Ont L5L 1C6, Canada

Michael M WATKINS, Centre for Space Research, University of Texas at Austin, WRW 402, AUSTIN Tx 78712, USA

Jack A WEIGHTMAN, Molibus, 44 Avton Road, Pollockshields, GLASGOW G41 5MN, United Kingdom

Prof Hans-Georg WENZEL, Geodetic Institute, University of Karlsruhe, Englerstrasse 7, D-7500 KARLSRUHE 1, Fed Rep Germany

Jane WESTROP, Institute of Engineering Surveying & Space Geodesy, University Park, NOTTINGHAM NG7 2RD, United Kingdom

Stephen WHALLEY, Institute of Engineering Surveying & Space Geodesy, University Park, NOTTINGHAM NG7 2RD, United Kingdom

Dr Charles A. WHITTEN, 9606 Sutherland Road, SILVER SPRING Md 20901, USA

Dr George A WILKINS, 291 Kings Drive, EASTBOURNE, East Sussex BN21 2YA, United Kingdom

Gordon A WILKINSON, Geosurvey Holdings Ltd, PO Box 6079, RIYADH 11442, Saudi Arabia

Prof H S WILLIAMS, Department of Surveying, University of Cape Town, PO Private Bag, 7700 RONDEBOSCH, South Africa

Pascal WILLIS, Institut Geographique National, 2 ave Pasteur, F-94160 SAINT MANDE, France

Dr Ing Herbert WILMES, DGFI Abt 1, Marstallplatz 8, D-8000 MUNCHEN 22, Fed Rep Germany

Prof Clark R WILSON, Center for Space Research, University of Texas, AUSTIN Tx 78713-7909, USA

Dr-Ing Peter WILSON, Institut fur Andgewandte Geodasie, Richard Strauss Allee 11, D-6000 FRANKFURT am MAIN 70, Fed Rep Germany

Sydney WITIUK, Statistics Canada, Geocartographics Division, 2A6 Jean Talon Bldg, Tunney's Pasture, OTTAWA Ont K1A 0T6, Canada

Dr Detlef WOLF, c/o Dept of Geodesy, Uppsala University, Hallby, S-75592 UPPSALA, Sweden

Prof Otfried WOLFRUM, Geodetic Institute, TU Darmstadt, Petersenstrasse 13, D-6100 DARMSTADT, Fed Rep Germany

William WOODEN, Defence Mapping Agency, 8613 Lee Highway, FAIRFAX Va 22031-2137, USA

Dr Alan F WRIGHT, Global Surveys Ltd, 2 Ridgacre Lane, Quinton, BIRMINGHAM B32 1ES, United Kingdom

Prof YE ShuHua, Shanghai Astronomical Observatory, Chinese Academy of Sciences, 80 Nandau Road, SHANGHAI, China

Dr Koichi YOKOYAMA, National Astronomical Observatory, Division of Earth Rotation, MIZUSAWA-shi, Iwate-Ken 023, Japan

Yoshimitsu YOSHIMURA, Geographical Survey Institute, 1 Kitasato, TSUKUBA-City, Ibaraki 305, Japan

Prof Janusz B ZIELINSKI, Space Research Centre, Polish Academy of Sciences, ul Bartycka 18, 00716 WARSZAWA, Poland

Dr Victor ZLOTNICKI, Jet Propulsion Lab 300-323, 4800 Oak Grove Drive, PASADENA, California 91109, USA

Prof Hossein ZOMORRODIAN, Institute of Geophysics, Teheran University, Kargar Shomali Ave, TEHERAN, Iran

Symposium 102 Author Index